PROJECT SCHEDULING
Recent Models, Algorithms and Applications

INTERNATIONAL SERIES IN OPERATIONS RESEARCH & MANAGEMENT SCIENCE

Frederick S. Hillier, Series Editor
Department of Engineering-Economic Systems and Operations Research
Stanford University
Stanford, California

Saigal, Romesh
LINEAR PROGRAMMING: A Modern Integrated Analysis

Nagurney, Anna/ Zhang, Ding
PROJECTED DYNAMICAL SYSTEMS AND VARIATIONAL INEQUALITIES WITH APPLICATIONS

Padberg, Manfred/ Rijal, Minendra P.
LOCATION, SCHEDULING, DESIGN AND INTEGER PROGRAMMING

Vanderbei, Robert J.
LINEAR PROGRAMMING: Foundations and Extensions

Jaiswal, N.K.
MILITARY OPERATIONS RESEARCH: Quantitative Decision Making

Gal, Tomas / Greenberg, Harvey J.
ADVANCES IN SENSITIVITY ANALYSIS AND PARAMETRIC PROGRAMMING

Prabhu, N.U.
FOUNDATIONS OF QUEUEING THEORY

Fang, S.-C./Rajasekera, J.R./ Tsao, H.-S.J.
ENTROPY OPTIMIZATION AND MATHEMATICAL PROGRAMMING

Yu, Gang
OPERATIONS RESEARCH IN THE AIRLINE INDUSTRY

Ho, Teck-Hua / Tang, Christopher S.
PRODUCT VARIETY MANAGEMENT

El-Taha, Muhammad / Stidham, Jr., Shaler
SAMPLE-PATH ANALYSIS OF QUEUEING SYSTEMS

Miettinen, Kaisa M.
NONLINEAR MULTIOBJECTIVE OPTIMIZATION

Chao, Hung-po & Huntington, Hillard G.
DESIGNING COMPETITIVE ELECTRICITY MARKETS

PROJECT SCHEDULING
Recent Models, Algorithms and Applications

edited by

Jan Węglarz
Poznan University of Technology

Kluwer Academic Publishers
Boston/London/Dordrecht

Distributors for North, Central and South America:
Kluwer Academic Publishers
101 Philip Drive
Assinippi Park
Norwell, Massachusetts 02061 USA
Telephone (781) 871-6600
Fax (781) 871-6528
E-Mail <kluwer@wkap.com>

Distributors for all other countries:
Kluwer Academic Publishers Group
Distribution Centre
Post Office Box 322
3300 AH Dordrecht, THE NETHERLANDS
Telephone 31 78 6392 392
Fax 31 78 6546 474
E-Mail <orderdept@wkap.nl>

Electronic Services <http://www.wkap.nl>

Library of Congress Cataloging-in-Publication Data

Project scheduling : recent models, algorithms, and applications / edited by Jan Weglarz.
p. cm. -- (International series in operations research & management science)
Includes bibliographical references and index.
ISBN 0-7923-8268-4
1. Production scheduling--Mathematical models. I. Weglarz, Jan. II. Series.
TS157.5.P75 1998
658.5'3--dc21 98-30712
CIP

Printed on acid-free paper.

Printed in the United States of America

TABLE OF CONTENTS

Part I - Classical Models – Classification, Exact Algorithms

Part II - Classical Models – Heuristics, Benchmark Instances, Software Evolution

Part IV - Extensions and Applications

PREFACE

Project scheduling problems are, generally speaking, the problems of allocating scarce resources over time to perform a given set of activities. The resources are nothing other than the arbitrary means which activities complete for. Also the activities can have a variety of interpretations. Thus, project scheduling problems appear in a large spectrum of real-world situations, and, in consequence, they have been intensively studied for almost fourty years.

Almost a decade has passed since the multi-author monograph: R. Słowiński, J. Węglarz (eds.), Advances in Project Scheduling, Elsevier, 1989, summarizing the state-of-the-art across project scheduling problems, was published. Since then, considerable progress has been made in all directions of modelling and finding solutions to these problems. Thus, the proposal by Professor Frederick S. Hillier to edit a handbook which reports on the recent advances in the field came at an exceptionally good time and motivated me to accept the challenge. Fortunately, almost all leading experts in the field have accepted my invitation and presented their completely new advances often combined with expository surveys. Thanks to them, the handbook stands a good chance of becoming a key reference point on the current state-of-the-art in project scheduling, as well as on new directions in the area.

The contents are divided into four parts. The first one, dealing with classical models - exact algorithms, is preceded by a proposition of the classification scheme for scheduling problems. Although the notation proposed is not strictly observed in the rest of the handbook, it is helpful in ordering the material.

Notice that by "classical models" we understand it to mean the deterministic problems involving discrete resources only and evaluating schedules on the basis of a single criterion. Nevertheless, this term covers a wide variety of problems, including (discrete) resource-duration interaction, i.e. mutli-mode models.

Part II deals with heuristic algorithms for these models and is complemented by chapters treating such important problems as benchmark instances, interval capacity consistency tests, and the evolution of software quality for (classical) project scheduling problems.

In Part III the most important new or "non-classical" models are considered. These include probabilistic and fuzzy models, continuous resources as well as multi-criteria formulations and knowledge-based systems.

The last part contains some important extensions and/or applications of the project scheduling problems considered in the previous chapters. The handbook is supplemented with a name index containing the names of the authors of all the referred papers, including the authors of the chapters.

The monograph is addressed primarily to research workers (including Ph.D. and graduate students), specialists and professional people in the fields of: production and operations management, system analysis, business administration, and applied mathematics. However, specialists in other disciplines, like civil, computer, and industrial engineering, in which problems of allocating scarce resources of different types are of vital importance, can also benefit from it.

I would like to thank all the authors for their contributions and their fruitful co-operation. Thanks are also due to Joanna Józefowska and Jarosław Nabrzyski for their help in preparing the manuscript, and to Izabella Tkocz for helping me in regarding correspondence.

Jan Węglarz

1 A CLASSIFICATION SCHEME FOR PROJECT SCHEDULING

Willy Herroelen[1]
Erik Demeulemeester[1]
Bert De Reyck[2]

1. Katholieke Universiteit Leuven (Belgium)
2. Erasmus University Rotterdam (The Netherlands)

1.1. Introduction

The basic concern of scheduling is commonly described as the allocation of limited resources to tasks over time (Lawler et al. 1993, Pinedo 1995). The resources and tasks may take many forms. In project scheduling the tasks refer to the activities belonging to one or more projects. The execution of project *activities* may require the use of different types of resources (money, crews, equipment, ...). The scheduling *objectives* may also take many forms (minimizing project duration, minimizing project costs, maximizing project revenues, optimizing due date performance, ...). The result is a wide and steadily growing variety of problem types which motivates the introduction of a systematic notation that can serve as a basis for a classification scheme.

A classification scheme for project scheduling may serve a variety of objectives. First, a classification scheme would greatly facilitate the presentation and discussion of project scheduling problems. Intensive research efforts over the past few years have greatly expanded the variety of project scheduling problems under study (for recent reviews we refer to Icmeli et al. 1993, Elmaghraby 1995, Herroelen and Demeulemeester 1995, Özdamar and Ulusoy 1995, Herroelen et al. 1997, 1998). These problems are often identified within the project scheduling community in a non-standardized manner by a rather confusing set of acronyms, most often consisting of a simple concatenation of characters. Examples are *RCPSP* for the resource-constrained project scheduling problem, *MRCPSP* for the multi-mode resource-constrained project scheduling problem, *RCPSP-GPR* for the resource-constrained project scheduling problem with generalized precedence relations, just to cite a few. A concise and rigorous classification scheme immediately highlights the fundamental problem characteristics and avoids the use of these lengthy and often ambiguous character concatenations. In addition, it saves both authors and speakers the repetitive use of lengthy, verbal, introductive statements about the precise characteristics and assumptions of the project scheduling problem under study.

Second, a comprehensive classification scheme allows for the immediate identification of viable areas of research through the identification of interesting open problems which have remained unstudied or largely ignored by the researchers in a fastly growing field. It helps in identifying the common characteristics of project scheduling problems and reveals the important fact that certain problems are in fact subproblems of more generic ones.

Third, a classification scheme simplifies the assessment of problem complexity. It reveals the close relationships between the various project scheduling problems through the use of reduction graphs which show the various interrelations among the different values of the particular classification parameters. As such it helps in identifying the fundamental characteristics which account for the inherent complexity of the problem under study.

Last, but not least, a classification scheme facilitates the match of solution procedures to problem settings and as such facilitates the preparation of problem surveys and literature reviews.

It is common practice to classify deterministic machine scheduling problems by a standard three-field notation proposed by Graham et al. (1979) and

Blazewicz et al. (1983). The extensive scheme we propose in this chapter resembles the standard scheme for machine scheduling problems in that it is also composed of three fields $\alpha|\beta|\gamma$. In machine scheduling problems (Blazewicz et al. 1994, Brucker 1995) the first field α describes the machine environment. This field allows for the identification of single machine problems, various types of parallel machine problems, flow shops, job shops, general shops, open shops, mixed shops and multiprocessor task systems. The second field β is used to describe the task and resource characteristics. This field includes parameter settings for characterizing the possibility for preemption, the precedence constraints, ready times, deadlines, task processing times, batching and additional resources. The characterization of the additional resources (Blazewicz et al. 1986) is done using a field parameter which takes the value of the empty symbol ° to denote the absence of additional resources and $res\lambda\sigma\rho$ to specify the resource constraints. λ, σ, $\rho \in \{\cdot, k\}$ respectively denote the number of additional resource types, resource limits and resource requests. If λ, σ, $\rho = \cdot$, then the number of additional resource types, resource limits and resource requests are arbitrary, and if λ, σ, $\rho = k$, then the number of additional resource types is equal to k, each resource is available in the system in the amount of k units and the resource requests of each task are at most equal to k units. The third field γ denotes a performance measure.

The characterization of the additional resources in the $\alpha|\beta|\gamma$ notation of the deterministic machine scheduling classification scheme, however, does not allow for the precise characterization of the wide variety of problems which manifest themselves in the specific and much more complex environment of *project* scheduling. This motivates the introduction of a specific classification scheme for project scheduling problems. The scheme proposed in this chapter is also based on a three-field notation, but the composition of the fields and the precise meaning of the various parameters, however, are mostly new and specific to the field of project scheduling. It should be understood from the outset, however, that our objective is not to build an extremely rigid classification scheme which attempts to create futile classification holes to accommodate any possible project scheduling problem. The proposed scheme tries to combine rigidity with flexibility. It provides sufficient detail to allow for a concise taxonomy of the project scheduling field which covers the majority of the project scheduling problems described in the literature and at the same time offers sufficient degrees of freedom to the user in the specification of the various parameters.

The organization of this chapter is as follows. The classification scheme is presented in the next section, in which we also discuss the simple reductions between the various project scheduling problems and problem parameters. We present the reduction graphs which can be defined on the field parameters of the classification scheme. In Section 1.3 we illustrate the potential use of the scheme by using the various parameters in an effort to characterize and classify the most important project scheduling problems commonly discussed in the literature. The last section is then reserved for overall conclusions.

1.2. Classification of project scheduling problems

The classification scheme is composed of three fields $\alpha | \beta | \gamma$. The meaning of the three fields is explained below.

1.2.1. Field α: resource characteristics

The resource characteristics of a project scheduling problem are specified by a set α containing at most three elements α_1, α_2 and α_3. Let ° denote the empty symbol which will be omitted when presenting specific problem types. Parameter $\alpha_1 \in \{°, 1, m\}$ denotes the number of resource types:

α_1 = °: no resource types are considered in the scheduling problem,
$\alpha_1 = 1$: one resource type is considered,
$\alpha_1 = m$: the number of resource types is equal to m.

Parameter $\alpha_2 \in \{°, 1, T, 1T, v\}$ denotes the specific resource types used. In the project scheduling literature a common distinction is made (Blazewicz et al. 1986) between renewable resources, nonrenewable resources and doubly-constrained resources. *Renewable resources* (e.g. manpower, machines, tools, equipment, space, ...) are available on a period-by-period basis, that is, the available amount is renewed from period to period. Only the total resource usage at every time instant is constrained. *Nonrenewable* resources (e.g. money, raw materials, energy, ...), on the contrary, are available on a total project basis, with a limited consumption availability for the entire project. *Doubly-constrained resources* are constrained per period (e.g. per period cash flow) as well as for the overall project (e.g. total expenditures, overall pollution limits, ...). Recently, researchers (Böttcher et al. 1996, Schirmer and Drexl 1996, Drexl 1997) have introduced the concept of *partially (non)renewable resources* referring to resources the availability of which is defined for a specific time interval (subset of periods). For each resource type there are a number of subsets of periods, each characterized by a specific (nonrenewable) availability of the resource type. Essentially, partially (non)renewable resources can be viewed as a generic resource concept in project scheduling, as they include both renewable and nonrenewable (and, hence, also doubly-constrained) resources. A partially renewable resource with a specified availability for a time interval equal to the unit duration period (identified below by the parameter setting equal to 1) is essentially a renewable resource. A partially renewable resource with a specified availability for a time interval equal to the project horizon (identified below by a parameter setting equal to T) is essentially a nonrenewable resource. Partially renewable resources with a specified availability on both a unit duration and a total project horizon basis (denoted in the parameter setting by $1T$) can be interpreted as doubly-constrained resources. As a result we can use the partially renewable resource concept in our classification scheme in a generic way which allows for a straightforward identification of the various resource categories considered:

$\alpha_2 = \circ$: absence of any resource type specification,
$\alpha_2 = 1$: renewable resources, the availability of which is specified for the unit duration period (e.g. hour, shift, day, week, month, ...),
$\alpha_2 = T$: nonrenewable resources, the availability of which is specified for the entire project horizon T,
$\alpha_2 = 1T$: both renewable and nonrenewable resources (also called doubly-constrained resources the availability of which is specified on both a unit duration period and a total project horizon basis),
$\alpha_2 = v$: partially (non)renewable resources the availability of which is renewed in specific time periods, i.e. they are nonrenewable in variable time intervals (e.g. in total 5 units are available in the set of periods 1 up to 5, 7 units are available in period 6, in total 0 units are available in the set of periods 7 up to 9, ...).

Parameter $\alpha_3 \in \{\circ, va\}$ describes the resource availability characteristics of the project scheduling problem. Some scheduling problems assume that renewable resource availabilities are a given constant while others assume that resource availability varies over time. We assume that the availability of partially renewable resources may vary over the various time intervals. The following parameter specifications are used in our classification scheme:

$\alpha_3 = \circ$: (partially) renewable resources are available in constant amounts,

$\alpha_3 = va$: (partially) renewable resources are available in variable amounts.

Stochastic settings for the resource availability characteristics are discussed in Section 1.3.3.

The simple reductions between the various resource parameters are shown by the reduction graphs in Figure 1:1. The nodes in the graph represent particular assumptions made about the parameters. The directed arcs show the direction of polynomial transformations.

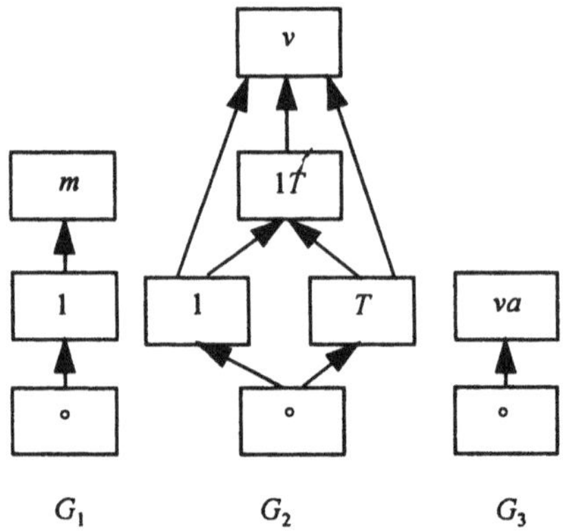

Figure 1:1. Simple reductions for the resource parameters

First consider the reduction graph G_1. If we replace ° by 1 in the specification for α_1, we get a simple reduction because the problem without any resource constraints is a special case of the problem with a single resource type. In a similar fashion, replacing 1 by m yields a simple reduction as we move from a problem with a single resource type to a problem which uses m resource types. The reduction graph G_2 specifies the reductions for α_2. Replacing ° by 1 or T yields a simple reduction because we move from a problem without any resource specification to a problem which involves either renewable or nonrenewable resources. Both 1 and T reduce to $1T$, since both the renewable and nonrenewable resources are a special case of a doubly-constrained resource. The case where the partially renewable resources are renewed in specific time periods constitutes a special case of 1, T and $1T$. The reduction graph G_3 denotes the simple reduction from the case of constant resource availabilities to the situation where resources are available in time varying amounts.

1.2.2. Field β: activity characteristics

The second field β specifies the activity characteristics of a project scheduling problem. It contains at most nine elements β_1, β_2, β_3, β_4, β_5, β_6, β_7, β_8 and β_9. Parameter $\beta_1 \in \{°, pmtn, pmtn\text{-}rep\}$ indicates the possibility of activity preemption:

$\beta_1 = °$: no preemption is allowed,
$\beta_1 = pmtn$: preemptions of the preempt-resume type are allowed,
$\beta_1 = pmtn\text{-}rep$: preemptions of the preempt-repeat type are allowed.

Preemption (activity splitting) implies that the processing of an activity may be interrupted and resumed at a later time (preempt-resume). The situation where activities may be interrupted but cannot be resumed at the point of interruption, i.e., must be completely redone (preempt-repeat), can be accommodated by setting $\beta_1 = pmtn\text{-}rep$.

Parameter $\beta_2 \in \{°, cpm, min, gpr, prob\}$ reflects the precedence constraints:

$\beta_2 = °$: no precedence constraints (the activities are unordered),
$\beta_2 = cpm$: the activities are subject to strict finish-start precedence constraints with zero time lag, as used in the basic PERT/CPM model,
$\beta_2 = min$: the activities are subject to precedence diagramming constraints of the type start-start, finish-start, start-finish and finish-finish with *minimal* time lags,
$\beta_2 = gpr$: the activities are subject to generalized precedence relations of the type start-start, finish-start, start-finish and finish-finish with both *minimal* and *maximal* time lags.

We deem it necessary to make an explicit distinction between the use of minimal and maximal time lags. A minimal time lag specifies that an activity can only start (finish) when the predecessor activity has already started (finished) for a certain time period. A maximal time lag specifies that an activity should be started (finished) at the latest a certain number of time periods beyond the start (finish) of another activity. A minimal time lag is essentially a generalized precedence relation which can be transformed into a non-negative start-start time lag, with the additional assumption that no cycles may occur in the network. The introduction of maximal time lags offers a wide amount of relevant and practical modelling capabilities (De Reyck 1995) which are well beyond the scope of minimal time lags. The introduction of both minimal and maximal time lags, however, also shifts the problem setting to a much higher complexity level (quite often, the feasibility problem already becomes NP-complete). The fact that many commercial project planning software packages do not allow for the use of maximal time lags does not come as a surprise.

β_2 = *prob:* the activity network is of the probabilistic type where the evolution of the corresponding project is not uniquely determined in advance.

This category encompasses generalized activity networks (Elmaghraby 1977) such as GERT (Neumann and Steinhardt 1979). This field leaves the user the freedom to be very specific about the precise type of probabilistic network in use. As such, the β_2-parameter can be set to *gert* to specify a GERT network, to *deor* to specify GERT networks with exclusive-or node entrance and deterministic node exit, to *steor* to specify GERT networks with exclusive-or node entrance and stochastic node exit, etc.

The third parameter $\beta_3 \in \{°, \rho_j\}$ describes ready times:

$\beta_3 = °$: all ready times are zero,
$\beta_3 = \rho_j$: ready times differ per activity.

Parameter $\beta_4 \in \{°, cont, d_j{=}d\}$ describes the duration of the project activities:

$\beta_4 = °$: activities have arbitrary integer durations,
β_4 = *cont*: activities have arbitrary continuous durations,
$\beta_4 = (d_j{=}d)$: all activities have a duration equal to d units.

The specification of stochastic activity durations is discussed in Section 1.3.3.

Parameter $\beta_5 \in \{°, \delta_j, \delta_n\}$ describes deadlines:

$\beta_5 = °$: no deadlines are assumed in the system,
$\beta_5 = \delta_j$: deadlines are imposed on activities,
$\beta_5 = \delta_n$: a project deadline is imposed.

It should be noted that the specification $\beta_5 = \delta_j$ also allows deadlines to be imposed on *events* in the network. Events are identified as the initial or terminal point of one or several activities and can easily be accommodated by the network logic. Moreover, we have opted not to specify the use of activity (event) or project *due dates* in the β_5-field. A *deadline* indicates that activities (events) or projects must finish not later than the deadline and may not be violated. The eventual use of due dates which may be violated at a specific cost should be apparent from the specification of the corresponding due-date based performance criterion in the γ-field of the classification scheme.

Parameter $\beta_6 \in \{°, vr, disc, cont, int\}$ denotes the nature of the resource requirements of the project activities. A common assumption is that activities request their resources in constant amounts or in variable amounts over their periods of execution. Some models (e.g. Weglarz 1980) assume that the resource requests of the tasks are continuous, i.e. concern resource amounts which are arbitrary elements of given intervals. For example, if a resource request of a task is characterized by the interval $(0,N]$, it means that an arbitrary amount of this resource greater than 0 and not greater than N can be used for processing the task at any moment. Such resource requests concern continuously-divisible resources like electric current (or power), fuel flow, etc. Other models study the simultaneous requirement of discrete and continuous resources (e.g. Jozefowska and Weglarz 1994, Weglarz and Jozefowska 1997). Still other models (Hackman and Leachman 1989, Leachman 1983, Leachman et al. 1990) assume that there is a feasible range of *intensity* of resource assignments to each activity, resulting in a range of possible durations. These models assume that all the different types of resources which are required by an activity are applied proportionally throughout the activity execution, i.e. each activity utilizes a constant mix of resources as it progresses exactly proportional to the mix of total resource requirements to complete the activity. The rates of applications of different types of resources to an activity can be indexed in terms of one rate which is called the *intensity of the activity*. The activity intensity is assumed to be continuously variable within given upper and lower limits. The rate of progress of an activity is assumed to be proportional to its intensity. Both types of models essentially boil down to similar problem environments in which the resource requirement of an activity is a function of time. Our classification scheme uses the following specifications for the resource requirements:

β_6 = °: activities require the resources in a constant discrete amount (e.g. a number of units for every time period of activity execution),

β_6 = *vr*: activities require the resources in variable discrete amounts (e.g. a number of units which varies over the periods of activity execution).

Stochastic resource requirement settings are discussed in Section 1.3.3.

For those cases where the activity durations have to be determined by the solution procedure on the basis of a resource requirement function, the following settings are used:

β_6 = *disc*: the activity resource requirements are a discrete function of the activity duration,
β_6 = *cont*: the activity resource requirements are a continuous function of the activity duration,
β_6 = *int:* the activity resource requirements are expressed as an intensity or rate function.

We leave it up to the user to be more specific in the specification of the resource requirement function. If so desired, the setting β_6 = *cont* can be made more specific: β_6 = *lin* can be used to specify that activity resource requirements are a linear funtion of the activity duration, the setting β_6 = *conc* can be used to specify that the activity resource requirements are a concave function of the activity duration, and β_6 = *conv* may denote the fact that resource requirements are a convex function of the activity duration.

The type and number of possible execution modes for the project activities is described by parameter $\beta_7 \in \{°, mu, id\}$. Most problems assume a single execution mode per activity. Various problems assume time/cost, time/resource and/or resource/resource trade-offs which give rise to various possible execution modes for the activities of the project. Recently, researchers (Salewski 1996, Salewski and Lieberam-Schmidt 1996, Salewski et al. 1997) have started to study project scheduling problems which generalize multiple activity modes to so-called mode identity constraints in which the set of activities is partitioned into disjoint subsets. All activities in a subset must then be executed in the same mode. Both the time and cost incurred by processing a subset of activities depend on the resources assigned to it. The following parameter settings are used:

β_7 = °: activities must be performed in a single execution mode,
β_7 = *mu*: activities have multiple prespecified execution modes,
β_7 = *id*: the activities are subject to mode identity constraints.

Parameter $\beta_8 \in \{°, c_j, per, sched\}$ is used to describe the financial implications of the project activities. In most models with cash flows, the cash flow amounts are assumed to be known and are either associated with network activities or network events. Both situations can be represented by associating the cash flows with the nodes in an activity-on-the-node network. Other models assume that the cash flows are periodic in that they occur at regular time intervals or with a known frequency. Still other models assume that both the amount and the timing of the cash flows have to be determined (Herroelen et al. 1997). The following settings are used in the classification scheme:

$\beta_8 = °$: no cash flows are specified in the project scheduling problem,
$\beta_8 = c_j$: activities have an associated arbitrary cash flow,
$\beta_8 = c_j^+$: activities have an associated positive cash flow,
$\beta_8 = per$: periodic cash flows are specified for the project (e.g. payments at regular intervals),
$\beta_8 = sched$: both the amount and the timing of the cash flows have to be determined.

In Section 1.3.3. a brief discussion is given on the use of stochastic cash flow specifications.

A common assumption in project scheduling is that change-over times (i.e. setup times or transportation times) are sequence-independent and included in the activity durations. As recognized by several authors (e.g. Kaplan 1991, Kolisch 1995, Dodin and Elimam 1997, Bartsch et al. 1997), however, sequence-dependent change-over times may be very important in project settings. Examples include equipping excavators with different types of scoops, the travel of workers between jobs and the transportation of heavy equipment to different construction sites. Parameter $\beta_9 = \in \{°, s_{jk}\}$ is used to denote change-over times:

$\beta_9 = °$: no change-over (transportation) times,
$\beta_9 = s_{jk}$: sequence-dependent change-over times.

Mode-dependent change-over times are modelled implicitly.

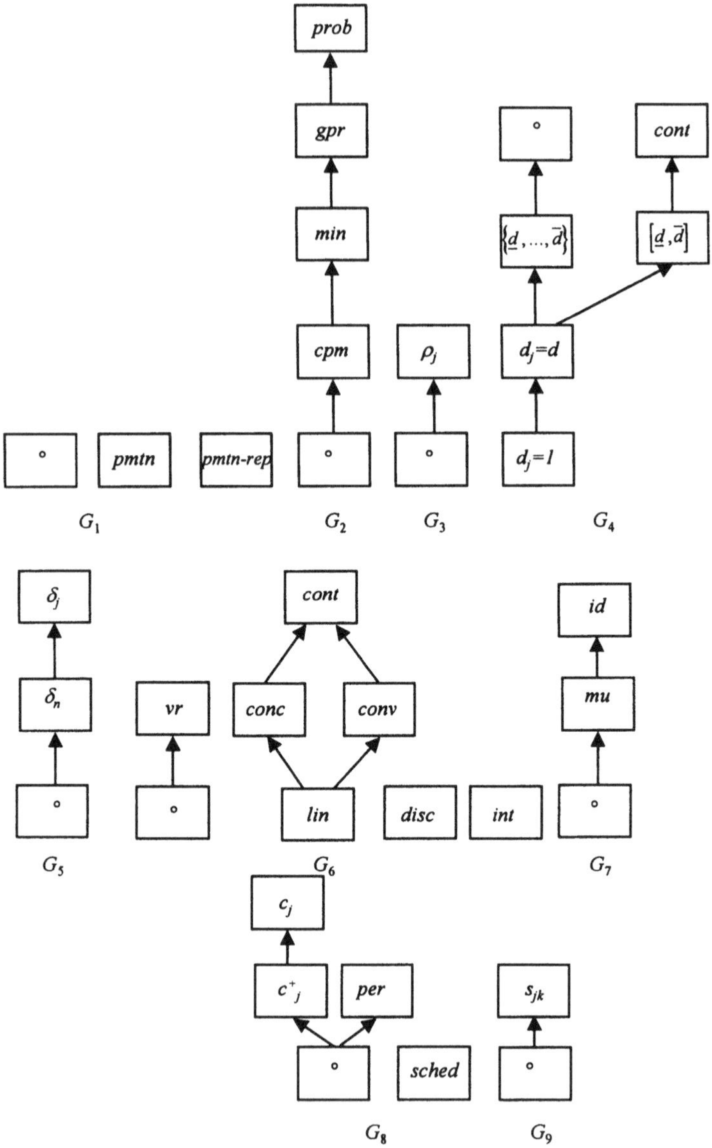

Figure 1:2. Simple reductions for the activity parameters

The reduction graphs in Figure 1:2 show the simple reductions between the various activity parameters. Reduction graph G_1 shows the case of a project scheduling problem in which preemptions are not allowed and the environment which does allow for preemption.

Reduction graph G_2 shows the simple reduction from a problem with unordered activities to a problem with finish-start precedence constraints with zero time lag, to a problem with precedence constraints with minimal time lags, to a problem with generalized minimal and maximal time lags. The latter reduces to the case where the network is probabilistic.

Reduction graph G_3 shows the simple reduction from a project scheduling problem in which all activities have a ready time (release date) equal to zero to a problem setting which specifies ready times which differ per activity. The simple reductions in reduction graph G_4 are for the duration parameters in the discrete and continuous case. A scheduling problem in which all activity durations are equal to 1 reduces to a scheduling problem with all activities having equal durations. This problem in turn reduces to a problem in which the given activity durations are bounded from below and above or ($\{\underline{d},\ldots,\overline{d}\}$ or $[\underline{d},\overline{d}]$), which can be reduced to a problem with arbitrary integer or continuous durations, respectively. The deadline reductions are shown in graph G_5. These reductions indicate that a problem without activity deadlines is a subproblem of a problem in which a deadline on the last activity (project duration) is imposed. The latter problem in turn reduces to a problem with individual activity deadlines.

Reduction graph G_6 describes the interrelations between the resource requirement characteristics. Problems in which activities have constant discrete resource requirements are subproblems of problems in which activities require the resources in variable discrete amounts. Problems with linear resource requirement functions are subproblems of problems in which the resource requirement functions are concave or convex. These in turn are subproblems of problems in which general resource functions are used. Reduction graph G_7 explains the reductions between the execution mode parameters. Single mode problems are subproblems of multi-mode problems, which reduce to mode-identity constrained problems.

Reduction graph G_8 shows the reduction between the cash flow parameters. A project scheduling problem in which no cash flows are specified for the activities (nodes in the network) reduces to a problem where activities have positive cash flows associated with the activities and a problem in which cash flows are specified on a periodic basis. The former reduces to the problem with arbitrary cash flows associated with the activities.

Finally, reduction graph G_9 shows the simple reduction from a project scheduling problem with no or sequence-independent change-over times to a problem where change-over times are sequence-dependent.

1.2.3. Field γ. performance measures

The third field γ is reserved to denote optimality criteria (performance measures). The performance measures are either *early completion measures* or *free completion measures*. *Early completion measures* (commonly denoted in the scheduling literature as r*egular performance measures*) involve penalty functions which are nondecreasing in activity completion times (Conway et al. 1967). Common examples of early completion criteria often used in project scheduling are the minimization of the project duration (makespan), the minimization of the project lateness or tardiness and the minimization of project costs. A less common example used in project scheduling with discounted cash flows is the maximization of the project net present value under the assumption that only positive cash flows are considered (see Herroelen et al. 1997). There are, however, many applications in which *free completion measures* (commonly denoted as *nonregular performance measures* in the scheduling literature) are appropriate. A typical example is the maximization of the net present value of a project characterized by arbitrary cash flow values or the minimization of the weighted earliness-tardiness.

The following settings are used:

$\gamma = reg$: the performance measure is any early completion (regular) measure,
$\gamma = nonreg$: the performance measure is any free completion (nonregular) measure.

Obviously the list of possible performance measures is almost endless. We provide the user with sufficient degrees of freedom to introduce suitable measures through a proper setting of the parameter value or through the specification of the mathematical expression of the objective function(s). The following is a nonexhaustive list of example settings:

makespan:

$\gamma = C_{max}$: minimize the project makespan.

flow time:

$\gamma = \bar{F}$: minimize the average flow time over all subprojects or activities.

due date performance:

$\gamma = L_{max}$: minimize the project lateness (i.e. the maximum of the lateness of the subprojects or activities),
$\gamma = T_{max}$: minimize the project tardiness (i.e. the maximum of the tardiness of the subprojects or activities),
$\gamma =$ *early/tardy*: minimize the weighted earliness-tardiness of the project,
$\gamma = n_T$: minimize the number of tardy activities.

levelling:

$\gamma = \Sigma sq.dev.$: minimize the sum of the squared deviations of the resource requirements from the average,
$\gamma = \Sigma jump$: minimize the weighted jumps in resource usage for all resource types over all time periods,
$\gamma = \Sigma abs.dev.$: minimize the sum of the absolute deviations of the resource requirements from the average,
$\gamma = av$: minimize the resource availability in order to meet the project deadline,
$\gamma = rac$: minimize the resource availability costs, i.e. the weighted availability of each resource type,
$\gamma = curve$: determine the complete time/cost trade-off curve.

financial:

$\gamma = npv$: maximize the net present value of the project.

stochastic:

$\gamma =$ E[.]: optimize the expected value of a performance measure,
$\gamma = cdf$: determine the cumulative density function of the project realization

date,

$\gamma = ci$: determine the criticality index of an activity (the probability that it will be on the critical path) or of a path (the probability that it will be critical),

$\gamma = mci$: determine the most critical path(s) or activities based on the criticality index values.

The above examples refer to the use of *single objective functions*. Obviously *multiple objectives* may be used (see for example Weglarz 1990, Nabrzyski and Weglarz 1994, Slowinski et al. 1994). We suggest the setting $\gamma = multi$ to specify the multi-objective case where different objectives are weighted or combined and the setting $\gamma = multicrit$ to specify multicriteria functions. Again we leave it up to the user to be more specific in the specification of the multiple objectives used.

In the above examples no distinction is made between single and multi-project scheduling. *Multi-project scheduling* problem settings can easily be accommodated. The network logic allows for the combination of the activity networks of the various projects into a single network and the user is given the freedom to specify the proper performance measures used. In practice *hybrid multi-project programs* occur which are made up of several classes of projects, each possibly having its own distinctive characteristics. In such cases, we suggest to identify the overall multi-project scheduling problem by the parameter settings corresponding to the most general case. The various reductions among the resource and activity parameters discussed earlier may prove to be very useful in this respect. The eventual use of multiple objectives can be specified along the lines indicated above.

1.3. Use of the classification scheme

The purpose of this section is to illustrate the potential use of the classification scheme in the characterization and classification of the project scheduling problems encountered in the literature.

1.3.1. Scheduling in the absence of resource constraints

1.3.1.1. Time analysis of activity networks. The classical problem of computing the longest path (critical path length) in a PERT/CPM project with finish-start precedence relations with zero time lag and known activity durations is represented as $cpm|C_{max}$. The same problem in precedence diagramming networks, i.e. networks with finish-start, start-finish, start-start and finish-finish relations with minimal time lags only is denoted as $min|C_{max}$. The problem of computing the critical path in a project with generalized precedence relations with minimal and maximal time lags is denoted as $gpr|C_{max}$.

1.3.1.2. Maximizing the net present value in project networks. The problem of scheduling project activities subject to finish-start precedence constraints with zero time lag in order to maximize the net present value of the project (often referred to as the unconstrained max-npv problem; for a review see Herroelen et al. 1997) is represented as $cpm,c_j|npv$. This notation implies that the cash flows are associated with the network activities and that the cash flow amounts are assumed to be given. The same problem in networks with minimal time lags is denoted as $min,c_j|npv$. For networks with generalized precedence relations the problem is denoted as $gpr,c_j|npv$. The so-called payment scheduling problem (Dayanand and Padman 1997) which involves the simultaneous determination of both the amount and timing of progress payments in an unconstrained max-npv environment can be denoted as $cpm,sched|npv$.

1.3.2. Project scheduling under resource constraints

The literature on project scheduling under various types of resource constraints has expanded drastically over the past few years. For a review of the recent developments we refer the reader to Herroelen et al. (1998).

1.3.2.1. The resource-constrained project scheduling problem. The resource-constrained project scheduling problem (often denoted as RCPSP) involves the scheduling of project activities subject to finish-start precedence constraints with zero time lag and constant renewable resource constraints in order to minimize the project duration. Activities have a single execution mode with a fixed integer duration, preemption is not allowed and renewable resource requirements are constant throughout the duration of an activity. This problem is denoted as $m,1|cpm|C_{max}$. The problem is known to be a generalization of the job shop scheduling problem.

1.3.2.2. The preemptive resource-constrained project scheduling problem. The preemptive resource-constrained project scheduling problem (often denoted as PRCPSP) extends the $m,1|cpm|C_{max}$ in that it allows for activity preemption at integer points in time. The problem can be denoted as $m,1|pmtn,cpm|C_{max}$.

1.3.2.3. The generalized resource-constrained project scheduling problem. The generalized resource-constrained project scheduling problem (often referred to as GRCPSP) extends problem $m,1|cpm|C_{max}$ to the case of minimal time lags, activity release dates and deadlines and variable resource availabilities. The proposed classification scheme classifies the problem as $m,1,va|min,\rho_j,\delta_j|C_{max}$.

1.3.2.4. The resource-constrained project scheduling problem with generalized precedence relations. The resource-constrained project scheduling problem with generalized precedence relations (often denoted as RCPSP-GPR or RCPSP/max) extends problem $m,1|cpm|C_{max}$ in that it allows for start-start, finish-start, start-finish and finish-finish precedence constraints with both minimal and maximal time lags. This basic extension can be denoted as $m,1|gpr|C_{max}$. The use of minimal and maximal time lags also allows, however, for the modelling of ready times and activity deadlines, variable resource requirements and availabilities. Procedures have been developed which allow for the use of any regular objective function. This general problem setting is denoted as $m,1,va|gpr,\rho_j,\delta_j,vr|reg$.

1.3.2.5. Time/cost trade-off problems. The classical time/cost trade-off problem in CPM-networks (described in project management textbooks such as Elmaghraby 1977, Moder et al. 1983, Wiest and Levy 1977, Shtub et al. 1994) basically assumes that resources are available in infinite amounts and hence does not *explicitly* take resource decisions into account. A direct activity cost function is used instead, representing the direct activity costs as a function of the activity duration. Activity durations are bounded from below by the crash duration (corresponding to a maximum allocation of resources) and bounded from above by the normal duration (corresponding to the most efficient resource allocation). Given a project deadline, the objective is to determine the activity durations and to schedule the activities in order to minimize the project cost; i.e. the sum of the direct costs of the activities. Essentially, the project costs correspond to a requirement for a nonrenewable resource, the total requirement of which is to be minimized. This corresponds to minimizing the (required) availability of the nonrenewable resource. Hence, the problem can be denoted as $1,T|cpm,\delta_n,lin,mu|av$ when the activity cost functions are linear. The notation $1,T|cpm,\delta_n,conc,mu|av$ is used for the case of concave activity cost functions while the notation $1,T|cpm,\delta_n,conv,mu|av$ is used in case the activity cost functions are convex. The notation $1,T|cpm,\delta_n,cont,mu|av$ denotes the time/cost trade-off problem for the case of general continuous activity cost functions.

The discrete time/cost trade-off problem (for a review see De et al. 1995) assumes a single nonrenewable resource. The duration of an activity is a discrete, nonincreasing function of the amount of a single resource allocated to it. An activity assumes different execution modes according to the possible resource allocations. When a limit on the total availability of the single nonrenewable resource is specified, the problem is to decide on the vector of activity durations that completes the project as early as possible under the limited availability of the single nonrenewable resource type. This problem is denoted by our classification scheme as $1,T|cpm,disc,mu|C_{max}$. Obviously, the notation becomes $1,T|min,disc,mu|C_{max}$ and $1,T|gpr,disc,mu|C_{max}$ for the case of minimal and both minimal and maximal time lags, respectively. A second objective function reverses the problem formulation. Now a project deadline is specified and the minimization is over the sum of the resource use over all activities. The notation for this problem is $1,T|cpm,\delta_n,disc,mu|av$, where again the parameter *cpm* can be changed into *min* and *gpr* to denote the corresponding type of precedence relations. It should be noted that the only difference with the classical time/cost trade-off problem in CPM-networks lies in the use of the *disc* parameter specification, referring to the use of a discrete resource requirements function. In some studies room is made for a third objective which involves the computation of the complete time/cost trade-off function for the total project. Exploiting the degrees of freedom allowed by our classification scheme, this problem could be represented as $1,T|cpm,disc,mu|curve$, where the value *curve* is given to the parameter γ.

1.3.2.6. Discrete time/resource trade-off problems. The discrete time/resource trade-off problem (often referred to as DTRTP) assumes that the duration of an activity is a discrete, non-increasing function of the amount of a single *renewable* resource committed to it. Given a specified work content for an activity, all its efficient execution modes are determined based on time/resource trade-offs. An activity when performed in a specific mode has a duration and a resource requirement during each period it is in progress, such that the resource-duration product is at least equal to the specified work content. The single resource has a constant per period availability. The objective is to schedule each activity in one of its modes, subject to the precedence and the renewable resource constraints, under the objective of minimizing the project makespan.

The discrete time/resource trade-off problem can be denoted as $1,1|cpm,disc,mu|C_{max}$. Obviously, *min* or *gpr* can be used instead of *cpm* to denote the proper type of precedence relations. Moreover, the problem resembles the discrete time/cost trade-off problem $1,T|cpm,disc,mu|C_{max}$ which studies time/cost trade-offs for a single nonrenewable resource. The only difference in the notation lies in the second parameter of the α-field: a T is used for the time/cost trade-off problems while a 1 is used for the time/resource trade-off problems.

It is also possible to define a kind of dual problem to the discrete time/resource trade-off problem which involves the minimization of the resource availability subject to the project deadline. The corresponding notation is $1,1|cpm,\delta_n,disc,mu|av$.

1.3.2.7. Multi-mode resource-constrained project scheduling problems. The multi-mode resource-constrained project scheduling problem (sometimes referred to as MRCPSP) includes time/resource and resource/resource trade-offs, multiple renewable, nonrenewable and doubly-constrained resources and a variety of objective functions. In the basic problem setting activities have to be scheduled in one of their possible execution modes subject to renewable and nonrenewable resources. Under the minimum makespan objective the general problem, including renewable and nonrenewable resources, can be denoted as $m,1T|cpm,disc,mu|C_{max}$ for projects with finish-start precedence constraints with zero time lag. It should be noted that algorithms have been developed to deal with multi-mode problems involving deadlines, variable availabilities and any type of regular objective function. The problem tackled by such procedures would be denoted as $m,1T,va|cpm,\delta_j,disc,mu|reg$. The problem with *mode identity constraints* can be denoted as $m,1T|cpm,disc,id|C_{max}$. The multi-mode problem with partially renewable resource constraints can be denoted as $m,v|cpm,disc,mu|C_{max}$.

1.3.2.8. Resource levelling problems. Various types of resource levelling problems have been studied in the project scheduling literature. The classical *resource levelling problem* (denoted by some authors as *RLP*) is somewhat the dual of the resource-constrained project scheduling problem. Instead of minimizing the project duration subject to renewable resource constraints, the problem now is to schedule the activities in order to level the resource use subject to a project deadline. An explicit resource availability constraint is not taken into account. Various levelling objectives have been used in the literature. In essence, they can be considered as resource cost functions. The resource levelling problem can be denoted according to the rules of our classification scheme as $m,1|cpm,\delta_n|\Sigma sq.dev.$, if the objective is to minimize the squared deviations of the resource requirements around the average resource requirement, which is equivalent to minimizing the sum of squares of the resource requirements for each period in the project schedule as such. If the objective is to minimize the absolute deviations of the resource requirements, the notation becomes $m,1|cpm,\delta_n|\Sigma abs.dev.$ The γ-field in the notation becomes $\Sigma jump$ for the objective of minimizing the weighted jumps in resource consumption for each resource type over all time periods.

Instead of taking the view of finding the minimum project length which does not violate the precedence and resource constraints, the *resource availability cost problem* (*RACP*) takes the view that the individual resource availabilities determine the cost of executing the schedule. It aims at determining the cheapest resource availability amounts for which a feasible project schedule exists that does not violate the project deadline. Resource costs are to be determined under the assumption of a discrete, non-decreasing cost function of the constant availability of the renewable resource types. The notation for this problem is $m,1|cpm,\delta_n|rac$.

1.3.2.9. Resource-constrained project scheduling with discounted cash flows. The basic problem of maximizing the net present value of a project subject to finish-start precedence and renewable resource constraints is denoted as $m,1|cpm,c_j|npv$. This problem setting assumes that the cash flow amounts, c_j, are known and associated with the network activities. If the known cash flows occur at regular intervals or with a known frequency, the corresponding problem can be denoted as $m,1|cpm,per|npv$. In the case where both the amount and timing of the cash flows have to be determined the so-called payment scheduling problem results, which can be denoted as $m,1|cpm,sched|npv$. Obviously, the parameter setting *min* can be used in the β-field to denote the case with minimal time lags and the setting *gpr* can be used for the case of both minimal and maximal time lags. In a similar fashion ready times and deadlines can be introduced in the notation through the parameter settings ρ_j and δ_j.

1.3.3. Classification of stochastic problem settings

The above classifications mainly apply to a deterministic problem setting. The classification of the stochastic problem equivalents should pose no major problems as the degrees of freedom present in our classification scheme allow for proper classification in those cases. Setting the activity duration parameter $\beta_4 = \tilde{d}_j$ to denote that the activity durations are stochastic, the problem involving the determination (estimation) of the cumulative probability density function of the project realization date in projects with stochastic activity durations can be represented as $cpm, \tilde{d}_j \mid cdf$. The problem of computing the expected duration of a GERT network can be denoted as $gert, \tilde{d}_j \mid \mathrm{E}[C_{max}]$.

In most cases, the stochastic characteristics apply to the activity durations and the amount of the cash flows. Setting the cash flow parameter $\beta_8 = \tilde{c}_j$ in order to indicate that the cash flows are stochastic, and using $E[npv]$ to indicate the objective of maximizing the *expected* net present value of the project, the stochastic resource-constrained project scheduling problem with discounted cash flows subject to finish-start precedence constraints and renewable resource constraints is denoted as $m,1 \mid cpm, \tilde{d}_j, \tilde{c}_j \mid E[npv]$.

It is also possible to specify stochastic resource availabilities and requirements. Setting $\alpha_3 = \tilde{a}$ denotes a stochastic resource availability which remains constant over time. The setting $\alpha_3 = v\tilde{a}$ denotes a stochastic resource availability which varies over time. In a similar fashion the setting $\beta_6 = \tilde{r}$ denotes a stochastic discrete resource requirement which is the same for every time period of activity execution. The setting $\beta_6 = v\tilde{r}$ denotes a stochastic resource requirement of a discrete number of units which varies over the periods of activity execution.

1.4. Conclusions

Over the past few years the variety of project scheduling problems studied in the literature has been drastically increased. The conception of new problem types and the research of existing basic models under more realistic problem assumptions has created the need for a detailed classification scheme which allows for a precise and unambiguous classification of the problems under study.

In this chapter a classification scheme is introduced which is composed of three fields $\alpha \mid \beta \mid \gamma$, denoting the resource characteristics, activity characteristics and performance measures, respectively. Precise settings for the various parameters are introduced together with the corresponding reduction graphs which allow the description of the interrelations between the various problem settings. The scheme is illustrated by applying it to the most common project scheduling problems which have been studied in the literature.

The classification scheme proves to possess sufficient rigidity *and* flexibility to accommodate in an unambiguous way the full spectrum of problem type characteristics. It offers sufficient detail to allow for a *unique* codification of the relevant project scheduling problems studied by the project scheduling community. Obviously, this imposes a rather high degree of rigour on the various parameter specifications. At the same time, however, we tried to keep the scheme *flexible* and *workable*, i.e. as concise and simple as possible, with sufficient degrees of freedom to be used by the individual user in order to specify individual "desires".

We are convinced that the scheme is usable. It allows for the unique codification of the overwhelming variety of project scheduling problems under study, as illustrated by the sample we provide in Section 1.3. Obviously, the scheme takes some time to digest. It must be studied before it can be used. Our experience is that once one tries it out on a few problem settings, the smoke curtain which initially may hide the logic behind the three field parameter settings steadily disappears.

We do hope that the scheme gains wide acceptance within the project scheduling community and we hope that (a) it facilitates the presentation and discussion of project scheduling problems, (b) it relieves authors and presenters of "wasting" a lot of time in preparing lengthy descriptions of problem assumptions and characteristics, (c) it allows for immediate problem identification and simplifies the assessment of problem complexity, (d) it allows for the identification of viable, relevant areas of research, (e) it provides additional insight in the ideosyncracies of the various problem settings studied in the literature by showing problem interrelations, dependencies, common characteristics and complicating factors.

Obviously there may always be a project setting that does not fit exactly in the "pigeon holes" we provide. We have done our best to combine "rigour" and "freedom" and have tried to provide sufficient generality in the categories, which offers the user sufficient degrees of specification freedom.

Acknowledgement

The development of the classification scheme presented in this chapter was inspired by discussions held at the Workshop on Scheduling & Heuristic Search, held on May 8, 1997 at the A. Gary Anderson Graduate School of Management, University of California, Riverside. We are grateful to the constructive comments and criticism received from many colleagues within the project scheduling community. Especially the detailed and constructive remarks of Richard Deckro (Air Force Institute of Technology), Bajis Dodin (University of California at Riverside), Salah Elmaghraby (North Carolina State University at Raleigh), Concepcion Maroto (Universidad Politécnica de Valencia), Jim Patterson (Indiana University) and Jan Weglarz (Poznan University of Technology) are deeply appreciated. Their suggestions to remove ambiguity and modify various parameter settings have been crucial in making the scheme to what it is now.

References

BARTSCH, T., R. NISSEN AND F. SALEWSKI. 1997. Genetic Algorithms for Resource-Constrained Project Scheduling with Changeover Times. Research Report, Christian-Albrechts-Universität zu Kiel.

BLAZEWICZ, J., W. CELLARY, R. SLOWINSKI AND J. WEGLARZ. 1986. *Scheduling under Resource Constraints - Deterministic Models*. Baltzer, Basel.

BLAZEWICZ, J., K. H. ECKER, G. SCHMIDT AND J. WEGLARZ. 1994. *Scheduling in Computer and ManufacturingSystems*. Springer-Verlag, Berlin.

BLAZEWICZ, J., J. K. LENSTRA AND A. H. G. RINNOOY KAN. 1983. Scheduling Subject to Resource Constraints: Classification and Complexity. *Discrete Appl. Math.* **5**, 11-24.

BÖTTCHER, J., A. DREXL AND R. KOLISCH. 1996. A Branch-and-Bound Procedure for Project Scheduling with Partially Renewable Resource Constraints. *Proceedings of the Fifth Workshop on Project Management and Scheduling*, Poznan, April 11-13, 48-51.

BRUCKER, P. 1995. *Scheduling Algorithms*. Springer-Verlag, Berlin.

CONWAY, R. W., W. L. MAXWELL AND L. W. MILLER. 1967. *Theory of Scheduling*. Addison-Wesley, Reading, Mass.

DAYANAND, N. AND R. PADMAN. 1997. On Modelling Payments in Projects. *J. Opnl. Res. Soc.*, to appear.

DE, P., E. J. DUNNE, J. B. GOSH AND C. E. WELLS. 1995. The Discrete Time/Cost Trade-Off Problem Revisited. *Eur. J. Opnl. Res.* **81**, 225-238.

DE REYCK, B. 1995. Project Scheduling under Generalized Precedence Relations - A Review: Part 1 and Part 2. Research Reports 9517-9518, Department of Applied Economics, Katholieke Universiteit Leuven.

DODIN, B. AND A. A. ELIMAM. 1997. Audit Scheduling with Overlapping Activities and Sequence-Dependent Setup Costs. *Eur. J. Opnl. Res.* **97**, 22-33.

DREXL, A. 1997. Local Search Methods for Project Scheduling under Partially Renewable Resource Constraints. Paper presented at the INFORMS San Diego Spring Meeting, May 4-7, 1997.

ELMAGHRABY, S. E. 1977. *Activity Networks - Project Planning and Control by Network Models*. Wiley Interscience, New York.

ELMAGHRABY, S. E. 1995. Activity Nets: A Guided Tour through Some Recent Developments. *Eur. J. Opnl. Res.* **82**, 383-408.

GRAHAM, R. L., E. L. LAWLER, J. K. LENSTRA AND A. H. G. RINNOOY KAN. 1979. Optimization and Approximation in Deterministic Sequencing and Scheduling Theory: A Survey. *Annals Discrete Math.* **5**, 287-326.

HACKMAN, S. T. AND R. C. LEACHMAN. 1989. An Aggregate Model of Project-Oriented Production. *IEEE Trans. Systems, Man and Cybernetics.* **19**, 220-231.

HERROELEN, W. AND E. DEMEULEMEESTER. 1995. Recent Advances in Branch-and-Bound Procedures for Resource-Constrained Project Scheduling Problems, Chapter 12 in *Scheduling Theory and Its Applications* (Chrétienne, Ph. et al. (Eds)), John Wiley & Sons, Chichester.

HERROELEN, W., E. DEMEULEMEESTER AND B. DE REYCK. 1998. Resource-Constrained Project Scheduling - A Survey of Recent Developments. *Comput. and O.R.* **25**, 279-302.

HERROELEN, W. S., P. VAN DOMMELEN AND E. L. DEMEULEMEESTER. 1997. Project Network Models with Discounted Cash Flows: A Guided Tour through Recent Developments. *Eur. J. Opnl. Res.* **100**, 97-121.

ICMELI, O., S. ERENGÜÇ AND J. C. ZAPPE. 1993. Project Scheduling Problems: A Survey. *Int. J. Prod. Opns. Mgmt.* **13**, 80-91.

JOZEFOWSKA, J. AND J. WEGLARZ. 1994. Approximation Algorithms for Some Discrete-Continuous Scheduling Problems. *Proceedings of the Fourth International Workshop on Project Management and Scheduling*, Leuven, July 12-15, 62-63.

KAPLAN, L. 1991. Resource-Constrained Project Scheduling with Setup Times. Research Report, University of Tenessee.

KOLISCH, R. 1995. *Project Scheduling under Resource Constraints.* Physica-Verlag, Heidelberg.

LAWLER, E. L., J. K. LENSTRA, A. H. G. RINNOOY KAN AND D. B. SHMOYS. 1993. Sequencing and Scheduling: Algorithms and Complexity, Chapter 9 in *Logistics of Production and Inventory*, Handbooks in Operations Research and Management Science, Volume 4 (S.C. Graves et al. (Eds.)), North-Holland, Amsterdam, 445-522.

LEACHMAN, R. C. 1983. Multiple Resource Leveling in Construction Systems through Variation in Activity Intensities. *Nav. Res. Log. Quart.* **30**, 187-198.

LEACHMAN, R. C., A. DINCERLER AND S. KIM. 1990. Resource-Constrained Scheduling of Projects with Variable-Intensity Activities. *IIE Trans.* **22**, 31-40.

MODER, J. J., C. R. PHILLIPS AND E. W. DAVIS. 1983. *Project Management with CPM, PERT and Precedence Diagramming.* Van Nostrand Reinhold Company, Third Edition.

NABRZYSKI, J. AND J. WEGLARZ. 1994. A Knowledge-Based Multiobjective Project Scheduling System. *Revue des systèmes de décision*, **3**, 185-200.

NEUMANN, K. AND U. STEINHARDT. 1979. *GERT Networks and the Time-Oriented Evaluation of Projects.* Lecture Notes in Economics and Mathematical Systems, 172, Springer-Verlag, Berlin.

ÖZDAMAR, L. AND G. ULUSOY. 1995. A Survey on the Resource-Constrained Project Scheduling Problem. *IIE Transactions*, 27, 574-586.

PINEDO, M. 1995. *Scheduling - Theory, Algorithms and Systems*. Prentice-Hall, Inc., Englewood Cliffs, New Jersey.

SALEWSKI, F. 1996. Tabu Search Algorithms for Project Scheduling under Resource and Mode Identity Constraints. Paper presented at the IFORS 14th Triennial Meeting, Vancouver, B.C., July 8-12.

SALEWSKI, F. AND S. LIEBERAM-SCHMIDT. 1996. Greedy Look Ahead Methods for Project Scheduling under Resource and Mode Identity Constraints, *Proceedings of the Fifth International Workshop on Project Management and Scheduling*, Poznan, April 11-13, 207-211.

SALEWSKI, F., A. SCHIRMER AND A. DREXL. 1997. Project Scheduling under Resource and Mode Identity Constraints - Model, Complexity, Methods and Application. *Eur. J. Opnl. Res.* **102, 88-110**.

SCHIRMER, A. AND A. DREXL. 1996. Partially Renewable Resources - A Generalization of Resource-Constrained Project Scheduling. Paper presented at the IFORS Triennial Meeting, Vancouver, B.C., July 8-12.

SHTUB, A., J. F. BARD AND S. GLOBERSON. 1994. *Project Management: Engineering, Technology, And Implementation*. Prentice Hall International, Inc., Englewood Cliffs.

SLOWINSKI, R., B. SONIEWICKI AND J. WEGLARZ. 1994. DSS for Multiobjective Project Scheduling. *Eur. J. Opnl. Res.* **79**, 220-229.

WEGLARZ, J. 1980. Control in Resource Allocation Systems. *Foundations of Control Engineering*, **5**, 159-180.

WEGLARZ, J. 1990. Synthesis Problems in Allocating Continuous, Doubly Constrained Resources among Dynamic Activities. *Proceedings of the IFORS 12th Triennial Meeting*, 715-724.

WEGLARZ, J. AND J. JOZEFOWSKA. 1997. Parallel Machine Scheduling with Additional Continuous Resource: Mean Completion Time Results. Paper presented at the INFORMS San Diego Spring Meeting, May 4-7.

WIEST, J. D. AND F. K. LEVY. 1977. *A Management Guide to PERT/CPM: with GERT/PDM/DCPM and Other Networks*. Prentice-Hall, Inc., Englewood Cliffs.

2 SOLVING LARGE-SIZED RESOURCE-CONSTRAINED PROJECT SCHEDULING PROBLEMS

Peter Brucker

Sigrid Knust *

Universität Osnabrück

2.1 INTRODUCTION

The **resource-constrained project scheduling problem (RCPSP)** may be formulated as follows:

Given are n activities $1, \ldots, n$ and r renewable resources $1, \ldots, r$. A constant amount of R_k units of resource k is available at any time. Activity i must be processed for p_i time units, where preemption is not allowed. During this time period a constant amount of r_{ik} units of resource k is occupied. The values R_k, p_i and r_{ik} are supposed to be non-negative integers. Furthermore, finish-

*Supported by the Deutsche Forschungsgemeinschaft, Project 'Komplexe Maschinen-Schedulingprobleme'

start precedence relations are defined between the activities. The objective is to determine starting times S_i for the activities $i = 1, \ldots, n$ in such a way that

- at each time t the total resource demand is less than or equal to the resource availability for each resource type,
- the given precedence constraints are fulfilled, and
- the makespan $C_{\max} = \max_{i=1}^{n} C_i$ is minimized, where $C_i := S_i + p_i$ is the completion time of activity i.

A schedule is called **active** if the starting times of the activities are such that no activity can be started earlier without delaying some other activity or violating the constraints. If not otherwise stated we assume that all schedules to be considered are active.

The problem is presented by an activity-on-the-node network with dummy starting and end nodes 0 and $n + 1$, respectively, and $p_0 = p_{n+1} = 0$. We assume $S_0 = 0$ and identify S_{n+1} with the $C_{\max}$-value. We denote by V the set of all activities (including the dummy activities).

We present a branch-and-bound algorithm from Brucker et al. (1997) which generalizes branch-and-bound methods that have been developed for the job-shop scheduling problem (Brucker et al. 1994) and for the multiprocessor task scheduling problem (Krämer 1997). Both problems are special cases of the RCPSP. Furthermore, we describe a tabu-search procedure for the RCPSP. These algorithms are based on schedule schemes which allow a generalization of immediate selection procedures. Immediate selection is a kind of constraint propagation which has already been used in connection with the job-shop scheduling problem.

We also discuss methods for calculating lower bounds for the RCPSP using a linear programming approach with delayed column generating techniques.

In the next section schedule schemes are introduced which are the basis of immediate selection techniques discussed in Section 2.3. In Section 2.4 a tabu-search algorithm is presented. Lower bound calculations are described in Section 2.5 while the branch-and-bound algorithm is presented in Section 2.6. Computational results can be found in the final section.

2.2 SCHEDULE SCHEMES

In this section we introduce the notion of schedule schemes which represent sets of feasible schedules. We will use schedule schemes to represent the nodes of a branch-and-bound algorithm. Schedule schemes will also be used as a crucial component in connection with local search methods.

To motivate schedule schemes we consider a feasible schedule $S = (S_i)$ defined by the starting times S_i of all activities $i = 0, \ldots, n+1$. S induces **conjunctions** $i \rightarrow j$ and **parallelity relations** $i \parallel j$ for activities i, j. $i \rightarrow j$ stands for $C_i \leq S_j$, i.e. activity j does not start before activity i is finished. $i \parallel j$ means that activities i and j are processed in parallel for at least one time unit. This is the case if and only if $S_i < C_j$ and $S_j < C_i$. Note that for two activities i and j we either have $i \rightarrow j$ or $j \rightarrow i$ or $i \parallel j$.

To create sets of schedules we replace conjunctions by **disjunctions** $i - j$ where $i - j$ means that either $i \rightarrow j$ or $j \rightarrow i$ holds. $i - j$ is the logical negation of $i \parallel j$. Furthermore, we may relax $i - j$ or $i \parallel j$ by a so-called **flexibility relation** $i \sim j$. $i \sim j$ means that it has not yet been decided which of the relations $i \parallel j$ or $i - j$ (or even $i \rightarrow j$ or $j \rightarrow i$) hold. Note that the relations $i - j$, $i \parallel j$, and $i \sim j$ are symmetric while $i \rightarrow j$ is not. Two activities i and j are called **incompatible** if $i - j$, or $i \rightarrow j$, or $j \rightarrow i$ holds. Let C be a set of conjunctions, D a set of disjunctions, N a set of parallelity relations, and let F be a set of flexibility relations. Then (C, D, N, F) is called a schedule scheme if

- the relation sets C, D, N, F are disjoint, and
- for arbitrary activities $i \neq j$ either $i \rightarrow j \in C$ or $j \rightarrow i \in C$ or $i - j \in D$ or $i \parallel j \in N$ or $i \sim j \in F$ holds.

The schedule scheme (C, D, N, F) represents the set $\mathcal{S}(C, D, N, F)$ of all active schedules not violating the relations in C, D, N, F. A schedule scheme may represent infeasible schedules as well as feasible ones because the resource constraints may be violated. The subset of feasible schedules that are represented by (C, D, N, F) is denoted by $\mathcal{S}_f(C, D, N, F)$.

Let C_0 be the set of conjunctions which are defined by the original precedence relations and let D_0 be an initial set of disjunctions that is induced by the resource constraints for pairs of activities (i.e. we set $i - j \in D_0$ if and only if $r_{ik} + r_{jk} > R_k$ for some $k = 1, \ldots, r$). Then $(C_0, D_0, \emptyset, F_0)$ with $F_0 = \{i \sim j \mid i \rightarrow j \notin C_0,\ j \rightarrow i \notin C_0,\ i - j \notin D_0\}$ represents all feasible active schedules.

The branch-and-bound algorithm to be presented in Section 2.6 uses schedule schemes for the enumeration process. It starts with the schedule scheme $(C_0, D_0, \emptyset, F_0)$ representing all feasible schedules. Branching is done by turning a flexibility relation $i \sim j$ either into the parallelity relation $i \parallel j$ or into the disjunction $i - j$.

Next, we show that for a scheduling scheme $(C, D, N, \emptyset)$ we can either recognize that no feasible schedule exists which corresponds with $(C, D, N, \emptyset)$ or we can calculate a feasible schedule dominating all feasible schedules which correspond with $(C, D, N, \emptyset)$. Furthermore, both can be done in polynomial time.

Thus, schedule schemes of the form $(C, D, N, \emptyset)$ can be treated as leaves of the enumeration tree and the branch-and-bound procedure solves the problem after a finite number of branchings.

A schedule scheme $(C', \emptyset, N, \emptyset)$ is called a **transitive orientation** of $(C, D, N, \emptyset)$ if

- $(C', \emptyset, N, \emptyset)$ is derived from $(C, D, N, \emptyset)$ by turning all disjunctions $i - j \in D$ into conjunctions $i \rightarrow j$ or $j \rightarrow i$, and
- C' is transitive, i.e. for all $i \rightarrow j$, $j \rightarrow k \in C'$ we have $i \rightarrow k \in C'$.

Obviously, a transitive orientation is acyclic, i.e. the corresponding graph (V, C') is acyclic. Note that the set of solutions $\mathcal{S}(C, \emptyset, N, \emptyset)$ represented by a transitive orientation contains at most one element because only active schedules are considered. Of course, such a schedule may be infeasible with respect to the resource constraints.

Given a transitive orientation $(C, \emptyset, N, \emptyset)$ of some schedule scheme we construct a schedule by the following procedure. Consider the directed graph (V, C) associated with C. Calculate for each activity $i \in V$ the length r_i of a longest path from the starting activity 0 to i (the length of a path to i equals the sum of all processing times of vertices in this path, i excluded). Schedule each activity $i \in V$ at time r_i. The schedule defined by these starting times $S_i = r_i$ is called the **earliest start schedule** associated with $(C, \emptyset, N, \emptyset)$ and is denoted by $S_{es}(C)$. The earliest start schedule associated with $(C, \emptyset, N, \emptyset)$ may not be represented by $(C, \emptyset, N, \emptyset)$ because only the set C of conjunctions is considered to construct $S_{es}(C)$, i.e. some parallelity relations may be violated. Due to results of Golumbic (1980) and Möhring (1985) an earliest start schedule $S_{es}(C')$ associated with a transitive orientation $(C', \emptyset, N, \emptyset)$ of a schedule scheme $(C, D, N, \emptyset)$ always has the same makespan, regardless which transitive orientation is chosen.

The following theorem shows that a schedule scheme $(C, D, N, \emptyset)$ can be treated as a leaf of the enumeration tree.

Theorem 1 *Let $(C', \emptyset, N, \emptyset)$ be an arbitrary transitive orientation of a schedule scheme $(C, D, N, \emptyset)$ and let $S_{es}(C')$ be the corresponding earliest start schedule. If $S_{es}(C')$ is feasible, it dominates all feasible schedules represented by $(C, D, N, \emptyset)$. Otherwise, the set $S_f(C, D, N, \emptyset)$ of feasible solutions is empty.*

Clearly, a feasible schedule always corresponds to a transitive orientation. Due to this observation and Theorem 1 we can proceed as follows when the set F of flexibility relations is empty. First we decide whether there exists a transitive orientation of $(C, D, N, \emptyset)$ and determine an arbitrary one in the case of a positive decision. (A polynomial algorithm for this problem can be found in Korte and Möhring 1985). Given a transitive orientation $(C', \emptyset, N, \emptyset)$ we calculate the corresponding earliest start schedule $S_{es}(C')$. If $S_{es}(C')$ is feasible, then we compare this schedule with the best schedule found so far. Otherwise, we can ignore $(C, D, N, \emptyset)$.

2.3 IMMEDIATE SELECTION

By immediate selection we denote certain constrained propagation techniques which are based on schedule schemes and heads and tails of activities. A **head** r_i of an activity i is a lower bound for the earliest starting time of i. Symmetrically a **tail** q_i of activity i is a lower bound for the length of the time period between the completion of activity i and the optimal makespan. By immediate selection methods

- additional conjunctions, disjunctions, or parallelity relations are deduced,
- heads and tails are increased, or
- infeasibility is detected.

Immediate selection is used in connection with branch-and-bound methods or for lower bound calculations. A branch-and-bound algorithm usually provides an upper bound UB which is improved during the algorithm. Due to integrality of all data this upper bound is integral, too. Thus, we further search for solutions for which the makespan is bounded by $U := UB - 1$. In such a solution activity i must be processed within the time window $[r_i, d_i]$ where $d_i := U - q_i$. This temporal information is exploited by immediate selection methods.

To calculate a lower bound for the set of feasible schedules defined by a schedule scheme we may proceed as follows. For a fictive upper bound U we try to show that no feasible schedule exists satisfying this upper bound. If we succeed, then $U + 1$ is a valid lower bound.

In the following sections we will describe different immediate selection techniques. Throughout this chapter we will assume that (C, D, N, F) is a schedule scheme and $S = (S_i)$ is a corresponding feasible schedule which is defined by the starting times S_i of all activities.

2.3.1 *Transitivity Conjunctions and Positive Cycles*

A feasible schedule S has the following basic properties:

- $i \rightarrow j$ holds if and only if $S_i + p_i \leq S_j$ or equivalently $S_j - S_i \geq p_i$ is satisfied.

- If $i \parallel j$ is given, then $S_j - S_i \geq -(p_j - 1)$ holds.

To prove the second property assume that $S_j - S_i < -(p_j - 1)$ is fulfilled. Then we have $S_j + p_j < S_i + 1$ which implies that $C_j \leq S_i$ holds because all data are integer causing a contradiction to $i \parallel j$.

Thus, if we define a **distance matrix** $D_S = (d_{ij})_{i,j \in V}$ associated with a schedule scheme (C, D, N, F) by

$$d_{ij} = \begin{cases} 0 & \text{if } i = j \\ p_i & \text{if } i \rightarrow j \\ -(p_j - 1) & \text{if } i \parallel j \\ -\infty & \text{otherwise,} \end{cases} \tag{2.1}$$

then for each pair (i, j) $(i, j = 0, \ldots, n + 1)$ the value d_{ij} is a lower bound for the difference $S_j - S_i$.

Recall that we have an upper bound U for the C_{max}-value which implies the constraint

$$S_{n+1} \leq U. \tag{2.2}$$

Since we have $S_i \geq 0$ for each activity i we may derive

$$S_i + p_i \leq S_{n+1} \leq U \leq U + S_j \text{ or } S_j - S_i \geq p_i - U$$

for an arbitrary pair of activities i and j. Thus, in (2.1) we may replace $-\infty$ by $p_i - U$. Additionally, we may include heads and tails in D_S by setting $d_{0i} := r_i$ and $d_{i,n+1} := p_i + q_i$ for $i = 1, \ldots, n$ because $r_i \leq S_i = S_i - 0 = S_i - S_0$ and $S_i + p_i + q_i \leq C_{\max} = S_{n+1}$, respectively.

The relation

$$S_j - S_i \geq d_{ij} \tag{2.3}$$

has the following transitivity property:

$$\text{If } S_j - S_i \geq d_{ij} \text{ and } S_k - S_j \geq d_{jk}, \text{ then } S_k - S_i \geq d_{ij} + d_{jk}. \tag{2.4}$$

Due to the transitivity property (2.4) we may replace the distance matrix $D_S = (d_{ij})_{i,j \in V}$ by its transitive closure $\overline{D}_S = (\overline{d}_{ij})_{i,j \in V}$. $\overline{D}_S$ can be calculated in $O(n^3)$ time by applying the **Floyd-Warshall algorithm** to D_S.

From $\overline{D}_S$ we may derive the following conclusions:

- If

  $$\overline{d}_{ij} \geq p_i \tag{2.5}$$

  holds, then we may fix the conjunction $i \rightarrow j$, and

- if

$$\bar{d}_{ii} > 0 \tag{2.6}$$

holds for some activity i, then no feasible schedule exists.

A relation $i \to j$ added to C by condition (2.5) is called a **transitivity conjunction**. Note that condition (2.6) indicates a positive cycle of the graph (V, C).

2.3.2 Symmetric Triples and Extensions

The concepts presented in this section are extensions of dominance rules commonly applied in the field of resource-constrained scheduling and line balancing (see, e.g. Johnson 1988).

A triple of activities (i, j, k) is called **symmetric** if $k \parallel i$ and $k \parallel j$ hold and i, j and k cannot be processed simultaneously because of resource constraints. Thus, we may introduce a disjunction $i - j \in D$. Figure 2:1 illustrates a symmetric triple.

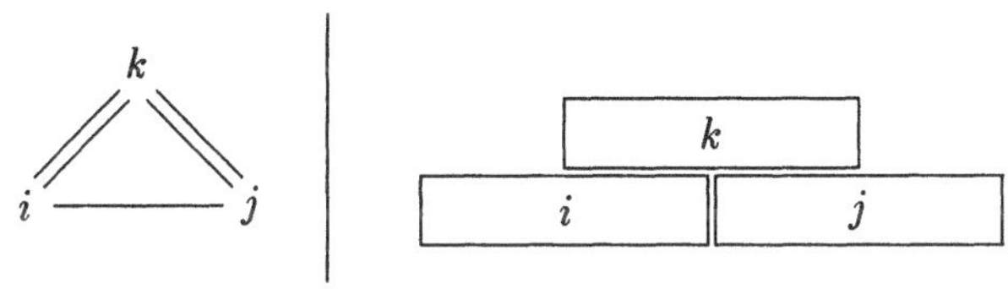

Figure 2:1. Symmetric triple

A procedure with time complexity $O(rn|N|)$ that calculates all symmetric triples (i, j, k) and updates $D := D \cup \{i - j\}$ is easily implemented. This can be seen as follows: We scan each parallelity relation $k \parallel i \in N$ which takes $O(|N|)$ time. Each activity k of some relation $k \parallel i \in N$ has at most $O(n)$ further parallelity relations $k \parallel j \in N$. Finally, checking the feasibility of (i, j, k) with respect to the resource capacities is done in $O(r)$ time.

We extend our procedure by incorporating additional conditions that force additional schedule scheme modifications in order to save feasibility. Provided with a symmetric triple (i, j, k) and another arbitrary activity l six cases based on additional conditions may be described as follows:

(1) Let the parallelity relation $l \parallel i$ be given and assume that j, k and l cannot be processed simultaneously. Then we can update $D := D \cup \{l - j\}$, see Figure 2:2.
Additionally, let the precedence relation $i \to j$ or $j \to i$ be given. Then we may even fix the conjunction $l \to j \in C$ or $j \to l \in C$, respectively.

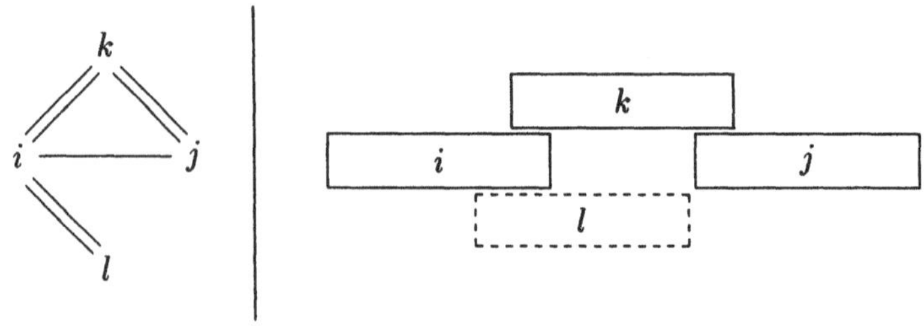

Figure 2:2. Extension of a symmetric triple - Condition 1

(2) Suppose that the conditions $p_k - 1 \leq p_i$, $p_k - 1 \leq p_j$, and $p_k - 1 \leq p_l$ hold. Furthermore, assume that i, k and l as well as j, k and l cannot be processed simultaneously. Then we can update $D := D \cup \{k - l\}$ since the parallel processing of k and l is excluded, see Figure 2:3.

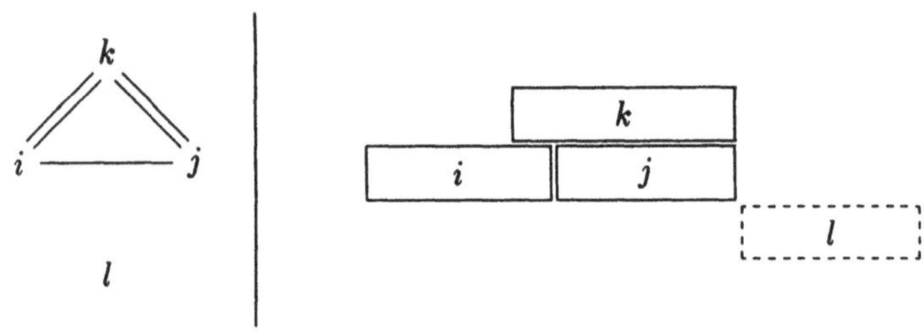

Figure 2:3. Extension of a symmetric triple - Condition 2

(3) Suppose that the conditions $p_k - 1 \leq p_i$ and $p_k - 1 \leq p_l$ hold. Furthermore, assume that the precedence relations $l \to j$ and $i \to j$ are given and i, k and l cannot be processed in parallel. Then the additional conjunction $l \to k \in C$ can be fixed, see Figure 2:4.

(4) Suppose that the conditions $p_k - 1 \leq p_j$ and $p_k - 1 \leq p_l$ hold. Furthermore, assume that the precedence relations $i \to j$ and $i \to l$ are given and j, k and l cannot be processed in parallel. Then the additional conjunction $k \to l \in C$ can be fixed, see Figure 2:5.

(5) Suppose that the conditions $p_k - 1 \leq p_i$ and $p_k - 1 \leq p_j$ hold. Furthermore, assume that the precedence relations $l \to i$ and $l \to j$ ($i \to l$ and $j \to l$) are given. Then we may fix the conjunction $l \to k \in C$ ($k \to l \in C$), see Figure 2:6.

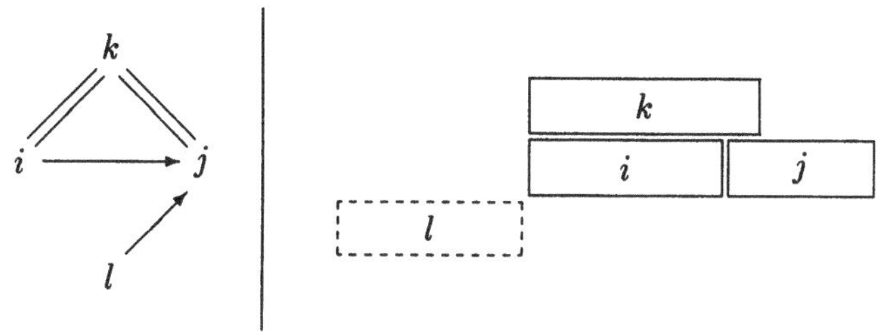

Figure 2:4. Extension of a symmetric triple - Condition 3

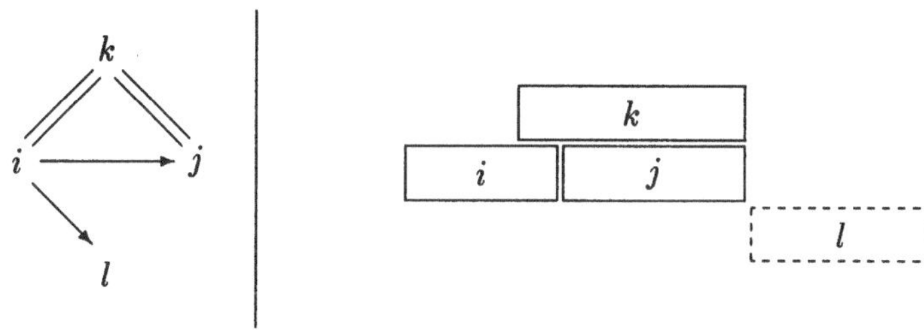

Figure 2:5. Extension of a symmetric triple - Condition 4

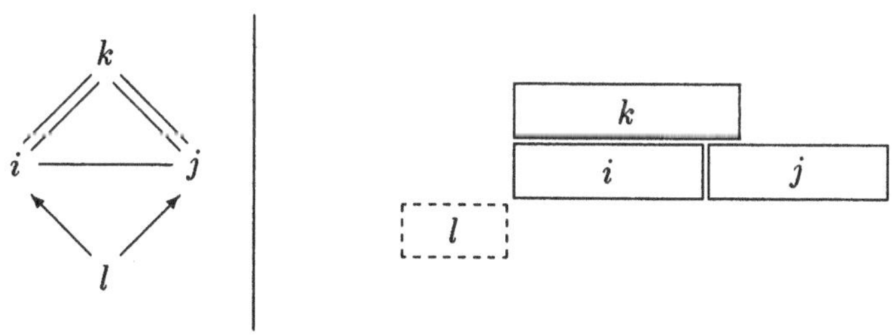

Figure 2:6. Extension of a symmetric triple - Condition 5

(6) Let the parallelity relations $l \parallel i$ and $l \parallel j$ be given. Then the additional parallelity relation $l \parallel k$ can be fixed, see Figure 2:7.

It is not difficult to prove that it is correct to fix relations as indicated and that all conditions can be fixed in $O(rn^2 \,|N|)$ time (see Schoo 1996).

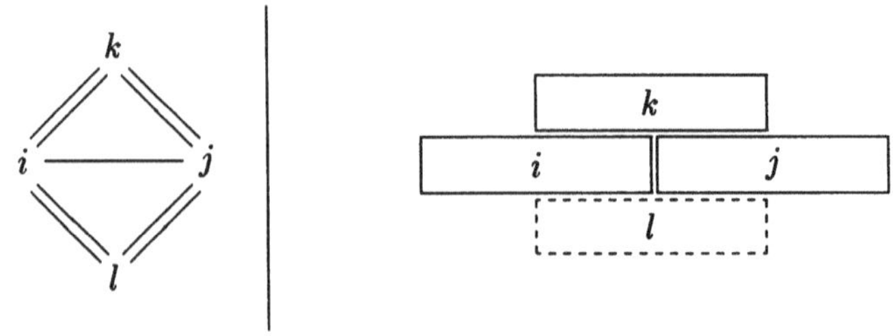

Figure 2:7. Extension of a symmetric triple - Condition 6

2.3.3 Additional Conjunctions and Parallelity Relations

Based on the set C of conjunctions heads and tails can be calculated in $O(|C|)$ time using the formulae

$$r_i := \max\{r_j + p_j \mid j \rightarrow i \in C\} \tag{2.7}$$

and

$$q_i := \max\{p_j + q_j \mid i \rightarrow j \in C\}, \tag{2.8}$$

respectively. These recursive definitions require initializations which are given by $r_0 = q_{n+1} = 0$.

Heads and tails provide time windows $[r_i, d_i]$ with $d_i = U - q_i$. Each activity i must be processed within its time windows $[r_i, d_i]$. Let a disjunction $i - j \in D$ be given. If the condition

$$p_i + p_j > d_i - r_j \tag{2.9}$$

holds, then j cannot be processed before i and thus we may fix the additional conjunction $i \rightarrow j$.

By setting $d_{0i} = r_i$ and $d_{i,n+1} = p_i + q_i$ heads and tails may be included in the distance matrix $D = (d_{ij})$.

Now let $D_S = (d_{ij})$ be the distance matrix defined by (2.1) where $-\infty$ is replaced by $p_i - U$. If the condition

$$d_{ij} \geq -(p_j - 1) \tag{2.10}$$

is fulfilled, then $S_j - S_i \geq d_{ij} \geq -(p_j - 1)$ or equivalently $C_j > S_i$ holds. Thus, if (2.10) holds for the activities of a disjunction $i - j \in D$ we may fix the additional conjunction $i \rightarrow j \in C$ which is called a **direct conjunction**. In Schoo (1996) it is shown that each conjunction fixed by condition (2.9) is also embraced by condition (2.10). A procedure with time complexity $O(|D|)$ for fixing additional conjunctions due to (2.10) is easily implemented.

Condition (2.10) is also useful to determine additional parallelity relations. If both conditions

$$d_{ij} \geq -(p_j - 1) \quad \text{and} \quad d_{ji} \geq -(p_i - 1) \tag{2.11}$$

are fulfilled, we clearly may introduce the additional parallelity relation $i \parallel j \in N$. Since each pair of activities i and j has to be considered the computational effort is $O(n^2)$.

2.3.4 *Clique Conjunctions*

Provided with heads and tails we may apply classical concepts of immediate selection. The basic idea of these concepts is to improve heads and tails of activities belonging to a clique in the undirected graph $G_{SS} = (V, E_{SS})$ associated with a schedule scheme $SS = (C, D, N, F)$. The set of edges E_{SS} consists of all pairs of incompatible activities, i.e. $[i, j] \in E_{SS}$ if and only if $i \to j \in C$, or $j \to i \in C$, or $i - j \in D$. Heads and tails can be derived from the distance matrix $\overline{D}_S$. The improved heads and tails are retransformed into distances. This may lead to additional conjunctions due to conditions (2.5) or (2.10). The whole process will be repeated until no more additional conjunctions are derived.

To improve heads and tails of activities belonging to a clique we use a well-known procedure which has been applied in connection with the job-shop problem (see, e.g. Brucker 1995, p.202).

It remains to describe how suitable cliques are calculated. This is done in a heuristical manner. We calculate cliques $I_1, \ldots, I_q$ in such a way that each I_ν is maximal and $I_1 \cup \ldots \cup I_q = \{1, \ldots, n\}$. To build I_1 we start with some activity i_1 and add an activity i_2 which is incompatible with i_1. Then we add some activity i_3 which is incompatible with i_1 and i_2, etc. I_1 is completed if there exists no activity incompatible with all activities in I_1. To build I_2 we start with some activity not belonging to I_1 and apply the same procedure. Note that activities $i \in I_1$ may be added to I_2. If $I_1, \ldots, I_k$ are built and there is some activity $i \notin I_1 \cup \ldots \cup I_k$, then we start I_{k+1} with i. The procedure stops if all activities are covered by the constructed cliques.

The classical immediate selection procedure is applied to each of these cliques.

2.3.5 *Updating Distances and Applying Immediate Selection*

As a foundation of immediate selection concepts we need the distance matrix $\overline{D}_S$ that provides transitive distances. We apply an extended version of the **Floyd-Warshall algorithm** denoted as the procedure **Calculate Transitive Distances** to the initialized matrix. Within this extension we fix transitivity conjunctions according to (2.5) and additional parallelity relations according to

(2.11). Furthermore, we detect cycles according to (2.6). This extended version still runs in $O(n^3)$ time. The computational cost to update the distances in connection with a branching step or a single immediate selection decision can be reduced from $O(n^3)$ to $O(n^2)$ (see Bartusch et al. 1988).

Next we give a full description of the procedure **Immediate Selection**. Besides the subprocedure **Calculate Transitive Distances** it uses the subroutines **Symmetric Triples, Fix Direct Conjunctions**, and **Process Cliques**. **Symmetric Triples** fixes relations by investigating symmetric triples and their extensions as described in Section 2.3.2. **Fix Direct Conjunctions** fixes conjunctions as described in 2.3.3. **Process Cliques** processes the generated cliques as described in the previous section and sets *success* := true if additional conjunctions are fixed.

The following procedure is applied to a schedule scheme (C, D, N, F).

```
PROCEDURE Immediate Selection
Calculate Transitive Distances;
IF infeasibility is not detected THEN BEGIN
    Symmetric Triples;
    Fix Direct Conjunctions;
    REPEAT
        success := Process Cliques;
    UNTIL success = FALSE;
END;
```

2.4 HEURISTIC METHODS

In this section we will develop a tabu-search heuristic based on schedule schemes for the RCPSP (cf. Baar et al. 1997). The search is guided by operators which are defined on schedule schemes in the following way. Given a schedule scheme $SS = (C, D, N, F)$ a so-called parallelity heuristic H first calculates a feasible schedule $S = H(SS)$. Depending on S a set of operators is defined which transform SS into new schedule schemes.

In the next subsection we describe the parallelity heuristic which is also used in connection with the branch-and-bound algorithm. In Section 2.4.2 the operators for the tabu-search procedure are introduced, while in Section 2.4.3 further details of the implementation are given.

2.4.1 A Parallelity Heuristic

The problem of deciding whether a schedule exists which belongs to $\mathcal{S}_f(C, D, N, F)$ has been shown to be $\mathcal{NP}$-hard by Krämer (1995). Thus, it is unlikely to find such a schedule in polynomial time. For these reasons, in this section we will develop a heuristic procedure H which, for a given schedule scheme $SS = (C, D, N, F)$, calculates a schedule $S = H(SS)$. This schedule

- satisfies all conjunctions C and disjunctions D,
- satisfies a "large" number of parallelity relations N, and
- has a "good" $C_{\max}$-value.

The basis of this heuristic is the concept of parallelity components. A **parallelity component** is a connectivity component in the undirected graph induced by the set N of parallelity relations. Let P_1 and P_2 be two different parallelity components. Then a precedence relation $P_1 \rightarrow P_2$ is introduced if and only if activities $i \in P_1$ and $j \in P_2$ exist such that $i \rightarrow j \in C$ holds (cf. Figure 2:8). These parallelity components and the precedences between components have a high impact on the priority of an activity.

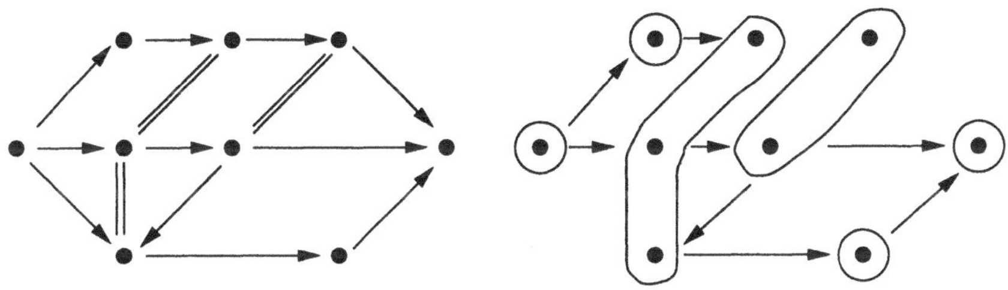

Figure 2:8. Example for parallelity components ∘ for a graph with vertices •

The heuristic constructs a feasible schedule by adding one activity to the current partial schedule in each step. It can be described as follows:

(1) Calculate the parallelity components and the corresponding precedence constraints between these components.

(2) If no parallelity component without predecessor components exists, then go to (5). Otherwise, one of the parallelity components without predecessor is chosen which contains a job with the smallest $(p_i + q_i)$-value. The component P is now scheduled until it is empty in the following way:

(3) Choose an activity $i \in P$ according to a specific priority rule and calculate an earliest starting time r_i at which i can be started if it is added to the partial schedule. More precisely, let

$$h_i := \max\{\max_{j \rightarrow i}\{S_j + p_j \mid j \text{ planned}\}, \min_{k \| i}\{S_k - (p_i - 1) \mid k \text{ planned}\}.$$

Calculate the earliest starting time $r_i \geq h_i$ such that activity i can be scheduled without violating the disjunctions and resource constraints and schedule i at time r_i. Eliminate i from P, update the resource profile of the partial schedule and repeat Step (3) until P is empty.

(4) When P becomes empty, eliminate this component with its precedences and go to (2).

(5) A cycle between some parallelity components exists. Choose an activity i without an unfinished conjunctive predecessor using the same criterion as in (3). Schedule i and update the parallelity components with its precedences. If a component without predecessors now exists, go to (2). Otherwise, repeat Step (5).

This procedure can be implemented in $O(n^3r)$.

2.4.2 *The Neighborhood*

We introduce four different types of operators which are defined on a schedule scheme $SS = (C, D, N, F)$.

The first operator move $(i \sim j)$ is defined for $i \sim j \in F$.

- move $(i \sim j)(SS)$ transforms $i \sim j \in F$ into a parallelity relation $i \parallel j$, i.e.

$$\text{move } (i \sim j)(SS) := (C, D, N \cup \{i \parallel j\}, F \backslash \{i \sim j\}).$$

The next two operators are defined for $i \parallel j \in N$ and the parallelity component P to which i, j belong. These operators eliminate parallelity relations with the objective that the parallelity relation $i \parallel j$ may be satisfied by the heuristic H.

- move $(P\backslash i, j)$ eliminates all parallelity relations $k \parallel i$ with $k \neq j$, i.e.

$$\text{move } (P\backslash i, j)(SS) := (C, D, N\backslash\{k \parallel i \mid k \neq j\}, F \cup \{k \sim i \mid k \parallel i, k \neq j\})$$

- move $(P\backslash i\backslash j)$ is the concatenation of move $(P\backslash i, j)$ and move $(P\backslash j, i)$, i.e. it eliminates all parallelity relations $k \parallel i$ and $k \parallel j$ except $i \parallel j$.

Finally, we have an operator which cancels all parallelity relations in a parallelity component. Then it may be possible that H can satisfy other parallelity relations.

- move (P) eliminates all parallelity relations in a parallelity component P, i.e.

$$\text{move } (P)(SS) := (C, D, N\backslash\{k \parallel l \mid k \parallel l \in P\}, F \cup \{k \sim l \mid k \parallel l \in P\}).$$

The neighborhood defined by these operators is far too large. We have to restrict the operators. This is done by constructing a feasible schedule $S = H(SS)$ using the parallelity heuristic H and calculating a parallel critical path in S.

A sequence $0 = i_0, i_1, \ldots, i_l = n+1$ of activities is called a **parallel critical path** $CP_{||}(S)$ if for $i_\nu, i_{\nu+1}$, $(\nu = 0, \ldots, n)$ one of the following conditions holds:

- activity $i_{\nu+1}$ starts when activity i_ν finishes, or
- activity i_ν starts earlier than $i_{\nu+1}$ and both activities are processed in parallel for at least one time unit.

The makespan of a schedule S can only be shortened if more of the activities on $CP_{||}(S)$ are processed in parallel. Therefore we restrict the first three operators to activities i, j on the parallel critical path. More specifically, the neighborhood $NBH_{||}$ is defined by $NBH_{||}(SS) := N_1 \cup N_2 \cup N_3 \cup N_4$ with

$$\begin{aligned} N_1 :=\ & \{\text{move } (i \sim j)(SS) \mid i, j \in CP_{||}(S), i \sim j \in F\}, \\ N_2 :=\ & \{\text{move } (P\backslash i, j)(SS) \mid i, j \in CP_{||}(S), i \parallel j \in N\}, \\ N_3 :=\ & \{\text{move } (P\backslash i\backslash j)(SS) \mid i, j \in CP_{||}(S), i \parallel j \in N\} \text{ and} \\ N_4 :=\ & \{\text{move } (P_u)(SS) \mid P_u \text{ is a parallelity component with at} \\ & \text{least two activities}\}. \end{aligned}$$

To keep the parallelity neighborhood $NBH_{||}$ small we additionally choose a parallel critical path $CP_{||}(S)$ where the number of relations $i \parallel j$ and $i \sim j$ with $i, j \in CP_{||}(S)$ is small.

Practical tests with this $O(n^2)$-neighborhood have shown that it is still too large. Therefore we considered a subneighborhood $NBH_{||}^{CritRes}$ of $NBH_{||}$ which is constructed as follows.

Instead of choosing all pairs $i, j \in CP_{||}(S)$ for move-operators of the four types described above, we only consider a limited number of operators in each set. These operators are chosen in such a way that we get some operators that only disturb the current solution slightly, some that change it intensively and others which cause changes in the medium range. We estimate the magnitude of these changes as follows.

For $SS' =$ move $(i \sim j)(SS)$, $SS' =$ move $(P\backslash i, j)(SS)$ or $SS' =$ move $(P\backslash i\backslash j)(SS)$ we define Δ_{ij} to be the minimal number of time units we have to move activity i such that i and j are processed in parallel, i.e.

$$\Delta_{ij} = \max\{S_j - C_i + 1, S_i - C_j + 1, 0\},$$

where S_j is the starting time of activity j and C_i is the finishing time of activity i in the schedule $S = H(SS)$. Let $\Delta_{\max}(\Delta_{\min})$ be the maximal (minimal) value of these numbers Δ_{ij} for all $i, j \in CP_{||}(S)$. For $SS' =$ move $(P_u)(SS)$ we consider the number of parallelity relations in P_u. Let $\alpha_{\max}(\alpha_{\min})$ be the maximal (minimal) value of these numbers.

We split $NBH_{||}(SS)$ into nine disjoint sets $M_1, \ldots, M_9$. We divide both intervals $[\Delta_{\min}, \Delta_{\max}]$ and $[\alpha_{\min}, \alpha_{\max}]$ into three parts of equal length. According to this partition we divide N_1 into sets M_1, M_2, M_3 with small, medium, and large Δ_{ij}-values. In the same way, $N_2 \cup N_3$ is partitioned into sets M_4, M_5, M_6. N_4 is partitioned into sets M_7, M_8, M_9 with respect to the α-values.

Now in each of the nine sets all schedule schemes are chosen which have the lowest priority value. For schedule schemes $SS' \in M_1 \cup \ldots \cup M_6$ this priority value is given by $CritRes(SS') = -\frac{r_{ik}+r_{jk}}{R_k}$ where k is the index of a critical resource. A **critical resource** is a resource k which induces the largest number of resource conflicts $u - v$ due to $r_{uk} + r_{vk} > R_k$.

For $SS' =$ move $(P_u)(SS)$ the priority value is equal to the number of parallelity relations in P_u.

The neighborhood $NBH_{||}^{CritRes}(SS)$ consists of all $SS' \in M_i$ $(i = 1, \ldots, 9)$ which have the lowest priority value:

$$NBH_{||}^{CritRes}(SS) = \bigcup_{i=1}^{9} \{SS' \mid SS' \in M_i \text{ has the lowest priority}\}.$$

2.4.3 Implementation

In this section the main strategies of the tabu-search procedure are presented.

We use the neighborhood $NBH_{||}^{CritRes}$. In each iteration we have a schedule scheme SS and calculate the corresponding heuristic solution $S = H(SS)$. The next schedule scheme SS' is chosen from $NBH_{||}^{CritRes}(SS)$ which is defined on the basis of a parallel critical path $CP_{||}(S)$ of the schedule S.

There are situations where an easy test shows that an operator move $(i \sim j)$ leads to a schedule scheme $SS' =$ move $(i \sim j)(SS)$ with no feasible solution. This situation occurs if in $SS = (C, D, N, F)$ we have for some activities k, l the relations $i \to k$, $k \parallel l$, and $l \to j$. These relations imply $i \to j$. Thus, move $(i \sim j)$ leads to a schedule scheme with only infeasible schedules. In order to apply move $(i \sim j)$, all parallelity relations $k \parallel l$ which induce $i \to j$ or $j \to i$ are eliminated.

Other features of the tabu-search procedure are:

- The tabu-status is marked by storing the relations $i \parallel j$ which are added or eliminated. If $i \parallel j$ is in the tabu-list, then all operators involving $i \parallel j$ are tabu.

- In Step (3) of the heuristic H we used the following priority rule. For each activity $i \in P$ an earliest starting time r_i is calculated as described and an $i \in P$ with the smallest r_i-value is chosen. In the case of ties, an activity with the largest $\left(\sum_{k=1}^{r} \frac{r_{ik}}{R_k}\right)$-value is determined.
- Experiments have shown that dynamic tabu-lists of minimal length 3 and maximal length 6 gave the best results.
- After each iteration which improves the current upper bound immediate selection is applied.
- An aspiration criterion is used: If an operator which is tabu leads to a solution with a better $C_{\max}$-value than the best solution value found so far, then the tabu-status of the operation is cancelled.
- The tabu-search is stopped if in the last 250 iterations no solution is found which improves the best solution calculated so far.
- We allowed two restarts, i.e. when stopping the tabu-search after 250 iterations we start the search again with the best solution found so far and choose a new neighbor randomly.
- To get a good starting solution we applied different priority-based heuristics and selected the best solution found (see Alvarez-Valdes and Tamarit 1989).

2.5 LOWER BOUNDS

Several methods for calculating lower bounds for the RCPSP are available. We tested different methods in connection with our branch-and-bound algorithm. The main result of these tests was that the lower bound LB_2 from Mingozzi et al. (1996) turned out to be the best.

LB_2 is the optimal value of a linear program which partially relaxes the precedence constraints and allows preemption. The columns of this LP correspond to so-called nondominated feasible subsets. Feasible sets X are sets of activities which may be processed simultaneously, i.e. there are no conjunctions or disjunctions between any pair of activities $i, j \in X$ and all resource constraints are respected. Such a set is called nondominated if it is not a proper subset $X \subset Y$ of some feasible set Y.

We denote all nondominated feasible sets by $X_1, X_2, \ldots, X_q$. With each set X_j we associate an incidence vector $a_j \in \{0,1\}^n$ defined by

$$a_{ij} := \begin{cases} 1 & \text{if } i \in X_j \\ 0 & \text{otherwise.} \end{cases}$$

Furthermore, let x_j be a variable denoting the number of time units where al activities in X_j are processed simultaneously. Then a LP-formulation of the relaxation may be written as follows:

$$\text{MIN} \sum_{j=1}^{q} x_j \tag{2.12}$$

$$\text{s.t.} \quad \sum_{j=1}^{q} a_{ij} x_j \geq p_i \quad (i = 1, \ldots, n) \tag{2.13}$$

$$x_j \geq 0 \quad (j = 1, \ldots, q). \tag{2.14}$$

Conditions (2.13) ensure that each activity i is processed for at least p_i time units. To calculate all incidence vectors which correspond to nondominated feasible sets we enumerate all incidence vectors $a = (a_1, \ldots, a_n)$ which correspond to feasible sets and check for each a whether the corresponding set is dominated.

Unfortunately, the cardinality of all nondominated feasible sets grows exponentially with the number n of activities. For $n = 60$ we have approximately 300 000 columns, for $n = 90$ even 8 000 000. For this reason Mingozzi et al. (1996) used a different weaker lower bound derived from the dual LP.

However, if delayed column generation techniques are applied, then it is not necessary to generate and store all these columns. We can work with only a few of them at a time and only generate new feasible sets when they are really needed. In the following we describe the details of our approach (cf. Baar et al. 1997). We work iteratively with a set of columns which contains a basis and the n slack-variables corresponding to (2.13).

In the general step of the column generation procedure we solve a linear program to optimality with a current set of columns which is a subset of all columns. Then we have to calculate new entering columns or to show that no improving columns exist. New columns are calculated in the following way. If we denote the dual-variables corresponding to conditions (2.13) by y_i, then we have to find a column $a \in \{0, 1\}^n$ with

$$\sum_{i=1}^{n} y_i \cdot a_i > 1 \tag{2.15}$$

$$\sum_{i=1}^{n} r_{ik} \cdot a_i \leq R_k \quad (k = 1, \ldots, r) \tag{2.16}$$

$$a_i \cdot a_j = 0 \quad (i - j \in D, i \to j \in C \text{ or } j \to i \in C). \tag{2.17}$$

Conditions (2.16) and (2.17) ensure that a column a corresponds to a feasible subset (respecting the resource constraints, the disjunctions D, and the conjunctions C). If (2.15) holds for a column a, then the current objective value is improved when a enters the basis.

Columns which satisfy conditions (2.15) - (2.17) are generated by a branch-and-bound procedure. If no such columns are found, the current basic solution is an optimal solution for the LP (2.12) - (2.14) and the process is stopped.

Our branch-and-bound procedure works as follows: Since we have solved the current LP to optimality and the slack-variables are kept in the working set, for all dual-variables we have $y_i \geq 0$. Because of (2.15) and (2.16) we only have to consider activities i with $y_i > 0$ $(i = 1, \ldots, n)$.

At first these activities are sorted according to nonincreasing dual-variables $y_1 \geq y_2 \geq \ldots \geq y_{n_0} > 0$, $y_{n_0+1} = \ldots = y_n = 0$. To calculate incidence vectors a which correspond to nondominated feasible sets we define a branching tree in which the vertices represent partial strings $(a_1, \ldots, a_l)(1 \leq l \leq n_0)$. Such a string has at the most two sons. $(a_1, \ldots, a_l, 0)$ is the right son. The left son $(a_1, \ldots, a_l, 1)$ only exists if the set which corresponds to $(a_1, \ldots, a_l, 1, 0, \ldots, 0)$ is feasible.

We undertake a depth-first search in the branching tree where leaves correspond to feasible subsets. The sets are generated in lexicographic order. We generate a subset of nondominated feasible columns which satisfy (2.15) - (2.17). Each time a column a^* which satisfies these conditions has been found, we only search for further columns a with

$$\sum_{i=1}^{n_0} y_i a_i > \sum_{i=1}^{n_0} y_i a_i^*. \tag{2.18}$$

This ensures that all sets are nondominated (for proof see Baar et al. 1997).

When we have found a suitable leaf a^*, we fill this column successively with all activities $n_0+1, \ldots, n$ which do not violate feasibility. Then the corresponding column is also nondominated and we backtrack to depth n_0.

We included the following simple bounding rule into our branch-and-bound procedure: If we set $a_k = 0$ in step k and

$$\sum_{\lambda=1}^{k-1} y_\lambda \cdot a_\lambda + \sum_{\lambda=k+1}^{n_0} y_\lambda \leq \sum_{i=1}^{n_0} y_i \cdot a_i^*$$

holds, the best column a^* found so far cannot be improved and therefore we can backtrack.

The search process stops if one of the following conditions holds:

- we have found a given number of columns, or

- we have found at least one column and the quotient of the number of generated leaves and the number of leaves satisfying (2.18) exceeds a certain limit.

The generated columns are added to the working set of columns and the LP is resolved to optimality. If the number of columns in the working set exceeds a given size, some arbitrary columns are deleted.

Unfortunately, for problems with a larger number of activities the quality of this lower bound worsens. The reason is that the possibility of preemption allows us to split the activities into many pieces and to put them appropriately into different feasible subsets yielding low objective values. Therefore, we strengthen the lower bound by additionally taking into account time windows. Since each activity i has to be processed in its time window $[r_i, d_i]$, we restrict the generation of feasible sets to certain time intervals. Furthermore, instead of calculating constructively a lower bound by combinatorial considerations we use a "destructive improvement" technique proposed by Klein and Scholl (1997). Starting with a hypothetical upper bound U we try to prove infeasibility. If some method provides the result that no feasible schedule with $C_{\max} \leq U$ exists, then $U + 1$ is a valid lower bound value. One method for proving infeasibility is to use the immediate selection techniques presented in Section 2.3. Another method may be described as follows. Given a threshold value U, we consider the problem of finding a preemptive schedule with $C_{\max} \leq U$ such that all activities are processed within their time windows and all resource constraints are respected.

For this method we assume that the threshold value U is taken into account in the time windows $[r_i, d_i]$ by setting $d_{n+1} = U$ and by modifying the other due dates appropriately.

Let $z_0 < z_1 < \ldots < z_\tau$ be the ordered sequence of all different r_i- and d_i-values. For $t = 1, \ldots, \tau$ we consider the intervals $I_t := [z_{t-1}, z_t]$ of length $z_t - z_{t-1}$. With each interval I_t we associate a set F_t of all activities i which can be planned in this interval, i.e. $r_i \leq z_{t-1}$ and $z_t \leq d_i$.

We denote all nondominated feasible subsets in an interval I_t by X_{jt} ($j = 1, \ldots, q_t$) and associate with each set X_{jt} an incidence vector $a^{jt} \in \{0, 1\}^n$ defined by

$$a_i^{jt} = \begin{cases} 1, & \text{if } i \in X_{jt} \\ 0, & \text{otherwise.} \end{cases}$$

Furthermore, let x_{jt} be a variable denoting the number of time units where all activities in X_{jt} are processed simultaneously. Then the preemptive relaxation may be written as follows:

$$\sum_{t=1}^{\tau}\sum_{j=1}^{q_t} a_i^{jt} x_{jt} \geq p_i \qquad (i = 1, \ldots, n) \tag{2.19}$$

$$\sum_{j=1}^{q_t} x_{jt} \leq z_t - z_{t-1} \qquad (t = 1, \ldots, \tau) \tag{2.20}$$

$$x_{jt} \geq 0 \qquad (t = 1, \ldots, \tau;\ j = 1, \ldots, q_t) \tag{2.21}$$

Due to restrictions (2.19) we process all activities i for at least p_i time units. Conditions (2.20) ensure that the number of time units scheduled in interval I_t do not exceed the length $z_t - z_{t-1}$ of this interval.

By introducing artificial variables u_t for $t = 1, \ldots, \tau$ in conditions (2.20) the feasibility problem can be formulated as a linear program:

$$\text{MIN} \sum_{t=1}^{\tau} u_t \tag{2.22}$$

s.t.

$$\sum_{t=1}^{\tau}\sum_{j=1}^{q_t} a_i^{jt} x_{jt} \geq p_i \qquad (i = 1, \ldots, n) \tag{2.23}$$

$$-\sum_{j=1}^{q_t} x_{jt} + u_t \geq -z_t + z_{t-1} \qquad (t = 1, \ldots, \tau) \tag{2.24}$$

$$x_{jt} \geq 0 \qquad (t = 1, \ldots, \tau;\ j = 1, \ldots, q_t) \tag{2.25}$$

$$u_t \geq 0 \qquad (t = 1, \ldots, \tau) \tag{2.26}$$

There exists a solution for the preemptive relaxation if and only if the linear program has the optimal solution value zero.

This linear program can again be solved with column generation by extending the approach presented above. To find the largest lower bound U we apply binary search.

2.6 A BRANCH-AND-BOUND ALGORITHM

In this section a branch-and-bound algorithm is described which is based on the concepts of schedule schemes and all other techniques presented before.

The root of the search tree is associated with the schedule scheme $(C_0, D_0, \emptyset, F_0)$ which represents all feasible schedules. Branching is done by choosing some flexibility relation $i \sim j \in F$ and turning it either into a disjunction $i - j$ or into a parallelity relation $i \parallel j$. This branching process is repeated until scheduling schemes with $F = \emptyset$ are reached which can be treated as leaves due to Theorem 1.

By this branching rule the set of feasible schedules corresponding to a search tree node is split into two disjoint sets creating a binary search tree. The flexibility relation $i \sim j \in F$ chosen for branching is selected heuristically according to the following priority rule. For each pair $i \sim j \in F$ we calculate values w_{i-j} and $w_{i\|j}$ which may be considered as approximations for lower bounds for the two possible descendants of the current search tree node. These values are obtained from the LP-solution after the lower bound calculation for the current search tree node. We choose a flexibility relation $i \sim j \in F$ such that the sum $w_{i-j} + w_{i\|j}$ is maximized. If this priority rule fails to determine a unique flexibility relation, we specify $i \sim j \in F$ randomly.

Finally, we describe the basic organization of our branch-and-bound method. In the root of the search tree we calculate a good initial solution by applying the tabu-search algorithm presented in Section 2.4. The branching rule explained above is imbedded in a depth-first search algorithm. The immediate successors of a search tree node are examined according to non-decreasing lower bound values.

However, before branching we apply some other procedures. First we apply immediate selection. The purpose is to detect successor nodes of the current node that can be excluded from consideration in order to find an optimal solution. Therefore, immediate selection is an important tool in our branch-and-bound algorithm which reduces the number of search tree nodes. Furthermore, it may influence lower and upper bound calculations because the solution set becomes more restricted. If the set F is not yet empty, we continue with the calculation of bounds. At first we calculate the lower bound LB_2. If this value exceeds the makespan of the best solution found so far, we may drop the current node. Otherwise, we calculate heuristically a feasible solution providing an upper bound. This is done by the heuristic H presented in Section 2.4. Each time an improvement is found the global upper bound UB is updated and a further comparison with the lower bound of the current node is indicated.

2.7 COMPUTATIONAL RESULTS

In this section we present some computational results for the algorithms developed in the previous sections.

We investigated the performance of our algorithms with respect to various benchmark instances generated by the project generator PROGEN published in Kolisch et al. (1995) and Kolisch and Sprecher (1997).

They specified a large number of parameter constellations and generated problem groups, each group consisting of 10 instances. For $n = 30, 60, 90$ and $r = 4$ they obtained 48 groups.

problem set	$n = 60$	$n = 90$
number of instances	480	480
verified instances	318	322
average $\Delta(LB)$	3.9%	4.3%
maximal $\Delta(LB)$	25.4%	27.0%
average computation time	41	61
maximal computation time	306	479

Table 2:1. Results for large PROGEN-instances

The results presented in this section were obtained using a Sun Ultra 2 workstation with operating system Solaris 2.5 and 320 MB general storage (167 MHz). The algorithms were coded in C. We used CPLEX as an LP-solver to provide the lower bounds.

For all instances with $n = 30$ activities the optimal values are known. The tabu-search procedure presented in Section 2.4 solves 389 of the 480 instances to optimality with an average (maximal) computation time of 15 (108) seconds. The average percentage deviation of the tabu-search value TS from the optimal value for all instances is 0.4 %, the maximal deviation 6.9 %.

For the test sets with $n = 60, 90$ activities not all optimal values are known. Therefore we used the lower bounds described in Section 2.5 to obtain bounds for the quality of the calcuated solutions. With a binary search procedure we calculated the largest lower bound LB where infeasibility could be proved. To detect infeasibility we first used immediate selection techniques. If this failed to prove infeasibility, then the linear programming formulation with possibly improved time windows (after immediate selection) was used for another feasibility check. Table 2:1. contains computational results for the large test sets. The number of instances for which the tabu-search value TS could be verified to be optimal is indicated. Furthermore, we calculated the average (maximal) percentage deviation $\Delta(LB) = \frac{TS-LB}{LB}$ from the lower bound LB for all instances. The computation times (in seconds) include the time needed for calculating the starting heuristic and applying immediate selection within the tabu-search procedure. Not included are the times for calculating the lower bounds.

For instances with a large number of activities most of the branch-and-bound algorithms known for the RCPSP do not perform well because faster implementations would require hundreds of MB storage. The main advantage of the algorithm presented in Section 2.6 is its relatively small storage requirement. Using column generation techniques for the lower bound calculations our implementation uses at most 25 MB for problems with 120 activities. 322 of the 480 test problems with 60 activities could be solved within a time limit of 10 minutes. Unfortunately, for the test sets with $n = 90$ the branch-and-bound algorithm only improves the tabu-search values in a small number of cases and verifying optimality is a very time-consuming process.

Detailed results of our algorithms can be found under the address `http://www.mathematik.uni-osnabrueck.de/research/OR/software.html`.

References

ALVAREZ-VALDES, R., J.M. TAMARIT. 1989. Heuristic algorithms for resource-constrained project scheduling: a review and an empirical analysis. In *Advances in project scheduling*. R. Slowinski, J. Weglarz (eds.). Elsevier Science Publ., Amsterdam, 113-134.

BAAR, T., P. BRUCKER, S. KNUST. 1997. Tabu-search algorithms and lower bounds for the resource-constrained project scheduling problem. To appear in *Proceed. of the 2nd International Conference on Metaheuristics - MIC97*, Sophia-Antipolis, France.

BARTUSCH, M., R.H. MÖHRING, F.J. RADERMACHER. 1988. Scheduling project networks with resource constraints and time windows. *Annals of Operations Research* **16**, 201-240.

BRUCKER, P. 1995. Scheduling algorithms, Springer Verlag, Berlin.

BRUCKER, P., B. JURISCH, B. SIEVERS. 1994. A branch & bound algorithm for the job-shop problem. *Discrete Applied Mathematics* **49**, 107-127.

BRUCKER, P., S. KNUST, A. SCHOO, O. THIELE. 1997. A branch & bound algorithm for the resource-constrained project scheduling problem. To appear in *Europ. Journal of Oper. Res.*.

GOLUMBIC, M.C. 1980. Algorithmic graph theory and perfect graphs. Academic Press, New York.

JOHNSON, R. 1988. Optimally balancing large assembly lines with FABLE. *Management Science* **34** (2), 240-253.

KLEIN, R., A. SCHOLL. 1997. Computing lower bounds by destructive improvement - an application to resource-constrained project scheduling. Schriften zur Quant. BWL, TH Darmstadt, to appear in *Europ. Journal of Oper. Res.*

KRÄMER, A. 1995. Scheduling multiprocessor tasks on dedicated processors. Dissertation, Fachbereich Mathematik/Informatik, Universität Osnabrück.

KRÄMER, A. 1997. Branch and bound methods for scheduling problems with multiprocessor tasks on dedicated processors. *OR Spektrum* **19**, 219-227.

KOLISCH, R., A. SPRECHER, A. DREXL. 1995. Characterization and generation of a general class of resource-constrained project scheduling problems. *Management Science* **41**, 1693-1703.

KOLISCH, R., A. SPRECHER. 1997. PSPLIB - A project scheduling problem library. *Europ. Journal of Oper. Res.* **96**, 205-216.

KORTE, N., R.H. MÖHRING. 1985. Transitive orientation of graphs with side constraints. In *Proceedings 11th international workshop on graph theoretic concepts in computer science WG '85.* H. Noltemeier (ed.) Trauner Verlag, Linz 1985, 143-160.

MINGOZZI, A., V. MANIEZZO, S. RICCIARDELLI, L. BIANCO. 1996. An exact algorithm for project scheduling with resource constraints based on a new mathematical formulation. Technical report no. 32, Department of Mathematics, University of Bologna, Italy, to appear in *Management Science.*

MÖHRING, R.H. 1985. Algorithmic aspects of comparability graphs and interval graphs. In *Graphs and order: the role of graphs in the theory of ordered sets and its applications.* I. Rival (ed.). NATO Advanced Science Institute Series, Series C., Mathematical and Physical Sciences, 41-101.

SCHOO, A. 1996. Untere Schranken und Immediate Selection für das Resource-Constrained-Project Scheduling Problem. Diplomarbeit, Fachbereich Mathematik/Informatik, Universität Osnabrück.

3 LOWER BOUNDS IN DIFFERENT PROBLEM CLASSES OF PROJECT SCHEDULES WITH RESOURCE CONSTRAINTS

Erwin Pesch

University of Bonn
Faculty of Economics, BWL 3
Adenauerallee 24-42, D-53113 Bonn, Germany

3.1. Introduction

The worldwide opening of the markets, hence the continuously increasing number of competitors, the higher customers standards and shorter product innovation cycles require a cost- or time-minimizing manufacturing of customer orders or projects, preferably just in time, under limited availability of resources. Manufacturing planning can no longer be the result of experience.

Thus for many make-to-order companies computer supported project scheduling is not only the application of PERT or CPM techniques, based upon the assumption, that all resources necessary in order to perform the project are available in unlimited quantities. Resource-constrained project scheduling is devoted to the timing of project activities such that precedence constraints between activity pairs are observed and in each time period t for any resource k its required amount of resource units never exceeds its availability. Scheduling of projects has many faces, e.g. the design of production facilities, maintenance projects, new product development and introduction, installation of hardware or software etc.. They are frequently considered under time-cost tradeoff objectives such as maximize the net-present-value, minimize the project duration (makespan), minimize mean (or maximum) lateness or earliness, etc.. Moreover, many scheduling problems can be considered as project scheduling problems under resource constraints, among others assembly line balancing and, as a particular qualification there is the job shop (Blazewicz et al. 1996). Obviously, the problem is NP-hard.

The most popular way to tackle project scheduling applications are by far priority rules, cf. Drexl (1991).

Here we discuss project scheduling problems comprised of jobs or activities which once started, cannot be interrupted (non-preemption). Resources are non-renewable, i.e. over the whole project duration the available resource amount is fixed, or resources are renewable, i.e. in each time interval t the available amount of the resource is fixed (which is independent of the resource usage in previous intervals). The project can be illustrated by an acyclic activity-on-node network where activities are topologically labeled. Activities can be performed in one among different modes, each of which is specified by its processing time, the required type and resource usage (in order to perform the activity), i.e. an activity is assumed to be scheduled in one mode which requires its specific amount of renewable and non-renewable resources. Project duration, i.e. the makespan has to be minimized. Our method is based on the exact algorithm presented by Talbot (1982). The quality of an enumeration procedure heavily depends on the considered ordering of the variables and the subsequent value assignment to these variables. This however is also the basic part of the first of the two steps from Talbot's algorithm. Hence, the first step consists of finding a most advantageous sequencing of activities, activity modes and resources in order to minimize the required computation time within the second step, i.e. the solution of the problem. Thus the order of activity selection within the enumeration step has been fixed in advance. An activity is tentatively scheduled in it's fastest mode in order to calculate the earliest possible completion. In case the resource constraints are violated a backtrack step leads to a reassignment of the previous activity either to a different (later) activity completion time or to the next mode, both with respect to free the most restrictive resources. Whenever a feasible solution is generated the whole procedure restarts with reduced latest completion times for each activity. We consider a number of possible modifications of Talbot's procedure where different variable and mode orderings are applied as well as look-ahead procedures in order to detect extensive resource requirements. A crucial aspect of any backtrack search is its quality of lower bounds. We describe modifications of Talbot's algorithm as top-down or bottom-up backtrack search. The former applies in the sense of a constraint satisfaction, i.e. it provides a feasible solution with respect to an upper bound. The latter proves a constraint non-satisfaction, i.e. an pretended upper bound is proved to be a lower bound of the makespan of an optimal solution. A particular contribution of this chapter is the application of a truncated backtrack search in order to provide goof lower bounds quickly. Different problem characteristics are elaborated based on the structure of precedence constraints and the resource requirements and availability. The performance of our algorithm is demonstrated in large number of runs on a huge test bed under severe time constraints.

3.2. The model

The general multi-mode resource-constrained project scheduling problem can be formulated as an integer programming problem where binary variables x_{jtm} accept value 1 if activity j operating in mode m completes at the end of time period t;

otherwise x_{jtm} has value 0. Precedence constraints (i,j) between activity pairs i and j are described by means of a project network G = (V, E) where V defines the activity set and E the set of ordered pairs (i,j). A simple (CPM-)forward calculation associates with an activity j from the set V := {1, ..., n} of all activities an earliest finish time E_j. Similarly, a simple backward calculation associates with an activity j a latest finish time L_j. The calculation starts with a makespan upper bound obtained e.g. from a heuristic, i.e. a time horizon of t = 1, ..., T time periods. Both E_j and L_j are calculated using the minimum processing time modes for all activities. Thus, for any feasible schedule with makespan at most T the completion time period t of activity j is always within the interval $[E_j, L_j]$. For convenience, we may assume that there is one last activity n such that each activity j precedes n. This can easily be achieved if n defines a dummy activity with processing time and resource demands equal to zero. Thus, the makespan upper bound T equals L_n.

For any activity $j \in V$, let M_j be the number of different processing modes and let p_{jm} be the processing time of activity j in mode m. We assume that the modes are ordered with respect to a non-decreasing processing time sequence, The first mode of a activity is the one having shortest processing time. Let r_{jkm} (and w_{jkm}) be the (per time period constant) resource consumption in each period t of activity j of the renewable (and non-renewable) resource k if j is performed in mode m; let R_{kt} be the number of available units of the renewable resource k in time period t. Let us assume that the resource consumption of an activity, for any resource type, is proportional to the activity processing time. W_k is the number of available units of the non-renewable resource k (for performing the whole project). Finally define $\mathcal{R}$ and $\mathcal{N}$ to be the set of renewable and non-renewable resources, respectively. With the introduced notation we obtain the following model (cf. Talbot 1982):

$$\text{Minimize} \sum_{m=1}^{M_n} \sum_{t=E_n}^{L_n} t\, x_{ntm} \tag{1}$$

s.t.

$$\sum_{m=1}^{M_j} \sum_{t=E_j}^{L_j} x_{jtm} = 1 \qquad j = 1, \ldots, n \tag{2}$$

$$-\sum_{m=1}^{M_i} \sum_{t=E_i}^{L_i} t\, x_{itm} + \sum_{m=1}^{M_j} \sum_{t=E_j}^{L_j} (t - p_{jm})\, x_{jtm} \geq 0 \text{ for all } (i,j) \in E \tag{3}$$

$$\sum_{j=1}^{n} \sum_{m=1}^{M_j} \sum_{q=t}^{t+p_{jm}-1} r_{jkm}\, x_{jqm} \leq R_{kt} \qquad \text{for all } k \in \mathcal{R};\ t = 1, \ldots, T \tag{4}$$

$$\sum_{j=1}^{n} \sum_{m=1}^{M_j} \sum_{t=E_j}^{L_j} w_{jkm}\, x_{jtm} \leq W_k \qquad \text{for all } k \in \mathcal{N} \tag{5}$$

The objective (1) is to minimize the project completion time, i.e. the makespan. The set (2) constraints insure that each activity is completed exactly once. The constraint set (3) insures that precedence relationships are maintained. In each period t the renewable resource k is available with R_{kt} units. Constraint set (4) insures that the resource requirements r_{jkm} of resource k of all activities which are performed in time period t does not exceed the availability of k in t, i.e. R_{kt}. Non-renewable resource limitations are imposed by (5).

There are different effective branch and bound algorithms in order to solve problems of the aforementioned type (e.g. Talbot / Patterson, 1978, Demeulemeester / Herroelen, 1992, Patterson et al. 1990, Sprecher, 1994, Sprecher / Drexl, 1996). The efficiency of these procedures depends on the location of the makespan of an optimal solution within a calculated interval defined by a lower and upper bound. We are going to consider different branch and bound strategies in order to calculate good upper bounds and / or to generate useful lower bounds. The latter is based on an incomplete propagation of an insufficient 'upper bound' which has been imposed on the problem from the beginning. As comparable approaches led to a characterization and differentiation of easy and difficult job shops (see Pesch and Tetzlaff, 1996), we try to find similar qualifications for a differentiation of project scheduling problems different from the characterizations of Kolisch (1995) and Kolisch et al. (1995). Therefore, all data sets are kept easily comprehensible clear and similar.

3.3. Talbot's algorithm

Talbot's algorithm (1982) solves the problem (1) to (5) in two stages. In stage 1 activities, modes and resources are ordered in order to accelerate the solution procedure of stage 2. The specified activity order of stage 1 defines how the activities will be scheduled during stage 2 of the algorithm. As the activities order of stage 1 is fixed for stage 2, it is crucial for the enumeration part in stage 2 to keep the algorithm's run time small. In stage 2 activities are scheduled in that order at their earliest possible completion time without violating any of the constraints. All activity orders have to be topological and are obtained through a set of different priorities. Activities are ordered with respect to the following criteria:

- their average processing time $\sum_{m=1}^{Mj} p_{jm}/M_j$;
- their maximum processing time $\max \{p_{jm} \mid m = 1, ..., M_j\}$;
- their latest completion time L_j ;
- their average resource demand $(\sum_{m=1}^{Mj} p_{jm} / M_j)(\sum_{m=1}^{Mj}\sum_{k\in\mathcal{R}} r_{jkm} / (M_j |\mathcal{R}|)$
- their earliest completion time E_j;
- their latest start time, i.e. $L_j - \min \{p_{jm} \mid m = 1, ..., M_j\}$;

- randomly;
- their latest completion time L_j reduced by the average processing time, i.e. $L_i - \sum_{m=1}^{Mj} p_{jm}/M_j$.

Modes of an activity j are sorted by increasing processing time, i.e. $p_{j1} \leq p_{j2} \leq ... \leq p_{jM_j}$. The sorting of the renewable resources tries to detect 'critical' resources. These are resources the constraints of which tend to be violated first when an newly considered activity has been scheduled. The idea is as follows: For each activity j and resource k an average resource availability $\overline{R}_{jk} := \sum_{t=E_j}^{L_j} R_{kt} / (L_j - E_j + 1)$ is calculated. Then Talbot defines an index $I_{jkm} := p_{jm}$ if $r_{jkm} / \overline{R}_{jk} = \max \{ r_{jk'm} / \overline{R}_{jk'} \mid k' \in \mathcal{R} \}$, and $I_{jkm} := 0$ for other resources k. These indices are summed over all activities and modes and the resources are sorted in non-increasing order of these indices $\sum_{j=1}^{n} \sum_{m=1}^{Mj} I_{jkm}$. The idea of this sorting is to identify resource infeasibility as early as possible.

In stage two of the algorithm a backtrack search generates schedules by considering activities for assignment to time periods in increasing numerical order. The procedure attempts to assign the activities 1, 2, 3, etc. to time periods (observing precedence constraints) in order to calculate their earliest possible completion time. Activities are performed in their fastest (first) mode possible such that precedence and resource constraints are observed. If the assignment of mode m of activity j is prevented because of resource infeasibility, mode m + 1 is considered. An assigned activity j causes an update of the resource availability. For the non-renewable resource k W_k is to reduce by w_{jkm} and in each processing period the renewable resource k amount R_{kt} is to reduce by r_{jkm} if j is performed in mode m. When an activity i has been assigned, the earliest possible completion time E_j of successor activities j of i might increase. If for any activity j the renewable resource availability is insufficient (for all non-renewable resource-feasible modes) in order to complete j latest at time L_j a backtrack is initiated. The most recently assigned activity, say j - 1, will be rescheduled at a later time or in a less resource consuming mode in order to free resources for use by the unassigned activity j. If there is no rescheduling of activity j - 1 possible another backtrack occurs to activity j - 2, etc.. Whenever a backtrack from activity j to activity j - 1 occurs an update of the resource availability is induced, i.e. for a non-renewable resource k W_k is to increase by w_{j-1km} and in each recent processing period of j - 1 the renewable resource k amount R_{kt} is to increase by r_{jkm} if j - 1 was supposed to be performed in mode m. This process is applied to all activities and their modes until an improved solution is found (or the current solution is proved to be optimum; or the upper bound T on the makespan, if not obtained through a heuristic, is recognized as infeasible, i.e. it provides a lower bound). An improved solution is found if activity n is assigned a completion L_n which is smaller than the makespan of the currently best solution known. The algorithm terminates if the makespan of a feasible solution coincides with a lower bound or an attempt is made to backtrack beyond activity 1.

3.4. The data

We introduce two parameters $\mathcal{R}_k$ and $\mathcal{N}_k$ in order to evaluate ex ante a project's shortage of renewable or non-renewable resource k, respectively. The domains of both parameters are minimum-/ maximum-/ and median-values '$\mathcal{R}_{min}$', '$\mathcal{R}_{max}$', '$\mathcal{R}_{med}$', '$\mathcal{R}_{med1}$', '$\mathcal{R}_{med2}$' and '$\mathcal{N}_{min}$', '$\mathcal{N}_{max}$', '$\mathcal{N}_{med}$', '$\mathcal{N}_{med1}$', '$\mathcal{N}_{med2}$', respectively, with respect to the data sets in consideration (cf. Drexl, 1991, Patterson, 1984 and Russel, 1986). The intention of these parameters is to achieve a comparability of test data sets with respect to resource shortage.

For the non-renewable resource k the values are:

$$\mathcal{N}_{max} := \sum_{j=1}^{n} w_{jk1} \text{ and } \mathcal{N}_{min} := \sum_{j=1}^{n} \min\{W_{jkm} \mid m = 1, ..., M_j\};$$

$$\mathcal{N}_{med} := \lceil (\mathcal{N}_{min} + \mathcal{N}_{max})/2 \rceil;$$

$$\mathcal{N}_{med1} := \lceil (\mathcal{N}_{min} + \mathcal{N}_{med})/2 \rceil \text{ and } \mathcal{N}_{med2} := \lceil (\mathcal{N}_{med} + \mathcal{N}_{max})/2 \rceil;$$

Parameter $\mathcal{N}_k$ is set to $\mathcal{N}_{max}$ if $\mathcal{N}_{max} = W_k$ and $\mathcal{N}_k$ is set to $\mathcal{N}_{min}$ if $\mathcal{N}_{min} = W_k$. If $\mathcal{N}_{max} = W_k$ then there are sufficiently many units of resource k available in order to perform each activity in its fastest mode. If $\mathcal{N}_{min} > W_k$ then there is no feasible schedule for the project.

For the renewable resource k the values are:

$\mathcal{R}_{max} := \max_t \sum_{j=1}^{n} r_{jk1} \cdot z_{jt}$ where $z_{jt} = 1$ if activity j belongs to a critical path in the underlying CPM-network when the project duration has been minimized without resource limitations, i.e. we have $L_j = E_j$ and activity j is supposed to be in process in period t. Hence, $\mathcal{R}_{max}$ measures the maximum resource usage of time critical activities.

$$\mathcal{R}_{min} := \max\{\min\{r_{jkm} \mid m = 1, ..., M_j\} \mid j = 1, ..., n\};$$

$$\mathcal{R}_{med} := \lceil (\mathcal{R}_{min} + \mathcal{R}_{max})/2 \rceil;$$

$$\mathcal{R}_{med1} := \lceil (\mathcal{N}_{min} + \mathcal{R}_{med})/2 \rceil \text{ and } \mathcal{R}_{med2} := \lceil (\mathcal{R}_{med} + \mathcal{R}_{max})/2 \rceil;$$

Parameter $\mathcal{R}_k$ is set to $\mathcal{R}_{max}$ if $\mathcal{R}_{max} \leq R_{kt}$ and $\mathcal{R}_k$ is set to $\mathcal{R}_{min}$ if $\mathcal{R}_{min} \geq R_{kt}$. If $\mathcal{R}_{max} \leq R_{kt}$ then there are sufficiently many units of resource k in period t available in order to perform each activity in its fastest mode. If $\mathcal{R}_{min} > R_{kt}$, for all t, then there is no feasible schedule for the project. Value $\mathcal{R}_{min}$ is defined by the resource consumption of the least resource consuming activity-mode configuration.

In order to keep the data sets comparable and in order to be able to solve the problem instances within a time bound of at most 3 minutes, we decided to consider instances of 15 activities each of which can be processed in two different modes. The last activity is a dummy. An activity to mode assignment requires two different types of non-renewable resources and two types of renewable resources. We considered 6 different activity duration schemes as described in the following Table 3:1.

Table 3:1. Activity processing times in two different modes

Activity	*1*	*2*	*3*	*4*	*5*	*6*	*7*	*8*	*9*	*10*	*11*	*12*	*13*	*14*	*15*
	scheme No 0 (short activities)														
p_{i1} / p_{i2}	2/4	1/3	2/3	2/3	1/3	2/4	1/3	3/4	2/4	1/3	2/3	1/3	2/4	2/3	0/0
	scheme No 1 (medium activities)														
p_{i1} / p_{i2}	1/3	5/7	2/4	5/6	2/5	1/4	6/7	5/8	3/4	1/3	4/6	2/4	1/5	5/8	0/0
	scheme No 2 (long early / short late activities)														
p_{i1} / p_{i2}	6/9	4/8	5/9	4/8	5/7	3/5	2/4	3/4	2/5	3/4	2/3	1/2	1/3	2/3	0/0
	scheme No 3 (short early / long late activities)														
p_{i1} / p_{i2}	2/3	1/3	1/2	2/3	3/4	2/5	3/4	2/4	3/5	5/7	4/8	5/9	4/8	6/9	0/0
	scheme No 4 (many short / few long activities)														
p_{i1} / p_{i2}	5/6	2/3	1/3	1/3	2/3	1/2	6/8	2/3	2/3	4/6	1/2	5/6	2/3	1/2	0/0
	scheme No 5 (many long / few short activities)														
p_{i1} / p_{i2}	6/8	1/3	5/7	2/4	6/9	5/7	2/3	1/3	6/8	5/7	4/6	3/6	6/9	2/4	0/0

There are 8 resource schemes described in the next Table 3:2. Figures 3:1 to 3:5 show the different network structures A to J of the problem instances. A network is described by means of its incidence matrix. As the networks are acyclic we can use a matrix to represent two networks, one by the upper the other one by the lower triangle matrix. Thus a matrix entry at position (r,s), $r < s$, indicates that r is an immediate predecessor of s; a matrix entry at position (r,s), $r > s$, indicates that r is an immediate successor of s. Transitivities are not shown in a matrix.

3.5. The strategies

We consider different modifications of Talbot's base algorithm. These modifications mainly differ in their search strategies, i.e. either they depart from a feasible solution or any upper bound and continuously reduce this bound until the optimum is reached, or they depart from a lower bound and continuously increase this bound until the optimum is found. Moreover different resource feasibility tests are introduced in order to detect an early resource infeasibility. The different modifications are applied on the same data sets in order to judge on the quality of the strategies.

Any of the different algorithms will be described by means of a character sequence, the entries of which have the following meaning:

T1 T1 is the implementation of Talbot's algorithm (as introduced aforementioned) as a backtrack search departing from an upper bound L_n for the project duration. Whenever a new solution with a makespan $\overline{L_n}$ is found a new upper bound $\overline{L_n}$ - 1 will be imposed for a restart of the backtrack search. The order in which the activities are considered is the topological order that corresponds to the numbering of the activities in the networks A to J.

T2 T2 is the implementation of Talbot's algorithm (as introduced aforementioned) as a backtrack search departing from a lower bound E_n for the project duration. Whenever E_n is confirmed as a lower bound, i.e. there exists no feasible solution of that makespan, a new lower bound $\overline{E_n} + \delta$ will be imposed for a restart of the backtrack search. If $\delta = 1$ then the first feasible solution is optimum, otherwise, in some iteration, $\overline{E_n} + \delta$ provides a close to the optimum upper bound and T1 will be invoked with an upper bound of $\overline{E_n} + \delta - 1$.

B Whenever the next activity j is supposed to be processed in mode m and is assigned its completion time in a partial schedule, parameter B refers to an implementation of Talbot's algorithm where the amount of remaining non-renewable resources is checked. For any non-renewable resource k it must be at least w_{jkm} added to the sum of non-renewable resource demands in the least resource k consuming mode of each activity not scheduled right now (i.e. not included into the partial schedule). If the remaining resource amount is insufficient for at least one resource k a backtrack is initiated to activity j - 1 in case $m = M_j$, otherwise, in case $m < M_j$ the test is repeated for the next mode m + 1 in sequence of activity j.

E Whenever the next activity j is supposed to be processed in mode m and is assigned its completion time, say t*, in a partial schedule, parameter E refers to an implementation of Talbot's algorithm where the amount of remaining renewable resource k in any period t in the interval $[t^*-p_{jm}+1; t^*]$ is checked. If the resource amount for some resource k is not sufficient in period $t' \in [t^*-p_{jm}+1; t^*]$ then activity j cannot be completed before $t' + p_{jm}+1 > t^*$. Hence any period in the interval $[t^*; t' + p_{jm}]$ need not be considered as a possible completion time of j. If $L_j < t' + p_{jm}+1$ then the test is repeated with respect to mode m + 1 of activity j. If $m = M_j$ a backtrack to activity j - 1 is initiated.

Table 3:2. Activity resource demands in two different modes

Activity	*1*	*2*	*3*	*4*	*5*	*6*	*7*	*8*	*9*	*10*	*11*	*12*	*13*	*14*	*15*
	scheme No 0 (equal small resource demand)														
w_{i11} /	4/2	3/1	3/2	3/2	3/1	4/2	3/1	4/3	4/2	3/1	3/2	3/1	4/2	3/2	0/0
w_{i21} /	4/2	3/1	3/2	3/2	3/1	4/2	3/1	4/3	4/2	3/1	3/2	3/1	4/2	3/2	0/0
r_{i11} / r_{i12}	4/2	3/1	3/2	3/2	3/1	4/2	3/1	4/3	4/2	3/1	3/2	3/1	4/2	3/2	0/0
r_{i21} / r_{i22}	4/2	3/1	3/2	3/2	3/1	4/2	3/1	4/3	4/2	3/1	3/2	3/1	4/2	3/2	0/0
	scheme No 1 (mixed resource demand)														
w_{i11} /	3/1	4/2	6/3	2/1	5/3	8/2	4/3	3/1	5/4	2/1	4/3	3/2	5/2	3/1	0/0
w_{i21} /	4/2	2/1	4/2	2/1	4/2	6/1	2/1	4/2	6/5	5/3	3/1	6/4	3/1	3/1	0/0
r_{i11} / r_{i12}	4/1	5/2	5/4	2/1	3/2	4/2	3/1	4/1	5/2	2/1	6/3	5/4	5/3	5/3	0/0
r_{i21} / r_{i22}	3/2	4/1	6/2	2/1	4/1	3/1	4/2	5/2	4/1	4/3	2/1	6/2	6/1	2/1	0/0
	scheme No 2 (mixed / equal resource demand)														
w_{i11} /	3/1	4/2	6/3	2/1	5/3	8/2	4/3	3/1	5/4	2/1	4/3	3/2	5/2	3/1	0/0
w_{i21} /	4/2	2/1	4/2	2/1	4/2	6/1	2/1	4/2	6/5	5/3	3/1	6/4	3/1	3/1	0/0
r_{i11} / r_{i12}	3/1	4/2	3/1	3/2	4/2	3/1	4/3	3/2	3/1	4/2	3/2	3/1	4/2	3/2	0/0
r_{i21} / r_{i22}	3/1	4/2	3/1	3/2	4/2	3/1	4/3	3/2	3/1	4/2	3/2	3/1	4/2	3/2	0/0
	scheme No 3 (equal / mixed resource demand)														
w_{i11} /	3/1	4/2	3/1	3/2	4/2	3/1	4/3	3/2	3/1	4/2	3/2	3/1	4/2	3/2	0/0
w_{i21} /	3/1	4/2	3/1	3/2	4/2	3/1	4/3	3/2	3/1	4/2	3/2	3/1	4/2	3/2	0/0
r_{i11} / r_{i12}	4/1	5/2	5/4	2/1	3/2	4/2	3/1	4/1	5/2	2/1	6/3	5/4	5/3	5/3	0/0
r_{i21} / r_{i22}	3/2	4/1	6/2	2/1	4/1	3/1	4/2	5/2	4/1	4/3	2/1	6/2	6/1	2/1	0/0
	scheme No 4 (increasing resource demand)														
w_{i11} /	2/1	3/2	2/1	3/1	4/2	4/3	5/3	5/4	6/2	6/3	7/3	7/5	8/4	8/5	0/0
w_{i21} /	2/1	3/2	2/1	3/2	4/2	4/2	4/3	6/4	6/3	7/4	6/5	7/5	7/4	8/4	0/0
r_{i11} / r_{i12}	3/2	2/1	3/1	3/2	3/2	4/3	5/4	4/2	6/3	5/4	7/4	6/4	7/5	8/5	0/0
r_{i21} / r_{i22}	3/2	3/1	2/1	2/1	5/2	5/3	6/3	5/3	5/4	6/4	6/4	7/4	8/6	7/4	0/0
	scheme No 5 (decreasing resource demand)														
w_{i11} /	8/5	8/4	7/5	7/3	6/3	6/2	5/4	5/3	4/3	4/2	3/1	2/1	3/2	2/1	0/0
w_{i21} /	8/4	7/4	7/5	6/5	7/4	6/3	6/4	4/3	4/2	4/2	3/2	2/1	3/2	2/1	0/0
r_{i11} / r_{i12}	8/5	7/5	6/4	7/4	5/4	6/3	4/2	5/4	4/3	3/2	3/2	3/1	2/1	3/2	0/0
r_{i21} / r_{i22}	7/4	8/6	7/4	6/4	6/4	5/4	5/3	6/3	5/3	5/2	2/1	2/1	3/1	3/2	0/0
	scheme No 6 (increasing / decreasing resource demand)														
w_{i11} /	2/1	3/2	2/1	3/1	4/2	4/3	5/3	5/4	6/2	6/3	7/3	7/5	8/4	8/5	0/0
w_{i21} /	2/1	3/2	2/1	3/2	4/2	4/2	4/3	6/4	6/3	7/4	6/5	7/5	7/4	8/4	0/0
r_{i11} / r_{i12}	8/5	7/5	6/4	7/4	5/4	6/3	4/2	5/4	4/3	3/2	3/2	3/1	2/1	3/2	0/0
r_{i21} / r_{i22}	7/4	8/6	7/4	6/4	6/4	5/4	5/3	6/3	5/3	5/2	2/1	2/1	3/1	3/2	0/0
	scheme No 7 (decreasing / increasing resource demand)														
w_{i11} /	8/5	8/4	7/5	7/3	6/3	6/2	5/4	5/3	4/3	4/2	3/1	2/1	3/2	2/1	0/0
w_{i21} /	8/4	7/4	7/5	6/5	7/4	6/3	6/4	4/3	4/2	4/2	3/2	2/1	3/2	2/1	0/0
r_{i11} / r_{i12}	3/2	2/1	3/1	3/2	3/2	4/3	5/4	4/2	6/3	5/4	7/4	6/4	7/5	8/5	0/0
r_{i21} / r_{i22}	3/2	3/1	2/1	2/1	5/2	5/3	6/3	5/3	5/4	6/4	6/4	7/4	8/6	7/4	0/0

C Consider a partial schedule, and let V_1 be the set of already assigned activities in this partial schedule. Let $V_2 := V \setminus V_1$. Whenever the next activity j is supposed to be processed in mode m consider the interval IP $:= [\min \{E_i - p_{im}+1 \mid i \in V_2\}, L_n]$ of the earliest possible start period of any activity i from V_2 and the currently latest completion period (or possibly an upper bound of it) of the last activity. Parameter C refers to an implementation of Talbot's algorithm where the whole amount of renewable resource k within this interval, i.e. $\sum_{t \in IP} R_{kt}$, is compared to the necessary amount. The latter is calculated as the sum of amounts of resource k required by the already assigned activities in V_1 and the minimum amounts of the non-assigned activities from V_2 except j plus the required resource amount of j in mode m, i.e. $\sum_{i \in V_1} p_{im} \cdot r_{ik.} + \sum_{i \in V_2 \setminus j} \min \{ p_{im} \cdot r_{ikm} \mid m = 1, ..., M_j\} + p_{jm} \cdot r_{jkm}$. If for some resource k the remaining amount is insufficient the least required amount then activity j is tried to schedule in a next less resource consuming mode. In case there is none a backtrack is initiated and activity j-1 is rescheduled at its next possible starting period.

D Talbot's algorithm with parameters C or D are very alike. The difference being that in case of D the required resource amount of renewable resource k is estimated by $\sum_{i \in V_1} p_{im} \cdot r_{ik.} + \sum_{i \in V_2 \setminus j} p_{im} \cdot \min \{ r_{ikm} \mid m = 1, ..., M_j\} + p_{jm} \cdot r_{jkm}$. If for some resource k the remaining amount is insufficient then activity j is tried to schedule in its next mode. In case there is none a backtrack is initiated and activity j-1 is rescheduled in its next mode. This implementation need not necessarily yield an optimal solution but yields good lower bounds.

H Talbot's algorithm with parameters H refers to the implementation with parameter C except that the number of activities in V_2 is restricted to the next 5 in sequence. Furthermore, the right hand side of the interval is reduced to $\max \{L_i \mid i \in V_2 \wedge \#(V_2) \leq 5\}$.

I Parameter I refers to an implementation with parameter H except that the left hand side of the interval is increased to $\min_i \max_j \{t \cdot x_{jtm} \mid j \rightarrow i; j \in V_1; i \in V_2, \#(V_2) \leq 5\}$ where $j \rightarrow i$ defines a precedence constraint between j and i.

The following parameters refer to implementations of Talbot's algorithm branch and bound where the ordering of the activities may differ from the topological ordering of the vertices in the networks.

F Parameter F refers to an activity ordering in Talbot's algorithm according to their earliest completion time E_j. The activities are ordered in increasing E_j.

F′ Parameter F′ refers to an activity ordering in Talbot's algorithm according to their currently earliest possible start time which will be updated whenever a predecessor is scheduled. Hence, the activity i considered next in Talbot's algorithm is that one where $\max_j \{t \cdot x_{jtm} \mid j \rightarrow i; j \in V_1; i \in V_2\}$ is minimum.

Different set sizes for the number of activities in V_2 are considered. V_2 contains only those activities of which all predecessors have been scheduled already, i.e. their completion time is known. Whenever a new trial is started based on a different imposed upper bound to generate either a feasible solution or to show that the imposed upper bound is to restrictive, the ordering of the previous solution or solution effort is kept until some earliest start times are updated. Ties are broken with respect to the previous solution or, in case there is no assignment of activities in a previous solution, they are solved randomly.

G Talbot's algorithm with parameters G refers to the implementation where each activity tries to be processed from the last (here: the second) to the first mode, i.e. less resource consuming modes are preferred to short processing times. This parameter might accelerate the algorithm in cases where the quantities of resources available are small.

S Parameter S refers to an activity ordering in Talbot's algorithm according to their latest completion time L_j. The activities are ordered in increasing L_j.

F'S A combination of both parameters F' and S means that the first activity ordering is generated using parameter S and then the parameter F' applies.

3.6. The results

Any instance is described through its network, the processing time scheme as well as the resource demand scheme. Hence we use a three character code to identify instances: the first character refers to the network, the second one to the processing time scheme, and the last one to resource demand scheme, e.g. A35 for network A and schemes No 3 and No 5. Columns 'R_1' and 'R_2' ('W_1' and 'W_2') contain the available amount of renewable (same in each period) (non-renewable) resource 1 and 2. Column 'C_{max}' contains the makespan of an optimal solution. Column 'run' contains the number of restarts with a decreased (in case of T1) or increased (in case of T2) upper bound. Parameters B or E etc. refer to the aforementioned implementations and parameter '1' in T2BE1 defines the step size in time periods in algorithm T2. Columns $\mathcal{R}$ and $\mathcal{N}$ refer to the resource tightness as introduced above. The tightness of both resources of each type (renewable or not) is assumed to be similar. E_n is the earliest possible completion time in the CPM-network where resource constraints are neglected while L_n is an upper bound calculated as sum of all processing times of all activities in their slowest mode. Columns 'sec' contain the run time of the particular implementation. A series of tests were considered. All runs are performed on an Intel-Pentium processor 133 MHz. The algorithms are coded in Pascal.

Table 3:3. Resource consistency check

	resources					*T1*		*T1BE*		*T2BE1*					
	W_1	W_2	R_1	R_2	C_{max}	run	sec	run	sec	run	sec	n	$\mathcal{R}$	E_n	L_n
F31	52	52	18	20	18	5	37	5	36	7	7	$>n_{med2}$	$\mathcal{R}_{max}$	18	74
G11	50	48	11	10	19	9	36	9	33	3	3	n_{med2}	$\mathcal{R}_{med}$	19	47
H41	40	40	7	7	22	5	71	4	38	36	36	$<n_{med}$	$\mathcal{R}_{med1}$	20	53
I21	36	34	17	21	22	2	71	2	2	1	1	n_{med1}	$\mathcal{R}_{max}$	18	74
J57	58	59	19	20	27	6	1735	6	378	26	26	$>n_{med}$	$\mathcal{R}_{med2}$	27	84

If instances of the networks with parameters '.00'are used and resources are available in amounts $>n_{med2}$, n_{med2} , n_{med} and $\mathcal{R}_{max}$, $\mathcal{R}_{med2}$ $\mathcal{R}_{med}$ for the two non-renewable and the two renewable resource types, respectively, then T1, T1BE and T2BE1 behave very similar. The first mode is most often an optimal one. Run times are below a second. Table 3:3 of test 1 shows that in different networks or more restrictive resource availability T2BE1 outperforms T1. This however, is probably also an effect of the good lower bounds. Hence, T2BE1 quickly finds a first feasible solution which is optimum. Obviously a resource consistency check pays off where the competition for scarce resources is high, see row 'J57'.

In the second test we considered the influence of the network structure if resource consumption and activity process times are kept the same while provided resource quantities are similar. It turned out that problem instances underlying network E and scarce resource quantities ($\mathcal{R}_{med}$ /n_{med}) are the most difficult ones. Hence, a network structure were many activities have almost the same predecessors and successors competing simultaneously, i.e. within the same time periods, for the resources increases run times. A network structure A seems to be a simple one because at most two activities share the same predecessor or successor.

Table 3:4 shows for test 3 the influence of the activity processing times provided network structures, resource consumption and available resource quantities are kept similar. T2BE1 was in all cases superior to T1 and T1BE because the lower bounds are close to the optimal solution. In general, we can say that instances with short (duration) activities are easier to handle. Moreover, instances where short activities are supposed to be scheduled last while long activities are those which have to be scheduled at the beginning, are in general easier to solve. In particular, those instances turned out to be difficult where renewable resources are scarce (parameter constellation $\sim\mathcal{R}_{med}$ and n_{med2}), network E defines the precedence constraints, and, in particular a time consuming lower bound calculation does not go far beyond values of the simple bound E_n. The processing times reflect that there are many long and a few short activities which are insufficient to smooth the schedule to a good one. This are probably cases where good upper bounds for T1** can give better results.

Table 3:4. Influence of the activity processing times

(C_{max} column contains the optimum solution or the best lower bound found with T2BE1 when a time bound 300_{TB} of 300 seconds was imposed. An entry '-'says that C_{max} (i.e. an optimal solution) could not be found within 300 seconds.)

	Resources					*T1*	*T1BE*	*T2BE1*				
	W_1	W_2	R_1	R_2	C_{max}	sec	sec	sec	$\mathcal{N}$	$\mathcal{R}$	E_n	L_n
A01	43	41	12	11	11	8	5	2	$\mathcal{N}_{med}$	$\mathcal{R}_{med}$	9	47
A11	43	41	12	11	20	71	24	0	$\mathcal{N}_{med}$	$\mathcal{R}_{med}$	20	74
A21	43	41	12	11	17	2	1	0	$\mathcal{N}_{med}$	$\mathcal{R}_{med}$	13	74
A31	43	41	12	11	18	-	116	1	$\mathcal{N}_{med}$	$\mathcal{R}_{med}$	17	74
A41	43	41	12	11	14	0	0	0	$\mathcal{N}_{med}$	$\mathcal{R}_{med}$	14	53
A51	43	41	12	11	24	-	192	168	$\mathcal{N}_{med}$	$\mathcal{R}_{med}$	22	84
	W_1	W_2	R_1	R_2	C_{max}	sec	sec	sec	$\mathcal{N}$	$\mathcal{R}$	E_n	L_n
B01	43	41	9	8	15	6	4	2	$\mathcal{N}_{med}$	$\mathcal{R}_{med}$	11	47
B11	43	41	9	8	24	-	600_{TB}	10	$\mathcal{N}_{med}$	$\mathcal{R}_{med}$	22	74
B21	43	41	9	8	23	9	5	1	$\mathcal{N}_{med}$	$\mathcal{R}_{med}$	18	74
B31	43	41	9	8	26	-	-	300_{TB}	$\mathcal{N}_{med}$	$\mathcal{R}_{med}$	22	74
B41	43	41	9	8	19	7	2	3	$\mathcal{N}_{med}$	$\mathcal{R}_{med}$	19	53
B51	43	41	9	8	31	-	345	67	$\mathcal{N}_{med}$	$\mathcal{R}_{med}$	27	84
	W_1	W_2	R_1	R_2	C_{max}	sec	sec	sec	$\mathcal{N}$	$\mathcal{R}$	E_n	L_n
E01	50	48	22	21	9	1	1	0	$\mathcal{N}_{med2}$	$\mathcal{R}_{med}$	9	47
E11	50	48	22	21	14	-	56	0	$\mathcal{N}_{med2}$	$\mathcal{R}_{med}$	13	74
E21	50	48	22	21	14	-	58	0	$\mathcal{N}_{med2}$	$\mathcal{R}_{med}$	13	74
E31	50	48	22	21	17	-	88	1	$\mathcal{N}_{med2}$	$\mathcal{R}_{med}$	17	74
E41	50	48	22	21	14	9	5	1	$\mathcal{N}_{med2}$	$\mathcal{R}_{med}$	14	53
E51	50	48	22	21	20	-	27	1	$\mathcal{N}_{med2}$	$\mathcal{R}_{med}$	20	84
	W_1	W_2	R_1	R_2	C_{max}	sec	sec	sec	$\mathcal{N}$	$\mathcal{R}$	E_n	L_n
A44	63	63	15	16	14		-	0	$\mathcal{N}_{med2}$	$\mathcal{R}_{med2}$	14	53
A54	63	63	15	16	22	-	-	0	$\mathcal{N}_{med2}$	$\approx\mathcal{R}_{med2}$	22	84
A44	55	56	15	16	14	-	-	0	$\mathcal{N}_{med}$	$\mathcal{R}_{med2}$	14	53
A54	55	56	15	16	22	-	-	297	$\mathcal{N}_{med}$	$\approx\mathcal{R}_{med2}$	22	84
A44	53	53	15	16	14	-	-	0	$<\mathcal{N}_{med}$	$\mathcal{R}_{med2}$	14	53
A54	53	53	15	16	23	-	-	300_{TB}	$<\mathcal{N}_{med}$	$\approx\mathcal{R}_{med2}$	22	84
	W_1	W_2	R_1	R_2	C_{max}	sec	sec	sec	$\mathcal{N}$	$\mathcal{R}$	E_n	L_n
B44	63	63	15	16	19	-	0	0	$\mathcal{N}_{med2}$	$\mathcal{R}_{med2}$	19	53
B54	63	63	15	16	27	-	0	0	$\mathcal{N}_{med2}$	$\approx\mathcal{R}_{med2}$	27	84
B44	59	59	15	16	19	-	5	5	$<\mathcal{N}_{med2}$	$\mathcal{R}_{med2}$	19	53
B54	59	59	15	16	27	-	48	45	$<\mathcal{N}_{med2}$	$\approx\mathcal{R}_{med2}$	27	84
B44	58	58	15	16	19	-	60	26	$>\mathcal{N}_{med}$	$\mathcal{R}_{med2}$	19	53
B54	58	58	15	16	28	-	-	300_{TB}	$>\mathcal{N}_{med}$	$\approx\mathcal{R}_{med2}$	27	84

	Resources					*T1*	*T1BE*	*T2BE1*				
	W_1	W_2	R_1	R_2	C_{max}	sec	sec	sec	$\mathcal{N}$	$\mathcal{R}$	E_n	L_n
E44	63	63	24	26	14	-	0	0	$\mathcal{N}_{med2}$	$\mathcal{R}_{med2}$	14	53
E54	63	63	24	26	20	-	28	26	$\mathcal{N}_{med2}$	$\approx\mathcal{R}_{med2}$	20	84
E44	60	60	24	26	14	-	0	0	$<\mathcal{N}_{med2}$	$\mathcal{R}_{med2}$	14	53
E54	60	60	24	26	21	-	-	207	$<\mathcal{N}_{med2}$	$\approx\mathcal{R}_{med2}$	20	84
E44	57	57	24	26	14	-	1	1	$>\mathcal{N}_{med}$	$\mathcal{R}_{med2}$	14	53
E54	57	57	24	26	21	-	-	257	$>\mathcal{N}_{med}$	$\approx\mathcal{R}_{med2}$	20	84
E44	55	56	24	26	14	-	3	0	$\mathcal{N}_{med}$	$\mathcal{R}_{med2}$	14	53
E54	55	56	24	26	21	-	-	300_{TB}	$\mathcal{N}_{med}$	$\approx\mathcal{R}_{med2}$	20	84
E44	53	53	24	26	14	-	61	52	$<\mathcal{N}_{med}$	$\mathcal{R}_{med2}$	14	53
E54	53	53	24	26	21	-	-	300_{TB}	$<\mathcal{N}_{med}$	$\approx\mathcal{R}_{med2}$	20	84
E44	58	58	24	26	14	-	1	1	$>\mathcal{N}_{med}$	$\approx\mathcal{R}_{med2}$	14	53
E54	58	58	24	26	21	-	-	294	$>\mathcal{N}_{med}$	$\mathcal{R}_{med2}$	20	84
E44	55	56	24	26	14	-	20	19	$\mathcal{N}_{med}$	$\approx\mathcal{R}_{med2}$	14	53
E54	55	56	24	26	21	-	-	300_{TB}	$\mathcal{N}_{med}$	$\mathcal{R}_{med2}$	20	84

Table 3:5 contains the influence of resource consumption on the run time from test 4 whereby processing times and resource availability are kept fixed. The resources are available in amounts $\mathcal{R}_{med}$ and $\mathcal{N}_{med}$ for the two renewable and the two non-renewable resource types, respectively. While Table 3:5a contains the different bounds and optimal values, Table 3:5b shows the run times and the number of restarts. The influence of the resource consumption seems to be less important on the run time than the processing times of the activities. However, a continuously decreasing consumption (with an increasing activity order) seems to lead to shorter run times than an increasing resource consumption. In contrast to Table 3:4 that indicates that processing time differences are more influential on run times than network structures, the network structure considered in Tables 5a and 5b more likely imposes run time difference. This probably could also be explained through the indices $\mathcal{R}$ and $\mathcal{N}$ because of their choice an increased resource demand leads to an increased availability. Furthermore, a large amount of resource usage over time seems to be less restrictive as long as this is not limited to a small set of activities.

The results of test 5 are shown in Table 3:6. The first five columns describe the problem instance with respect to processing times, network structure, resource consumption and resource availability. Columns 'OPT' contain the makespan C_{max} of an optimum solution of the particular problem instance as well as the run time to generate a (optimum) solution when C_{max} is imposed as an upper bound to T2BE1 or T1BE. The run times for T1BE and T2BE1 to find an prove C_{max} to be the makespan of an optimum solution are shown in columns 'T1BE' and 'T2BE1', respectively. The intial upper bound for T1BE is L_n while T2BE1 used E_n as the first upper bound (which, whenever no feasible solution could be generated turned out to be a lower bound).

Table 3:5. Influence of the resource consumption

(C_{max} column contains the optimum solution found with T2BE1 when a time bound 300_{TB} of 300 seconds was imposed. An entry '-' says that C_{max} was not found and confirmed within 300 seconds.)

Table 3:5a.

	resources								*resources*				
	W_1	W_2	R_1	R_2	C_{max}	E_n	L_n		R_1	R_2	C_{max}	E_n	L_n
A11	43	41	12	11	20	20	74	B11	9	9	24	22	74
A12	43	41	10	10	20	20	74	B12	7	7	23	22	74
A31	43	41	12	11	18	17	74	B31	9	9	-	22	74
A32	43	41	10	10	18	17	74	B32	7	7	26	22	74
A21	43	41	12	11	17	13	74	B21	9	9	22	18	74
A23	36	36	12	11	16	13	74	B23	9	9	22	18	74
A41	43	41	12	13	14	14	53	B41	9	9	19	19	53
A44	55	56	12	13	15	14	53	B44	12	13	19	19	53
A45	55	56	19	20	14	14	53	B45	13	14	19	19	53
A56	55	56	19	20	24	22	84	B56	13	14	31	27	84
A57	55	56	12	13	-	22	84	B57	12	12	29	27	84

	resources								*resources*				
	W_1	W_2	R_1	R_2	C_{max}	E_n	L_n		R_1	R_2	C_{max}	E_n	L_n
D11	43	41	9	8	28	26	74	E11	22	21	14	13	74
D12	43	41	7	7	26	26	74	E12	19	19	14	13	74
D31	43	41	11	11	-	20	74	E31	22	21	18	17	74
D32	43	41	9	9	24	20	74	E32	17	17	29	18	74
D21	43	41	9	9	24	22	74	E21	22	21	16	14	74
D23	36	36	9	9	23	22	74	E23	22	21	16	14	74
D41	43	41	9	9	21	20	53	E41	22	21	14	14	53
D44	55	56	12	13	21	20	53	E44	24	26	14	14	53
D45	55	56	11	12	21	20	53	E45	28	31	14	14	53
D56	55	56	12	13	29	24	84	E56	28	31	-	20	84
D57	55	56	15	16	28	24	84	E57	25	26	20	20	84

Table 3:5b.

	network A		*network B*		*network D*		*network E*	
A/B/C/D	run	sec	run	sec	run	sec	run	sec
11	1	0	3	0	3	173	2	9
12	1	0	2	0	2	0	2	8
31	2	1	-	300_{TB}	-	300_{TB}	2	57
32	2	1	5	12	5	107	65	300
21	5	1	5	1	3	1	3	18
23	4	1	5	3	2	0	3	20
41	1	0	1	3	2	31	1	7
44	2	17	1	21	2	65	1	0
45	1	0	1	0	2	0	1	0
56	3	103	5	71	6	28	-	300_{TB}
57	-	300_{TB}	3	43	5	183	1	13

Whenever T1BE generated a feasible solution with makespan t_n (possibly smaller than L_n) then T1BE is restarted with the new upper bound t_n - 1 and either proves $t_n = C_{max}$, i.e. the solution is optimal, or generates an improved feasible solution. If T2BE1 recognizes that there is no feasible solution to an upper bound t_n then T2BE1 restarts with an increased upper bound $t_n + 1$, which is the currently best lower bound of the underlying problem instance. Hence, while T1BE does not necessarily check all time periods from L_n to C_{max}-1 T2BE1 checks each period from E_n to C_{max}.

If no information about the optimum makespan is available the test could not prove that one approach is definitely superior (faster) than the other one. Run times are very similar. If T1BE is provided with a better initial upper bound it seems to be faster; moreover it generates feasible solutions. However, we cannot confirm that as larger the difference L_n - C_{max} as easier it is to find a feasible solution to L_n. For instance, in order to find a first feasible solution for J45 the algorithm runs 105 seconds with an upper bound of 26, while an upper bound of 27 required 1412 seconds to find the same feasible solution. Only less than 1 second is necessary to find a feasible solution to upper bounds 28 or 29. Other instances are even worse. An always only one period increasing upper bound (above C_{max} and below L_n) for D12 caused run times from 4, 17, 66, 4, 18, 99, 1017, 119, 3 and finally less than 1 seconds. Thus, the run time of T1BE heavily depends on the upper bounds and the impression that close to optimum bounds are more likely to increase the run time than very weak bounds could in general not be confirmed. The behavior of T2BE1 was different. This algorithm could easily prove infeasibility for bounds far below C_{max} while close to optimum bounds are more time consuming to prove to be infeasible. Surprisingly, it turned out, that cases where E_n provides a reasonable good bound were the most difficult ones.

Table 3:6. Efficiency of T1BE to T2BE1

	resources				*OPT*		*T1BE*	*T2BE1*		
	W_1	W_2	R_1	R_2	C_{max}	sec	sec	sec	E_n	L_n
A00	27	27	6	6	16	3	269	315	9	47
D44	39	42	15	15	28	0	83	83	20	53
H56	50	52	16	16	22	5	739	761	20	84
D44	55	56	12	12	21	0	48	47	20	53
D11	53	53	10	9	27	24	68	68	26	74
E56	52	52	22	22	16	13	50	50	14	48
H12	35	35	6	6	28	47	74	74	25	74
J45	43	45	17	18	25	12	157	171	20	53
A12	35	35	6	6	24	37	63	63	20	74
D12	35	35	6	6	29	147	171	191	26	74
B57	58	58	14	14	28	26	29	28	27	84

Table 3:7. Lower bounds / upper bounds / interval lengths obtained within 180 sec

	$\mathcal{R}$ / $\mathcal{N}$				
	$\mathcal{R}_{min}$ / $\mathcal{N}_{min}$	$\mathcal{R}_{min}$ / $\mathcal{N}_{med}$	$\mathcal{R}_{med1}$ / $\mathcal{N}_{med1}$	$\mathcal{R}_{med2}$ / $\mathcal{N}_{min}$	$\mathcal{R}_{med2}$ / $\mathcal{N}_{med2}$
A46	36/40/4	opt	opt	33/35/2	opt
B11	38/51/13	31/39/8	27/40/13	35/53/18*	opt
C12	38/55/17	30/35/5	opt	34/49/15	opt
D41	32/39/7	opt	opt	29/38/9	opt
E57	33/65/32	23/29/6	21/32/11	30/67/37*	opt
F00	24/32/8	opt	opt	20/30/10	opt
F51	36/64/28	30/37/7	25/42/17	35/70/35*	25/28/3
G32	31/50/19	28/38/10	24/39/15	28/48/20	opt
G54	34/60/26	33/41/8	27/37/10	30/65/35*	opt
I21	43/47/4	opt	opt	40/48/8	opt

Table 3:7 contains results if severe time bounds of 180 seconds are imposed on the run time of both algorithms, T1BE and T2BE1. For 10 problem instances and different resource profiles $\mathcal{R}_{min}$ / $\mathcal{N}_{min}$, $\mathcal{R}_{min}$ / $\mathcal{N}_{med}$, $\mathcal{R}_{med1}$ / $\mathcal{N}_{med1}$, $\mathcal{R}_{med2}$ / $\mathcal{N}_{min}$, and $\mathcal{R}_{med2}$ / $\mathcal{N}_{med2}$. The left column entry contains the best lower bound (i.e. the last upper bound for which T2BE1 proved infeasibility increased by one) obtained where the initial upper bound for T2BE1 is E_n. The second column entry contains the best upper bound for which T1BE could generate a feasible solution. The right column entry is just the length of the interval (right bound - left bound) that contains the optimum. The run times for T1BE and T2BE1 were limited to 180 seconds each. Entries 'opt' indicate that left and right bound coincide and an optimum has been found by either T1BE or T2BE1 or both. In some cases it turned out that in order to improve the bounds beyond those of Table 3:7 the required run times increased rapidly. This happened in particular if the optimum was close to the upper bound. For example, in order to show that there is no feasible solution with a makespan of 46 for I21 and resource profile $\mathcal{R}_{min}$ / $\mathcal{N}_{min}$ (i.e. to show that 47 is optimal) the run time increased to about 40 minutes. About 20 minutes were necessary to prove that there is no solution with makespan 39 to A46 ($\mathcal{R}_{min}$ / $\mathcal{N}_{min}$), hence the upper bound of 40 is optimal. For problem instances with '*' entries in column '$\mathcal{R}_{med2}$ / $\mathcal{N}_{min}$' a scarce resource profile (column '$\mathcal{R}_{min}$ / $\mathcal{N}_{min}$') accelerated to find even better bounds. Hence, scarce resource profiles can help to detect infeasibility faster. One particular example is F51 with resource profile $\mathcal{R}_{med2}$ / $\mathcal{N}_{min}$: while there is a feasible solution of less than 66 time unit, the algorithm could find one within 80 minutes. Once again it seems to be true that the ease of the network structure A is responsible for the high quality solution with respect to A46. It also turned out again that a network structures E and F are more difficult ones.

Table 3:8a. A comparison of lower bounds calculated within 1 second

	T2BE	*T2BDE*	*identical*
No. of times the algorithm found the shortest interval	6	24	20
Sum of all interval lengths	741	423	-
No. of times the algorithm found a better upper bound	0	0	50
No. of times the algorithm found a better lower bound	6	24	
Resource parameters $\mathcal{R}$ / $\mathcal{N}$			
$\mathcal{R}_{min}$ / $\mathcal{N}_{min}$	0/0	10/140	0
$\mathcal{R}_{min}$ / $\mathcal{N}_{med}$	5/5	0/0	5
$\mathcal{R}_{med1}$ / $\mathcal{N}_{med1}$	1/1	4/13	5
$\mathcal{R}_{med2}$ / $\mathcal{N}_{min}$	0/0	10/161	0
$\mathcal{R}_{med2}$ / $\mathcal{N}_{med2}$	0/0	0/0	10
Sum	6/6	24/314	20

Table 3:8b. A comparison of lower bounds calculated within 180 second

	T2BE	*T2BDE*	*Identical*
No. of times an optimum solution is found	18	17	-
Sum of all interval lengths	460	263	-
No. of times a better upper bound is found	2	0	48
No. of times a better lower bound is found	10(12)	20(210)	20
Resource parameters $\mathcal{R}$ / $\mathcal{N}$			
$\mathcal{R}_{min}$ / $\mathcal{N}_{min}$	2/4	8/92	0
$\mathcal{R}_{min}$ / $\mathcal{N}_{med}$	6/6	0/0	4
$\mathcal{R}_{med1}$ / $\mathcal{N}_{med1}$	1/1	3/13	6
$\mathcal{R}_{med2}$ / $\mathcal{N}_{min}$	1/1	9/105	0
$\mathcal{R}_{med2}$ / $\mathcal{N}_{med2}$	0/0	0/0	9
sum	10/12	20/210	20

Right now only two parameters were responsible for checking the resource sufficiency, namely parameters B and E for non-renewable and renewable resources. In the next tests we consider additional termination criteria with respect to renewable resources. They are implemented in the algorithm modifications C, D, I, and J. Moreover, up to now the activity ordering for branching was fixed from the beginning. It corresponds to the numbering of the activities in the networks A to J. We are going to consider four different activity orders indicated by the aforementioned parameters F, F′, S, and F′S.

The results of test 6 are described in Tables 8a,b. They are based on the algorithms T2BE and T2BDE, both in order to generate useful lower bounds within a run time limit of 1 second (Tables 8a) and 180 seconds (Tables 8b). For each of the ten instances A46, B11, C12, D41, E57, F00, F51, G32, G54, I21 five different resource profiles are generated.

Table 3:8a gives an overview of the interval lengths which is the difference of upper and lower bound, obtained from both algorithms T2BE and T2BDE. In all 50 cases the upper bounds were the same, but in 24 cases T2BDE calculated better lower bounds while T2BE reached a better lower bound only in 6 cases. For 20 instances both algorithms terminated with the same lower bound. The sum of all 50 interval lengths was 423 for T2BDE and 741 for T2BE. In the bottom part of Table 3:8a the resource profiles of the parameters $\mathcal{R}$ / $\mathcal{N}$ are shown. For each resource profile the first entry gives the number of times the particular algorithm achieved a better lower bound. The second entry gives the sum of the time periods the lower bounds of the first entry saved. In the last column the number of cases where both lower bounds (that one of T2BE and T2BDE) were identical. From the last row we can see that in case severe time bounds are imposed T2BDE could generate much better lower bounds than the exact procedure T2BE. We found similar results in the case where the run time was increased to 180 seconds. Again T2BDE outperformed T2BE. The results are shown in Table 3:8b. Twice algorithm T2BE found a better upper bound, i.e. in these cases the upper bound was proved to be the makespan of an optimal solution. While T2BE generated in only 10 cases a better lower bound (the overall improvement over all 10 cases, i.e. the number of time periods saved, is shown in brackets) saving 12 periods of time, T2BDE saved 210 periods distributed over 20 cases where the lower bound was superior to the one of T2BE. We also compared the T2BDE lower bounds to those generated by T2BCE, T2BEI, T2BEJ. In all cases the latter algorithms were unable to find lower bounds on the quality level of T2BDE. The average distance of the lower bounds generated by T2BCE, T2BEI, T2BEJ from the T2BDE lower bounds was 4.5, 7.5, and 10.7 time periods, respectively.

We wonder if in case of scarce resources an initial assignment of the least resource consuming mode (parameter G) can accelerate the calculation of good lower and upper bounds. Over the 50 problem instances T1BCEG and T1BEG found the same upper bounds which were on average almost one period below the T1BE upper bounds. In total T1BE needed 656 seconds to calculate all upper bounds (within a run time limit of 180 seconds per instance). T1BCEG used 252 seconds and T1BEG used only 87 seconds in order to calculate the 50 best possible upper bounds. When the computation time was limited to 1 second per instance the upper bound of T1BCEG was on average two periods below the T1BE bound. The lower bounds of T2BEG however were very weak, unfortunately close to E_n. The lower bounds of T2BCEG were competitive (on average 4 periods above the T2BE bound) but still below those that can be obtained through T2BCE. Finally, we would like to mentioned that, in particular, for resource profiles of $\mathcal{R}_{med2}$ / $\mathcal{N}_{min}$ parameter G could improve the upper bound calculation.

Finally, we considered tests which include activity orderings based on the parameters F, F′ and S as well as combinations of them. In 41 out of all 50 cases T2BES gave the best lower bounds among the 4 procedures of Table 3:9. Even the run time was smallest for T2BES about 1/3 of the run time of T2BE. In 37 out of 50 cases T1BEF′S gave the best upper bounds among the procedures of Table 3:9. A summary of the results is given in Table 3:9. T* indicates T2 or T1 for lower or upper bound calculation.

Table 3:9. The quality of different activity orderings

	*T*ABE*	*T*ABEF'*	*T*ABES*	*T*ABEF'S*
Sum of the lower bounds (180 sec)	1349	1339	1403	1394
Sum of the upper bounds (180 sec)	1818	1818	1799	1787
Sum of the lower bounds (1 sec)	1176	1181	1258	1246
Sum of the upper bounds (1sec)	1908	1902	1836	1837

3.7. Conclusions

The network structure, i.e. the number of predecessors or successors of activities, different processing times (e.g. many long, few short ones) and the resource profile with respect to the parameters $\mathcal{R}$ and $\mathcal{N}$ seem to have a high influence on the performance of a branch and bound algorithm. Surprisingly, different priority based mode or activity sequences are less influential on an algorithms performance; they are more a matter of fine tuning. Hence, we cannot recommend to implement one procedure in order to solve resource constrained project scheduling problems efficiently. Most probably there is none. Different implementations are necessary and they are to apply to the different classes of problem instances, i.e. in order to be more effective it seems to be save not to invest only time in the construction of a general problem solver (the branch and bound code) but, in particular, to create instance classes within the same problem in consideration. A classification of this classes is necessary and this paper might be a first step in that way. The problem generator of Kolisch et al. (1995) is another one, however, their results cannot be applied to a classification and differentiation of difficult and easy job shop instances.

Moreover, it seems save to claim that the difficulty an algorithm has to solve an instance of a problem will be reflected in its subproblems. That means, whenever subproblem optimal solutions have makespans far from the subproblem makespan of an optimum problem solution (which is induced to the subproblem) then the problem is difficult to solve. These are also the cases where consistency tests (see the survey by Dorndorf et al., 1998) in branch and bound procedures are less effective. Problems seem to be more challenging if the matrix representation of the network contains many non-diagonal subproblem parts, e.g. network E. Similar results could be confirmed for job shops, see Pesch and Tetzlaff (1996). These are cases where lower bounds are hard to prove because the different solutions of subproblems (i.e. partial schedules at different branches) do not fit to each other and require increased lower bounds. Consistency tests are insufficient to recognize contradictions in time, however, a propagation of the subproblem imposed constraints could accelerate the lower bound contradiction. Hence, as many constraint propagation approaches on the complete problem are either too weak or too time consuming, a propagate and backtrack approach seems to be more effective.

Finally, we would like to mention that branch and bound procedures can be used approaching an optimal solution from both sides. From the upper bound side it is to consider as a method in order to guarantee feasibility. From the lower bound side a limited branching strategy of the same procedure might deliver useful lower bounds. In our case this was achieved through parameter D.

Acknowledgment

I would like to thank Fabian Dreibholz and Carl Scharffenorth for their support in the algorithms implementation and testing.

Remark

The Code can be obtained from the author.

References

BLAZEWICZ J., K.H. ECKER, E. PESCH, G. SCHMIDT AND J. WEGLARZ. 1996. *Scheduling Computer and Manufacturing Processes*. Springer, Berlin.

DEMEULEMEESTER, E. AND W. HERROELEN. 1992. A branch-and-bound procedure for the multiple resource-constrained project scheduling problem. *Management Science* **38**, 1803-1818.

DORNDORF U., T. PHAN HUY, AND E. PESCH. 1998. A survey of interval capacity consistency tests for time- and resource-constrained scheduling. in: *Recent Advances in Project Scheduling*, Chapter 10, Kluwer.

Drexl A. 1991. Scheduling of project networks by job assignment. *Management Science* **37**, 1590-1602.

KOLISCH R. 1995. *Project Scheduling under Resource Constraints - Efficient heuristics for Several Problem Classes*. Physica, Heidelberg.

KOLISCH R, A. SPRECHER AND A. DREXL. 1995. Characterization and generation of a class of resource-constrained project scheduling problem. *Management Science* **41**, 1693-1703.

PATTERSON J.H. 1984. A comparison of exact approaches for solving the multiple constrained resource project scheduling problem. *Management Science* **30**, 854-867.

PATTERSON, J.H., F.B. TALBOT, R. SLOWINSKI AND J. WEGLARZ. 1990, Computational experience with a backtracking algorithm for solving a general class of precedence and resource-constrained scheduling problems. *European Journal of Operational Research* 49, 68-79.

PESCH E. AND U. TETZLAFF. 1996. Constraint propagation based scheduling of job shops. *INFORMS Journal on Computing* **8**, 144-157.

RUSSEL R.A. 1986. A comparison of heuristics for scheduling projects with cash-flows and resource restrictions. *Management Science* **32**, 1291-1300.

SPRECHER A. 1994. *Resource-Constrained Project Scheduling - Exact Methods for the Multi-Mode Case*. Springer, Berlin.

SPRECHER A. AND A. DREXL. 1996, Solving multi-mode resource-constrained project scheduling problems by a simple, general and powerful sequencing algorithm. *European Journal of Operational Research*, to appear.

TALBOT F.B. 1982. Resource-constrained project scheduling with time-resource tradeoffs: the nonpreemptive case. *Management Science* **28**, 1197-1210.

TALBOT F.B. AND J.H PATTERSON. 1978. An efficient integer programming algorithm with network cuts for solving resource constrained scheduling problems. *Management Science* **24**, 1163-1174.

	1	2	3	4	5	6	7	8	9	10	11	12	13	14	15
1						a									
2						a	a								
3							a	a							
4	b							a	a						
5		b							a						
6			b							a					
7				b						a	a				
8					b						a	a			
9						b						a			
10							b						a		
11								b					a	a	
12									b					a	
13										b	b				a
14												b	b		a
15														b	

Figure 3:1. Networks A (entries a) and B (entries b)

	1	2	3	4	5	6	7	8	9	10	11	12	13	14	15
1												c			
2						c									
3	d					c	c								
4	d						c								
5		d											c		
6			d					c							
7				d				c							
8						d	d		c	c	c				
9					d									c	
10								d						c	
11								d						c	
12								d						c	
13								d	d					c	
14										d	d	d	d		c
15														d	

Figure 3:2. Networks C (entries c) and D (entries d)

	1	*2*	*3*	*4*	*5*	*6*	*7*	*8*	*9*	*10*	*11*	*12*	*13*	*14*	*15*
1		e	e	e	e	e	e	e	e	e	e				
2												e			
3												e			
4		f										e			
5			f									e			
6				f								e			
7				f									e		
8					f								e		
9					f								e		
10	F												e		
11						f	f						e		
12									f					e	
13									f					e	
14								f				f	f		e
15										f	f			f	

Figure 3:3. Networks E (entries e) and F (entries f)

	1	*2*	*3*	*4*	*5*	*6*	*7*	*8*	*9*	*10*	*11*	*12*	*13*	*14*	*15*
1					g										
2					g										
3						g									
4															g
5	h	h					g								
6				h				g	g						
7					h					g					
8						h								g	
9			h								g	g			
10							h		h					g	
11									h				g		
12								h					g		
13								h						g	
14										h	h				g
15												h	h	h	

Figure 3:4. Networks G (entries g) and H (entries h)

	1	*2*	*3*	*4*	*5*	*6*	*7*	*8*	*9*	*10*	*11*	*12*	*13*	*14*	*15*
1					i										
2					i										
3									i						
4	J					i									
5	J	j					i								
6				j				i							
7				j						i					
8						j						i	i		
9						j	j			i	i				
10					j		j							i	
11					j									i	
12								j	j						i
13										j	j				i
14												j	j		i
15			j											j	

Figure 3:5. Networks I (entries i) and J (entries j)

4 ALGORITHMS FOR SCHEDULING PROJECTS WITH GENERALIZED PRECEDENCE RELATIONS

Bert De Reyck[1]
Erik Demeulemeester[2]
Willy Herroelen[2]

1. Erasmus University Rotterdam (The Netherlands)
2. Katholieke Universiteit Leuven (Belgium)

4.1. Introduction

The problem of scheduling projects under various types of resource constraints constitutes an important and challenging problem which has received increasing attention during the past several years. The bulk of the models and procedures designed for coping with these problem types aim at scheduling project activities to minimize the project duration subject to constant availability constraints on the required set of resources and precedence constraints that indicate that activities can only be started when all of their predecessors have already been finished. However, real-life project scheduling applications often involve more complicated types of precedence relations such as arbitrary minimal and maximal time lags between the starting and completion times of the activities, and require more sophisticated regular and nonregular objective functions. Over the past few years, considerable progress has been made in the use of exact solution procedures for this problem type and its variants. We will review the fundamental logic and report new computational experience with solution procedures for optimally solving resource-constrained project scheduling problems in which such generalized precedence relations and objective functions can be explicitly considered.

We distinguish between four types of generalized precedence relations (GPRs): start-start (SS), start-finish (SF), finish-start (FS) and finish-finish (FF). These relations specify a minimal or maximal time lag between a pair of activities. A minimal time lag specifies that an activity j can only start (finish) when its predecessor i has already started (finished) for a certain time period. A maximal time lag specifies that an activity should be started (finished) at the latest a specific number of time periods beyond the start (finish) of activity i.

GPRs enhance the capabilities of project scheduling models because they can be used to model a wide variety of real-life problem characteristics. Next to the straightforward use of GPRs, namely allowing activity overlaps (which will often lead to a substantial decrease of the project makespan) and ensuring a maximal delay between the execution of specific activities (useful, for instance, when dealing with perishable products or chemical operations), GPRs can be used to model a wide variety of specific problem characteristics, including activity release dates and deadlines, activities that have to start or terminate simultaneously, non-delay execution of activities, several types of mandatory activity overlaps, fixed activity start times, time-varying resource requirements and availabilities, set-up times, overlapping production activities (process batches, transfer batches) and assembly line zoning constraints.

The first comprehensive treatment of GPRs is due to Kerbosch and Schell (1975), based on the pioneering work of Roy (1962). Other studies include Crandall (1973), Elmaghraby (1977), Wiest (1981), Moder et al. (1983), Bartusch et al. (1988), Elmaghraby and Kamburowski (1992), Zhan (1994), Neumann and Zhan (1995), Schwindt (1996), Brinkmann and Neumann (1996), Schwindt and Neumann (1996), Franck and Neumann (1996), De Reyck and Herroelen (1996b, 1997, 1998ab), Neumann and Schwindt (1997), Demeulemeester and Herroelen (1997b) and De Reyck (1998).

The remainder of this chapter is organized as follows. Section 4.2 elaborates on the concept of GPRs and clarifies the terminology and project representation used. Section 4.3 briefly reviews the temporal analysis of activity networks with GPRs. A distinction is made between the precedence diagramming case, i.e. start-start, start-finish, finish-start and finish-finish relations with minimal time lags for minimizing the project makespan (defined as $min|C_{max}$ using the classification scheme of Herroelen et al. 1998), and the case of minimal and maximal time lags ($gpr|C_{max}$). In Section 4.4, the resource analysis required by the introduction of additional resource constraints is discussed. In Section 4.5, we discuss the fundamentals of a branch-and-bound procedure for optimally solving resource-constrained project scheduling problems with generalized precedence relations of the precedence diagramming type. This problem type, when extended to cope with activity release dates and deadlines as well as variable resource availabilities, is also referred to as the *generalized resource-constrained project scheduling problem* (Demeulemeester and Herroelen 1997b) and is denoted as $m,1,va|min,\rho_i,\delta_i|C_{max}$ using the classification scheme of Herroelen et al. (1998). Subsequently in Section 4.6, we review several branch-and-bound procedures for the case of minimal and maximal time lags ($m,1|gpr|C_{max}$), and demonstrate how the solution methodology can be extended to cope with variable resource availabilities and requirements as well as other real-life project characteristics ($m,1,va|gpr,\rho_i,\delta_i,vr|C_{max}$), and with other regular ($m,1,va|gpr,\rho_i,\delta_i,vr|reg$) and nonregular ($m,1,va|gpr,\rho_i,\delta_i,vr|nonreg$) objective functions. In Section 4.7, we briefly describe the modifications that the original algorithms have undergone since their development. In Section 4.8, computational experience is reported using a set of randomly generated problem instances. Section 4.9 is reserved for our conclusions.

4.2. Terminology and representation

Assume a project represented in activity-on-the-node format by a directed graph $G = \{V, E\}$ in which V is the set of vertices or activities, and E is the set of edges or GPRs. The non-preemptable activities are numbered from 1 to n, where the dummy activities 1 and n mark the beginning and the end of the project. The duration of an activity is denoted by $d_i (1 \le i \le n)$, its start time by $s_i (1 \le i \le n)$ and its finish time by $f_i (1 \le i \le n)$. There are m renewable resource types, with r_{ikx} $(1 \le i \le n, 1 \le k \le m, 1 \le x \le d_i)$ the resource requirements of activity i with respect to resource type k in the xth period it is in progress and a_{kt} $(1 \le k \le m; 1 \le t \le T)$ the availability of resource type k in time period $]t$-1, $t]$ (T is an upper bound on the project length). If the resource requirements and availabilities are not time-dependent, they are represented by r_{ik} $(1 \le i \le n, 1 \le k \le m)$ and a_k $(1 \le k \le m)$ respectively. The minimal and maximal time lags between two activities i and j have the form:

$$s_i + SS_{ij}^{\min} \leq s_j \leq s_i + SS_{ij}^{\max}; \qquad s_i + SF_{ij}^{\min} \leq f_j \leq s_i + SF_{ij}^{\max}$$
$$f_i + FS_{ij}^{\min} \leq s_j \leq f_i + FS_{ij}^{\max}; \qquad f_i + FF_{ij}^{\min} \leq f_j \leq f_i + FF_{ij}^{\max}$$

where $SS_{ij}^{\min}$ represents a minimal time lag between the start time of activity i and the start time of activity j (similar definitions apply for $SS_{ij}^{\max}$, $FS_{ij}^{\min}$,...). The various time lags can be represented in a *standardized form* by transforming them to minimal start-start precedence relations, using the transformation rules given in Bartusch et al. (1988). Consequently, all GPRs are consolidated in the expression $s_i + l_{ij} \leq s_j$, where l_{ij} denotes a minimal start-start time lag.

A path $<i_s, i_k, i_l, ..., i_t>$ is called a *cycle* if $s = t$. With 'path' we mean a *directed* path, and with 'cycle' we mean a *directed* cycle. The *length* of a path (cycle) is defined as the sum of the lags associated with the arcs belonging to that path (cycle). To ensure that the dummy start and finish activities correspond to the start and the completion of the project, we assume that there exists at least one path with nonnegative length from node 1 to every other node and at least one path from every node i to node n which is equal to or larger than d_i. If there are no such paths, we can insert arcs (1,i) or (i,n) with weight zero or d_i respectively. $P(i) = \{j \mid (j,i) \in E\}$ is the set of all *immediate predecessors* of node i, $Q(i) = \{j \mid (i,j) \in E\}$ is the set of all its *immediate successors*.

4.3. Temporal analysis

The (resource-unconstrained) project scheduling problem with GPRs under the minimum makespan objective ($gpr|C_{max}$) can be mathematically formulated as follows:

Minimize s_n [1]

Subject to

$s_i + l_{ij} \leq s_j \qquad (i,j) \in E$ [2]

$s_i \in N \qquad i \in V$ [3]

where N denotes the set of natural numbers. The objective function [1] minimizes the project duration (makespan), determined by the completion time (or start time, since $d_n = 0$) of the dummy end activity n. Constraints [2] represent the GPRs. Constraints [3] ensure that all start times assume nonnegative integer values. Solving this problem can be accomplished by finding a precedence-feasible *earliest start schedule* (*ESS*), i.e. the minimum start times (es_1, es_2, ..., es_n) satisfying [2] and [3]. The earliest start of an activity i can be computed by finding the longest path from node 1 to node i.

4.3.1. The precedence diagramming case

The CPM analysis for project networks with zero-lag finish-start precedence relations ($cpm|C_{max}$) can easily be extended to cope with minimal time lags ($min|C_{max}$). A forward computation step $es_i = \max\left\{es_j + l_{ji} | \forall j \in P(i)\right\}$ yields an *ESS* (assuming that es_1 = 0). A backward computation step $ls_i = \min\left\{ls_j - l_{ij} | \forall j \in Q(i)\right\}$ (assuming that $ls_n = es_n$) yields a latest start schedule (*LSS*) which can be used for float calculations and activity criticality analysis.

4.3.2. The case of generalized precedence relations

When maximal time lags are introduced ($gpr|C_{max}$), there may not be a schedule that satisfies all of the GPRs. There exists a precedence-feasible schedule for G iff G has no cycle of positive length (Bartusch et al. 1988). Therefore, if we compute the matrix $D = [d_{ij}]$, where d_{ij} denotes the longest path from node i to node j, a positive path length from node i to itself (d_{ii}>0) indicates the existence of a cycle of positive length and, consequently, the non-existence of a precedence-feasible schedule. The computation of the matrix D can be done by the Floyd-Warshall algorithm in time $O(n^3)$ (see Lawler 1976). The ESS = (es_1, es_2, ..., es_n) is given by (d_{11}, d_{12}, ..., d_{1n}).

4.4. Resource analysis

When we introduce additional renewable resource constraints, we obtain problems $m,1|min|C_{max}$ and $m,1|gpr|C_{max}$, which can both be conceptually formulated as follows:

Minimize s_n [4]

Subject to

$s_i + l_{ij} \le s_j \quad \forall (i,j) \in E$ [5]

$\sum_{i \in S(t)} r_{ik} \le a_k \quad k = 1, 2, \ldots, m \quad t = 1, 2, \ldots, T$ [6]

$s_i \in N \quad i = 1,2,\ldots,n$ [7]

where $S(t)$ is the set of activities in progress in time period $]t\text{-}1, t]$ and T is an upper bound on the project duration, for instance $T = \sum_{i \in V} \max\left\{d_i, \max_{j \in Q(i)}\left\{l_{ij}\right\}\right\}$. The objective function [4] minimizes the project duration. The GPRs are denoted in standardized form by constraints [5]. Constraints [6] represent the resource constraints and constraints [7] ensure that the activity start times assume nonnegative integer values.

In the precedence diagramming case, the l_{ij} values are restricted to nonnegative values. Consequently, an activity can never start before its predecessor. Activity release dates ρ_i need not be specified separately because they can be modelled using standardized time lags of the type $l_{1i} = \rho_i$. The algorithm of Demeulemeester and Herroelen (1997b) for the generalized resource-constrained project scheduling problem (GRCPSP; $m,1,va|min,\rho_i,\delta_i|C_{max}$), also deals with deadlines δ_i and variable resource availabilities.

In the case of GPRs, the l_{ij} values may assume arbitrary integer values. In that case also, there is no need to specify activity deadlines separately, because they can be modelled using negative standardized time lags of the type $l_{i1} = d_i - \delta_i$. Also time-varying resource availabilities and requirements need not be specified explicitly (Bartusch et al. 1988). Time-varying resource availabilities can be handled by creating dummy activities which absorb a certain amount of each resource type for which a constant availability (equal to the maximum availability over time of that resource type) can then be assumed. These dummy activities should then be assigned a fixed start time using appropriate minimal and maximal time lags. Time-varying resource requirements can be modelled by splitting up the activities in a number of subactivities with a different *constant* resource requirement for each of the resource types. The subactivities should then be connected with appropriate minimal and maximal time lags which ensure a non-delay execution of all the subactivities of each activity. Therefore, problem $m,1,va|gpr|C_{max}$ and problem $m,1,va|gpr,\rho_i,\delta_i,vr|C_{max}$ can be solved using the same algorithm.

Problems $m,1|min|C_{max}$ and $m,1|gpr|C_{max}$ are known to be strongly NP-hard. For problem $m,1|gpr|C_{max}$ or problem $m,1|min,\delta_n|C_{max}$ with an imposed project deadline, even the decision problem of detecting problem (in)feasibility is NP-complete (Bartusch et al. 1988). To the best of our knowledge, the only exact solution procedures presented in the literature are the branch-and-bound algorithms of Bartusch et al. (1988), Demeulemeester and Herroelen (1997b) and De Reyck and Herroelen (1998a). Because of the extreme complexity of problem $m,1|gpr|C_{max}$, quite a number of heuristics have been developed (Zhan 1994, Neumann and Zhan 1995, Brinkmann and Neumann 1996, Franck and Neumann 1996, Schwindt and Neumann 1996).

In the next section, we will review two exact solution procedures for project scheduling problems with GPRs. Again, we distinguish between the precedence diagramming case $(m,1,va|min,\rho_i,\delta_i|C_{max})$, for which the procedure of Demeulemeester and Herroelen (1997b) will be reviewed, and the case of generalized precedence relations $(m,1,va|gpr,\rho_i,\delta_i,vr|C_{max})$, for which the fundamentals of the procedure developed by De Reyck and Herroelen (1998a) will be discussed.

4.5. A branch-and-bound procedure for the GRCPSP

4.5.1. The search tree

The branch-and-bound procedure of Demeulemeester and Herroelen (1997b, further referred to as GDH) is an extension of the DH-algorithm presented in Demeulemeester and Herroelen (1992,1997a) for the resource-constrained project scheduling problem with zero-lag finish-start precedence constraints $(m,1|cpm|C_{max})$. It is based on a depth-first solution strategy in which nodes in the search tree represent resource- and precedence-feasible partial schedules. Branches emanating from a parent node correspond to exhaustive and subset-minimal combinations of activities, the delay of which resolves a resource conflict at the parent node (referred to as *minimal delaying alternatives*). The search process closely resembles the one used in the procedure of Demeulemeester and Herroelen (1992) for the RCPSP $(m,1|cpm|C_{max})$. The modifications involve a different definition of the decision point, a different process of delaying temporarily scheduled activities, a different definition of the cutset activities, and a modified backtracking scheme. In addition, the procedure relies on a different set of dominance rules and bounding calculations.

The nodes in the search tree correspond to partial schedules in which finish times temporarily have been assigned to a subset of the activities of the project. Scheduling decisions are temporary in the sense that scheduled activities may be delayed as the result of decisions made at later stages in the search process. Partial schedules are constructed by semi-active timetabling[1]; i.e. each activity is started as soon as possible within the

[1] Sprecher and Drexl (1996) correctly claim that the procedure of Demeulemeester and Herroelen (1992) does not only generate semi-active schedules. The same applies to the algorithm described here. Only if the left-shift dominance rule is extended to examine also local left-shifts prior to the current decision point, the schedules are guaranteed to be semi-

precedence and resource constraints. A precedence-based lower bound is calculated by adding the maximal remaining critical path length of any of the activities that belong to a delaying alternative to the current delaying point. The delaying alternative with the smallest lower bound is chosen for further branching. When a complete schedule is constructed or when a partial schedule can be dominated using one of the node fathoming rules described below, the procedure backtracks to a previous level in the search tree. The procedure is completed upon backtracking to level 0. Activity deadlines are coped with through a standard critical path-based backward pass computation starting from the deadlines.

4.5.2. Node fathoming rules

Three dominance rules are used to prune the search tree. Proofs can be found in Demeulemeester and Herroelen (1997b).

THEOREM 1. *In order to resolve a resource conflict it is sufficient to consider only minimal delaying alternatives (which do not contain other delaying alternative as a subset).*

THEOREM 2. *Consider a partial schedule PS_t (the set of scheduled or completed activities) at level p of the search tree in which activity i is started at time t. If activity i was delayed at level p-1 of the search tree, and if this activity can be left-shifted without violating the precedence or resource constraints, then the partial schedule PS_t is dominated.*

THEOREM 3. *Consider a cutset C_t (the set of unscheduled activities for which at least one direct predecessor belongs to PS_t) which contains the same activities as a cutset $C_{t'}$, which was previously saved during the search of another path in the search tree, and which was considered during the same resource interval. If $t' \leq t$, if all activities in progress at time t' did not finish later than the maximum of t and the finish time of the corresponding activities in PS_t, and if the earliest possible start time of every activity in $C_{t'}$ is smaller than or equal to the earliest start time of the corresponding activity in C_t, then the current partial schedule PS_t is dominated.*

Remark here that the definition of the cutset differs slightly from the one formulated in Demeulemeester and Herroelen (1997b). In the GRCPSP, contrary to the RCPSP, it is possible that an unscheduled activity is precedence constrained by an already scheduled activity, while some of its

---

active. However, the branching scheme and the dominance rules are based on the *principle* of semi-active timetabling.

other predecessors are not yet scheduled. A subtle change in the definition of the cutset is therefore needed. This fact was overlooked in the original implementation of the algorithm. Based on new computational experience presented in this chapter, we discovered one project example (out of the many examined) for which the optimal solution was missed because of this flaw. As will be indicated in Section 4.8.2.3, the correction of the flaw only slightly affects the computational results.

4.6. A branch-and-bound procedure for the RCPSP-GPR

4.6.1. The search tree

Essentially, the algorithm of De Reyck and Herroelen (1998a, further referred to as DRH) is a hybrid depth-first / laser beam search branch-and-bound algorithm. The nodes in the search tree represent the initial project network, described by the longest path matrix $D = [d_{ij}]$, extended with extra zero-lag finish-start precedence relations to resolve a number of resource conflicts, which results in an *extended* matrix $D' = [d'_{ij}]$. Nodes which represent precedence-feasible but resource-infeasible project networks and which are not fathomed by any node fathoming rules described below lead to a new branching. Resource conflicts are resolved using the concept of minimal delaying alternatives. However, contrary to the GDH procedure, each of these minimal delaying alternatives is delayed (enforced by extra zero-lag finish-start precedence relations $i \prec j$, implying $s_i + d_i \le s_j$) by each of the remaining activities also belonging to the *conflict set* $S(t^*)$, the set of activities in progress in period $]t^*-1, t^*]$ (the period of the *first* resource conflict). Consequently, each minimal delaying alternative can give rise to several *minimal delaying modes*.

In general, the *delaying set DS*, i.e. the set of all minimal delaying alternatives, is equal to

$$DS = \left\{ D_d \,\middle|\, \begin{array}{l} D_d \subset S(t^*) \text{ and } \forall \text{ resource } k: \sum\limits_{i \in S(t^*)} r_{ik} - \sum\limits_{i \in D_d} r_{ik} \le a_k \\ \text{and } \forall\, D_{d'} \in \mathrm{DS} \setminus \{D_d\} : D_{d'} \not\subset D_d \end{array} \right\}$$

The set of minimal delaying modes equals:

$M = \left\{ M_m \,\middle|\, M_m = \{k \prec D_d\}, k \in S(t^*) \setminus D_d, D_d \in DS \right\}$. Activity k is called the *delaying activity*: $k \prec D_d$ implies that $k \prec l$ for all $l \in D_d$.

***THEOREM* 4.** *The delaying strategy which consists of delaying all minimal delaying alternatives* D_d *by each activity* $k \in S(t^*) \setminus D_d$ *will lead to the optimal solution in a finite number of steps.*

PROOF. See De Reyck and Herroelen (1998a).

Each minimal delaying mode is then examined for precedence-feasibility and evaluated by computing the critical path-based lower bound lb_0. Each precedence-feasible minimal delaying mode with a lower bound $lb_0<T$ is then considered for further branching, which occurs from the node with smallest lb_0. If the node represents a project network in which a resource conflict occurs, a new branching occurs. If it represents a feasible schedule, the upper bound T is updated and the procedure backtracks to the previous level in the search tree. Branching occurs until at a certain level in the tree, there are no delaying modes left to branch from. Then, the procedure backtracks to the previous level in the search tree and reconsiders the other delaying modes (not yet branched from) at that level. The procedure stops upon backtracking to level 0.

The fact that semi-active timetabling cannot be applied when dealing with both minimal and maximal time lags results in a different solution strategy employed by the GDH and DRH procedures. In the GDH procedure, *partial* schedules based on the precedence relations are constructed, until a resource conflict is observed. The remainder of the schedule need not be computed. A resource conflict is resolved through the delay of activities participating in the conflict, which corresponds to the addition of precedence relations. The search then advances through time to subsequent resource conflicts until the dummy end activity is scheduled. Upon finding such a complete (precedence- and resource-feasible) schedule, the procedure backtracks, which corresponds to a partial destruction of the schedule. Contrary, in the DRH procedure, in each node of the search tree, *complete* (precedence-based) schedules are evaluated. The first resource conflict in that schedule is subsequently resolved using additional precedence relations. However, contrary to the GDH procedure, the procedure cannot proceed through time, because in a newly derived schedule, obtained by resolving a resource conflict at time t, a new conflict can occur at a time instant $t'<t$. When a precedence- and resource-feasible schedule is encountered, the procedure backtracks to the previous level in the search tree which corresponds to removing precedence relations from the project network.

4.6.2. Node fathoming rules

Nodes are fathomed if they represent a precedence-infeasible project network or when lb_0 exceeds (or equals) T. Four additional node fathoming rules (three dominance rules and a new lower bound) and a procedure which reduces the solution space and which can be executed as a pre-processing rule are added. These rules are given below. Additional information and proofs can be found in De Reyck and Herroelen (1998a).

THEOREM 5. *If there exists a minimal delaying alternative D_d with activity $i \in D_d$ but its real successor $j \notin D_d$ $(d_{ij} \geq 0)$, we can extend D_d with activity j. If the resulting delaying alternative becomes non-minimal as a result of this operation, it may be eliminated from further consideration.*

THEOREM 6. *When a minimal delaying alternative D_d gives rise to two minimal delaying modes M_{m_1} and M_{m_2} with delaying activities i and j respectively, M_{m_2} is dominated by M_{m_1} if $d_{ij} + d_j \geq d_i$. If $d_{ij} + d_j = d_i$ and $d_{ji} + d_i = d_j$, either delaying mode M_{m_1} or M_{m_2} can be dominated.*

Mingozzi et al. (1998) have developed five new lower bounds for the RCPSP, namely lb_1, lb_2, lb_p, lb_x and lb_3, derived from different relaxations of a new mathematical formulation. They compute lb_3 using a heuristic for the maximum weighted independent set problem. Demeulemeester and Herroelen (1997a) have incorporated another version of lb_3 in their GDH algorithm, which can be extended to the RCPSP-GPR as follows. For each activity $i \in V$, its *companions* are determined (the activities with which it can be scheduled in parallel, respecting both the precedence ($d_{ij} < d_i$ and $d_{ji} < d_j$) and resource constraints ($\forall k \leq m$: $r_{ik} + r_{jk} \leq a_k$)). All activities i are then entered in a list L in non-decreasing order of the number of companions (non-increasing duration as tie-breaker). The following procedure then yields a lower bound, lb_3^g. For each activity, we define a remaining duration d_j^r. Initially, all d_j^r are equal to d_j and $lb_3^g = 0$.

while L not empty **do**
    take the first activity (activity i) in L and remove it from L
    $lb_3^g = lb_3^g + d_i^r$
    for every companion j of i **do**
        if $d_{ij} > 0$ **then** $d_j^r = d_j^r - (d_i - d_{ij})$

else if $d_{ji} > 0$ **then** $d_j^r = d_j^r - \min\{d_j - d_{ji}, d_i\}$

else $d_j^r = d_j^r - d_i$

endif

if $d_j^r \leq 0$, remove activity *j* from *L*

enddo

enddo

THEOREM 7. *lb_3^g is a valid lower bound for problem $m, 1, va|gpr, \rho_i, \delta_i, vr|C_{\max}$.*

lb_3^g is used to fathom nodes for which $lb_3^g \geq T$. However, whereas lb_0 is calculated immediately upon the creation of a node, the calculation of lb_3^g is deferred until a decision has been made to actually branch from that node. The rationale behind this is that (a) lb_3^g is more difficult to compute than lb_0, and (b) calculating lb_3^g implies calculating the entire matrix *D*. We defer the calculation of lb_3^g and *D* until the node is actually selected for branching. As a result, only lb_0 is used as a branching criterion.

THEOREM 8. *If the set of added precedence constraints which leads to the project network in node x contains as a subset another set of precedence constraints leading to the project network in a previously examined node y in another branch of the search tree, node x can be fathomed.*

THEOREM 9. *If $\exists i, j \in V$ and $k \leq m$ for which $r_{ik} + r_{jk} > a_k$ and $-d_j < d_{ij} < d_i$, we can set $l_{ij} = d_i$.*

4.6.3. Extensions to other objective functions

In real-life project scheduling applications, the minimization of the project length is undoubtedly the most popular objective function. Minimizing the project makespan implies that the resources tied up in the activities of the project are released as soon as possible, thereby making them available for other projects in the future. Also, minimizing the project length releases tied-up capital because in many projects, the majority of income payments occur at the end of a project (Kolisch 1996).

Nevertheless, for many actual project scheduling applications, minimizing the project length may be a misrepresentation of the actual conditions, in which considerations such as cost minimization, tardiness minimization and revenue maximization may be much more relevant. In the literature, a rich variety of objective functions has been the subject of extensive study. These objective functions can be classified into two distinct classes: regular and nonregular measures of performance. A regular measure of performance (which is to be minimized) is a nondecreasing function of the activity completion times. When not imposed by resource, precedence or temporal constraints, it will not be advantageous to delay activities solely to obtain an improved performance under a regular measure of performance. For a nonregular objective function, the condition above does not hold. This implies that delaying activities may improve the performance of the schedule, even if such a delay is not imposed by any constraints.

4.6.3.1. Regular performance measures. Practical applications of regular measures of performance often take the form of a cost function based on the activity completion times. Such cost functions may take the following form:

- Minimizing project costs determined by a weighted function of the tardiness of the activities with respect to pre-set due dates $(m,1,va|gpr,\rho_i,\delta_i,vr|\sum w_i T_i)$, where $T_i = \max\{f_i - \overline{\delta_i}, 0\}$, $\overline{\delta_i}$ represents the due date of activity i (not to be confused with the activity deadlines δ_i which can also be present and which constitute hard constraints that cannot be violated) and w_i denotes the weight (penalty) associated with an additional delay (of one time period) of activity i beyond its due date. We may also want to minimize the number of tardy activities $(m,1,va|gpr,\rho_i,\delta_i,vr|n_T)$ or the maximal activity tardiness $(m,1,va|gpr,\rho_i,\delta_i,vr|T_{max})$.
- Minimize the mean flow time: $m,1,va|gpr,\rho_i,\delta_i,vr|\overline{F}$, where $\overline{F} = \frac{1}{n}\sum_{i=1}^{n}(f_i - \rho_i)$ and ρ_i denotes the release date of activity i.

In the DRH procedure, we evaluate the project networks in each node of the search tree by computing an *ESS* (by means of a longest path matrix *D*), which yields a critical path-based lower bound lb_0. If we optimize any other *regular* measure of performance $(m,1,va|gpr,\rho_i,\delta_i,vr|reg)$, we can still use the *ESS* to evaluate the project networks and simply replace the calculation of lb_0 by the regular performance measure under consideration. The branching strategy based on minimal delaying modes (Theorem 4) can also be used when dealing with other regular performance measures. Also Theorems 5, 6, 8 and 9 are still applicable. Therefore, only two slight modifications are needed to extend the procedure. First, we need to replace lb_0 by the new measure and use the resulting value as a lower bound. Second, the lower bound lb_3^g can no longer be used as a node fathoming rule since it is based on the minimum makespan objective.

4.6.3.2. Nonregular performance measures. If we optimize a *nonregular* measure of performance ($m,1,va|gpr,\rho_i,\delta_i,vr|nonreg$), the branching strategy based on minimal delaying modes to resolve resource conflicts (Theorem 4) can still be used. However, we *cannot* use the *ESS* anymore to compute the objective function value and replace the calculation of lb_0 by the performance measure under consideration. Rather, the project networks in each node of the search tree should be optimized using the nonregular objective function while discarding the resource constraints ($gpr|nonreg$). Also resource-feasibility should be checked against the schedule obtained by optimizing the nonregular objective function for the resource-unconstrained project network.

A popular nonregular performance measure is the maximization of the net present value (*npv*) of the project, in which positive and negative cash flows are associated with the activities ($m,1,va|gpr,\rho_i,\delta_i,vr,c_i|npv$): Maximize $\sum_{i=1}^{n} c_i e^{-\alpha f_i}$ with $c_i = \sum_{t=1}^{d_i} g_{it}\, e^{\alpha(d_i - t)}$, the cash flow (positive or negative) associated with each activity compounded up to its completion.

For a review of project scheduling problems in which financial considerations are explicitly included, we refer the reader to Herroelen et al. (1997). When maximizing the project *npv*, the evaluation and optimization of the project networks in each node should be accomplished by maximizing the *npv* of the corresponding (precedence-feasible, but not necessarily resource-feasible) project without taking the resource constraints into account ($gpr,\delta_n,c_i|npv$). Algorithms for the unconstrained max-*npv* project scheduling problem ($cpm,\delta_n,c_i|npv$) can be found in Russell (1970), Grinold (1972), Elmaghraby and Herroelen (1990), Herroelen and Gallens (1993) and Herroelen et al. (1996). Unfortunately, none of these algorithms can cope with GPRs ($gpr,\rho_i,\delta_i,c_i|npv$). De Reyck and Herroelen (1996b) have developed an exact recursive enumeration procedure for optimizing the *npv* in project networks with GPRs ($gpr,\rho_i,\delta_i,c_i|npv$), which will be briefly reviewed in the next section.

4.6.3.3. Maximizing the net present value of projects: the resource-unconstrained case. The algorithm of De Reyck and Herroelen (1996b) for problem $gpr,\rho_i,\delta_i,c_i|npv$ is based on the intuitive idea that activities carrying positive cash flows should be executed as early as possible, whereas activities with a negative cash flow should be delayed as much as possible. The procedure consists of 3 steps. In STEP 1, the longest path matrix D is computed. If the project is precedence-feasible, the *early tree* is computed, which spans all activities scheduled at their earliest start time. For every activity i, a predecessor j is determined for which $d_{1,j} + d_{j,i} = d_{1,i}$, upon which activities j and i are linked.

The *current tree* is computed in STEP 2 of the algorithm by delaying all activities i with a negative cash flow c_i and no successor in the early tree as much as possible within the early tree, i.e. without affecting the start times of the successor activities in the constraint digraph. This results in a local optimum which cannot be improved by delaying single activities and will reduce the number of recursions required in STEP 3 of the procedure which examines the simultaneous delay of activities. If any activity i has been delayed while computing the current tree, STEP 2 is repeated. After STEP 2 has been repeated a sufficient number of times, the procedure enters a recursive search in STEP 3, in which partial trees PT (with a negative *npv*) are identified that can be disconnected from the current tree and shifted forwards in time in order to increase the *npv* of the project. When such a partial tree is found, the algorithm computes the maximal shift of the partial tree by identifying the maximal possible increase in the start times of the activities belonging to the partial tree without violating any of the precedence constraints, keeping all activities not belonging to PT at their current start times. Therefore, a new arc is determined with minimal displacement, i.e. an arc (k,l) ($k \in PT, l \notin PT$) with minimal value for $d_{1,l} - d_{1,k} - d_{k,l}$. We disconnect the partial tree from the remainder of the current tree and we add the arc (k,l) to the current tree, thereby relinking the forward-shifted partial tree to the current tree. Then, we update the completion times of the activities in the partial tree as follows: $\forall\, j \in PT$: $d_{1,j} = d_{1,j} + \min_{\substack{k \in PT \\ l \notin PT}} \left\{ d_{1,l} - d_{1,k} - d_{k,l} \right\}$. If a shift has been found and implemented, the recursive procedure is restarted until no further shift can be accomplished. Then, the optimal schedule with its corresponding *npv* is reported.

4.6.3.4. Maximizing the net present value of projects: the resource-constrained case. De Reyck and Herroelen (1998b) have developed a branch-and-bound procedure for problem $m,1,va|gpr,\rho_i,\delta_i,vr,c_i|npv$ based on the DRH algorithm for the minimum makespan case, using the recursive search procedure described above for the calculation of a bound on the project *npv*. Each time a node in the branch-and-bound tree is chosen to branch from, the corresponding longest path matrix D is computed and the schedule which optimizes the project *npv* is computed, yielding an upper bound on the *npv*. However, in the DRH procedure, when a number of nodes are created at a certain level of the search tree (not yet chosen to be branched from), the matrices D are not yet computed. Therefore, it is not possible to use the algorithm for the unconstrained *npv* maximization to compute an upper bound on the project *npv*. Therefore, the computation of the upper bound *ub* on the project *npv* is not made upon creation of a node, but is deferred until a decision has been made to actually branch from it. As a result, another criterion (a myopic criterion based on the cash flows of the delayed activities) is used in order to select the node to branch from at a certain level.

Another often encountered example of a nonregular performance measure is the minimization of the weighted earliness-tardiness of the activities in a project, in which a due date, earliness penalties as well as tardiness penalties are associated with the activities ($m,1,va|gpr,\rho_i,\delta_i,vr|early/tardy$): $\sum_{i=1}^{n} w_i \sum_{t=es_i}^{ls_i} \left| t + d_i - \overline{\delta_i} \right| x_{it}$. In that case, the project network in each node of the search tree should be optimized such that a minimum penalty value due to earliness or tardiness of the activities is obtained, while the activities are subject to GPRs only ($gpr,\rho_i,\delta_i|early/tardy$). Exact solution procedures for optimizing due date performance in project networks are sparse. To the best of our knowledge, if the precedence relations among the activities are allowed to be GPRs, no solution procedure is available at all for minimizing earliness-tardiness-based objective functions. This constitutes a promising area for future research.

4.6.3.5. Multiple objective functions. From the discussion above, it is clear that project management has the choice between a wide variety of performance measures. These measures may pertain to the makespan of the project, the tardiness of activities or subprojects, the activity flow times, the levelness of the resource profile(s) and may even include financial considerations. In many situations, several of these objective functions may be relevant at the same time. Often, however, the relevant objectives are in conflict. In that case, a trade-off will be present in the sense that the project manager will have to decide which performance measure is the most important, in which order they should be considered or which weights should be assigned to each of the measures.

This gives rise to the problem of scheduling projects under multiple objectives. We distinguish between the case where (a) multiple objectives are considered in a pre-specified order or have been assigned a weight determining the trade-off between the measures, and (b) where the solution method should present a series of alternative solutions from which the decision maker should select a solution based on his/her perspectives of the relation between the performance measures. In the former case, the solution procedure can unambiguously determine the optimal solution because the multiple objectives can be merged into a single objective function. In the latter case, the solution procedure cannot determine an optimal schedule because a trade-off between the various performance measures has not been firmly established. The procedure should then present a number of efficient solutions, from which the decision maker can select a schedule.

If multiple regular performance measures are considered, each one given a weight to determine its importance vis-à-vis the other measures (or a rank order), the DRH procedure can still be used. In that case, the *ESS* can still be used to evaluate the resource-unconstrained project networks in each node of the search tree. If, however, nonregular performance measures are considered, the problem becomes much more complex. In that case, the resource-unconstrained projects should be optimized taking into account the weighted nonregular (and regular) objective functions. Also when no weights or strict order can be assigned to the measures, the solution approach should be modified rather extensively. In that case, the procedure should present multiple viable alternatives and allow the user to determine which schedule he/she prefers based on the associated values for the various objective functions.

4.7. Modifications to the original algorithms

The GDH and DRH procedures have been recoded and compiled using Microsoft Visual C++ 4.0 under Windows NT for use on a Dell Pentium Pro-200Mhz personal computer. The GDH code requires 91 Kb, whereas the data structures are allowed to use up to 16 Mb. This memory is mainly allocated to the application of the cutset dominance rule. For the DRH code, which requires 90 Kb, only 400 Kb should be reserved for storing the data.

The GDH procedure has undergone some modifications since its development. First, we have corrected the application of the cutset dominance rule as explained in Section 4.5.2. Secondly, we now only consider efficient cutsets when applying the cutset dominance rule: every time a new cutset is saved, all cutsets that are dominated by it are removed, resulting in a significantly smaller set of cutsets and a substantial speed-up of the dominance rule. The codes of both algorithms have been modified in order to take full advantage of modern 32-bit compiler architecture. This results in a significant efficiency gain. The major change in the GDH procedure involves a new coding scheme for the cutset dominance rule. Similar adjustments have been described for the case with zero-lag finish-start precedence relations (see Demeulemeester and Herroelen 1997a). Other changes involve merging different resource types into one global resource type (using 32-bit integers). Additional code polishing also leads to an increase in performance. For technical details we refer the reader to Demeulemeester and Herroelen (1997a).

4.8. Computational experience

4.8.1. Previous computational experience

4.8.1.1. The GRCPSP ($m,1,va|min,\rho_i,\delta_i|C_{max}$). Computational experience with the GDH procedure is reported by Demeulemeester and Herroelen (1997b) on the problem set consisting of the 110 RCPSP instances assembled by Patterson (1984). The results were very promising and indicated that the algorithm was, on the average, only 2.5 times slower than the similar procedure designed for the RCPSP. For the Simpson problem set, consisting of the same 110 RCPSP instances, extended with variable resource availabilities by Simpson et al. (1992), the procedure also performed substantially better than the procedure of Simpson et al. (1992). A third problem set consisted of ten problem instances based on Patterson problem 72 in which ready times, deadlines and precedence diagramming constraints were introduced. All ten instances could be solved very quickly (in at most 32.08 seconds).

The changes in the coding of the GDH procedure result in a dramatic decrease in the computation times for these problem instances. Using a Pentium Pro-200Mhz computer, the average computation time for the Simpson problem set decreases by a factor of more than one thousand (0.009 seconds versus 9.273 seconds). For the ten problems based on Patterson problem 72, the speed-up factor is only 12 (0.434 seconds versus 0.035 seconds).

4.8.1.2. The RCPSP-GPR ($m,1,va|gpr,\rho_i,\delta_i,vr|C_{max}$). De Reyck and Herroelen (1998a) report computational results on three different problem sets in order to validate the DRH procedure against the serial and parallel heuristics developed by Franck and Neumann (1996). These heuristics improve upon the procedures developed by Neumann and Zhan (1995), Zhan (1994) and Brinkmann and Neumann (1996). All three data sets have been generated using the random problem generator ProGen/max developed by Schwindt (1996). The first problem set consists of 1,080 100-activity instances. The second set consists of 1,440 100-activity problem instances. The third set consists of 7,200 instances with 10 up to 100 activities.

Table 4:1 shows some computational results on the first problem set. These results are obtained using Microsoft Visual C++ 2.0 under Windows NT for a Digital Venturis Pentium-60 personal computer with 16Mb of internal memory. The branch-and-bound procedure is truncated after a specific amount of running time (1, 10 and 100 seconds). The results include the number of problems solved to optimality (for which the optimum was found *and verified*), the number of problems for which the optimal solution is obtained (but *not necessarily verified*), the number of problems for which the best known solution is obtained, the number of unsolved problems (for which a feasible solution could not be determined and neither infeasibility of the instance could be proven), the average deviation from a lower bound and the average deviation from the best known solution. The lower bound lb used to compute the deviations, is the maximum of the critical path-based lower bound lb_0, the resource-based lower bound $lb_r = \max_{k=1}^{m}\left\{\left\lceil \sum_{i=1}^{n} d_i r_{ik} / a_k \right\rceil\right\}$ and lb_3^g (computed in the root node of the search tree after pre-processing). The column labelled F&N in Table 4:1 contains the results obtained by Franck and Neumann (1996), which are obtained by running a collection of 44 different heuristics which rank among the best currently available. The best known solution referred to in Table 4:1 is the best of the solutions obtained with various versions of the DRH algorithm running for up to 1 hour per problem and with the heuristic (F&N) solutions, and can therefore be considered as near-optimal.

Table 4:1. The results on problem set I

	F&N	DRH 1 sec	DRH 10 sec	DRH 100 sec
Solved to optimality	196 (18%)	543 (50%)	592 (55%)	609 (56%)
Optimal solution is found	220 (20%)	578 (54%)	596 (55%)	609 (56%)
Best known solution is found	378 (35%)	606 (56%)	652 (60%)	682 (63%)
Unsolved problems	21 (2%)	205 (19%)	86 (8%)	68 (6%)
Average deviation from lb	17.02%	5.99%	9.77%	10.00%
Average deviation from best solution	7.20%	2.20%	2.54%	2.31%

The DRH procedure manages to solve more than 50% of the 100-activity problem instances to optimality within 1 second of computation time. However, increasing the allowed computation time from 1 second to 100 seconds leads to an increase of only 12% in the problem instances solved to optimality (from 543 to 609). The average deviation from the best known solution (lower bound) never exceeds 2.54% (10.00%), whereas the F&N heuristics result in an average deviation of 7.20% (17.02%). Less reassuring, however, is that, especially for small time limits, a relatively large number of problems remains unsolved. The F&N heuristics do a better job on this issue.

This inspired us to develop another approach which is based on finding a feasible solution first, rather than going immediately for the optimal solution. When no feasible solution has been obtained yet, this approach uses a new criterion (referred to as time window slack *TWS*) to decide on which node to branch from, based on feasibility criteria. The node that entails the highest chance of leading to a feasible solution is selected first, regardless of its lower bound (which is only used as a tie-breaker). When a feasible solution is obtained, again the lower bound is used as a branching criterion. Using this new approach, the number of unsolved problems decreases to 27 (2.5%), 8 (0.7%) and 6 (0.6%) for the three time limit settings, whereas the number of problems solved to optimality does not significantly differ from the original approach. The average deviation from the best known solution (lower bound) increases somewhat, but never exceeds 4.5% (14%), thereby still outperforming the heuristics. More details can be found in De Reyck and Herroelen (1998a).

4.8.2. New computational results

4.8.2.1. Benchmark problem set. In this section, we present new computational experience with the two enhanced branch-and-bound procedures on a new benchmark problem set consisting of 1,620 randomly generated instances. The parameters used to generate the new problem set are given in Table 4:2. The indication $[x,y]$ means that the corresponding value is randomly generated in the interval $[x,y]$, whereas x; y; z means that three settings for that parameter were used in a full factorial experiment. For each combination of parameter values, 10 instances have been generated.

The resource factor *RF* (Pascoe 1966) reflects the average portion of resources requested per activity. If *RF*=1, then each activity requests all resources. *RF*=0 indicates that no activity requests any resource: $RF = \frac{1}{nK}\sum_{i=1}^{n}\sum_{k=1}^{K}\begin{cases}1, & \text{if } r_{ik} > 0\\ 0, & \text{otherwise}\end{cases}$. The *resource strength RS* (Cooper 1976) is redefined by Kolisch et al. (1995) as $\left(a_k - r_k^{\min}\right)/\left(r_k^{\max} - r_k^{\min}\right)$, where a_k is the total availability of renewable resource type k, $r_k^{\min} = \max_{i=1,\ldots,n} r_{ik}$ (the maximum resource requirement for each resource type), and $r_k^{\max}$ is the peak demand for resource type k in the precedence-based early start schedule. The resource availability is assumed to be constant over time.

Table 4:2. The parameter settings of the new problem sets

Control parameter	Values
number of activities	10; 20; 30
activity durations	[2,10]
number of resource types	4
minimum / maximum number of resources used per activity	1 / 4
activity resource demand	[1,10]
resource factor, *RF* (Pascoe 1966)	0.50; 1.00
resource strength, *RS* (Kolisch et al. 1995)	0.00; 0.25; 0.50
number of initial and terminal activities	[2,4]
maximum number of predecessors and successors	3
order strength, *OS* (Mastor 1970)	0.35; 0.50; 0.65
% maximal time lags	0%; 10%; 20%
number of cycle structures (Brinkmann and Neumann 1996)	[1,10]
minimum / maximum number of nodes per cycle structure	2 / 30
coefficient of cycle structure density (Schwindt 1996)	0.3
cycle structure tightness (Schwindt 1996)	0.5

The *order strength OS* is defined as the number of precedence relations, including the transitive ones, divided by the theoretical maximum of such precedence relations, namely $n(n-1)/2$, where n denotes the number of activities (Mastor 1970). Because *OS* only applies to acyclic networks, it is applied to the acyclic skeleton of the generated project networks obtained by ignoring all maximal time lags (for details see Schwindt 1996). For the definition of cycle structures and related measures, we refer the reader to Schwindt (1996).

4.8.2.2. Overall results. The problems without any maximal time lags correspond to instances of the GRCPSP, whereas the problems with 10% and 20% maximal time lags correspond to instances of the RCPSP-GPR. Therefore, we solved the former (540) instances with the GDH procedure, and the latter (1,080) instances with the DRH procedure. A time limit of 1,000 seconds is imposed. The overall results can be found in Table 4:3.

Table 4:3. Overall results

	0%		10%		20%		All problems	
n	Optimal	CPU-time	Optimal	CPU-time	Optimal	CPU-time	Optimal	CPU-time
10	180	0.001	180	0.00	180	0.00	540	0.00
20	180	0.005	180	0.11	180	0.12	540	0.08
30	180	0.021	177	36.37	178	34.27	535	23.56
All	540	0.009	537	12.16	538	11.46	1,615	7.88

The GRCPSP instances can be solved to optimality within very small CPU-times using the GDH procedure. However, the RCPSP-GPR instances, solved using the DRH procedure, require much more time. Five out of the 1,080 instances cannot be solved to (verified) optimality within 1,000 seconds. This illustrates the much higher complexity of the RCPSP-GPR versus the GRCPSP. The main reason for the difference in complexity is that for solving the RCPSP-GPR, semi-active timetabling cannot be applied. Consequently, most of the concepts developed for the minimal time lag case are not transferable to the GPR-case. The fundamental logic of the GDH procedure, namely its branching strategy, is based on the principle of semi-active timetabling and can therefore not be applied for the GPR case. Also the dominance rules, including the left-shift dominance rule and the powerful cutset dominance rule, which is mainly responsible for the efficiency of the GDH procedure, are not applicable anymore. Consequently, the DRH procedure is based on a different branching strategy and a new set of dominance rules and lower bounds.

The fact that the branching strategy and the dominance rules which are applicable for the GPR case are less powerful than those used for the precedence diagramming case, can be illustrated by comparing the performance of both algorithms on the problem instances with minimal time lags only. Whereas the GDH procedures solves all 540 instances to optimality with an average CPU-time of 0.009 seconds, the more general DRH procedure cannot solve 12 out of the 540 instances within 1,000 seconds. Naturally, when no maximal time lags are present, the DRH procedure is no longer efficient because it is designed for the inclusion of maximal time lags. In that case, the GDH procedure should be used. Similarly, when all precedence relations are of the zero-lag finish-start type, the procedure of Demeulemeester and Herroelen (1992, 1997a) should be used instead. The efficiency of the DRH procedure heavily depends on the relative number of maximal time lags in the problem instances. The more maximal time lags, the more effective the dominance and bounding rules, and the more efficient the DRH procedure.

4.8.2.3. The modified cutset dominance rule. We examined the impact of the new cutset dominance rule on the efficiency of the GDH procedure by implementing the original cutset dominance rule in the new procedure. The results indicate that the efficiency of the procedure does not substantially differ. For all but 1 instance, the optimal solution is obtained despite the use of the erroneous cutset dominance rule. The computation times using the modified cutset dominance rule are only slightly higher (0.009 versus 0.008 seconds).

4.8.2.4. Results with truncated procedures. Table 4:4 presents the results with a truncated version of the DRH procedure. We did not report any results with a truncated GDH procedure since it is able to solve all of the instances with minimal time lags to optimality with very little computational effort. Therefore, we report the results with the DRH procedure only (also for the instances without maximal time lags). The results reported are the number of instances for which the optimal solution is found (not necessarily verified, including the problems proven to be infeasible) and the average deviation from the best known solution (all but five of these solutions are known to be optimal). The deviations are only computed for the instances which are feasible and for which the truncated procedure was able to find a feasible solution.

Although the DRH procedure cannot solve all instances to optimality when the imposed time limit is rather small, the obtained heuristic solutions are of high quality, especially when the relative amount of maximal time lags is rather high. The results conform to the results of previous computational experiments (De Reyck and Herroelen 1998a), which show a similar performance of the truncated DRH procedure on instances with up to 100 activities (see also Table 4:1).

4.8.2.5. Impact of problem characteristics. In Table 4:5, the impact of the order strength *OS* on the complexity of the problem instances is examined. Clearly, *OS* has a negative impact on the problem complexity, measured by the number of problems solved to optimality and the required CPU-time. This result is in line with other results reported in the literature (De Reyck 1995, Schwindt 1996, De Reyck and Herroelen 1998a).

Table 4:4. Heuristic results

		0%		10%		20%		All problems	
		Optimal	% dev.	Optimal	% dev.	Optimal	% dev.	Optimal	% dev.
	$n = 10$	180	0.00%	180	0.00%	180	0.00%	540	0.00%
time limit	$n = 20$	168	0.26%	178	0.06%	179	0.01%	525	0.11%
1 SECOND	$n = 30$	112	2.23%	152	1.23%	148	1.68%	412	1.71%
	All	460	0.83%	510	0.43%	507	0.56%	1,477	0.61%
	$n = 10$	180	0.00%	180	0.00%	180	0.00%	540	0.00%
time limit	$n = 20$	177	0.05%	180	0.00%	180	0.00%	537	0.02%
10 SECONDS	$n = 30$	131	1.31%	167	0.59%	164	0.76%	462	0.89%
	All	488	0.46%	527	0.20%	524	0.25%	1,539	0.30%
	$n = 10$	180	0.00%	180	0.00%	180	0.00%	540	0.00%
time limit	$n = 20$	180	0.00%	180	0.00%	180	0.00%	540	0.00%
100 SECONDS	$n = 30$	151	0.65%	172	0.30%	173	0.14%	496	0.36%
	All	511	0.22%	532	0.10%	533	0.05%	1,576	0.12%

Table 4:5. Impact of *OS*

	$OS = 0.35$		$OS = 0.50$		$OS = 0.75$	
	Optimal	CPU-time	Optimal	CPU-time	Optimal	CPU-time
$n = 10$	180	0.00	180	0.00	180	0.00
$n = 20$	180	0.89	180	0.31	180	0.06
$n = 30$	166	136.22	175	58.29	177	22.38
All problems	526	45.71	535	19.53	537	7.48

The impact of the resource-based measures *RF* and *RS* is given in Tables 4:6 and 4:7. *RF* has a strong impact on the complexity of the problem. The higher *RF*, the harder the corresponding problem instances. These results conform to the conclusions drawn by other research for related problem types (Kolisch et al. 1995, De Reyck and Herroelen 1998a). The effect of *RS* is not monotonously increasing or decreasing. On the contrary, it is bell-shaped, the hardest instances corresponding to an intermediate *RS* setting (equal to 0.25). However, there is a clear difference between the complexity of problems with small or large *RS* values. Problems with small *RS* values (*RS*=0) are much more difficult than problems with a high *RS* value (*RS*=0.5). Therefore, the 'top' of the bell-shaped complexity curve is skewed towards *RS*=0. The impact of *RS* corresponds to the conjecture of Elmaghraby and Herroelen (1980) and the results of De Reyck and Herroelen (1996a) for the RCPSP.

Table 4:6. Impact of *RF*

	RF = 0.50		*RF* = 1.00	
	Optimal	CPU-time	Optimal	CPU-time
$n = 10$	270	0.00	270	0.78
$n = 20$	270	0.00	269	14.66
$n = 30$	270	0.06	249	129.94
All problems	810	0.02	788	48.46

Table 4:7. Impact of *RS*

	RS = 0.00		*RS* = 0.25		*RS* = 0.50	
	Optimal	CPU-time	Optimal	CPU-time	Optimal	CPU-time
$n = 10$	180	0.00	180	0.00	180	0.00
$n = 20$	180	0.98	180	0.21	180	0.07
$n = 30$	173	93.63	166	112.35	179	10.91
All problems	533	31.54	526	37.52	539	3.66

4.8.2.6. Variable resource availabilities. When the resource availabilities are allowed to vary over time, the complexity of the GRCPSP and the RCPSP-GPR increases. The GDH procedure needs to be explicitly equipped with the ability to handle such time-varying resource availabilities, which will result in an increased number of decision periods and nodes in the search tree. The DRH procedure need not be modified in order to be able to handle time-varying resource availabilities (or, for that matter, variable resource requirements). The introduction of dummy activities and appropriate time lags, as discussed in Section 4.4, will transform an instance with variable availabilities (and requirements) into an equivalent instance with constant availabilities (and requirements). Naturally, the increased number of activities in the project network will have a substantial effect on the efficiency of the DRH procedure.

In order to estimate the effect of introducing variable resource availabilities on the performance of the GDH procedure, we modified the

540 instances with minimal time lags as follows. The constant availabilities are replaced by variable availabilities which are constant for an interval equal to 5 time periods. The availability is varied from interval to interval by increasing, respectively decreasing the availability with 1 or 2 units (or by keeping it constant), each with equal probability. Each time the resource availability dropped below the maximal demand of any of the activities for that resource type, the availability was assigned that maximal demand. The computational results indicate that the performance of the GDH procedure does not suffer significantly from this relaxed assumption. The average computation time increases from 0.009 to 0.018 seconds, while the average number of nodes in the search tree increases from 781 to 1,492.

4.9. Conclusions

In this chapter, we review a number of algorithms for project scheduling problems with resource constraints and generalized precedence relations. These generalized precedence relations specify minimal and/or maximal time lags between the starting and completion times of activities, and allow to model various types of activity overlaps (either permissible or mandatory), and also allow to model a wide variety of characteristics of real-life project scheduling applications. Also several objective functions are dealt with, including all kinds of regular performance measures and the nonregular measure of maximizing the net present value of a project.

The algorithms are enhanced and recoded in order to gain computational efficiency taking full advantage of modern 32-bit compiler architecture. We report new computational results using these algorithms on a problem set consisting of randomly generated problem instances. A comparison with results reported in the literature reveals that the algorithms presented here constitute the state-of-the-art for project scheduling with generalized precedence relations. When the optimal solution cannot be guaranteed, a truncated version of the algorithms can be used to provide high-quality solutions at acceptable computational cost. The experiments also highlight the fundamental difference in complexity between the precedence diagramming case, i.e. the case with minimal time lags only, and the generalized precedence relations case, in which both minimal and maximal time lags are allowed.

An investigation into the relationship between the complexity of a problem instance, defined by the computational effort required for its solution, and its intrinsic characteristics, reveals that the network morphology as well as the resource-constrainedness of the problems significantly influence the required computational effort. The more dense the project network becomes, measured by an increase in the *order strength*, the easier it is to obtain the optimal solution. When more activities require the use of resources, measured by an increase in the *resource factor*, the harder the instances become. The resource-constrainedness, measured by the *resource strength*, has a bell-shaped impact on the computational complexity. Instances with a low or high resource-constrainedness are easier to solve than instances with an intermediate resource-constrainedness, although the most difficult problems are relatively highly resource-constrained.

Acknowledgements

We would like to thank Klaus Neumann and Christoph Schwindt from the University of Karlsruhe for providing us with the project generator ProGen/max and Jan Weglarz for providing the opportunity for contributing to the *Handbook on Recent Advances in Project Scheduling*.

References

BARTUSCH, M., R. H. MÖHRING, AND F. J. RADERMACHER. 1988. Scheduling Project Networks with Resource Constraints and Time Windows. *Annals O.R.* **16**, 201-240.

BRINKMANN, K. AND K. NEUMANN. 1996. Heuristic Procedures for Resource-Constrained Project Scheduling with Minimal and Maximal Time Lags: The Minimum Project-Duration and Resource-Levelling Problem. *J. Dec. Syst.* **5**, 129-156.

COOPER, D. F. 1976. Heuristics for Scheduling Resource-Constrained Projects: An Experimental Comparison. *Mgmt. Sci.* **22**, 1186-1194.

CRANDALL, K. C. 1973. Project Planning with Precedence Lead / Lag Factors. *Project Mgmt. Quart.* **4**, 18-27.

DE REYCK, B. 1995. On the Use of the Restrictiveness as a Measure of Complexity for Resource-Constrained Project Scheduling. Research Report 9535, Department of Applied Economics, Katholieke Universiteit Leuven.

DE REYCK, B. 1998. *Scheduling Projects with Generalized Precedence Relations - Exact and Heuristic Procedures.* Ph.D. Dissertation, Department of Applied Economics, Katholieke Universiteit Leuven.

DE REYCK, B. AND W. HERROELEN. 1996a. On the Use of the Complexity Index as a Measure of Complexity in Activity Networks, *Eur. J. Opnl. Res.* **91**, 347-366.

DE REYCK, B. AND W. HERROELEN. 1996b. An Optimal Procedure for the Unconstrained Max-*npv* Project Scheduling Problem with Generalized Precedence Relations. Research Report 9642, Department of Applied Economics, Katholieke Universiteit Leuven.

DE REYCK, B. AND W. HERROELEN. 1997. The Multi-Mode Resource-Constrained Project Scheduling Problem with Generalized Precedence Relations. Research

Report 9725, Department of Applied Economics, Katholieke Universiteit Leuven.

DE REYCK, B. AND W. HERROELEN. 1998a. A Branch-and-Bound Algorithm for the Resource-Constrained Project Scheduling Problem with Generalized Precedence Relations, *Eur. J. Opnl. Res.*, to appear.

DE REYCK, B. AND W. HERROELEN. 1998b. An Optimal Procedure for the Resource-Constrained Project Scheduling Problem with Discounted Cash Flows and Generalized Precedence Relations, *Comput. and O.R.* **25**, 1-17.

DEMEULEMEESTER, E. AND W. HERROELEN. 1992. A Branch-and-Bound Procedure for the Multiple Resource-Constrained Project Scheduling Problem. *Mgmt. Sci.* **38**, 1803-1818.

DEMEULEMEESTER, E. AND W. HERROELEN. 1997a. New Benchmark Results for the Resource-Constrained Project Scheduling Problem. *Mgmt. Sci.* **43**, 1485-1492.

DEMEULEMEESTER, E. AND W. HERROELEN. 1997b. A Branch-and-Bound Procedure for the Generalized Resource-Constrained Project Scheduling Problem. *Opns. Res.* **45**, 201-212.

ELMAGHRABY, S. E. 1977. *Activity Networks - Project Planning and Control by Network Models*. Wiley Interscience, New York.

ELMAGHRABY, S. E. AND W. HERROELEN. 1980. On the Measurement of Complexity in Activity Networks, *Eur. J. Opnl. Res.* **5**, 223-234.

ELMAGHRABY, S. E. AND W. HERROELEN. 1990. The Scheduling of Activities to Maximize the Net Present Value of Projects. *Eur. J. Opnl. Res.* **49**, 35-49.

ELMAGHRABY, S. E. AND J. KAMBUROWSKI. 1992. The Analysis of Activity Networks under Generalized Precedence Relations. *Mgmt. Sci.* **38**, 1245-1263.

FRANCK, B. AND K. NEUMANN. 1996. Priority-Rule Methods for the Resource-Constrained Project Scheduling Problem with Minimal and Maximal Time Lags - An Emprical Analysis. *Proceedings of the Fifth International Workshop on Project Management and Scheduling*, 11 - 13 April, Poznan, 88-91.

GRINOLD, R. C. 1972. The Payment Scheduling Problem, *Nav. Res. Log. Quart.* **19**, 123-136.

HERROELEN, W. AND E. GALLENS. 1993. Computational Experience with an Optimal Procedure for the Scheduling of Activities to Maximize the Net Present Value of Projects. *Eur. J. Opnl. Res.* **65**, 274-277.

HERROELEN, W., E. DEMEULEMEESTER, AND P. VAN DOMMELEN. 1996. An Optimal Recursive Search Procedure for the Deterministic Unconstrained Max-npv Project Scheduling Problem. Research Report 9603, Department of Applied Economics, Katholieke Universiteit Leuven.

HERROELEN, W., P. VAN DOMMELEN AND E. DEMEULEMEESTER. 1997. Project Network Models with Discounted Cash Flows: A Guided Tour through Recent Developments, *Eur. J. Opnl. Res.* **100**, 97-121.

HERROELEN, W., E. DEMEULEMEESTER AND B. DE REYCK. 1998. A Classification Scheme for Project Scheduling Problems, in: Weglarz J. (Ed.), *Handbook on Recent advances in Project Scheduling*, Kluwer Academic Publishers, this volume.

KERBOSH, J. A. G. M. AND H. J. SCHELL. 1975. Network Planning by the Extended METRA Potential Method. Report KS-1.1, University of Technology Eindhoven, Department of Industrial Engineering.

KOLISCH, R. 1996. Efficient Priority Rules for the Resource-Constrained Project Scheduling Problem. *J. Opns. Mgmt.* **14**, 179-192.

KOLISCH, R., A. SPRECHER AND A. DREXL. 1995. Characterization and Generation of a General Class of Resource-Constrained Project Scheduling Problems. *Mgmt. Sci.* **41**, 1693-1703.

LAWLER, E. L. 1976. *Combinatorial Optimization: Networks and Matroids*. Holt, Rinehart and Winston, New York.

MASTOR, A. A. 1970. An Experimental and Comparative Evaluation of Production Line Balancing Techniques. *Mgmt. Sci.* **16**, 728-746.

MINGOZZI, A., V. MANIEZZO, S. RICCIARDELLI, AND L. BIANCO. 1998. An Exact Algorithm for Project Scheduling with Resource Constraints Based on a New Mathematical Programming Formulation. *Mgmt. Sci.*, to appear.

MODER, J. J., C. R. PHILLIPS, AND E. W. DAVIS. 1983. *Project Management with CPM, PERT and Precedence Diagramming*. Van Nostrand Reinhold Company, Third Edition.

NEUMANN, K. AND C. SCHWINDT. 1997. Activity-on-Node Networks with Minimal and Maximal Time Lags and Their Application to Make-to-Order Production. *OR Spektrum*. **19**, 205-217.

NEUMANN, K. AND J. ZHAN. 1995. Heuristics for the Minimum Project-Duration Problem with Minimal and Maximal Time Lags under Fixed Resource Constraints. *J. Intelligent Manufacturing*. **6**, 145-154.

PASCOE, T. L. 1966. Allocation of Resources - CPM, *Rev.fr. Rech. Opér.* **38**, 31-38.

PATTERSON, J. H. 1984. A Comparison of Exact Procedures for Solving the Multiple Resource-Constrained Project Scheduling Problem. *Mgmt. Sci.* **30**, 854-867.

ROY, B. 1962. Graphes et Ordonnancement. *Rev. fr. Rech. Opér.* **25**, 323-333.

RUSSELL, A. H. 1970. Cash Flows in Networks. *Mgmt. Sci.* **16**, 357-373.

SCHWINDT, C. 1996. Generation of Resource-Constrained Project Scheduling Problems with Minimal and Maximal Time Lags. Report WIOR-489, Institut für Wirtschaftstheorie und Operations Research, Universität Karlsruhe.

SCHWINDT, C. AND K. NEUMANN. 1996. A New Branch-and-Bound-Based Heuristic for Resource-Constrained Project Scheduling with Minimal and Maximal Time Lags. *Proceedings of the Fifth International Workshop on Project Management and Scheduling*, 11 - 13 April, Poznan, 212-215.

SIMPSON III, W. P. AND J. H. PATTERSON. 1992. A Parallel Exact Solution Procedure for the Multiple Resource-Constrained Project Scheduling Problem. Research Report, Indiana University.

SPRECHER, A. AND A. DREXL. 1996. Minimal Delaying Alternatives and Semi-Active Timetabling in Resource-Constrained Project Scheduling. Research Report 426, Christian-Albrechts-Universität zu Kiel.

WIEST, J. D. 1981. Precedence Diagramming Methods: Some Unusual Characteristics and their Implications for Project Managers, *J. Opns. Mgmt.* **1**, 121-130.

ZHAN, J. 1994. Heuristics for Scheduling Resource-Constrained Projects in MPM Networks. *Eur. J. Opnl. Res.* **76**, 192-205.

5 AN EXACT SOLUTION PROCEDURE FOR MAXIMIZING THE NET PRESENT VALUE OF CASH FLOWS IN A NETWORK

Sami M. Baroum
James H. Patterson

Indiana University

5.1. Introduction

Projects, as a method of organizing large endeavors, have been a part of the human experience for thousands of years. Ancient civilizations utilized the project approach to control grand undertakings. Most of humanity's historic landmarks have been associated with projects, from Noah's Ark and the Tower of Babel to the conquests of Alexander the Great and the Viking's daring raids [Fangel (1988); Vila (1990)].

One of the earliest recorded achievements of the use of project planning, still standing today, is exemplified by the pyramids. The Egyptian master masons adroitly executed this spectacular construction project by breaking it down into different stages. Several of the management techniques introduced constructing the pyramids have survived and are still in use to this day Weeks (1971). Interestingly, modern techniques such as the Critical Path Method are now being used somewhat in reverse to estimate the length of time required to construct the pyramids, as well as estimate the size of the workforce required for the build in the time length indicated. This is being accomplished by providing modern Egyptian craftspeople with little more than known hand tools of that era (inclined planes, levers, hard cobble, and so forth) for constructing portions of the pyramids the "old fashioned way" in order to estimate individual activity times. Meanwhile, scientists, engineers, and historians are using modern day project management techniques to estimate the resulting build time and length of the construction schedule. This is being accomplished to help unravel one of the great mysteries of all time. In our modern age, the explosive growth in the rate of innovation, our ever increasingly ambitious ventures, and the dramatic and rapid transitions in the economy have all stimulated both practitioner and academic interest in project management, as a flexible and adaptive form of organizing and managing change. Many have argued for greater reliance on the techniques of project management in an environment in which the cost of computing has been declining so dramatically, while the general availability of both hardware and software has witnessed a near mirror increase.

In this chapter we describe an exact solution procedure for scheduling the activities of a project such that, when completed, the net present value or the *financial reward* of resulting cash flows is maximal among all possible schedules. Our procedure is based on a systematic search of the solution space representing potential activity schedules. Our procedure allows for multiple resources and limited per-period resource availabilities. The algorithm reliably solves MAXNPV scheduling problems consisting of networks with as many as fifty activities. This is a reasonable accomplishment given that for the duration minimization version of this problem (MINDUR), the most reliable solution procedures available cannot guarantee (within a reasonable amount of computation time) the optimality of the schedule obtained for projects consisting of more than fifty activities and requiring up to three scarce resource types. Further, given that the maximization of net present value problem subsumes the duration minimization version of this problem through the placement of all cash flows (positive) on the final activity of a project, the procedure described is a reasonable procedure for solving the duration minimization version of this problem as well. The specific problem addressed in this chapter is: $3,1|cpm,c_j|npv+$.

5.2. Review of the literature

Table 5:1 gives a brief review of the relevant literature for the maximization of net present value in a network problem. A more comprehensive review of the area (including various forms of contracting and establishing progress & final payment schedules) can be found in Herroelen et. al. (1997).

Table 5:1. Taxonomy of NPV maximization literature in project scheduling

	No Resource Constraints	Resource Constrained
Optimal Techniques	- Russell (1970) - Grinold (1972) - Erenguc et al. (undated) - Elmaghraby and Herroelen (1990) - Herroelen and Galens (1993) - Smith-Daniels (1986)	- Doersch and Patterson (1977) - Smith-Daniels and Smith-Daniels (1987) - Patterson et al. (1989) - Patterson et al. (1990) - Yang et al. (1992) - Icmeli and Erenguc (1996)
Heuristic Techniques	- No research needed in this area. The problem is amenable to solution using optimization based approaches	- Russell (1986) - Smith-Daniels and Aquilano (1987) - Baroum and Patterson (1996) - Padman et al. (1990

5.2.1 Solution of the NPV maximization problem in the absence of resource constraints

A.H. Russell (1970) was the first to introduce the objective of maximizing the net present value (NPV) of the cash flows in a project. Russell considers the unconstrained problem where both positive and negative payments occur as events in the project are completed. Russell transforms the nonlinear objective function of this problem to a linear one by approximation using the first linear term of the associated Taylor series expansion in the vicinity of a feasible schedule. The dual of the linearized model is a minimum cost, network flow problem, where the external flows correspond to discounted activity cash flows (according to a feasible schedule), and arc flow costs correspond to the negative of the activity durations. Grinold (1972) transforms the unconstrained problem formulated by Russell into an equivalent linear program. He then proves that the optimal NPV solution must correspond to a basic spanning tree in the project network. Grinold uses this result to limit his search to only basic feasible trees. Erenguc et al., develop an algorithm which employs the concepts first proposed by Grinold, and which further exploits the network structure in developing an efficient network programming based method for solving this problem.

Elmaghraby and Herroelen (1990) develop an intuitively appealing, simple algorithm for solving the non-resource constrained version of this problem. Their procedure bases its logic (search) on only the sign (-, 0, +) of select model coefficients. Using this information, their algorithm builds trees in a [activity-on-arc] network to solve this problem. Computational experience on 250 randomly generated problems with their approach is reported by Herroelen and Galens (1993). This latter work additionally clears-up a possible inconsistency in the interpretation of the manner in which their solution procedure is to be implemented. Computational experience with their procedure is extremely encouraging. Finally, Smith-Daniels (1986) builds a model to predict the NPV of a restricted category of the unconstrained cash flow scheduling problem where the only positive cash flows occur at completion (such as in projects under lump sum contracts). As could be easily shown in an optimal schedule for this problem category, activities falling on the critical path must be assigned to their critical path determined early start times, while those off of the critical path must be assigned to their late start times. A positive project NPV is a prerequisite for the validity of such an approach.

5.2.2 Solution to the NPV maximization problem with limitations on resource availability

Doersch and Patterson (1977) provide the first formulation of the constrained cash flow problem. Their formulation, which is a zero-one integer programming one and which is similar to the formulation given in Section 5.3.2, treats capital as the only limiting resource. This formulation additionally considers the impact of cash outflows and inflows on capital availability. Penalties (bonuses) are explicitly considered in both the objective function and in the constraints in this approach. Smith-Daniels and Smith-Daniels (1987) extend the Doersch and Patterson procedure to accommodate material management costs. Their formulation accounts for material investment costs as a component of the capital constraints. They demonstrate their method by a small illustrative example; no computational results are reported.

Patterson et al. (1989) develop a general purpose backtracking algorithm for solving RCPSP (the Resource-Constrained, Project Scheduling Problem). The algorithm has the capability to solve problems with an NPV maximization objective. Their general procedure can accommodate constraints on cumulative capital availability as well. In a subsequent study [Patterson et al. (1990)], the authors report that the ability of their general purpose, backtracking algorithm to obtain and verify an optimal solution decreases rapidly once an NPV based objective is introduced. Yang et al. (1992) modify the implicit enumeration algorithm of Talbot and Patterson (1978) for solving the duration minimization version of this problem to solve the MAXNPV problem. The fathoming technique employed in the original algorithm [network cut] to eliminate inferior partial schedules by comparing them to previously generated and saved "good" schedules (*comparison pruning*) is successfully modified to accommodate the NPV based objective in their approach. Icmeli and Erenguc (1996) present a branch-and-bound procedure for solving the MAXNPV problem. Their procedure (deep-sea-troll search [Johnson (1988)]) is similar to the approach of Demeulemeester and Herroelen (1992) for solving the duration minimization version of this problem. Beginning with the optimal solution to the unconstrained MAXNPV problem obtained using the fixed deadline algorithm of Grinold, their algorithm branches by constructing minimum delay alternatives to resolve resource conflicts. The node with the greatest objective function value at the current level of the solution tree is selected next. The algorithm terminates its search when all possible schedules have been either implicitly or explicitly considered, or after a solution is found within a specified tolerance of the optimal solution.

The solution approach of Yang et al. and of Icmeli and Erenguc can each be contrasted to the current solution procedure to highlight differences among these three approaches. Both of these earlier procedures solve the *less restrictive* $m,1|cpm,c_j|npv$ version of this problem. That is, both solve the MAXNPV problem for both positive and negative cash flow activities. Both additionally require the specification of a project deadline or maximum allowable schedule span. According to the computational results reported by Icmeli and Erenguc, their procedure is *much less* influenced by the proximity of the imposed project deadline to the MINDUR solution in terms of the CPU time required for problem solution than is the Yang et al. approach. In the event the optimal schedule span is greater than the deadline imposed by either of these two solution approaches, the procedures can be re-optimized with an increased project deadline (maximum allowable schedule span). The current procedure, on the other hand, requires no such specification on the deadline for project completion. As shown in Section 5.3.1, the MAXNPV schedule duration often exceeds the MINDUR schedule duration, and it can be difficult to specify in advance a project deadline for problem solution. Hence, the apparent trade-off in solution approaches. Ours does not allow for the inclusion of *loss-leader* or negative net cash flow activities. Nor does ours require the specification of a [sometimes arbitrary] project deadline as do these other solution approaches. We feel the stipulation that all of the activities of a project minimally break even in a cash flow sense (as would be the case with a cost plus percentage fee contract) is a small price to pay when developing a procedure to determine the optimal schedule span corresponding to a maximum net present value solution.

5.2.3 Heuristic solution of the NPV maximization problem with limitations on resource availability

Various heuristic procedures are available for solving the net present value maximization problem in the presence of resource constraints. Such procedures can be useful in the design of an exact solution procedure by providing a bound on the value of an NPV solution. R.A. Russell (1986) compares six heuristic decision rules for scheduling projects with cash flows and resource constraints. Three of Russell's heuristic rules are derived from information obtained from the optimal solution to the unconstrained version of this problem using A.H. Russell's approach. R.A. Russell's results are rather inconclusive. Under the majority of experimental conditions, none of the heuristic rules employed are statistically superior to a benchmark random rule. Smith-Daniels and Aquilano (1987) consider the resource-constrained, NPV maximization problem for the restricted case where only negative cash flows occur during the project, and a single positive cash payment is received upon project completion. Back-shifting of activities is employed to improve the results of scheduling in their approach. Baroum and Patterson (1996) extend the Ranked Positional Weight heuristic of Helgeson and Birnie (1961) [for solving the assembly line balancing problem] to the MAXNPV project scheduling problem with encouraging results. Their "Cash Flow Weight" procedure additionally allows for the selective shifting of activities. Finally, Padman et al. (1990) developed a heuristic solution procedure based on weights derived from the solution to the optimal unconstrained problem. Different from earlier approaches incorporating an optimization based, heuristic solution procedure, the unconstrained problem in their approach is re-solved whenever a scheduling conflict arises, and new weights are generated for each unassigned activity. Padman et al. report that their greedy, single-pass, optimization-guided heuristic provides higher NPV performance than do the heuristics included in the study by R.A. Russell.

5.3. The MAXNPV, Resource-Constrained, Project Scheduling Problem

5.3.1 Alternative Scheduling Objectives

A project may be characterized by the following attributes: a defined beginning and an end, a specified set of goals, a series of interrelated activities, and limited resources. It is the job of a project manager to insure that project goals are met within the limited resources available and without exceeding mandated project constraints (e.g., deadlines on the accomplishment of project milestones).

Figure 5:1 depicts an example project network where nodes correspond to activities and arcs correspond to precedence relationships (AON representation). This project has seven uniquely defined activities[1]. The example assumes that all activities require only one critical resource type which is available in the limited quantity of three units per period. We will use this example extensively in describing our approach.

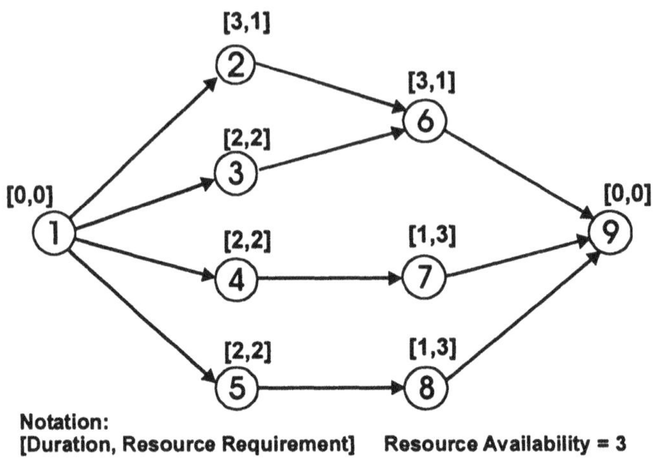

Figure 5:1. Example Project Network

Different optimal schedules may be generated for this example project depending on the assumptions made concerning the availability of resources and the objective sought in scheduling project activity. The type and complexity of the solution technique used in generating schedules also changes depending on the objectives of problem solution.

In the example of Figure 5:1, if we temporarily assume that the required single resource type is available in unlimited quantity each period, and that our objective is to complete each activity in the project as soon as technically possible, then the schedule shown in Figure 5:2 results (critical-path-determined, early start schedule).

When inspecting the resource profile of the schedule in Figure 5:2, we see that cumulative resource requirements in periods one through three exceed three units. It is apparent that this schedule would be infeasible if the per-period resource availability of three units per-period is enforced. If the desired objective is that of minimizing project duration, the resulting problem is the **resource-constrained, duration minimization project scheduling problem** (**MINDUR**).

A schedule for the project of Figure 5:1 which achieves a minimum feasible, resource-constrained duration is shown in Figure 5:3[2]. Several alternative schedules may be constructed, each of which will achieve an equivalent eight period project duration (schedule span). For example, exchanging activities four and five results in a resource feasible schedule having an eight period project duration.

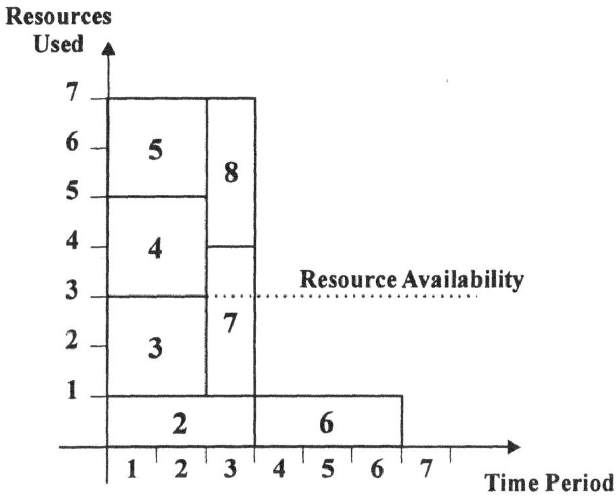

Figure 5:2. Early-Start, Time-Only Schedule of Example Project

Another, more reasonable objective in project scheduling maximizes the time-phased income from the project. When the objective of minimizing project duration is replaced with the objective of maximizing the net present value (NPV) of cash flows, **the resource-constrained, maximize NPV scheduling problem (MAXNPV)** results. The schedule that results when net present value is maximized is not necessarily the same as the minimum-duration schedule; some if not all of the activities' start times may be different [Bey et al. 1981].

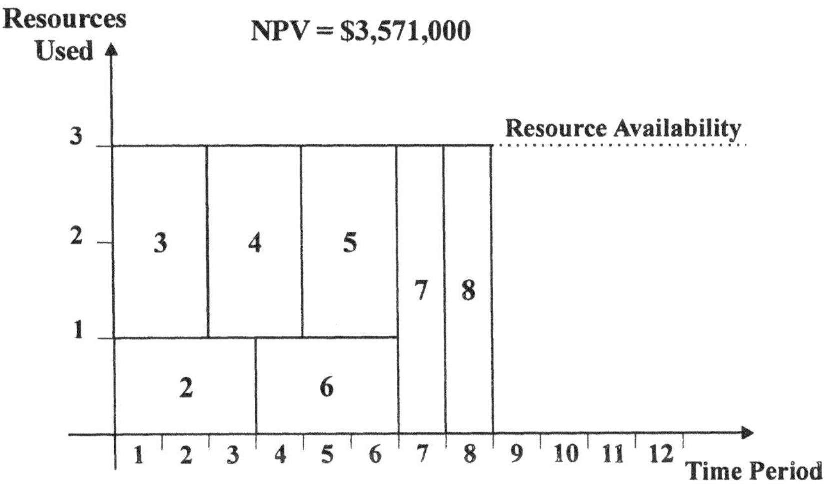

Figure 5:3. Optimal MINDUR Schedule of the Example Project

In the example project of Figure 5:1, we will assume that a positive cash flow is associated with the *completion* of each activity as indicated in Table 5:2. The corresponding schedule which produces the highest NPV amount is shown in Figure 5:4. The NPV of the schedule in Figure 5:4 is \$3,655,000—a significant improvement (\$84,000) over the NPV of the MINDUR schedule shown in Figure 5:3 (\$3,571,000). Note that the duration of the optimal MAXNPV schedule is twelve periods[3]—four periods longer than the MINDUR solution, and that this optimal project duration will most often not be known in advance of problem solution.

Table 5:2. Activity Progress Payment Cash Flows (000)

Activity	1	2	3	4	5	6	7	8	9
Cash Flow	100	100	100	110	120	100	1000	2000	200

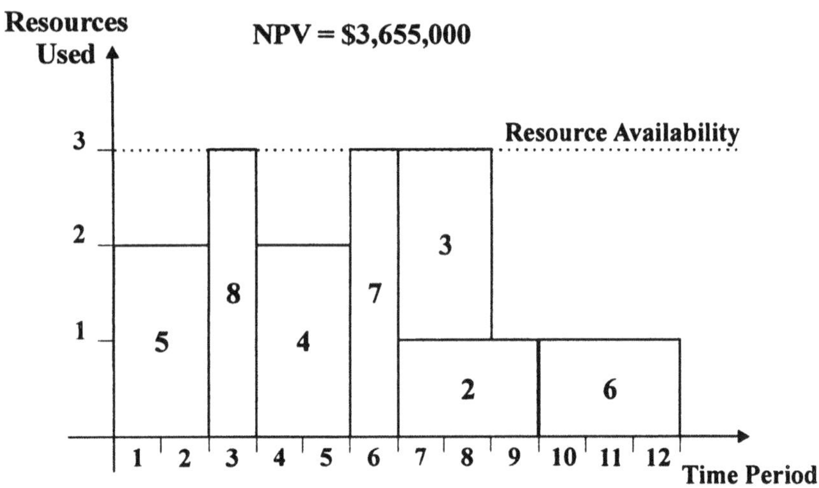

Figure 5:4. Optimal MAXNPV Schedule of the Example Problem

Even in the case where the optimal NPV schedule has a duration equivalent to an optimal MINDUR schedule, not all alternative optimal MINDUR schedules will achieve the largest or maximal NPV. In our example project, when a MAXNPV objective is assumed, the optimal MINDUR schedule shown in Figure 5:3 is suboptimal, even if we impose the additional constraint that the duration of the project may not exceed eight periods. A higher NPV may be achieved by an alternative MINDUR schedule generated by exchanging activities seven and eight in Figure 5:3, because activity eight has a higher cash flow associated with it than does activity seven. Also, since activity five has a higher cash flow than activity four, an improved NPV can be obtained by exchanging activities five and four (i.e., activity five is assigned earlier than activity four). Thus, of the four alternative optimal MINDUR schedules[4], only the schedule shown in Figure 5:5 achieves the highest NPV ($3,580,000) when a project completion of eight periods is additionally imposed. All of the other alternative optimal MINDUR schedules will result in inferior NPV amounts. Hence, unlike MINDUR, where the objective is to minimize the makespan of the project, MAXNPV's objective is evaluated by summing the project's discounted cash flows according to a given schedule. In addition to the expanded search effort introduced by the NPV objective, the nonlinear objective function exacts a heavy computational burden in solving the MAXNPV problem. Our solution approach takes cognizance of this distinction in generating the optimal NPV solution to a problem.

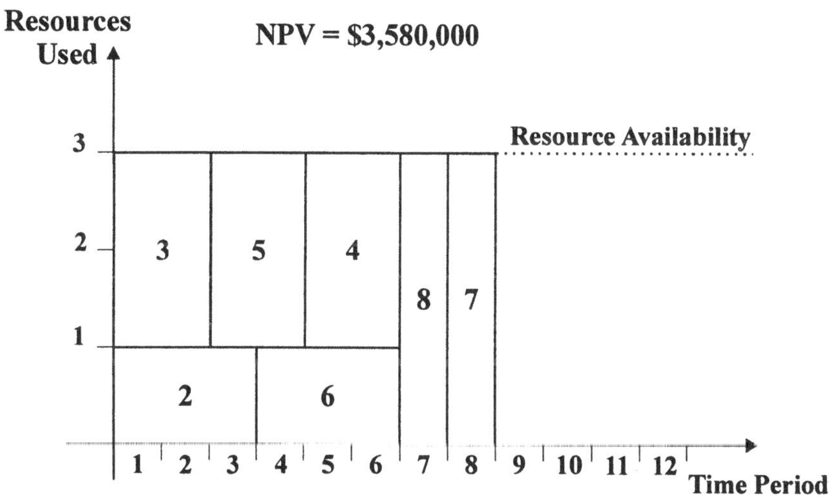

Figure 5:5. The Optimal MINDUR, Resource-Constrained Schedule Achieving The Highest NPV

5.3.2 The MAXNPV Model

We make the following assumptions [French (1982, p.8), adapted for the project scheduling case] in developing our procedure:

1. There is no uncertainty. This implies:
   a) activity durations are deterministic;
   b) the activities' per-period resource requirements and the project per-period resource availabilities are fixed and are known; and
   c) the cash flow associated with each activity is not subject to change.
2. Once an activity is begun, there is no pause in its execution (no preemption).
3. No cancellation is allowed. Each activity must be completed.
4. A single mode is available for completing each activity. Associated with that mode is a known activity duration, fixed resource requirements, and a fixed cash flow.
5. Net non-negative cash flows occur at the completion of an activity[5].

The resource-constrained, project scheduling problem under the above assumptions where the objective is to maximize the net present value of the project is formulated as a mathematical programming problem as follows:

$$\textit{Maximize:} \quad \sum_{i=1}^{N} \sum_{t=1}^{T} f_t^{\prime} \, G_i \, x_{it}$$

Subject to:

$$\sum_{t=E_i}^{L_i} x_{it} = 1 \qquad \textit{for} \quad i=1,2,\ldots,N$$

$$-\sum_{t=E_i}^{L_i} t x_{it} + \sum_{t=E_j}^{L_j} (t-d_j) x_{jt} \geq 0$$

$$\textit{for} \quad i \ll j, \quad \forall \quad (i,j) \in A$$

$$\sum_{i=1}^{N} \sum_{q=t}^{t+d_i-1} r_{ik} \, x_{iq} \leq R_{kt} \qquad t=1,2,\ldots,T^*; \; k=1,2,\ldots,K$$

$$x_{it} \in \{0,1\} \qquad i=1,2,\ldots,N, \; t=1,2,\ldots,T$$

Where:

$$x_{it} = \begin{cases} 1 \text{ if job i is completed in period t,} \\ 0 \text{ otherwise} \end{cases}$$

d_i = The duration of activity i.
E_i = Zero, or the critical-path-determined, earliest possible completion period for job i, whichever is larger.
L_i = The critical-path-determined, latest possible completion period for job i obtained from L_N.
L_N = The latest possible completion period for the dummy end activity, obtained from a heuristic solution or set equal to the sum of all activity durations .
N = The number of activities in the project
T^* = The time horizon for the project (obtained from L_N).
r_{ik} = Per period requirement of resource k by job i.
R_{kt} = Units of resource k available during time period t.
K = Number of resource types included in the project.
f_t = The discount factor over t periods when the internal rate of return of the company (α_i) is used for the discount rate (= $e^{-\alpha_i t}$).
G_i = The contribution of activity i upon its completion.
A = The project network.

The first constraint set represents the activity completion constraints. Satisfaction of these constraints insures that each activity is assigned a single completion period. The second set of constraints force each activity to start at a period which is later than the completion times of all of its immediate predecessor activities. The final constraint set ensures that the sum of the resource requirements by all activities active during any period t does not exceed resource availabilities in that period.

The objective function maximizes the sum of the present values of the activities' contributions discounted to the start of the project. The contribution of an activity is calculated as the sum of the future values of the cash flows associated with the different stages (time periods) of that activity, discounted to the end of the activity, minus the opportunity cost of the recoverable investment capital required by the activity. This value is given by:

$$G_i = \sum_{j=1}^{d_i} F_{ij}\, e^{\alpha(d_i - j)} - C_i(e^{\alpha d_i} - 1)$$

Where:

C_i = The recoverable capital investment required by activity i.
d_i = The duration of activity i.
F_{ij} = The cash flow in activity i at the end of period j, j = 0,1,.., d_i.

The first term on the right side of the above expression gives the sum of the cash flows associated with the activity (F_{ij}) multiplied by the appreciation factor to compound their values to the end of the activity ($e^{\alpha(d_i - j)}$). The term (d_i - j) gives the number of periods over which the cash flows occurring at the end of the jth period have to be compounded. The second term gives the value of the opportunity cost incurred by holding the initial investment (C_i) for the duration of the activity. Again we assume that each G_i is a positive quantity in our no loss-leader approach to this problem.

5.4. The Branch-and-Bound Solution Procedure for the MAXNPV Problem

Our developed solution procedure incorporates directly the economic consequences of activity sequencing decisions in algorithm design. That is, rather than begin with a solution procedure which was originally designed to satisfy another objective--minimizing project duration or maximizing the utilization of project resources--and modify it or incorporate elements of many such procedures to solve the net present value in a network problem, we begin instead by designing a procedure which incorporates directly the financial consequences of activity sequencing decisions.

The solution procedure we present here is a branch-and-bound based search algorithm. We have selected a **deep-sea-troll search** approach similar to Icmeli and Erenguc in preference to the LIFO laser search approach employed by Yang et al. for solving this problem. This choice has been motivated by the potential for exploiting the information derived from calculating NPV-based upper bounds in providing an

Table 5:3. Notation used in describing solution procedure

Symbol	Description
PS	A feasible partial schedule
S	The set of in **progress activities** in a partial schedule.
C	The set of **completed activities** corresponding to a partial schedule.
E	The set of **eligible activities** available for scheduling at the current decision point; i.e., whose predecessors have been completed by the current time period.
U	The set of activities yet to be scheduled (**unscheduled activities**) in a continuation of a partial schedule.
m	The **decision point** corresponding to the earliest clock time at which one or more in progress activities are completed in the partial schedule being considered.
$t_i(A)$	The start time assigned to activity i in partial schedule PS_A.
$f_i(A)$	The finish time assigned to activity i in partial schedule PS_A.
FS_A	A complete feasible schedule resulting from completing partial schedule PS_A.
$NPV_A(X)$	The NPV contribution of activities in the set X discounted using their scheduled assignments in PS_A.
D	The minimal set of in progress or eligible activities to delay in order to resolve a resource conflict; called a **delay alternative**.
p	Current branching level in the solution tree.
D(p)	The set of delay alternatives at branch level p.
$FNPV^*_A$	The NPV contribution of the maximum NPV feasible solution generated from PS_A

Table 5:4. Variable definitions used in algorithm description

Notation	Description
A	The project network.
t_i	The scheduled start time of activity i.
h_i	t_i - 1.
d_i	Duration of activity i.
f_i	The scheduled finish time of activity i (h_i+d_i).
CF_i	The cash flow associated with the completion of activity i
α	The continuous discount rate.
r_{ik}	Per-period requirement of activity i for resource k.
b_k	Amount of resource k available in each period
D_{parent}	The most recent delay alternative leading to the current decision point.
v_k	Amount of resource type k demanded by the in progress activities and all the eligible activities that is in excess of resource type k's per period availability (i.e., the magnitude of violation—or infeasibility—in resource constraint k).
UB_q	An upper bound on the NPV resulting from the schedule formed by delaying activities in D_q.
UB_{parent}	The upper bound on the NPV of the parent node of the node currently being explored.
LB_{npv}	The current highest NPV achieved from a complete feasible schedule.
CPNPV	Cumulative discounted cash flows of the completed tasks in the current partial schedule.
BFNPB	An upper bound on the NPV of the best feasible complete schedule formed by descendants of the current branch of the search tree.
FNPV(p)	The current highest upper bound on the NPV of any feasible schedule emanating from the partial schedule at branching level p.
PNPV(p)	Cumulative discounted cash flows of the completed activities in the partial schedule corresponding to the branching node at level p.
PS_i	The ith saved unfathomed partial schedule.
$SFNPV_i$	An upper bound estimate of the NPV contribution of the best complete feasible schedule emanating from saved partial schedule Psi.

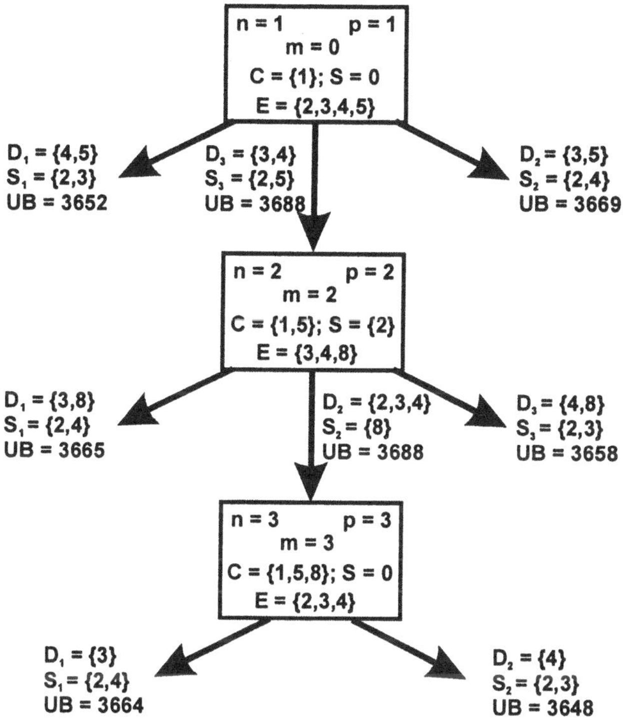

Figure 5:6. A portion of the branch and bound solution tree for our example problem

intelligent direction for the search to follow. In order to facilitate the discussion of our procedure, we present in Figure 5:6 a portion of the search tree generated from the solution of the example project shown in Figure 5:1 and in Table 5:1. Also, to clarify the notation needed in describing our approach, we provide definitions in Table 5:3 and Table 5:4.

5.4.1 Steps of the Branch-And-Bound Solution Procedure

The implementation steps of our approach necessary to derive and verify an optimal solution to the MAXNPV problem are as follows:

Step 1: Set an initial NPV lower bound on the project: LB_{npv}[6] = 0.0.
Initialize the project NPV: FNPV(0) = 0.
Initialize the branch level, and the decision point: p = 0, m = 0.
Initialize the cumulative NPV of the completed tasks in the partial schedule, CPNPV: CPNPV = 0.0.
Compute UB_0 from an early start, time-only schedule.
Set $UB_{parent} = UB_0$ and $D_{parent} = \varnothing$.
Initialize the completed activity set: C = Ø.
Schedule the required dummy start activity: set $h_1 = 0$; S = {1}.

Step 2: Advance to the next decision point:
For each task in progress, X, finishing at time m, $\{X \mid X \in S,\ h_X + d_X = m\}$:
Update S and C: $S = S - \{X\}$; $C = C + \{X\}$;
and $CPNPV = CPNPV + CF_X * e^{(-\alpha m)}$.
If the required dummy end activity has been completed ($N \in C$), a feasible schedule has been generated. Set: BFNPV = CPNPV.
If the newly generated full schedule is superior to the incumbent solution ($BFNPV > LB_{npv}$), replace the incumbent solution with the new complete schedule: set $LB_{npv} = BFNPV$. Go to step 6.
If $p > 0$, Perform comparison pruning:
If the current partial schedule is dominated:
Set $BFNPV = UB_{parent}$, go to step 6.
Else, save the current Completed Activity Set, CAS, the activities in progress, the decision point m, and the cumulative discounted cash flows of the activities in the current completed activity set (CPNPV).
Construct the set of eligible activities:
$E = \{j \mid (i,j) \in \mathbf{A};\ h_i + d_i = m;\ k \in C\ \forall\ (k,j) \in \mathbf{A}\}$.
If $E = \emptyset$: set $m = \min\{h_i + d_i,\ i \in S\}$; repeat step 2.

Step 3: For each resource k, compute the resource requirement when all the eligible activities are placed in progress. Calculate the resulting resource violation v_k:

$$v_k = \Sigma_{i \in S \cup E}\ r_{ik} - b_k$$

If for any resource k, $v_k > 0$, go to step 4.
Else, set $h_i = m\ \forall\ i \in E$, and update $S = S \cup E$.
Set $D_{parent} = \emptyset$. Determine the next decision point:
$m = \min\{h_i + d_i,\ i \in S\}$ Go to step 2.

Step 4: Establish the new node corresponding to the current conflict decision point:
Update the branch level: $p = p + 1$.
Set FNPV(p) = CPNPV, PNPV(p) = CPNPV.
Generate the delay alternative set D(p):

$$D(p) = \{D_q \mid D_q \subseteq S \cup E;\ \sum_{i \in D_q} r_{ik} \geq v_k\ \forall\ k;\ D_q \not\supseteq D_{q'}\ \forall\ D_{q'} \in D(p),\ q' \neq q\}$$

For every $D_q \in D(p)$:
Update $S_q = \{S \cup E\} - D_q$. Update start times for the newly started activities ($i \in E$, $i \notin D_q$): $h_i = m$.
Determine j'_q, the earliest finishing activity in S_q. Set $m_q = f_{j'_q}$.
Push back the start of the delayed activities: $h_i = m_q\ \ \forall\ i \in D_q$.
Check for left shift dominance in alternative D_q. If D_q is left shift dominated (LSD), update $D(p) = D(p) - D_q$.
Set $FNPV(p) = \max\{FNPV(p), UB_{parent}\}$.
If delaying alternative D_q is not left shift dominated, compute the updated UB_q applying PUSH routine. If $UB_q < LB_{npv}$, drop delay alternative D_q.

$D(p) = D(p) - D_q$.
Update $FNPV(p) = \max\{FNPV(p), UB_q\}$.
If the delay alternative is not dominated, store (updated) activity start times (h_i), UB_q, set of completed activities C_q, set of in progress activities S_q, the delay alternative D_q, and next decision point, m_q.
If all the delay alternatives have been evaluated and there is at least one alternative which has not been dominated ($D(p) \neq \emptyset$):
Select the delaying alternative $D_q^* \in D(p)$ having the highest UB_q. Place D_q^* at the end of the candidate list. Drop parent node.
If all the delay alternatives have been pruned (i.e., $D(p) = \emptyset$):
Set $BFNPV = FNPV(p)$, $p = p - 1$, go to step 6.

Step 5: Restore the node at the end of the candidate list.
Restore the completed activities' NPV: Set $CPNPV = PNPV(p)$.
Initialize the Best full NPV of the new branch to the NPV of the completed activities in the parent node: $BFNPV = PNPV(p)$
Restore S, C, m and the start time vector (h_i) to the values stored in the selected node. Set UB_{parent} to the value of UB of the selected node. Set D_{parent} to the delay alternative of the selected node. Drop the restored alternative from the alternative set at the current branching level, $D(p) = D(p) - D_q^*$.
Go to step 2.

Step 6: Backtrack: If branching level $p = 0$, stop. The incumbent solution is the optimal one.
Else, update the full schedule NPV for saved partial schedules: For each CAS saved while branching at level p, set $SFNPV = BFNPV$.
Update the full schedule NPV for the current node at level p, $FNPV(p)$: $FNPV(p) = \max\{BFNPV, FNPV(p)\}$.
For each delay alternative remaining at the current branching level ($D_q \in D(p)$):
If $UB_q < LB_{npv}$: drop D_q ($D(p) = D(p) - D_q$) and set $FNPV(p) = \max\{FNPV(p), UB_q\}$.
If any unfathomed delay alternatives remain at the current branching level ($D(p) \neq \emptyset$): select $D_q^* \in D(p)$ having the highest UB_q^*; Place selected alternative D_q^* at the end of the candidate list. Go to step 5.
Else, backtrack to a higher branching level if all delay alternatives have been evaluated (i.e., $D(p) = \emptyset$): set $BFNPV = FNPV(p)$, $p = p - 1$, and repeat step 6.

5.4.2 Developing the Branch-and-Bound Solution Tree

Each node in the branch-and-bound solution tree generated in our approach corresponds to a resource and precedence feasible partial schedule. The search procedure confines its search only to instances where a scheduling decision has to be made; i.e., time periods in which one or more activities are completed. Each one of these **decision points, m**, corresponds to a **partial schedule, PS**. Each partial schedule, PS, corresponds to three collectively exhaustive and mutually exclusive sets of project activities. The first is the set of completed activities, the **completed set C**; the second is the set of activities that have been started, but are not yet finished, the **in progress set** S; and the third is the set of **unscheduled activities**, U. For example, Node 2 (n = 2) in Figure 5:6 corresponds to the partial schedule generated at the completion time of activity five at period two (i.e., m = 2). The partial schedule corresponding to this partial solution divides the activities of the project into three distinct sets: C = {1,5}, S = {2}, and U = {3,4,6,7,8,9}.

Since the problem under study is one where positive net present values are associated with each activity, any schedule in which an activity is delayed while it is possible for it to be scheduled at an earlier time period is an inferior schedule. Higher NPV amounts would result from advancing the occurrence of the delayed activity. Consequently, we confine the search for an optimal solution to semi-active partial schedules where each activity is started as soon as possible within precedence and resource constraints. This is accomplished by attempting to schedule every candidate activity as soon as it becomes **precedence feasible**, i.e., at the instance in which its last immediate predecessor has been completed. When it becomes infeasible to simultaneously schedule all eligible activities due to exceeding current resource availabilities, a decision has to be made as to which activities to start, or to remain in progress, and which activities to delay. Corresponding to each decision point where there exists such a resource conflict, a node is formed in the branch-and-bound solution tree. The branches emanating from the node correspond to the set of possible subsets of activities to delay in order to allow a feasible continuation of the partial schedule. It is sufficient to delay a set of activities that would free sufficient resources to eliminate the resource conflict (infeasibility). The set of activities selected for delaying, which is called the **delay alternative**, D, can consist of a combination of eligible or candidate activities (E), as well as in progress activities (S). Only minimal delaying sets of activities are considered in our approach. A **non-minimal delay alternative** D_1 would result in the delay of activities that could have been started earlier using a **minimal alternative** set, D_2 ($D_2 \subset D_1$). The schedule resulting from D_1 will lead to an inferior NPV since it causes the lowering of the discounted cash flows of the delayed activities which could have been started sooner (i.e., $i \in D_1 - D_2$). In Figure 5:6, there are three minimal delay alternatives emanating from Node 2: D_1 = {3,8}, D_2 = {2,3,4}, and D_3 = {4,8}. An example of a nonminimal delay alternative would be D = {3,4,8} —by including such an alternative, we will be wasting the resources that could have been utilized by starting activity 4 at the current decision point.

By allowing the delay of in progress activities, the procedure allows for backtracking to prior decisions to start an in progress activity at an earlier time period, when, in the optimal schedule, it is required to start at a later time period. Thus, the procedure does not confine the search only to **non-delay schedules**[7]. For example, the decision to start activity two in combination with activity five in Node 1 at m = 0 (corresponding to D_3 [at the same level] in Figure 5:6) is retracted in Node 2 at m = 2, and activity two is taken out of operation in order to allow activity eight to start. An alternative approach would be to enumerate all the possible combinations of in progress activities (both **maximal** and **non-maximal** sets) at the current decision point. In such an approach, an in progress activity once started may not be delayed. Stinson, Davis, and Khumawala (1978) adopt this approach for the MINDUR problem. The disadvantage of this approach is the tendency of the search tree to grow in width due to the larger number of possible alternative in progress sets generated at each decision point. Many of the non-minimal delay alternatives (non-maximal in progress sets) would result in inferior partial schedules where some of the activities that are to be started at succeeding decision points may have been feasiblely started at earlier periods.

The delaying decision is implemented through the introduction of auxiliary precedence relationships to postpone the activities in the selected delay alternative. The added relationships (called **delay arcs**) have as their origin the first activity to finish from among the activities that are in progress and left undelayed, or the eligible ones that are allowed to start. Each activity delayed will have an auxiliary arc terminating in it[8]. Again, corresponding to the branch at Node 1 formed by delay alternative $D_3 = \{3,4\}$ in Figure 5:6 (where both activities two and five are started), the earliest finishing activity is activity five, which gets completed in period two. The corresponding auxiliary precedence relationships to implement the delay decision are thus (5,3) and (5,4).

Only a decision point where resource conflicts arise results in a new node in our approach. The only decision that can be made at a decision point where no conflicts arise—the resources available can accommodate all of the eligible activities—is to start all the eligible activities and move on to the next decision point. We conjecture that the strategy of confining the search to only those decision points having resource conflicts will prove superior to a typical zero-one or LIFO laser based solution approach where all of the combinations of possible start times for the activities are enumerated.

A feasible partial schedule at any point is thus the result of a series of scheduling decisions based on resolving conflicts on resource usage. Each scheduling decision corresponds to the selection of a delay alternative from among the available alternatives at the corresponding node. Each decision point is indexed by its position (or chronological order) in the chain of decisions made. This index, denoted by **p**, corresponds to the **branching level** in the search tree. The **delay alternative set** emanating from the node corresponding to the pth decision is labeled: **D(p)**. For example the delay set corresponding to level one in the solution tree of Figure 5:6, D(1), is: $\{\{4,5\}, \{3,5\}, \{3,4\}\}$.

In selecting the delay alternative to evaluate first, the algorithm employs an NPV based branching criteria. An estimate of the NPV achievable under each delay alternative is formed by summing the NPV of the current partial schedule and the NPV of a resource-relaxed continuation of the current partial schedule discounted to the current scheduling period, m. This result is achieved with the addition of delay arcs corresponding to the evaluated delay alternatives. The delay alternative with the maximum decision bound directs the search. Thus, at each branching level p in the branch-and-bound search tree, the delay alternative generating the highest NPV upper bound from among the delay alternatives available at that level ($D \in D(p)$) is selected to branch on next. The NPV directed choice—we conjecture—increases the probability that a good incumbent solution will be reached early in the enumeration process.

The computation of the upper bound for each alternative at a given level implies that branches descending from an open node must be partially explored. However, at each branching level, the search fully explores only one branch at a time. This allows for a more intelligent search than a LIFO laser strategy of the depth-first variety, at the additional expense of generating potentially many more new upper bounds at each level. For example, in Node 1 of Figure 5:6, the NPV upper bounds corresponding to the three delay alternatives in D(1) are: $3,652, $3,669 and $3,688, [000] respectively. Of those, the upper bound generated by D_3, $3,688, is the highest. Therefore, the search continues by evaluating the branch corresponding to D_3 first. An added benefit of this search is the potential early discovery of a good feasible schedule resulting in a reduction in search effort through the fathoming of inferior partial schedules (i.e. those having inferior upper bounds) early in the search process. Unlike a breadth-first search strategy, a deep-sea-troll search of the type investigated here will also have only one node open at any branching level. Thus, relatively moderate storage overhead is required, and much of the required information may be regenerated incrementally when moving from a parent node to a lower level node.

The NPV of a partial schedule is updated at each decision point by augmenting the discounted cash flows of the activities that are completed at that decision point. The augmentation process continues until a full feasible schedule is achieved. A full feasible schedule is one in which the dummy end activity has been assigned. When such a schedule is achieved, its NPV is compared to that of the stored incumbent schedule; if it is higher, then the new schedule is taken as the incumbent[9]. When backtracking occurs during the search, the NPV corresponding to activities in the parent schedule is restored.

The search backtracks whenever it can be proven that the partial schedule corresponding to the current branch will only lead to an inferior schedule. Backtracking also occurs when a full feasible schedule (**FS**) is achieved. Backtracking is implemented by restoring the conditions prevailing at an unexplored sibling delay alternative. The search continues by constructing a new branch corresponding to the new delay decision. If sibling alternatives have all been enumerated, the algorithm backtracks to the branch that led to the parent node and continues the search at this higher branching level. When the algorithm attempts to backtrack to level 0, the solution space of the problem has been fully enumerated and the algorithm terminates with the current incumbent solution being the optimal one.

The general framework described so far is basically a full enumeration of all undominated partial schedules. However, the search process may be further reduced through the implementation of various fathoming rules that will identify and eliminate inferior portions of the search tree.

5.4.3 Fathoming Partial Schedules

In order to reduce the amount of search effort expended in evaluating nodes of the branch-and-bound solution tree representing partial solutions to the MAXNPV problem, two fathoming procedures are employed. The first fathoming procedure is based on the identification and elimination of partial schedules that are left-shift dominated. We call this **left-shift pruning**. The second procedure employs **comparison pruning** to identify current partial schedules that are inferior to previously generated and saved partial schedules.

5.4.3.1 Left Shift Pruning. Left shift pruning essentially operates by identifying delay alternatives that will lead to schedules where one or more activities may be left shifted (advanced) within the project's resource and precedence constraints, without delaying any of the other activities in the schedule. Such non left-shifted schedules are obviously dominated since a higher NPV may be obtained by simply advancing the set of activities that are candidates for left shifting, while at the same time, maintaining the original assignment of the other activities. Our implementation attempts to identify and prune schedules that are not semi-active (i.e., schedules where one or more activities may be feasibley left-shifted without changing the assignments of the rest of the activities in the schedule). Because we consider only activities with positive net present values, any schedule that is left-shift dominated is also NPV dominated.

Within our procedure, it is possible to generate left shift dominated partial schedules in one situation. If an activity j that has been delayed at an earlier decision point, m_1, is allowed to start at the current conflict decision point, m_2; if the delay alternative allowing the start of j causes the delay of one or more activities (i) that are still in progress at time m_2 ($D_S = \{i \mid i \in D \cap S\}$); and if the resources released by activities in D_S are greater than or equal to the resource requirements of activity j; then, activity j can be feasiblely started earlier than time period m_2, i.e, it can be left-shifted.

If the unshifted partial schedule is further augmented until it becomes a complete schedule, two possible cases arise in any feasible completion. In the first case, activity j (the activity eligible for left shifting) is not delayed in succeeding decision points. In this case, a completion of the partial schedule in which activity j has been left shifted and every other activity has been assigned identical start times as in the corresponding complete schedule generated from the unshifted partial schedule will result in a higher NPV solution. Thus the unshifted schedule is dominated. In the second case, activity j is further delayed in future delay alternatives. In such a case, any feasible completion generated from the unshifted partial schedule may be matched by identical activity assignments in the corresponding completion of the shifted partial schedule. This is due to the fact that when activity j is further delayed in both partial schedules, identical solution spaces are generated beyond m_2. Thus, the unshifted schedule is redundant and may be eliminated from further consideration.

5.4.3.2 Comparison Pruning. The second fathoming technique that we implement is completed activity set (CAS) pruning. It may be categorized as a comparison pruning technique (Stinson, 1976). It operates by comparing the current partial schedule to previously saved partial schedules.

Comparison pruning has been previously suggested by researchers for pruning the search tree in solving the resource-constrained project scheduling problem. Johnson

(1967) first introduced it. Stinson applies it in his branch-and-bound solution procedure. Talbot and Patterson adapt this concept to their implicit enumeration procedure by introducing the notion of *network cuts*. Demeulemeester and Herroelen modify Stinson's implementation to fit their branch-and-bound solution procedure which utilizes delay arcs. Yang et al. extend the notion of comparison pruning one step further by adapting it to an NPV based objective in their implicit enumeration solution approach. They utilize the construct of the network cut of Talbot and Patterson to perform such an implementation.

Once a partial schedule has been proven to lead to a reduced solution space, two fathoming rules may be applied to ascertain whether it will also lead to an NPV dominated (fathomed) schedule. If the partial schedule meets the conditions set forth by any of the following two rules, it may be regarded as inferior and consequently need not be investigated further.

(CP1) Comparison Pruning Rule 1: The first fathoming procedure we employ recognizes that when the NPV contribution of the scheduled activities in a current partial schedule, PS_A, which has a smaller solution space than a previously saved partial schedule, PS_B, is less than the NPV contribution of the corresponding activities in the saved partial schedule, the current partial schedule is inferior. This result is based on the fact that the potential NPV contribution of the unscheduled activities in PS_A can never exceed the potential contribution of the unscheduled activities in PS_B. Since the NPV contribution of the activities in PS_A is smaller than the contribution of PS_B, the total potential contribution of the best complete schedule generated from PS_A is inferior to the total potential contribution of the best complete schedule generated from PS_B. Thus PS_A is inferior to PS_B.

(CP2) Comparison Pruning Rule 2: The second fathoming rule again assumes that the solution space of the current partial schedule, PS_A, is a subset of the solution space of the saved partial schedule, PS_B. Hence, the potential NPV contribution of the unscheduled activities in PS_A cannot exceed the contribution of these activities in the maximum NPV feasible completion of PS_B, $NPV_B^*(U)$. An upper bound on the contribution of the current partial schedule can thus be defined as the sum of the NPV contribution of the activities included in the current partial schedule, $NPV_A(C_A + S_A)$, and $NPV_B^*(U)$. If this upper bound is lower than the NPV of the incumbent solution, the current partial schedule is inferior and may be fathomed.

5.4.3.3 Adapting Comparison Pruning to Our Deep-Sea-Troll Search Procedure. In attempting to implement the above fathoming rules, the following observations are made. If the maximum NPV completion of a current partial schedule PS_A involves a delay of an in progress activity, i, in the full schedule (FS_A), the NPV contributions of the scheduled activities in PS_A, $NPV_A(C_A \cup S_A)$ will be reduced in FS_A. If the activity in progress, i, which is delayed in the full schedule FS_A, ($t_i'(A) > t_i(A)$) has the same start time in the saved partial schedule PS_B, ($t_i(A) = t_i(B)$) and is assigned an identical start time as its counterpart in FS_B, ($t_i'(A) = t_i'(B)$), then the reductions in the NPV of the partial schedules are equivalent and Rule 1 above may be applied without modifications. However, if activity i starts earlier in PS_B than in PS_A ($t_i(B) < t_i(A)$), but has an identical assignment in both FS_A and FS_B, ($t_i'(B) = t_i'(A)$), then the reduction in the contribution of activity i to the NPV of FS_B due to its delay is larger than the reduction in its contribution to the NPV of FS_A. In the worst case, all activities that are in progress in both PS_A and PS_B and which start earlier in PS_B may be delayed to start at identical time periods in the corresponding complete schedules. Thus, to ascertain that a saved partial schedule is superior, we have to reduce its NPV by the reductions induced by the delay of such activities. We may accomplish this by discounting the NPV contribution of each activity that is in progress in both PS_A and PS_B for the time difference between its scheduled assignments in PS_A and PS_B. Although this is possible, it does require substantial calculations in comparing partial schedules.

One way to avoid the additional computational burden is to consider only the NPV contributions of the completed activities (C) in each pair of compared partial schedules. Because the solution space in PS_B contains the solution space in PS_A, any feasible completion of PS_A can be feasibley augmented to PS_B. Each activity in S_A and U_A may be feasibley assigned an identical start time in a continuation of PS_B as they have been in a corresponding continuation of PS_A. The NPV contribution of the two identical completions will generate equal NPV contributions. If $NPV_B(C_A)$ is greater than $NPV_A(C_A)$, the total NPV of the completed schedule in A (FS_A) is less than that of FS_B.

Thus the condition for the dominance of PS_B over PS_A—in addition to the conditions stated earlier—is given as follows:

$$NPV_B(C_A) \geq NPV_A(C_A)$$

Where:

$NPV_B(C_A)$ = the NPV contribution of completed activities in PS_A ($i \in C_A$) discounted using the scheduled assignments of the corresponding activities in PS_B (i.e., $t_i(B)$).

The previous criterion uses the contribution of the activities completed in the current partial schedule PS_A, C_A. However, to determine the contribution of these activities for the saved partial schedule PS_B, we subtract the contribution of the activities that are completed in PS_B, but are not yet completed in PS_A, each time PS_A is compared to a saved partial schedule PS_B (i.e., $NPV_B(C_A) = NPV_B(C_B) - NPV_B(C_B - C_A)$)[10]. This requires that we know the finish times for the completed activities in the saved partial schedule PS_B, which in turn requires that we save the finish time of all completed activities with each partial schedule saved.

As the above discussion indicates, performing comparison pruning between schedules identified by their unscheduled sets (U) dictates an increased computational burden and requires a substantial amount of memory when a MAXNPV objective is introduced. However, with a minor redefinition of what constitutes comparable partial schedules, these additional requirements may be eliminated.

5.4.3.4 Use of the completed activity set (CAS) in comparing partial schedules. Comparison pruning under a MAXNPV objective is enhanced through the development of a new method for identifying comparable partial schedules. Demeulemeester & Herroelen and Stinson, Davis, & Khumawala define comparable partial schedules as those partial schedules possessing identical sets of unscheduled activities (DH), or where the unscheduled activities associated with the saved partial schedule is a subset of the corresponding set associated with the current partial schedule (SDK). The method we propose here identifies comparable schedules as those having identical sets of completed activities. Again, with each saved partial schedule PS—identified by its completed activity set, CAS—the activities in progress, their finish times, the decision point m, and the NPV contribution of the completed activities are all stored. One CAS may correspond to several partial schedules that have an identical set of completed activities, but differ in the set of activities in progress, their finish times, and the decision point, m.

Unlike when unscheduled activity sets are used for identifying matching partial schedules, the use of completed activity sets avoids the necessity of recalculating NPV contributions of the activities which are members of the set. The contribution of the completed activities is calculated as the partial schedule is constructed and is saved with the completed set information. Since partial schedules being compared have identical sets of completed activities, only the previously saved NPV contributions are needed to perform comparison operations.

5.5. Computational Results

In order to fully exploit the operational efficiencies of the data structures we use to implement our approach, the computer language used has to have two qualities: it has to be able to dynamically allocate data; and it has to have a facility to deal directly with memory addressing (i.e., pointers). We have made the choice to code our algorithm using the C programming language. It has an inherent data type corresponding to pointers. It allows for dynamic memory allocation and deallocation. The C language also has built in bitwise operators to perform bitwise boolean operations on whole computer words. It supports the creation of user defined complex data types from the basic data types provided by the language. This facility permits the definition of records embracing diverse data types which are used in our procedure.

In our study, networks of several sizes were investigated to estimate the size of the largest network solvable in less than the maximum CPU time allowable per observation (200 seconds per problem on an IBM RISC/6000). The project networks used in the study are adapted from the Patterson 110 problem data set [Patterson (1984)]. Problems are adapted for this experiment by sampling from an exponential distribution with a mean of 100 to generate the necessary activity cash flows. The exponential distribution is characterized by a relatively high coefficient of variation (CV = 1.0). It is also simple to generate using an elementary inverse transformation technique. Fifteen project networks were randomly selected from this data set. Five problems consist of networks with 22 activities each, five are 27 activity problems, and five are 51 activity problems.

Each activity requires the usage of three scarce resources during each period of its schedule duration.

The relative efficacy of the solution procedure developed is assessed by observing the **solution speed** and the **size** of the search tree resulting from application of the procedure. The measure used for solution speed is the computation (CPU) time required for problem solution, while the size of the solution tree is measured by counting the number of nodes explicitly enumerated. These two performance measures are employed to investigate whether there may be a tradeoff among the two. For example, while an upper bounding technique of the type investigated here may reduce the size of the search tree, the reduction may come at very high computational overhead, and the total computation time may exceed that of a procedure which does not employ upper bounding in developing a solution to the MAXNPV problem. Computational results using our procedure on the test problems for the two measures of solution efficacy *Number of Node Evaluations Required* and *CPU Time Required for Problem Solution* are given in Tables 5:5 and 5:6, respectively.

The first row depicting results in Tables 5:5 and 5:6 gives solution experience in the absence of upper bounding *and* comparison pruning (in essence, problem solution using a full enumeration and a LIFO laser search strategy). The second row reports similar experience where comparison pruning has been added. The third row gives results using the NPV based upper bound for directing the search, but in the absence of comparison pruning. Finally, the final row in these two tables report on experience with our procedure incorporating both the directional search using an NPV based upper bound, as well as comparison pruning to restrict the search effort. Caution should be exercised in interpreting these results when upper bounding is not used. The results reported are for eight solved problems (out of fifteen) when both upper bounding and comparison pruning are not included, and are based on fourteen solved problems when comparison pruning only is included. Were the unsolved problems allowed to run until an optimal solution was found and verified in each of these instances, the quantities quoted in the first two rows of results would be much greater than those reported in Tables 5:5 and 5:6.

What is readily apparent from the results reported in these two tables is the efficacy of upper bounding and of comparison pruning in solving the MAXNPV problem. The upper bounding procedure developed performs superior to a procedure employed in the absence of any bounds. This stronger upper bound is eliminating many of the inferior nodes that are not being fathomed in the absence of this bound. And while computation of this upper bound requires considerable CPU time at each node evaluated, it more than makes up for this increase in time through the reduction in processing effort it yields by not searching as extensively (explicitly) through the search tree. Upper bounding additionally reduces search effort when comparison pruning is used. The second comparison pruning rule we use requires an upper bound to operate. In the absence of upper bounding, CP2 is not able to fathom solutions from consideration.

The results reported in Tables 5:5 and 5:6 attest to the efficacy of using a strong upper bound and comparison pruning to solve the MAXNPV problem in the context of a deep-sea-troll search. Another group of test problems was also investigated for further testing the efficacy of our solution procedure. Various cash flow distribution profiles were used to generate cash flows for each of the 110 problems in the Patterson data set–exponential, uniform, equal, and a single cash payment on the final activity of each project. Note that problems in which there is a single cash flow payment on the final activity are equivalent to the MINDUR version of the RCPSP problem. Various discount amounts were investigated as well. In each instance, all problems were solved within seven-seconds CPU time on an IBM RISC/6000, indicating the procedure developed is

Table 5:5. Number of Node Evaluations Required (n = 15)

Solution Procedure	Mean	Standard Deviation	Minimum	Maximum
No NPV Upper Bound (LIFO Laser Search); No Comparison Pruning	132,724	114,840	18	394,923†
No NPV Upper Bound (LIFO Laser Search); With Comparison Pruning	28,459	53,396	14	178,599‡
NPV Upper Bounding; No Comparison Pruning	3,803	7,770	2	30,166
NPV Upper Bounding; With Comparison Pruning	433	384	2	1,183

†Seven (of fifteen) problems were terminated early due to time limit of 200 CPU seconds per problem being exceeded.
‡One problem was terminated early due to time limit of 200 CPU seconds per problem being exceeded.

Table 5:6. CPU Time Required for Problem Solution* (n = 15)

Solution Procedure	Mean	Standard Deviation	Minimum	Maximum
No NPV Upper Bound (LIFO Laser Search); No Comparison Pruning	108	91.2	0.031	200.01 †
No NPV Upper Bound (LIFO Laser Search); With Comparison Pruning	31	61.4	0.02	200.01 ‡
NPV Upper Bound; No Comparison Pruning	3	6.6	0.001	25.46
NPV Upper Bound; With Comparison Pruning	0.37	0.36	0.001	1.19

*IBM RISC System/6000
†Seven (of fifteen) problems were terminated early due to time limit of 200 CPU seconds per problem being exceeded.
‡One problem was terminated early due to time limit of 200 CPU seconds per problem being exceeded.

a reliable optimization procedure for both maximize net present value and minimize duration variations of these two problem instances for networks consisting of fifty or fewer activities. Further details of this experimentation is described in Baroum (1992).

5.6. Conclusions

Based upon the results reported, we conclude that the use of strong upper bounds and comparison pruning are robust procedures for solving the MAXNPV problem. Our developed procedure can reliably solve project networks of up to fifty activities in size with moderate computational effort. Its solid performance is maintained over diverse network and cash flow scenarios.

The efficiency of the MAXNPV algorithm makes it a useful tool for the practicing project manager. This is especially relevant if the algorithm is going to be used in the early phase of a project to evaluate strategic types of choices. The low processing time of the algorithm makes it quite useful as an interactive decision support tool as well.

The ability of the procedure to successfully solve projects of up to fifty activities in size indicates the potential of solving even larger problems. This allows a project manager to use a lower level of aggregation in the project network, thus allowing a richer model which is less prone to infeasibilities arising in the later disaggregation phase of a project plan.

Endnotes

1. Two additional activities have been added to this project, each of which has a duration of 0 periods and no resource requirements. The purpose of these two activities is to denote the start (activity 1) and the end (activity 9) of the project. They are referred to as **dummy activities**.
2. It is obvious that the makespan of eight periods is optimal. This can be deduced from the fact that the ratio of the total project work content (=24 unit resource periods) divided by the per-period resource availability (3 units) is equal to 8. That is, a schedule achieving an 8 period makespan utilizes every unit of the resource available for the duration of the project.
3. The optimal MAXNPV solution was derived using procedures described in Section 4.
4. There are 2 possible feasible permutations of activities 4 and 5 (2!) at the start of the project and 2 possible feasible permutations of activities 7 and 8 (2!) at the end of the project. Each combination of these permutations (2*2 = 4 combinations) results in a distinct MINDUR schedule.
5. Although condition 5 above appears to disallow positive cash outflows at intermediate (early) stages during the performance of an activity, this condition actually only dictates that cash flows occurring before the completion of an activity, when summed and compounded to the end of the activity, have a cumulative net future value that is positive, or is greater than zero.
6. LB_{npv} may be set to the value of a heuristic solution if one is known.
7. A non-delay schedule is one where activities are started as soon as they become eligible and are resource feasible. Non-delay schedules may not be optimal since a task which is started at an earlier decision point may delay the scheduling of another critical activity which will become eligible at a later time period when its predecessors are completed.
8. In our algorithm, the delay arcs are implicitly considered by setting the early start times for the delayed activities (the terminal nodes of the delay arcs) to the finish time of the earliest finishing in progress activity (the origin node).

9. Actually, once the optimal solution is reached and becomes the incumbent solution, the solution procedure will backtrack before any full schedule has been formed. This is due to the fact that upper bound fathoming will eliminate partial schedules before reaching a full schedule.

10. Assuming $C_A \subseteq C_B$.

REFERENCES

Baroum, Sami M. 1992. An Exact Solution Procedure For Maximizing The Net Present Value of Resource Constrained Projects. Unpublished PhD dissertation, Indiana University, Bloomington, Indiana.

______, and J. H. Patterson. 1996. The Development of Cash Flow Weight Procedures for Maximizing the Net Present Value of a Project. *Journal of Operations Management*, **14;3**, 209-228.

Bey, R.B., Doersch, R.H. and Patterson, J.H. 1981. The Net Present Value Criterion: Its Impact on Project Scheduling. *Project Management Quarterly*, **12;2**, 35-45.

Demeulemeester, E. and Herroelen, W. 1992. A Branch-and-Bound Procedure for the Multiple Resource-Constrained Project Scheduling Problem.. *Management Science*, **38;12**, 1803-1818.

____and _____. 1997. New Benchmark Results for the Resource-Constrained Project Scheduling Problem. *Management Science*, **43;11**, 1485-1492.

_____, _____, Simpson, W.P., Baroum, S., Patterson, J.H. and Yang, K.K. 1994. Comments on a Paper by Christofedes et al. for Solving the Multiple-Resource Constrained, Single Project Scheduling Problem. *European Journal of Operational Research*, **76**, 218-228.

Doersch, R.H. and Patterson, J.H. 1977. Scheduling a Project to Maximize its Present Value: A Zero-One Programming Approach. *Management Science*, **23;8**, 882-889.

Elmaghraby, S.E. and Herroelen, W.S. 1990. The Scheduling of Activities to Maximize the Net Present Value of Projects. *European Journal of Operational Research,* **49**, 35-49.

Erenguc, S.S., Sepil, C. and Tufekci, S. An Algorithm for a Project Scheduling Problem Using Discounted Cash Flows. Working paper, College of Business Administration, University of Florida.

Fangel, M. 1988. The Viking Approach to Project Management. *International Journal of Project Management,* **6;2**, 102-105.

French, S. 1982. *Sequencing and Scheduling: An Introduction to the Mathematics of Job Shop*. John Wiley & Sons, New York.

Grinold, R.C. 1972. The Payment Scheduling Problem. *Naval Research Logistics Quarterly*, **19;1**, 123-136.

Helgeson, W. B. and D. P. Birnie. 1961. Assembly Line Balancing Using Ranked Positional Weight Technique. *Journal of Industrial Engineering*, **12**, 161-172.

Herroelen, W. S. and E. Gallens. 1993. Computational Experience with an Optimal Procedure for the Scheduling of Activities to Maximize the Net Present Value of Projects. *European Journal of Operational Research*, **65**, 274-277.

_____, Willy S., Patrick Van Dommelen, and Erik L. Demeulemeester. 1997. Project Network Models with Discounted Cash Flows: A Guided Tour Through Recent Developments. *European Journal of Operational Research*, **100**, 97-121.

Icmeli, Oya and S. Selcuk Erenguc. 1996. A Branch and Bound Procedure for the Resource Constrained Project Scheduling Problem with Discounted Cash Flows. *Management Science*, **42;10**, 1395 - 1408.

Johnson, R.V. 1988. Efficient Modular Implementation of Branch-and-Bound Algorithms. *Decision Sciences*, **19**, 17-38.

Johnson, T.J.R. 1967. An Algorithm for the Resource-Constrained Project Scheduling Problem. Unpublished Ph.D. Thesis, School of Management, MIT.

Padman, R., Smith-Daniels, D.E. and Smith-Daniels, V.L. 1990. Heuristic Scheduling of Resource-Constrained Projects with Cash Flows: An Optimization Based Approach. Working Paper, College of Business, Arizona State University.

Patterson, J.H. 1984. A Comparison of Exact Approaches for solving the Multiple-Constrained Resource, Project Scheduling Problem. *Management Science*, **30;**7, 854-867.

_____, Slowinski, R., Talbot, F.B, and Weglarz, J. 1989. An Algorithm for a General Class of Precedence and Resource Constrained Scheduling Problem. *Studies in Production and Engineering Economics, (Advances in Project Scheduling)*, Slowinski, R. and Weglarz, J. (eds.), Elsevier, Amsterdam, The Netherlands, 9, 3-28.

_____, Talbot, F.B., Slowinski, R., and Weglarz, J. 1990. Computational Experience with a Backtracking Algorithm for Solving a General Class of Precedence and Resource Constrained Scheduling Problems. *European Journal of Operational Research*, **49**, 68-79.

Russell, A. H. 1970. Cash Flows in Networks. *Management Science*, **16;5**, 357-373.

Russell, R.A. 1986. A comparison of Heuristics for Scheduling Projects with Cash Flows and Resource Restrictions. *Management Science*, **32;10**, 1291-1300.

Smith-Daniels, D.E. 1986. Summary Measures for Predicting the Net Present Value of a Project. Working Paper, College of St. Thomas, St. Paul, Minnesota.

_____and Aquilano, N.J. 1987. Using a Late-Start Resource-Constrained Project Schedule to Improve Project Net Present Value. *Decision Sciences*, **18**, 617-630.

_____and Smith-Daniels, V.L. 1987. Maximizing the Net Present Value of a Project Subject to Materials and Capital Constraints. *Journal of Operations Management*, **7;1**, 33-45.

Stinson, J.P. 1976. A Branch and Bound Algorithm for a General Class of Multiple Resource-Constrained Scheduling Problems. Unpublished Ph.D. dissertation, University of North Carolina, Chapel Hill, North Carolina.

____, Davis, E.W. and Khumawala, B.M. 1978. Multiple Resource-Constrained Scheduling Using Branch and Bound. *AIIE Transactions*, **10;3**, 252-259.

Talbot, F. B. and Patterson, J.H. 1978. An Efficient Integer Programming Algorithm with Network Cuts for Solving Resource-Constrained Scheduling Problems. *Management Science*, **24;11**, 1163-74.
Vila, J.M. 1990. Reflections on a Complex Project. *Datamation*, **36**, 101-103.
Weeks, J. 1971. *The Pyramids*. Cambridge University Press, Great Britain.
Yang, K.K., Talbot, F.B., and J.H. Patterson. 1992. Scheduling a Project to Maximize its Net Present Value: An Integer Programming Approach. *European Journal of Operational Research*, **64;2**, 188-198.

6 SOLVING A PREEMPTIVE PROJECT SCHEDULING PROBLEM WITH COLORING TECHNIQUES

Lucio Bianco
Massimiliano Caramia
Paolo Dell'Olmo

University of Rome "Tor Vergata"

6.1. Introduction

In this paper we describe an exact algorithm to solve a preemptive project scheduling problem with scarce resources. As most of the scheduling problems also the studied one is *NP*-hard. Many different approaches, such as mathematical programming, heuristics, simulating annealing, etc., have been presented to solve this class of intractable problems. In our approach we consider a graph-theoretical method based on new coloring techniques. In particular, given a disjunctive graph we examine the correspondences between an assignment of time windows (i.e. activities starting time and finishing time) and an assignment of colors to the activities of a project, defining when an exact weighted coloring is an optimal preemptive schedule. We present computational results on randomly generated test problems of the branch and bound algorithm proposed. From the results obtained emerges that the possibility to manage hard problems using new coloring techniques seems to be more efficient than others, referring in particular to the performances of the algorithm proposed.

Scheduling problems are applied to many different areas, such as operating system, production management, nuclear power plant, transportation scheduling, etc. In this work we consider the area of project management where scheduling plays a very important role.

On behalf of the great interest in this problem, project scheduling has not been completely examined, referring in particular to some important aspects. In fact, although many papers in literature deal with project scheduling, and with exact procedures to solve it, indeed not many of these take into account the preemption mode. Preemption is not always a good assumption for representing real production systems especially for machine scheduling where tasks need to be processed by machines in processing times that are smaller than the working day. But, if we consider a project where we have a set of activities that need to be carried out in a given processing time (the duration of the activity) that often is at least some days or some weeks, we obviously have to consider the possibility to interrupt the activities at the end of the day and restart them the days after.

In general deadlines of projects are represented by a succession of discrete points of time, so we have focused on the problem of scheduling activities on pre-allocated renewable resources with precedence constraints and integer points preemption in which the aim is to minimize the time required to execute all the tasks.

In this paper, we present an exact branch and bound algorithm inspired on coloring techniques and on recent studies (Caramia and Dell'Olmo (1997) and (1998)) on some advances in graph theory and algorithms. Computational results are presented on randomly generated test problems.

At the best of our knowledge in the literature there is no exact algorithm for this problem thus we restrict the state of the art to a recent work (Bianco et al. 1994a) where preemption, of a fractional points of time type, is allowed over independent activities.

For references to the project scheduling problem the reader can see Bianco et al. (1994a), Bianco et al. (1994b), Herroelen et al. (1998), Weglarz (1977). For references to the coloring problem the reader can see Brelaz (1979), Kubale and Jackowski (1985), Cangalovic and Schreuder (1991), Mehrotra and Trick (1996), Sager and Lin (1993), Sewell (1993).

The paper is organized as follows: section 2 defines the project scheduling problem with integer points preemption; section 3 shows the analogies between the coloring problem and the preemptive scheduling problem; in section 4 we give the mathematical formulation of the analyzed scheduling problem; in section 5 the exact branch and bound algorithm is described and in section 6 computational results are reported. Finally, in the last section, we find the conclusions.

6.2. m,1|fix, pmtnZ, cpm|C_{max}: notations and definitions

In this paper we consider the problem of scheduling tasks on dedicated resources in which we want to minimize the time required to execute all the tasks. The problem can be stated as follows. Let be given a set $R=\{R_1,...., R_m\}$ of m renewable resources, and a set $T=\{T_1,....,T_n\}$ of n tasks to be carried out using these resources. A set of resources $\mu_j \subseteq R$ has been pre-allocated to each task $T_j \in T$ for a integer processing time

p_j. Resources can execute one task at a time, and task execution can be interrupted at integer points of time and restarted later. Precedence relations between tasks are also given: let $G=(T,A)$ be the digraph representing precedence constraints among tasks, such that, a task can start its execution only if all the tasks preceding it in G have been completed.

The problem consists of assigning a starting time to each task such that each resource is used by a single task at time, precedence constraints are satisfied and the time required to process all the tasks (i.e., the makespan) is minimized.

Using and extending the notation proposed by Herroelen et al. (1998), the problem can be denoted as a $m,1|fix, pmtnZ, cpm|C_{max}$ scheduling problem where m represents the number of different types of resources; 1 is the level of availability of each resource in each time period; *fix* define that a set of resources is pre-allocated to each task; *pmtnZ* means that the preemption mode, of the preempt-resume type, is allowed only at integer points of time; *cpm* stands for *critical path method* and means that activities are subject to strict finish-start precedence constraints with zero time lag, as used in the basic *pert/cpm* model. C_{max} is the object we want to minimize and corresponds to the minimum project makespan.

The problem is *NP*-hard (Ullman (1976)). Polynomial cases are: $2,1|fix, pmtnZ, cpm|C_{max}$ where there are only two resources and $m,1|fix, pmtnZ, cpm\text{-}tree|C_{max}$ where the precedence graph is a tree (Muntz and Coffman (1969) and (1970)). The best result in terms of approximation is $C_{max} \leq (2\text{-}2/m)C^*_{max}$ (Muntz and Coffman (1970)).

Without loss of generality, let task T_n be a dummy task, with $p_{n,1} = 0$, that can start only after the completion of tasks $T \setminus \{T_n\}$. In particular, let $\Gamma^{-1}(T_j) = \{T_k \in T \mid T_k \rightarrow T_j \in A\}$ be the set of (direct) predecessors of T_j, and $\Gamma(T_j) = \{T_k \in T \mid T_j \rightarrow T_k \in A\}$ be the set of (direct) successors of T_j. Moreover, let the digraph $G_c=(T,A_c)$ be the transitive closure of G, and $\Gamma_c^{-1}(T_j) = \{T_k \in T \mid T_k \rightarrow T_j \in A_c\}$, $\Gamma_c(T_j) = \{T_k \in T \mid T_j \rightarrow T_k \in A_c\}$ be the set of (direct and indirect) predecessors and the set of (direct and indirect) successors of T_j, respectively.

We define the *disjunctive graph* $G_A=(T, A \cup D_A)$ obtained inserting undirected edges $D_A=\{T_k — T_j \mid \mu_k \cap \mu_j \neq \emptyset\}$ into G. In other words, we insert an undirected edge between two vertices if the corresponding tasks cannot be processed simultaneously.

We say that a directed network G with a specified source node s and a specified sink node t is *layered* if we can partition its node set T into w *layers* $N_1, N_2, \ldots, N_w$ so that $N_1=\{s\}$, $N_w=\{t\}$ and for every arc $(k, j) \in A$, nodes k and j belong to adjacent layers (i.e., $k \in N_l$ and $j \in N_{l+1}$ for some $1 \leq l \leq w\text{-}1$).

6.3. The coloring problem: definitions ad analogies with the preemptive scheduling

Let $G=(V,E)$ be an undirected graph, with vertex (node) set V and edge set E. A subset $W \subseteq V$ of vertices is a *clique* if it induces a complete subgraph. A clique is *maximum* if there is no clique of G of larger cardinality. Let $\omega(G)$ be the number of vertices of a maximum clique of G. A *vertex coloring* of G is an assignment of colors to the vertices such that no two adjacent vertices have the same color. The aim of the coloring problem is to find the *minimum* number of colors that assures the specified conditions. The minimum number of colors needed is called the *chromatic number* of G and is denoted by $\chi(G)$. In the sequel for simplicity we write coloring instead of vertex coloring.

Now, let $p(j)=p_j \in Z^+$ be the weights assigned to the vertices $j \in V$. Let $p(V')=\Sigma_{j \in V'} p_j = p_{V'}$ with $V' \subseteq V$. The coloring problem transforms into a weighted coloring problem where the aim is to minimize the number of colors satisfying that each vertex has assigned a set of colors I_j such that $|I_j|=p_j$ and no two adjacent vertices j and k have $I_j \cap I_k \neq \emptyset$.

Consider the vertex set V to be the set T of tasks and assign to the vertices of the graph weights equal to the duration of the activities. First assume that there are no precedence relations between activities, so the graph G is undirected and obtained by considering resource conflicts, i.e. $E=D_A=\{T_j—T_k | \mu_j \cap \mu_k \neq \emptyset\}$. An assignment of admissible colors to the vertices of the disjunctive graph obtained is equivalent to an *admissible schedule* because it assigns time periods to the vertices such that no two adjacent vertices use simultaneously the same resources. The minimum number of colors assigned define the minimum completion time of the project.

Table 6:1. Sets of pre-allocated resources

	T_1	T_2	T_3	T_4	T_5	T_6	T_7
R_1	*	*					
R_2		*		*			
R_3			*		*		
R_4			*			*	
R_5				*	*		*
R_6						*	*
R_7	*		*				

Each * defines the resource required by each task

Here it follows an example of this analogies: consider seven activities and seven renewable resources, each activity requiring a set of pre-allocated resources (see for example table 6:1). As we assumed, there are no precedence relations and an incompatibility between activities arises only when two activities share the same resources. The graph represented in figure 6:1 is the incompatibility graph where the vertices are the activities of the project and an edge (j,k) occurs if $\mu_j \cap \mu_k \neq \emptyset$. Assume

also that preemption is allowed only at integer points of time. An admissible coloring is given by the assignment of colors in figure 6:2. This is also equivalent to the admissible schedule where each color defines the (unit) time period when the activity is processed. The optimal coloring (schedule) is given in figure 6:3. The considerations above leads us to formalize that:
Theorem 1. A feasible/optimal schedule of the $m,1|fix,\ pmtnZ|C_{max}$ is equivalent to a feasible/optimal weighted coloring of the resource incompatibility graph where integer duration have been assigned to the activities.

Let us recall the mathematical formulation of the weighted coloring problem: let x_{jh} be a binary variable that is equal to 1 if and only if to vertex j is assigned color h and x_{jh}=0 otherwise. Let M be the adjacency matrix of G and y_h be a binary variable that is equal to 1 if and only if the color h is used and y_h=0 otherwise. The weighted coloring problem can be formulated as follows:

$$\min \sum_{h=1}^{q} y_h \qquad (1)$$

s.t.

$$\sum_{h=1}^{q} x_{jh} = p_j \qquad j=1,\ldots,n; \qquad (2)$$

$$\sum_{k=1}^{n} M_{kj} x_{kh} \leq (1 \quad x_{jh})n \qquad j=1,\ldots,n;\ h=1,\ldots,q; \qquad (3)$$

$$\sum_{j=1}^{n} x_{jh} \leq L y_h \qquad h=1,\ldots,q; \qquad (4)$$

$$x_{jh} \in \{0,1\} \qquad j=1,\ldots,n;\ h=1,\ldots,q; \qquad (5)$$

$$y_h \in \{0,1\} \qquad h=1,\ldots,q; \qquad (6)$$

where L is a large integer, and q is an upper bound on the chromatic number.

Now, let us introduce the precedence relations and consider the disjunctive graph obtained. An admissible weighted coloring does not correspond to an admissible schedule because for all $(k, j) \in A$ might not be verified $max\{k \in I_k\} \leq min\{j \in I_j\}$. Thus an exact algorithm for the weighted vertex coloring have to be modified in order to be used for reaching on optimal solution for the $m,1|fix,\ pmtnZ,\ cpm|C_{max}$ problem such that when a precedence relation occurs then $max\{k \in I_k\} \leq min\{j \in I_j\}$ holds.
Now, referring to the previous example, let us suppose that precedence constraints among activities are represented by the graph G in figure 6:4. Each arc (k, j) implies that activity j can start only if all the activities k in the set $\Gamma_c^{-1}(T_j)=\{T_k \in T \mid T_k \rightarrow T_j \in A_c\}$ have finished their execution.

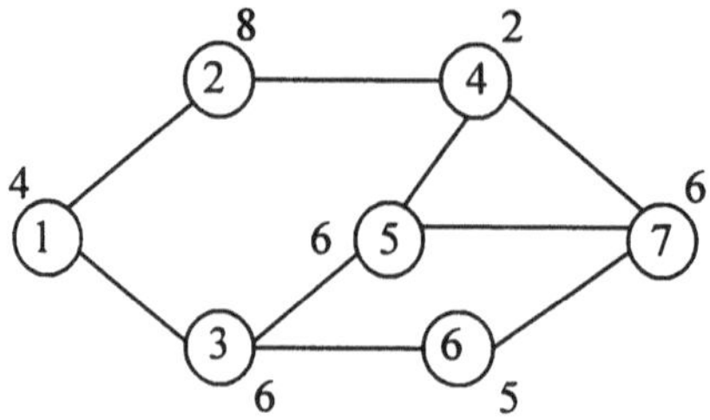

Figure 6:1

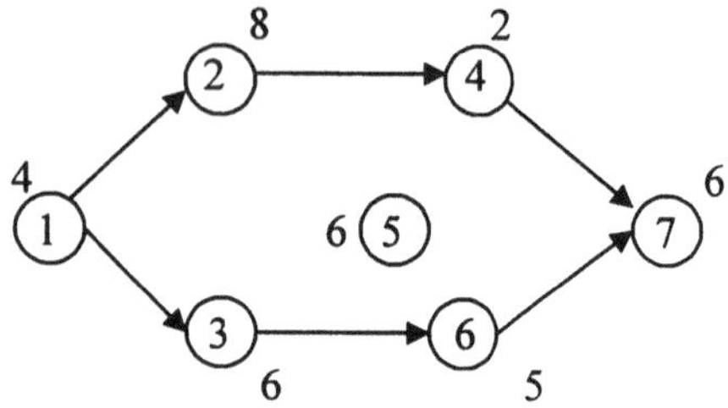

Figure 6:4

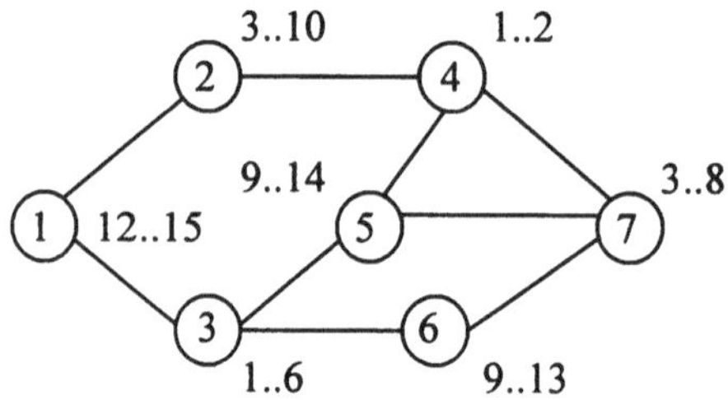

Figure 6:2

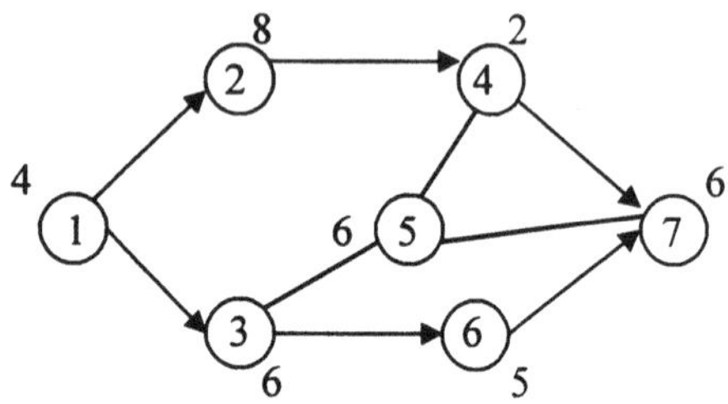

Figure 6:5

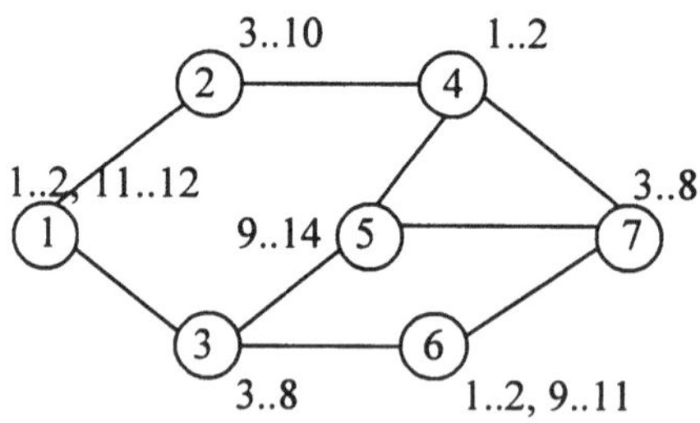

Figure 6:3

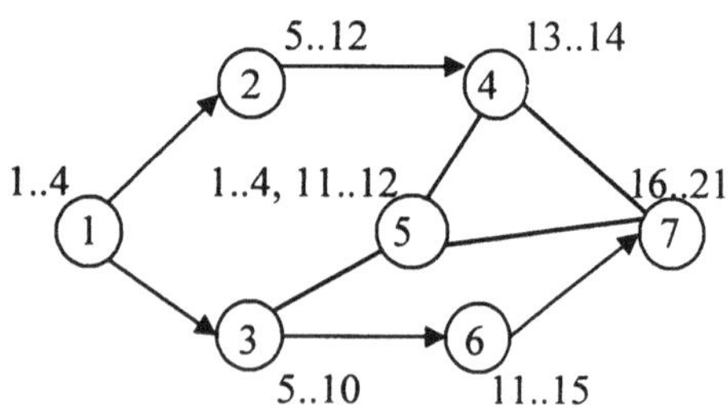

Figure 6:6

Consider the disjunctive graph G_A obtained inserting undirected edges into G (see figure 6:5). The optimal schedule is represented in figure 6:6. As for the above case we can state that:

Theorem 2. A feasible/optimal weighted coloring of the resource incompatibilities graph where integer duration have been assigned to the activities is equivalent to a feasible/optimal schedule of the $m,1|fix,\ pmtnZ,cpm|C_{max}$ iff $max\{k \in I_k\} \leq min\{j \in I_j\}$ for all $(k, j) \in A$.

So the weighted vertex coloring problem with precedence relations can be formulated as follows:

$$\min \sum_{h=1}^{q} y_j \tag{7}$$

s.t.

$$\sum_{h=1}^{q} x_{jh} = p_j \qquad j = 1,\dots,n; \tag{8}$$

$$\sum_{k=1}^{n} M_{kj} x_{kh} \leq (1 - x_{jh})n \qquad j = 1,\dots, n;\ h = 1,\dots, q; \tag{9}$$

$$\sum_{j=1}^{n} x_{jh} \leq L y_h \qquad h = 1,\dots, q; \tag{10}$$

$$x_{kh} + x_{jh} \leq 1 \qquad (k,j) \in A_c;\ h < h';\ h, h' = 1,\dots, q; \tag{11}$$

$$x_{jh} \in \{0,1\} \qquad j=1,\dots,n;\ h=1,\dots,q; \tag{12}$$

$$y_h \in \{0,1\} \qquad h=1,\dots,q; \tag{13}$$

6.4. Mathematical formulation

In this section, we recall the 0-1 integer programming formulation of the project scheduling problem. First, let us define some useful parameters. Let T be an upper bound on C^*_{max}, for example, $T = \Sigma_{j=1..n}\, p_j$. Moreover, let ES_j and LS_j be respectively the earliest starting time and the latest starting time of task T_j. ES_j and LS_j can be computed by standard forward and backward recursions considering each task T_j to be executed in p_j time, and considering T as the ending time of the last executed task in the backward recursion.

Let $a_{ji} = 1$ if resource $R_i \in \mu_j$, while $a_{ji} = 0$ otherwise and let $x_{j\tau}$ be a binary variable that is equal to 1 iff the task T_j starts at time τ. The problem can be formulated as follows:

$$\min \sum_{\tau=ES_n}^{LS_n} t x_{n\tau} \qquad (14)$$

s.t.

$$\sum_{\tau=ES_j}^{LS_j} x_{j\tau} = 1 \qquad j = 1,\ldots,n; \qquad (15)$$

$$\sum_{\tau=ES_k}^{LS_k} (t + p_k) x_{k\tau} \le \sum_{\tau=ES_j}^{LS_j} t x_{j\tau} \qquad (k,j) \in A; \qquad (16)$$

$$\sum_{j=1}^{n} a_{ji} \sum_{\tau=\max[t-p_j+1, ES_j]}^{\min[t, LS_j]} x_{j\tau} \le 1 \qquad t = 1,\ldots,T \qquad (17)$$

$$x_{j\tau} \in \{0,1\} \qquad j=1,\ldots,n;\ \tau=ES_j,\ldots,LS_j; \qquad (18)$$

Constraints [15] ensure that exactly one starting time is assigned to each task; constraints [16] take into account task precedence relations; constraints [17] guarantee that each resource is assigned to at most one task at a time and, finally, constraints [18] defines that $x_{i\tau}$ is a binary variable.

6.5. The branch and bound algorithm

The algorithm is obtained by adapting the exact algorithm for the weighted vertex coloring (Caramia and Dell'Olmo (1998)) WCDCOLOR. Before the execution of the algorithm an upper bound and a lower bound on the exact weighted coloring of G are computed.

6.5.1 Upper bound

Before the execution of the algorithm an upper bound *UB* is computed by the *Coloring Layered Network* (CLN) algorithm in the following way: CLN considers the layers N_i in the transitive closure of G and assigns intervals of colors to the activities N_l starting from the successive color of the maximum one assigned to the activities in

N_{l-1} predecessors of the considered one.

In the example in figure 6:4 we have four layers: $N_1=\{1\}$, $N_2=\{2,3\}$, $N_3=\{4,6\}$, $N_4=\{7\}$. The algorithm starts assigning colors $\{1..4\}$ to node 1 that is unique in N_1, thus the first color to assign in N_2 is $\{5\}$. CLN assigns to I_2 the interval $\{5..12\}$ and to $I_3=\{5..10\}$. Referring to N_5 we have $I_4=\{13..14\}$ and $I_6=\{11..15\}$. Finally $I_7=\{16..21\}$.

6.5.2 Lower bound

We adapt for our specific needs a branch and bound algorithm for the maximum weighted clique (Babel (1991)) W of the transitive closure G_c. p_W is a lower bound on the chromatic number of G.

In the example proposed in figure 6:4 p_W is 21 and corresponds also to the longest path from the activity 1 to the activity 7.

6.5.3 The branch and bound algorithm

The algorithm is based on an enumeration scheme where at the current subproblem the graph is partially colored with h colors 1,....,h. In each node of the tree the subproblem is to determine whether or not the coloring can be extended to the uncolored nodes. The algorithm expands at each subproblem the coloring by choosing the node with the fewest available colors. A tie breaking rule is computed selecting the node with the maximum degree in the uncolored graph.

The algorithm starts coloring the maximum clique of G_c and then begins to select activities to process. At each level l of the branch and bound tree corresponds a set of *schedulable activities* SA_l for which $\Gamma_c^{-1}(T_j)$ for all $j \in SA_l$ have already been processed. When an activity $j \in SA_l$ has been selected the algorithm, before expanding the tree and creating a new branching node, checks whether $\Gamma_c^{-1}(T_j) \neq \emptyset$: if so the subproblems relative to the activity j must have $min\{j \in I_j\} \geq max\{k \in I_k\}$ for each k in $\Gamma_c^{-1}(T_j)$; otherwise the algorithm checks resource incompatibilities and assigns colors. The set SA_l is updated at each level by deleting the activity processed and inserting all the activities in $\Gamma(T_j) = \{T_k \in T \mid T_j \rightarrow T_k \in A\}$.

The algorithm colors G pruning subproblems with more than UB colors. One important aspect of the algorithm so structured is that it works on the graph G and not on its transitive closure G_c reducing considerably the number of branching because of the lower density of G with respect to G_c. In fact, as can be also observed by the experimental behavior of other coloring algorithms the number of branches is higher for dense graphs.

6.6. Computational results

The algorithm has been implemented in C programming language and run on a DEC Alpha 400. We test the algorithm on a set of randomly generated problems. In all the generated instances, we consider m=10 resources. Different series of tests have been performed considering n=30, 60, 90, 120, 150 tasks. The resource sets are randomly chosen considering that the numbers of required resources μ_j are integer values uniformly distributed over the range [1,5]. Processing times p_j are integer values uniformly distributed over the range [1,10]. As for precedence relations, we have considered in all the generated instances that each task can have at most 3 predecessors. Moreover, for each value of n we have considered 5 test problems, and computed average result values.

In the table 6:2 we have listed the average solution values of the makespan produced by the algorithm ($\chi(G)$), the lower bound ($\omega(G)$), the upper bound *UB*, the average *CPU time* and the number of *branches*.

Table 6:2. All statistics are averages of 5 test problems solved on a DEC Alpha 400.

n	ω *(G)*	χ*(G)*	*UB*	*CPU*	*Branches*
30	110.4	125.4	146.8	0.0	25.8
60	220.6	248.6	281.2	2.2	2410.4
90	302.6	334.8	373.6	8.4	4089.2
120	406.4	460.4	500.4	27.2	11545.4
150	458.0	522.2	632.8	65.4	847742.4

Computational results

As expected, the number of branches grows exponentially with the size of the problem. However, the average time to solve instances with 150 activities is slightly more than one minute, which can be considered satisfactory for this problem. Note also that for all the instances $\omega(G)<\chi(G)$ and the number of branches is dependent both on this gap and on the problem size.

6.1. Conclusion

In this paper, we have studied the problem of scheduling tasks on dedicated resources, with precedence constraints, and preemption allowed at integer points of time, in which we want to minimize the time required to execute all the tasks.

For this problem, we have proposed an exact branch and bound algorithm obtained by adapting a weighted coloring algorithm through a procedure that, when a precedence relation between the activities k and j occurs, assigns colors such that the starting time of activity j is not lower than the finishing time of the activity k. The algorithm is experimentally evaluated on a set of randomly generated test problems.

References

BABEL, L. (1991). "Finding Maximum Cliques in Arbitrary and in Special Graphs" *Computing.* 46, 321-341.

BIANCO, L., BLAZEWICZ, J., DELL'OLMO, P., DROZDOWSKI, M. (1994a). "Scheduling Preemptive

Multiprocessor Tasks on Dedicated Processors" *Performance Evaluation* 20 (4), 361-371.

BIANCO, L., DELL'OLMO, P., SPERANZA, M. G. (1994b). "Nonpreemptive scheduling of independent tasks with prespecified processors allocation" *Naval Research Logistic* 41, 959-971.

BRELAZ, D. (1979). "New methods to color the vertices of a graph" *Communications of the ACM* 22, 251-256.

CANGALOVIC, M., SCHREUDER, J. A. M. (1991). "Exact coloring algorithm applied to timetabling problems with lectures of different lengths" *European Journal of Operational Research* 51, 248-258.

CARAMIA, M., DELL'OLMO, P. (1997). "An exact algorithm for the vertex coloring problem" *Tech. Report DISP University of Rome Tor Vergata 10.*

CARAMIA, M., DELL'OLMO, P. (1998). "An exact algorithm for the unweighted and weighted integer vertex coloring problem" *Tech. Report IASI-CNR.*

HERROELEN, W., DEMEULEMEESTER, E., DE REYCK, B. (1998). *A classification scheme for project scheduling*, chapter 1 of this book.

KUBALE, M., JACKOWSKI, B. (1985). "A generalized implicit enumeration algorithm for graph coloring" *Communications of the ACM* 28, 412-418.

MEHROTRA, A. AND TRICK, M.A. (1996). "A Column Generation Approach for Graph Coloring" *INFORMS Journal on Computing* 8, 344-354.

MUNTZ, R.R, COFFMAN JR., E.G. (1969). "Optimal Preemptive Scheduling on two processor Systems" *IEEE Trans. Comput.* C-18, 1014-1020.

MUNTZ, R.R, COFFMAN JR., E.G. (1970). "Preemptive scheduling of real time tasks on multiprocessor systems" *Journal of the Association Comput. Machinery*, 17, 324-338.

SAGER T.J., LIN S. (1993). "A Prunig Procedure for Exact Graph Coloring" *ORSA Journal on Computing* 3, 226-230.

SEWELL, E. C. (1993). "An Improved Algorithm for Exact Graph Coloring" *Second DIMACS Challenge (COLORING Papers)*, to appear in *DIMACS Series in Computer Mathematics and Theoretical Computer Science*.

ULLMANN, J. D. (1976). *Complexity of Sequencing Problems, Computer and Job Shop Scheduling Theory*, Wiley, New York, 139-164.

WEGLARZ, J., BLAZEWICZ, J., CELLARY, W., SLOWINSKI, R. (1977). "An Automatic Revised Simplex Method for Constrained Resource Network Scheduling" *ACM Trans. Math. Software*, 3 (3), 295-300.

7 HEURISTIC ALGORITHMS FOR THE RESOURCE-CONSTRAINED PROJECT SCHEDULING PROBLEM: CLASSIFICATION AND COMPUTATIONAL ANALYSIS

Rainer Kolisch
Sönke Hartmann

Christian–Albrechts–Universität zu Kiel

7.1 INTRODUCTION

The resource–constrained project scheduling problem (RCPSP) can be given as follows. A single project consists of a set $\mathcal{J} = \{0, 1, \ldots, n, n+1\}$ of activities which have to be processed. Fictitious activities 0 and $n+1$ correspond to the "project start" and to the "project end", respectively. The activities are interrelated by two kinds of constraints. First, precedence constraints force activity j not to be started before all its immediate predecessor activities comprised in the set $\mathcal{P}_j$ have been finished. Second, performing the activities requires resources with limited capacities. We have K resource types, given by the set $\mathcal{K} = \{1, \ldots, K\}$. While being processed, activity j requires $r_{j,k}$ units of resource type $k \in \mathcal{K}$ during every period of its non-preemptable duration p_j. Resource type k has a limited capacity of R_k at any point in time. The parameters p_j, $r_{j,k}$, and R_k are assumed to be deterministic; for the project start and end activities we have $p_j = 0$ and $r_{j,k} = 0$ for all $k \in \mathcal{K}$. The objective of the RCPSP is to find precedence and resource feasible completion times for all activities such that the makespan of the project is minimized. Figure 7:1 gives an example of a project comprising $n = 6$ activities which have to be scheduled subject to $K = 1$ renewable resource type with a capacity of 4 units. A feasible schedule with an optimal makespan of 13 periods is represented in Figure 7:2.

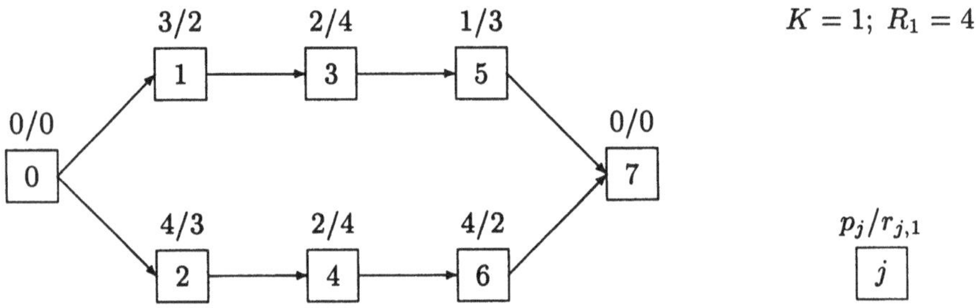

Figure 7:1 Project instance

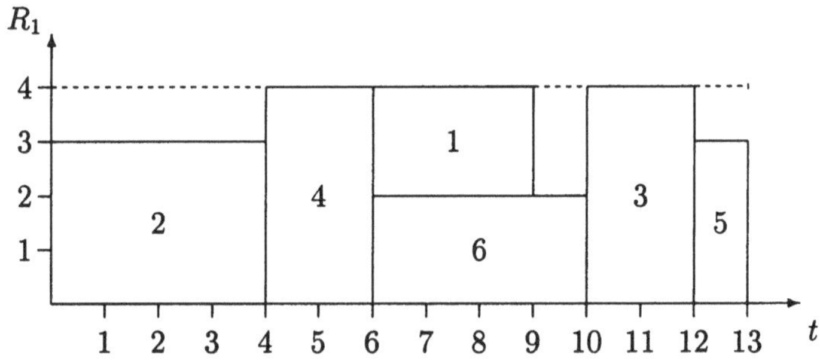

Figure 7:2 Example schedule

Let F_j denote the finish time of activity j. A schedule S is given by a vector of finish times $(F_1, F_2, \ldots, F_n)$. Let $\mathcal{A}(t) = \{j \in \mathcal{J} \mid F_j - p_j \leq t < F_j\}$ be the set of activities which are being processed (active) at time instant t. We now can provide the conceptual decision model (7.1) – (7.4) (cf. Christofides et al. 1987).

$$\text{Min } F_{n+1} \tag{7.1}$$

$$F_h \leq F_j - p_j \qquad j = 1, \ldots, n+1;\ h \in \mathcal{P}_j \tag{7.2}$$

$$\sum_{j \in \mathcal{A}(t)} r_{j,k} \leq R_k \qquad k \in \mathcal{K};\ t \geq 0 \tag{7.3}$$

$$F_j \geq 0 \qquad j = 1, \ldots, n+1 \tag{7.4}$$

The objective function (7.1) minimizes the finish time of the project end activity and thus the makespan of the project. Constraints (7.2) enforce the precedence constraints between activities, and constraints (7.3) limit for each resource type k and each time instant t that the resource demand of the activities which are currently processed does not exceed the capacity. Finally, (7.4) define the decision variables. (7.1) – (7.4) is a conceptual model since the sets $\mathcal{A}(t)$ are a function of the decision variables. Hence, the model cannot be solved with mixed integer programming (MIP) techniques. In order to solve the RCPSP with MIP–solvers such as CPLEX (cf. Bixby 1996), one has to employ the 0–1 problem formulation of Pritsker et al. (1969).

The RCPSP is denoted in Chapter 1 of this book (cf. Herroelen et al. 1998) as $m, 1/\text{cpm}/C_{\max}$ where a number of activities which are precedence related by finish–start relationships with zero time lags have to be scheduled on m renewable resource types such that $C_{\max}$, the maximal completion time of all activities is minimized.

It has been shown by Błażewicz et al. (1983) that the RCPSP as a generalization of the classical job shop scheduling problem belongs to the class of $\mathcal{NP}$–hard optimization problems. Therefore, heuristic solution procedures are indispensable when solving large problem instances as they usually appear in practical cases. Since 1963 when Kelley (1963) introduced a schedule generation scheme, a large number of different heuristic algorithms have been suggested in the literature.

The great number of optimal approaches (for a survey cf. Kolisch and Padman 1997) are mainly for generating benchmark solutions. Currently, the most competitive exact algorithms seem to be the ones of Brucker et al. (1998), Demeulemeester and Herroelen (1997), Mingozzi et al. (1998) and Sprecher (1996).

In what follows we will give an appraising survey of heuristic approaches for the RCPSP. We start in Section 7.2 with schedule generation schemes which are essential to construct feasible schedules. In Section 7.3 we show how these schemes are employed in priority rule based methods. Section 7.4 is devoted to metaheuristic algorithms such as simulated annealing, tabu search, and genetic algorithms. Heuristics which do neither belong to the class of priority rule based methods nor to metaheuristic approaches are treated in Section 7.5. In Section 7.6 we will report about a comparison of priority rule based and metaheuristic heuristics for the RCPSP. We end with a summary and an outlook on research opportunities in Section 7.7.

7.2 SCHEDULE GENERATION SCHEMES

Schedule generation schemes (SGS) are the core of most heuristic solution procedures for the RCPSP. SGS start from scratch and build a feasible schedule by stepwise extension of a partial schedule. A partial schedule is a schedule where only a subset of the $n+2$ activities have been scheduled. There are two different SGS available. They can be distinguished w.r.t the incrementation into activity– and time–incrementation. The so–called *serial* SGS performs *activity–incrementation* and the so–called *parallel* SGS performs *time–incrementation.*

7.2.1 Serial Schedule Generation Scheme

We begin with a description of the serial SGS. It consists of $g = 1,\ldots,n$ stages, in each of which one activity is selected and scheduled at the earliest precedence– and resource–feasible completion time. Associated with each stage g are two disjoint activity sets. The scheduled set $\mathcal{S}_g$ comprises the activities which have been already scheduled, the eligible set $\mathcal{D}_g$ comprises all activities which are eligible for scheduling. Note that the union of $\mathcal{S}_g$ and $\mathcal{D}_g$ does not give the set of all activities $\mathcal{J}$ because, generally, there are so–called ineligible activities, i.e. activities which have not been scheduled and can not be scheduled at stage g because not all of their predecessors have been scheduled. Let $\tilde{R}_k(t) = R_k - \sum_{j\in\mathcal{A}(t)} r_{j,k}$ be the remaining capacity of resource type k at time instant t and let $\mathcal{F}_g = \{F_j \mid j \in \mathcal{S}_g\}$ be the set of all finish times. Let further be $\mathcal{D}_g = \{j \in \mathcal{J} \setminus \mathcal{S}_g \mid \mathcal{P}_j \subseteq \mathcal{S}_g\}$ the set of eligible activities. We can now give the following description of the serial SGS.

Serial SGS

Initialization: $F_0 = 0, \mathcal{S}_0 = \{0\}$,

`For` $g = 1$ `to` n `do`

> Calculate $\mathcal{D}_g$, $\mathcal{F}_g$, $\tilde{R}_k(t)$ $(k \in \mathcal{K};\ t \in \mathcal{F}_g)$
> Select one $j \in \mathcal{D}_g$
> $EF_j = \max_{h\in\mathcal{P}_j} \{F_h\} + p_j$
> $F_j = \min\ \{t \in [EF_j - p_j, LF_j - p_j] \cap \mathcal{F}_g \mid$

$$r_{j,k} \leq \tilde{R}_k(\tau), k \in \mathcal{K}, \ \tau \in [t, t + p_j[\cap \mathcal{F}_g\} + p_j$$
$$\mathcal{S}_g = \mathcal{S}_{g-1} \cup \{j\}$$

$$F_{n+1} = \max_{h \in \mathcal{P}_{n+1}} \{F_h\}$$

The initialization assigns the dummy source activity $j = 0$ a completion time of 0 and puts it into the partial schedule. At the beginning of each step g, the decision set $\mathcal{D}_g$, the set of finish times $\mathcal{F}_g$, and the remaining capacities $\tilde{R}_k(t)$ at the finish times $t \in \mathcal{F}_g$ are calculated. Afterwards, one activity j is selected from the decision set. The finish time of j is calculated by first determining the earliest precedence feasible finish time EF_j and then calculating the earliest (precedence– and) resource–feasible finish time F_j within $[EF_j, LF_j]$. LF_j denotes the latest finish time as calculated by backward recursion (cf. Elmaghraby 1977) from an upper bound of the project's finish time T. Table 7:1 reports the serial SGS when generating the schedule given in Figure 7:2.

g	1	2	3	4	5	6
$\mathcal{D}_g$	{1,2}	{1,4}	{1,6}	{3,6}	{3}	{5}
j	2	4	1	6	3	5

Table 7:1 Example for serial SGS

The serial SGS generates always feasible schedules which are for the resource–unconstrained scheduling problem (7.1), (7.2), and (7.4) optimal. Kolisch (1996a) has shown that the serial SGS generates active schedules, that is schedules $S = (F_1, F_2, \ldots, F_n)$ where none of the activities can be started earlier without delaying some other activity. For scheduling problems with regular performance measure (for a definition of the latter cf. to Sprecher et al. 1995) such as makespan minimization, the optimal solution will always be in the set of active schedules. The time complexity of the serial SGS as given above is $\mathcal{O}(n^2 \cdot K)$ (cf. Pinson et al. 1994).

Let j_g denote the activity which is selected in iteration g. Then, an execution of the serial SGS can be recorded by a list $\lambda = \langle j_1, j_2, \ldots, j_n]$ which prescribes that activity j_g has been scheduled in iteration g. Note, that this list is precedence feasible, i.e. we have $\mathcal{P}_{j_g} \subseteq \{j_1, \ldots, j_{g-1}\}$ (cf. Hartmann 1998). The list for the above given example is $\lambda = \langle 2, 4, 1, 6, 3, 5]$. Given a list λ, we can now give a special case of the serial SGS, namely the serial SGS for activity lists.

Serial SGS for Activity Lists

Initialization: $F_0 = 0, \mathcal{S}_0 = \{0\}$,
For $g = 1$ to n do

Calculate $\mathcal{F}_g$, $\tilde{R}_k(t)$ $(k \in \mathcal{K};\ t \in \mathcal{F}_g)$
$j = j_g$
$EF_j = \max_{h \in \mathcal{P}_j} \{F_h\} + p_j$
$F_j = \min \{t \in [EF_j - p_j, LF_j - p_j] \cap \mathcal{F}_g \mid$
$r_{j,k} \leq \tilde{R}_k(\tau), k \in \mathcal{K},\ \tau \in [t, t + p_j[\cap \mathcal{F}_g\} + p_j$
$\mathcal{S}_g = \mathcal{S}_{g-1} \cup \{j\}$

$F_{n+1} = \max_{h \in \mathcal{P}_{n+1}} \{F_h\}$

The serial SGS for activity lists plays an important role in classical machine scheduling where it is referred to as list scheduling (cf. Kim 1995 and Schutten 1996). Since the serial SGS for activity lists is a special case of the serial SGS, it generates active schedules. Hence, there is always a list λ^* for which list scheduling will generate an optimal schedule when a regular measure of performance is considered.

7.2.2 *Parallel Schedule Generation Scheme*

The parallel scheduling scheme does time incrementation. For each iteration g there is a schedule time t_g. Activities which have been scheduled up to g are either element of the complete set $\mathcal{C}_g$ or of the active set $\mathcal{A}_g$. The complete set comprises all activities which have been completed up to t_g, i.e. $\mathcal{C}_g = \{j \in \mathcal{J} \mid F_j \leq t_g\}$ and the active set comprises all activities which are active at t_g, i.e. $\mathcal{A}_g = \mathcal{A}(t_g) = \{j \in \mathcal{J} \mid F_j - p_j \leq t < F_j\}$. The eligible set $\mathcal{D}_g$ comprises all activities which can be precedence– and resource–feasibly started at t_g, i.e. $\mathcal{D}_g = \left\{j \in \mathcal{J} \setminus (\mathcal{C}_g \cup \mathcal{A}_g) \mid \mathcal{P}_j \subseteq \mathcal{C}_g \wedge r_{j,k} \leq \tilde{R}_k(t_g)\ (k \in \mathcal{K})\right\}$. The remaining capacity at t_g is $\tilde{R}_k(t_g) = R_k - \sum_{j \in \mathcal{A}_g} r_{j,k}$. An algorithmic description of the parallel SGS can be given as follows:

Parallel SGS

Initialization: $g = 0, t_g = 0, \mathcal{A}_0 = \{0\}, \mathcal{C}_0 = \{0\}, \tilde{R}_k(0) = R_k$
While $\mid \mathcal{A}_g \cup \mathcal{C}_g \mid \leq n$ do

(1) $g := g + 1$
$t_g = \min_{j \in \mathcal{A}_g} \{F_j\}$
Calculate $\mathcal{C}_g$, $\mathcal{A}_g$, $\tilde{R}_k(t_g)$, $\mathcal{D}_g$

(2) `While` $\mathcal{D}_g \neq \emptyset$ `do`

Select one $j \in \mathcal{D}_g$
$F_j = t_g + p_j$
Calculate $\tilde{R}_k(t_g)$, $\mathcal{A}_g$, $\mathcal{D}_g$

$F_{n+1} = \max_{h \in \mathcal{P}_{n+1}} \{F_h\}$

The initilization sets the schedule time to 0, assigns the project start activity to the active and the complete set and sets the available capacity. Each iteration consists of two steps. (1) determines the next schedule time t_g, the associated activity sets $\mathcal{C}_g$, $\mathcal{A}_g$, $\mathcal{D}_g$ and the available capacity $\tilde{R}_k(t_g)$. (2) schedules a subset of the eligible activities to start at t_g. Table 7:2 reports the parallel SGS when generating the schedule given in Figure 7:2. Note that the parallel SGS might have less than n stages but that there are exactly n selection decisions which have to be made.

g	1	2	3	3	4	5	6
t_g	0	4	6	6	9	10	12
$\mathcal{D}_g$	{1,2}	{1,4}	{1,6}	{6}	{}	{3}	{5}
j	2	4	1	6		3	5

Table 7:2 Example for parallel SGS

As the serial, so does the parallel SGS always generate feasible schedules which are optimal for the resource–unconstrained case. It has been shown by Kolisch (1996a) that the parallel SGS constructs non–delay schedules. A non–delay schedule is a schedule where, even if activity preemption is allowed, none of the activities can be started earlier without delaying some other activity. The set of non–delay schedules is a subset of the set of active schedules. It thus has, on average, a smaller cardinality. But it has the severe drawback that it might not contain an optimal schedule with a regular performance measure. E.g., in Kolisch (1996a) it is shown that out of 298 problem instances from the set j30sm as given in Chapter 9 of this book (cf. Kolisch et al. 1998) only 175, i.e. 59.73 % have an optimal solution which is in the set of non–delay schedules. The time complexity of the parallel SGS is $\mathcal{O}(n^2 \cdot K)$.

7.3 PRIORITY RULE BASED HEURISTICS

Priority rule based heuristics employ one or both of the SGS in order to construct one or more schedules. The priority rule itself is used in order to select an activity j from the decision set $\mathcal{D}_g$. We first give a survey of different priority rules. Thereafter we show how scheduling schemes and priority rules can be combined in order to obtain different priority rule based heuristics.

7.3.1 Priority Rules

A priority rule is a mapping which assigns each activity j in the decision set $\mathcal{D}_g$ a value $v(j)$ and an objective stating whether the activity with the minimum or the maximum value is selected. In case of ties, one or several tie breaking rules have to be employed. The easiest ways to resolve ties is to choose the activity with the smallest activity label. There has been an overwhelming amount of research on priority rules for the RCPSP; cf. Alvarez–Valdés and Tamarit (1989a), Boctor (1990), Cooper (1976, 1977), Davies (1973), Davis and Patterson (1975), Elsayed (1982), Kolisch (1996a, 1996b), Lawrence (1985), Özdamar and Ulusoy (1994, 1996b), Patterson (1973, 1976), Shaffer et al. (1965), Thesen (1976), Thomas and Salhi (1997), Ulusoy and Özdamar (1989), Valls et al. (1992), and Whitehouse and Brown (1979).

Priority rules can be classified according to different criteria. W.r.t. to the type of information employed to calculate $v(j)$, we can distinguish network, time, and resource based rules (cf. Alvarez–Valdés and Tamarit 1989a and Lawrence 1985) as well as lower and upper bound rules. Lower bound rules calculate for each activity a lower bound of the objective function value, upper bound rules calculate for each activity an upper bound of the objective function value. W.r.t. the amount of information employed it can be distinguished in local or global rules. Local rules employ only information from the activity under consideration such as the processing time while global rules make use of a wider range of information. A distinction into static and dynamic rules is made w.r.t. the fact if the value $v(j)$ remains constant or changes during the iterations of the SGS. W.r.t. the SGS we can distinguish in rules which can be employed in the serial, the parallel or both SGS. Table 7:3 gives an overview of the well known priority rules GRPW (greatest rank positional weight), LFT (latest finish time), LST (latest start time), MSLK (minimum slack), MTS (most total successors), RSM (resource scheduling method), SPT (shortest processing time), and WCS (worst case slack). The MTS rule employs $\overline{\mathcal{S}}_j$, the set of all (direct and indirect) successors of activity j. The WCS and RSM rules employ $\mathcal{AP} = \{(i,j) \in \mathcal{D}_g \times \mathcal{D}_g \mid i \neq j\}$, the set of all activity pairs (i,j) which are in the decision set $\mathcal{D}_g$. Finally, the WCS rule uses $E(i,j)$, the earliest precedence– and resource–feasible start time of activity j if activity i is started

Rule	Ref.	$v(j)$
GRPW	Alvarez–Valdés, Tamarit (1989a)	$p_j + \sum_{i \in \mathcal{S}_j} p_i$
LFT	Davis, Patterson (1975)	LF_j
LST	Kolisch (1995)	$LF_j - p_j$
MSLK	Davis, Patterson (1975)	$LF_j - EF_j$
MTS	Alvarez–Valdés, Tamarit (1989a)	$\mid \overline{\mathcal{S}}_j \mid$
RSM	Shaffer et al. (1965)	$\max_{(i,j) \in \mathcal{AP}} \{0, t_g + p_j$ $-(LF_i - p_i)\}$
SPT	Alvarez–Valdés, Tamarit (1989a)	p_j
WCS	Kolisch (1996b)	$LF_j - p_j$ $- \max_{(i,j) \in \mathcal{AP}} \{E(i,j)\}$

Table 7:3 Priority Rules

at the schedule time t_g. Note that the RSM rule selects the activity with the lowest priority value.

7.3.2 Proposed Methods

Priority rule based heuristics combine priority rules and schedule generation schemes in order to construct a specific algorithm. If the heuristic generates a single schedule, it is called a single pass method, if it generates more than one schedule, is is referred to as multi pass method.

Single Pass Methods. The oldest heuristics for the RCPSP are *single pass methods* which employ one SGS and one priority rule in order to obtain one feasible schedule. Examples are the heuristics of Alvarez–Valdés and Tamarit (1989a), Boctor (1990), Cooper (1976, 1977), Davies (1973), Davis and Patterson (1975), Elsayed (1982), Kolisch (1996a, 1996b), Lawrence (1985), Patterson (1973, 1976), Thesen (1976), Valls et al. (1992), and Whitehouse and Brown (1979).

Recently, more elaborate priority rules have been proposed by Kolisch (1996b), Özdamar and Ulusoy (1996b) as well as Ulusoy and Özdamar (1994). Kolisch (1996b) developed, amongst other priority rules, the so–called worst case slack (WCS) rule for the parallel SGS which is given in Table 7:3. Özdamar and Ulusoy (1996b) as well as Ulusoy and Özdamar (1994) introduced the local constraint based analysis (LCBA). LCBA employs the parallel SGS and decides via feasibility checks and so–called essential conditions which activities have to be selected and which activities have to be delayed at the schedule time.

Multi Pass Methods. There are many possibilities to combine SGS and priority rules to a multi pass method. The most common ones are multi priority rule methods, forward–backward scheduling methods, and sampling methods.

Multi priority rule methods employ the SGS several times. Each time a different priority rule is used. Often, the rules are used in the order of descending solution quality. Boctor (1990) employed 7 different rules in his experimental study. Instead of using m priority rules in order to generate m schedules, a virtually unlimited number of schedules can be generated by using convex combinations of I priority rules $v(j) = \sum_{i=1}^{I} w_i \cdot v_i(j)$ with $w_i \geq 0$ for all i and $\sum_{i=1}^{I} w_i = 1$. Examples of such approaches are given by Ulusoy and Özdamar (1989) and Thomas and Salhi(1997). Ulusoy and Özdamar employed a convex combination of $I = 2$ rules in order to generate 10 different schedules.

Forward–backward scheduling methods employ an SGS in order to iteratively schedule the project by alternating between forward and backward scheduling. Forward scheduling is as outlined in Section 7.2. Backward scheduling applies one of the SGS to the reversed precedence network where the former end activity $n+1$ has become the new start activity. The priority values are usually obtained from the start or completion times of the lastly generated schedule. Forward–backward scheduling methods have been proposed by e.g. Li and Willis (1992) as well as Özdamar and Ulusoy (1996a, 1996b).

Sampling methods make generally use of one SGS and one priority rule. Different schedules are obtained by biasing the selection of the priority rule through a random device. Instead of a priority value $v(j)$ a selection probability $p(j)$ is computed. At selection decision g of the SGS, $p(j)$ is the probability that activity j from the decision set $\mathcal{D}_g$ will be selected. Dependent on how the probabilities are computed, one can distinguish random sampling, biased random sampling, and regret based biased random sampling (cf. Kolisch 1996a). *Random sampling (RS)* assigns each activity in the decision set the same probability $p(j) = 1/|\mathcal{D}_g|$. *Biased random sampling (BRS)* employs the priority values directly in order to obtain the selection probabilities. If the objective of the priority rule is to select the activity with the highest priority value, then the probability is calculated to $p(j) = v(j)/\left(\sum_{i \in \mathcal{D}_g} v(i)\right)$. Biased random sampling methods have been applied by Alvarez–Valdés and Tamarit (1989b) and Cooper (1976). Schirmer and Riesenberg (1997) propose a modification called normalized biased random sampling (NBRS) which essentially ensures that the selection probability of the activity with the smallest (highest) priority value is the same when seeking for the activity with the highest (smallest) priority value. *Regret based biased random sampling (RBRS)* uses the priority values indirectly via regret values. If the objective is again to select the activity with the highest priority value, the regret value $r(j)$ is the absolute difference between the priority value $v(j)$ of the activity under consideration and the worst priority value of all activities in the decision set, i.e. $r(j) = v(j) - \min_{i \in \mathcal{D}_g} v(i)$. Before calculating the selection probabilities based on the regret values, the latter can be modified by $r'(j) = (r(j) + \epsilon)^\alpha$ (cf. Drexl 1991). Adding the constant $\epsilon > 0$ to the regret value assures that the selection probability for each activity in the decision set is greater than zero and thus every schedule of the population can be generated. With the choice of the parameter α the amount of bias can be controlled. A high α will cause no bias and thus deterministic activity selection while an α of zero will cause maximum bias and hence random activity selection. Kolisch (1995) has found out that, in general, $\epsilon = \alpha = 1$ will provide good results. Drexl (1991) uses $\epsilon = \min_{i \in \mathcal{D}_g} v(i)$. Schirmer and Riesenberg (1997) propose a modified regret based biased random sampling (MRBRS) where ϵ is determined dynamically. Experimental comparisons performed by Kolisch (1995) as well as Schirmer and Riesenberg (1997) revealed (modified) regret based biased random sampling as the best sampling approach. Table 7:4 gives examplary different selection probabilities for $|\mathcal{D}_g| = 3$, $\epsilon = 1$, and $\alpha = 1$.

$j \in \mathcal{D}_g$ $v(j)$	1 11	2 13	3 20
Random Sampling	0.33	0.33	0.33
Biased Random Sampling	0.25	0.30	0.45
Regret Based Biased Random Sampling	0.07	0.21	0.72

Table 7:4 Selection Probabilities $p(j)$ for Different Sampling Methods

An adaptive multi pass approach has been proposed by Kolisch and Drexl (1996). The heuristic applies the serial SGS with the LFT–priority rule and the parallel SGS with the WCS–priority rule while employing deterministic and regret based sampling activity selection. The decision on the specific method is based on an analysis of the problem at hand and the number of iterations already performed. Partial schedules are discarded by the use of lower bounds. Schirmer (1998) has extended this approach by employing both schedule generation schemes together with four different priority rules (MTS,LFT,LST,WCS) and two different sampling schemes (MRBRS,RBRS).

Author(s)	SGS	Priority Rule	Sampling	Passes
Alvarez–Valdés/Tamarit (1989a)	p	several rules	—	single
Alvarez–Valdés/Tamarit (1989b)	p	several rules	BRS	multi
Boctor (1990)	p/s	several rules	—	multi
Cooper (1976)	s	several rules	BRS	multi
Davis/Patterson (1975)	p	LFT and other	—	single
Kolisch (1995)	p/s	several rules	RS/BRS/RBRS	multi
Kolisch (1996a)	p/s	several rules	RBRS	multi
Kolisch (1996b)	p	WCS and other	—	single
Kolisch/Drexl (1996)	p/s	LFT/WCS	RBRS	multi
Li/Willis (1992)	p	start/fin. times	—	multi
Özdamar/Ulusoy (1994)	p	LCBA	—	single
Özdamar/Ulusoy (1996a, 1996b)	p	LCBA	—	multi
Shaffer et al. (1965)	p	RSM	—	single
Schirmer/Riesenberg (1997)	p/s	several rules	several methods	multi
Schirmer (1998)	p/s	several rules	RBRS/MRBRS	multi
Thomas/Salhi (1997)	p	convex comb.	—	multi
Ulusoy/Özdamar (1989)	p	convex comb.	—	multi

Table 7:5 Survey of priority rule based heuristics for the RCPSP

Table 7:5 gives a survey of priority rule based heuristics for the RCPSP where with 'p' and 's' it is referred to the parallel and the serial SGS, respectively. Note that this convention holds for all tables in this chapter.

7.4 METAHEURISTIC APPROACHES

7.4.1 General Metaheuristic Strategies

Several metaheuristic strategies have been developed to solve hard optimization problems. The following summary briefly describes those general aproaches that have been used to solve the RCPSP.

Simulated Annealing. *Simulated Annealing (SA)*, introduced by Kirkpatrick et al. (1983), originates from the physical annealing process in which a melted solid is cooled down to a low–energy state. Starting with some initial solution, a so–called neighbor solution is generated by slightly perturbing the current one. If this new solution is better than the current one, it is accepted, and the search proceeds from this new solution. Otherwise, if it is worse, the new solution is only accepted with a probability that depends on the magnitude of the deterioration as well as on a parameter called temperature. As the algorithm proceeds, this temperature is reduced in order to lower the probability to accept worse neighbors.

Clearly, SA can be viewed as an extension of a simple greedy procedure, sometimes called *First Fit Strategy (FFS)*, which immediately accepts a better neighbor solution but rejects any deterioration.

Tabu Seach. *Tabu Search (TS)*, developed by Glover (1989a, 1989b), is essentially a steepest descent/mildest ascent method. That is, it evaluates all solutions of the neighborhood and chooses the best one, from which it proceeds further. This concept, however, bears the possibility of cycling, that is, one may always move back to the same local optimum one has just left. In order to avoid this problem, a tabu list is set up as a form of memory for the search process. Usually, the tabu list is used to forbid those neighborhood moves that might cancel the effect of recently performed moves and might thus lead back to a recently visited solution. Typically, such a tabu status is overrun if the corresponding neighborhood move would lead to a new overall best solution (aspiration criterion).

It is obvious that TS extends the simple steepest descent search, often called *Best Fit Strategy (BFS)*, which scans the neighborhood and then accepts the best neighbor solution, until none of the neighbors improves the current objective function value.

Genetic Algorithms. *Genetic Algorithms (GA)*, inspired by the process of biological evolution, have been introduced by Holland (1975). In contrast to the local search strategies above, a GA simultaneously considers a set or population of solutions instead of only one. Having generated an initial population, new solutions are produced by mating two existing ones (crossover) and/or by altering an existing one (mutation). After producing new solutions, the fittest solutions "survive" and make up the next generation while the others are deleted. The fitness value measures the quality of a solution, usually based on the objective function value of the optimization problem to be solved.

7.4.2 Representations

Once a metaheuristic strategy has been chosen to attack a given optimization problem, one has to select a suitable representation for solutions. Usually, metaheuristic approaches for the RCPSP rather operate on representations of schedules than on schedules themselves. Then, an appropriate decoding procedure must be selected to transform the representation into a schedule. Finally, operators are needed to produce new solutions w.r.t. the selected representation. A unary operator constructs a new solution from an existing one, that is, it makes up the neighborhood move in local search procedure such as SA and TS as well as the mutation in a GA. A binary operator constructs a new solution from two existing ones, as done by crossover in a GA.

This subsection summarizes five representations reported in the literature that have been used within metaheuristic approaches to solve the RCPSP. For each representation, we give the related decoding procedures and operators. In order to keep the description short, we will restrict the definition of binary operators to the one–point crossover type for the different representations. Roughly speaking, the one–point crossover splits two existing solutions and takes one part from the first and one part from the second solution in order to form a new one. Other general crossover types that are well known from the GA literature (such as two–point and uniform crossover) can be easily obtained from extending the one–point definitions given here, see e.g. Hartmann (1998).

Activity List Representation. In the activity list representation, a precedence feasible activity list

$$\lambda = \langle j_1, j_2, \ldots, j_n]$$

is given, in which each activity j_g must have a higher index g than each of its predecessors in $\mathcal{P}_{j_g}$. As shown in Section 7.2, the serial SGS can be used as a decoding procedure to obtain a schedule from an activity list. Note, however, that the parallel scheme cannot be applied without modification. As is easily verified, the serial SGS transforms example activity list

$$\lambda^E = \langle 2, 4, 6, 1, 3, 5]$$

for the project of Figure 7:1 into the schedule of Figure 7:2. The initial solution(s) can be generated by randomly selecting an activity from the decision set in each step of the SGS. To obtain better solution quality, one can also use a priority rule or priority rule based sampling scheme for choosing an eligible activity. In either case, recording the activities in the order of their selection results in a (precedence feasible) activity list.

Several unary operators have been proposed for the activity list representation, see e.g. Della Croce (1995). The so-called pairwise interchange is defined as swapping two activities j_q and j_s, $q, s \in \{1, \ldots, n\}$ with $q \neq s$, if the resulting activity list is precedence feasible. As a special case, the adjacent pairwise interchange swaps two activities j_q and j_{q+1}, $q \in \{1, \ldots, n-1\}$, that are adjacent in λ but not precedence related. Considering again the example project of Figure 7:1, we could apply the adjacent pairwise interchange for $q = 3$ to λ^E as given above and obtain

$$\lambda^N = \langle 2, 4, 1, 6, 3, 5] .$$

Furthermore, the simple shift operator selects some activity j_q and inserts it immediately after some other activity j_s, if the precedence constraints are not violated. In our example, shifting activity 6 immediately after activity 3 in λ^E results in neighbor activity list

$$\lambda^{N'} = \langle 2, 4, 1, 3, 6, 5] .$$

More sophisticated shift operators have been proposed by Baar et al. (1997) for the RCPSP. They make use of the schedule $S(\lambda)$ that is represented by the current activity list λ. The operators are based on the notion of a critical arc which is defined as a pair of successively scheduled activities (i, j), that is, $F_i + p_j = F_j$ in $S(\lambda)$. The underlying idea is that at least one critical arc must become non–critical to improve the current schedule. Hence, Baar et al. define three shift operators that may cancel a critical arc. They extend the simple shift by allowing more than one activity to be shifted. Without giving the formal definitions here, we illustrate such a shift operator on the critical arc (4,1) in the schedule $S(\lambda^E)$ shown in Figure 7:2: Shifting activity 4 and its successor activity 6 immediately after activity 1 leads to neighbor activity list

$$\lambda^{N''} = \langle 2, 1, 4, 6, 3, 5]$$

As a binary operator, i.e. as crossover for a GA, Hartmann (1998) used the following technique: Given two "parent" activity lists, a "mother" $\lambda^M = \langle j_1^M, \ldots, j_n^M]$ and a "father" $\lambda^F = \langle j_1^F, \ldots, j_n^F]$, the "child" activity list $\lambda^C = \langle j_1^C, \ldots, j_n^C]$ is defined as follows: After drawing a random integer q with $1 \leq q < n$, the positions $i = 1, \ldots, q$ are taken from the "mother" λ^M by setting $j_i^C := j_i^M$. The activity list of positions $i = q+1, \ldots, n$ is taken from the "father" λ^F. However, the jobs that have already been taken from the mother may not be considered again. We obtain $j_i^C := j_k^F$, where k is the lowest index such that $j_k^F \notin \{j_1^C, \ldots, j_{i-1}^C\}$. Choosing $q = 3$, this definition is illustrated for our project example by

$$\begin{aligned} \lambda^M &= \langle 1, 3, 2, 5, 4, 6] , \\ \lambda^F &= \langle 2, 4, 6, 1, 3, 5] , \\ \lambda^C &= \langle 1, 3, 2, 4, 6, 5] . \end{aligned}$$

As shown by Hartmann (1998), the "child" activity list λ^C resulting from two precedence feasible "parent" activity lists λ^M and λ^F is also precedence feasible.

Random Key Representation. Several researchers employed a representation which makes use of an array

$$\rho = (r_1, r_2, \ldots, r_n)$$

that assigns a (typically real-valued) number r_j to each activity j. Following Bean (1994), we will call such an encoding random key representation. It is similar to the priority value representation of Lee and Kim (1996) and Cho and Kim (1997) and the problem-space based representation of Storer et al. (1992), Leon and Ramamoorthy (1995), and Naphade et al. (1997), such that we will discuss these three approaches in a unified framework here.

For an initial solution, the random keys are usually chosen randomly (see e.g. Lee and Kim 1996) or computed by some priority rule (see e.g. Leon and Ramamoorthy 1995).

Both the parallel and the serial SGS can be used to derive a schedule from ρ: On each stage g, we can select activity j with the highest random key $r_j = \max\{r_i \,|\, i \in \mathcal{D}_g\}$ from the decision set $\mathcal{D}_g$ (clearly, if initialized with e.g. the latest finish time, one would select the activity with the minimum random key). In other words, the random keys play the role of priority values. Considering again our example project, we obtain the schedule of Figure 7:2 from applying either the parallel or the serial SGS to

$$\rho^E = (0.58, 0.64, 0.31, 0.87, 0.09, 0.34)\,.$$

An alternative approach is proposed by Naphade et al. (1997). Here, the random keys are used to perturb activity slacks which serve as priority values.

While both the parallel and the serial SGS as decoding procedures guarantee that only feasible schedules are found, only the serial one ensures the existence of at least one optimal schedule in the solution space, as discussed in Section 7.2. In order to overcome the drawback of possible exclusion of all optimal solutions by the parallel SGS, several researchers (see Cho and Kim 1997, Naphade et al. 1997, and Leon and Ramamoorthy 1995) introduced different modifications of the parallel SGS as decoding procedures for the random key representation. These essentially allow to delay a schedulable activity such that the search is not restricted to non–delay schedules.

As a unary operator, any pairwise interchange of r_j and r_i can be employed to the random key representation, including the adjacent pairwise interchange of r_j and r_{j+1}. Considering a pairwise interchange with $j = 2$ and $i = 4$, an example neighbor of ρ^E is

$$\rho^N = (0.58, 0.87, 0.31, 0.64, 0.09, 0.34)\,.$$

In an approach for the job shop problem, Storer et al. (1992) proposed the so–called problem–space based neighborhood which randomly reselects $r_j^{new} \in$

$[(1-\epsilon)\cdot r_j^{old}, (1+\epsilon)\cdot r_j^{old}]$ from a uniform distribution, where ϵ is a real-valued constant. For this neighborhood definition with $\epsilon = 0.1$ an example neighbor of ρ^E is given by

$$\rho^{N'} = (0.59, 0.62, 0.34, 0.89, 0.09, 0.33).$$

The random key representation allows the application of the standard one-point crossover as binary operator: Given a random integer q with $1 \le q < n$, a new random key array $\rho^C = (r_1^C, \ldots, r_n^C)$ is derived by taking the first q random keys from a "mother" array $\rho^M = (r_1^M, \ldots, r_n^M)$ and the remaining ones from a "father" array $\rho^F = (r_1^F, \ldots, r_n^F)$. We obtain $r_i^C = r_i^M$ for $i = 1, \ldots, q$ and $r_i^C = r_i^F$ for $i = q+1, \ldots, n$. An example for $q = 3$ is

$$\begin{aligned}
\rho^M &= (0.58, 0.64, 0.31, 0.87, 0.09, 0.34),\\
\rho^F &= (0.12, 0.43, 0.99, 0.65, 0.19, 0.22),\\
\rho^C &= (0.58, 0.64, 0.31, 0.65, 0.19, 0.22).
\end{aligned}$$

Priority Rule Representation. The priority rule representation, used by e.g. Dorndorf and Pesch (1994) for the job shop problem and adapted by Hartmann (1998) to the RCPSP, is based on a list of priority rules

$$\pi = \langle \pi_1, \pi_2, \ldots, \pi_n],$$

where each π_i is a priority rule. As decoding procedures, both the parallel and the serial SGS can be used by selecting the i-th activity to be scheduled according to priority rule π_i. For our example project, the schedule of Figure 7:2 can be obtained from e.g.

$$\pi^E = \langle \text{LST, GRPW, MTS, LST, MSLK, LFT}].$$

The commonly used unary operator randomly selects a new priority rule for some π_i. For the example priority list π^E as given above, a possible neighbor is

$$\pi^N = \langle \text{MTS, GRPW, MTS, LST, MSLK, LFT}].$$

The binary operator, i.e. the one-point crossover, follows again the standard definition. Given a random integer q with $1 \le q < n$, a new priority rule list π^C is derived by taking the first q priority rules from a "mother" list π^M and the remaining ones from a "father" list π^F. Consider as an example $q = 3$ and

$$\begin{aligned}
\pi^M &= \langle \text{LST, GRPW, MTS, LST, MSLK, LFT}],\\
\pi^F &= \langle \text{LFT, GRPW, MSLK, SPT, LFT, GRPW}],\\
\pi^C &= \langle \text{LST, GRPW, MTS, SPT, LFT, GRPW}].
\end{aligned}$$

Shift Vector Representation. The shift vector representation has been proposed by Sampson and Weiss (1993) for the RCPSP. A solution is represented by a shift vector

$$\sigma = (\sigma_1, \sigma_2, \ldots, \sigma_n),$$

where σ_j is a nonnegative integer. As a decoding procedure, an extension of the classical forward recursion (cf. Elmaghraby 1977) is used, in which the start time S_j of an activity j is calculated as the maximum of the finish times of its predecessors plus the shift σ_j of activity j, that is, $S_0 = 0$ and $S_j = \max\{S_h + p_h \mid h \in \mathcal{P}_j\} + \sigma_j$ for $j = 1, \ldots, n$. The following shift vector for our example project leads to the schedule of Figure 7:2:

$$\sigma^E = (6, 0, 1, 0, 0, 0)$$

As this decoding procedure does not consider the resource constraints, a schedule derived from a shift vector may be infeasible. This is illustrated by the following shift vector which forces activities 1 and 4 to be simultaneously in process, thus exceeding the resource capacity by 2 units:

$$\sigma^I = (4, 0, 1, 0, 0, 0)$$

Consequently, the objective function is extended by penalizing the violation of the resource constraints.

The neighborhood of a shift vector σ is given by those vectors that differ from σ in exactly one position and do not lead to a project makespan that exceeds some given upper bound. An example neighbor for shift vector σ^E above is

$$\sigma^N = (6, 0, 1, 0, 1, 0).$$

Schedule Scheme Representation. The schedule scheme representation has been introduced by Brucker et al. (1998) for a branch–and–bound algorithm for the RCPSP. In what follows we give a brief and rather informal description. A schedule scheme (C, D, N, F) consists of four disjoint relations. $(i, j) \in C$ implies that activity i must be finished before activity j can be started (conjunctions). $(i, j) \in D$ implies that activities i and j may not overlap (disjunctions). $(i, j) \in N$ implies that activities i and j must be processed in parallel in at least one period (parallelity relations). For activities i and j with $(i, j) \in F$ there are no restrictions (flexibility relations). A schedule scheme represents those (not necessarily feasible) schedules in which the related relations are maintained. As a decoding procedure, Baar et al. (1997) develop a heuristic that constructs a feasible schedule in which all relations of C and D and a "large" number of parallelity relations N are satisfied.

Baar et al. (1997) introduce a neighborhood definition which basically consists of moves that transform flexibility relations into parallelity relations and parallelity relations into flexibility relations. The neighborhood size is reduced by a critical path calculation and impact estimations for the moves.

We do not discuss this representation and its neighborhood definition in more detail here, referring the reader to Chapter 2 in this book (cf. Brucker and Knust 1998).

7.4.3 Proposed Methods

In this subsection, we briefly describe the metaheuristic approaches which are documented for the RCPSP. The references are listed in alphabetical order. We have, however, restricted the list to those papers that consider the RCPSP, while papers that cover its extensions (such as the multi-mode RCPSP) are not included. A summarizing overview is given in Table 7:6. For each reference, we denote the employed metaheuristic strategies together with the underlying representation, the SGS, and the used operator. In case of a GA, we only mention the crossover operator. The representation given in the third column is abbriaviated as follows: AL stands for activity list, RK for random key, PR for priority rule, SV for shift vector, and SS for schedule scheme.

Baar et al. (1997) develop two TS algorithms. The first one is based on the activity list representation in accordance with the serial SGS. The neighborhood is defined by three kinds of critical path based moves. Their second TS approach employs the schedule scheme representation with the related decoding procedure and neighborhood definition. Both TS algorithms use dynamic tabu lists as well as priority based start heuristics.

Bouleimen and Lecocq (1998) propose an SA procedure based on the activity list representation together with the serial SGS. For neighborhood moves, the shift operator is used.

Cho and Kim (1997) modify the SA algorithm of Lee and Kim (1996) (see below) by extending the random key representation in order to allow the delay of schedulable activities within an adapted parallel SGS.

Hartmann (1998) proposes a GA based on the activity list representation and compares it to GA's which make use of the random key and priority rule representations, respectively. All three approaches employ the serial SGS as well as two-point crossover operators related to the respective representation. In the activity list based GA, the regret based biased random sampling method with the serial SGS and the LFT-rule (see Section 7.3) is used to determine the initial generation.

Paper	Metah.	Rep.	SGS	Operator
Baar et al. (1997)	TS	AL	s	crit. path shifts
	TS	SS	related	related moves
Bouleimen, Lecocq (1998)	SA	AL	s	shift
Cho, Kim (1997)	SA	RK	mod. p	pairw. int.
Hartmann (1998)	GA	AL	s	two–point cr.
	GA	RK	s	two–point cr.
	GA	PR	s	two–point cr.
Kohlmorgen et al. (1998)	GA	RK	s	two–point cr.
Lee, Kim (1996)	SA	RK	p	pairw. int.
	TS	RK	p	pairw. int.
	GA	RK	p	one–point cr.
Leon, Ramamoorthy (1995)	FFS	RK	mod. p	problem–space
	BFS	RK	mod. p	problem–space
	GA	RK	mod. p	one–point cr.
Naphade et al. (1997)	BFS	RK	mod. p	problem–space
Pinson et al. (1994)	TS	AL	s	adj. pairw. int.
	TS	AL	s	pairw. int.
	TS	AL	s	shift
Sampson, Weiss (1993)	SA-var.	SV	ext. recurs.	related move

Table 7:6 Survey of metaheuristic strategies for the RCPSP

Kohlmorgen et al. (1998) develop a GA which employs the random key representation with standard two–point crossover. They test their approach on a massively parallel computer.

Lee and Kim (1996) propose an SA algorithm, a TS procedure and a GA. These three approaches are based on the random key representation with the parallel SGS as decoding procedure. While SA and TS make use of a restricted version of the pairwise interchange move, the GA employs the standard one–point crossover.

Leon and Ramamoorthy (1995) test an FFS and a BFS approach as well as a GA. They employ the problem–space based version of the random key representation. The random keys are initialized with values computed by a priority rule. A modified variant of the parallel SGS serves as decoding procedure. The unary operator is defined by the problem–space based neighborhood, and the binary one is the standard one–point crossover.

Naphade et al. (1997) use the BFS concept with the problem–space based variant of the random key representation. The random keys are initialized with the latest finish times of the activities and then modified according to the problem–space based neighborhood. As decoding procedure, they employ a modified parallel SGS, where the random keys are used to calculate slack based priority values.

Pinson et al. (1994) suggest several variants of a TS approach based on the activity list representation, the serial SGS, and a priority rule procedure for computing a start solution. The variants differ in the neighborhood definitions, using the adjacent pairwise interchange, the (general) pairwise interchange, and the shift move, respectively.

Sampson and Weiss (1993) suggest a local search procedure which can be viewed as a variant of SA. Their approach is based on the shift vector representation and the related neighborhood definition.

7.5 OTHER HEURISTICS

7.5.1 Truncated Branch and Bound Methods

Pollack–Johnson (1995) uses a so–called depth–first, jumptracking branch and bound search of a partial solution tree. The algorithm is essentially a parallel scheduling heuristic. Instead of scheduling the activity with the highest priority value it branches on certain occasions such that one branch has the activity with the highest priority value and the other branch has the activity with the second highest priority value, which is scheduled next. Note, due to use of the parallel SGS optimal solution might be excluded from the search space.

Sprecher (1996) employs his depth–first search branch and bound procedure as a heuristic by imposing a time limit. The enumeration process is guided by the so–called precedence tree which essentially branches on the activities in the decision set of the serial SGS. Via backtracking, all precedence feasible activity lists are (implicitly) enumerated. In order to obtain good solutions early in the search process (and thus within the time limit), priority rules are applied to select the most promising activity from the decision set for branching first.

7.5.2 Disjunctive Arc Based Methods

The basic idea of the disjunctive–arc–based approaches is to extend the precedence relations (the set of conjunctive arcs) by adding additional arcs (the disjunctive arcs) such that the minimal forbidden sets, i.e. sets of technologically independent activities which cannot be scheduled simultaneously due to resource constraints, are destroyed and thus the earliest finish schedule is feasible with respect to (precedence and) resource constraints.

Shaffer et al. (1965) restrict the scope, within their "resource scheduling method", to those forbidden sets for which all activities in the earliest finish schedule are processed at the same time. The disjunctive arc which produces the smallest increase in the earliest finish time of the unique sink is introduced and the earliest finish schedule is recalculated. The algorithm terminates as soon as a (precedence– and) resource–feasible earliest finish schedule is found. Note that this approach can be transformed into a single-pass priority rule method based on the parallel SGS, cf. Tables 7:3 and 7:5.

Alvarez–Valdés and Tamarit (1989a) propose four different ways of destroying the minimal forbidden sets. The best results were achieved by applying the following strategy: Beginning with the minimal forbidden sets of lowest cardinality, one set is arbitrarily chosen and destroyed by adding the disjunctive arc for which the earliest finish time of the unique dummy sink is minimal.

Bell and Han (1991) present a two–phase algorithm for this problem. The first phase is very similar to the approach of Shaffer et al. (1965). However, phase 2 tries to improve the feasible solution obtained by phase one as follows: after removing redundant arcs, each disjunctive arc that is part of the critical path(s) is temporarily cancelled and the phase 1 procedure is applied again.

7.5.3 Further Approaches

Integer programming based heuristics have been used by Oğuz and Bala (1994). The method employs the integer programming formulation originally proposed by Pritsker et al. (1969). The planning horizon is divided in T periods of equal length and the processing times p_j have to be given as discrete multiples of one period. The binary decision variable $x_{j,t}$ equals 1 if activity j is finished at the end of period t.

Mausser and Lawrence (1998) use block structures to improve the makespan of projects. They start by generating a feasible solution with the parallel SGS. Following this, they identify blocks which represent contiguous time spans that completely contain all activities processed within it. Each such block can be considered independent of the other blocks. The method essentially reschedules individual blocks in order to shorten the overall project length.

7.6 COMPUTATIONAL ANALYSIS

7.6.1 Test Design

This section reports on a computational comparison of several of the heuristics summarized above. As test instances we have employed the standard sets j30, j60, and j120 for the RCPSP which have been generated using ProGen (cf. Kolisch et al. 1995). The sets j30 and j60 consist of 480 projects with $n = 30$ and $n = 60$ activities, respectively. The set j120 consists of 600 projects, each with $n = 120$ activities. Details of these problem instances are given in Chapter 9 of this book (cf. Kolisch et al. 1998).

Each algorithm was tested by its author(s) using the original implementation. This allowed the authors to adjust the parameters in order to obtain good results. As a consequence, however, the tests were performed on different computer architectures and operating systems. Therefore, we could not impose a bound on the computation time to provide a basis for the comparison. Instead, we have chosen to limit the number of generated and evaluated schedules to 1,000 and 5,000, respectively. This decision is based on the assumption that the effort needed for generating one schedule is similar in the tested heuristics. With the exception of the schedule scheme representation based TS approach of Baar et al. (1997) all algorithms considered for our investigation make use of an SGS as described in Section 7.2. Hence, we found this assumption justified.

7.6.2 Results

Tables 7:7–7:11 display the results of the computational comparison. The heuristics are sorted according to descending performance with respect to 5,000 iterations. Table 7:7 summarizes the percentage deviations from the optimal makespan for the instance set j30. Since for the two instance sets j60 and j120 some of the optimal solutions are not known, we measured for these sets the average percentage deviation from an upper and a lower bound, respectively. The upper bound was set to the lowest makespan found by any of the tested heuristics while the lower bound was selected to be the critical path based lower bound (cf. Stinson et al. 1978). We employed the lower bound in order to allow researchers to compare their results with the ones obtained in this study. All lower and upper bounds can be obtained from the authors upon request. For the j60 set, the percentage deviations from the upper and lower bounds are reported in Tables 7:8 and 7:10, respectively. Note, that the schedule scheme based TS heuristic of Baar et al. (1997) was additionally run by allowing 2 trials where each trial was terminated after no improved solution was found after 250 iterations. This way, the deviation from the upper bound was lowered to 1.14 %. Finally, Tables 7:9 and 7:11 provide the respective deviations for the j120 set.

The heuristics that performed best in our study are the SA of Bouleimen and Lecocq (1998) and the GA of Hartmann (1998). While the procedure of Bouleimen and Lecocq performs best on the j30 set, the approach of Hartmann dominates on the instance sets with larger projects.

Generally, the results show that the best metaheuristic strategies outperform the best priority rule based sampling approaches. Increasing the number of schedules allowed to be computed from 1,000 to 5,000 further increases the

Algorithm	SGS	Reference	Iterations 1,000	5,000
SA – activity list	s	Bouleimen, Lecocq (1998)	0.38	0.23
GA – activity list	s	Hartmann (1998)	0.54	0.25
TS – sched. scheme	related	Baar et al. (1997)	0.86	0.44
sampling – adaptive	s/p	Kolisch, Drexl (1996)	0.74	0.52
sampling – LFT	s	Kolisch (1996a)	0.83	0.53
sampling – adaptive	s/p	Schirmer (1998)	0.71	0.59
sampling – WCS	p	Kolisch (1996b)	1.40	1.28
sampling – LFT	p	Kolisch (1996a)	1.40	1.29
GA – random key	mod. p	Leon, Ramamoorthy (1995)	2.08	1.59

Table 7:7 Average deviations from optimal solution — $J = 30$

Algorithm	SGS	Reference	Iterations 1,000	5,000
GA – activity list	s	Hartmann (1998)	0.99	0.45
SA – activity list	s	Bouleimen, Lecocq (1998)	1.17	0.49
sampling – adaptive	s/p	Schirmer (1998)	1.26	0.97
sampling – adaptive	s/p	Kolisch, Drexl (1996)	1.60	1.26
TS – sched. scheme	related	Baar et al. (1997)	1.79	1.54
sampling – LFT	s	Kolisch (1996a)	1.88	1.55
sampling – LFT	p	Kolisch (1996a)	1.83	1.56
sampling – WCS	p	Kolisch (1996b)	1.88	1.56
GA – random key	mod. p	Leon, Ramamoorthy (1995)	2.48	1.82

Table 7:8 Average deviations from best solution — $J = 60$

superiority of the metaheuristics. This is mainly because sampling procedures generate each schedule anew without considering any information given by already visited solutions while metaheuristic algorithms typically exploit the knowledge gained from the previously evaluated schedule(s).

A comparison of the results obtained from the metaheuristics shows that the choice of the underlying representation is crucial. The two best procedures make use of different metaheuristic paradigms while they both employ the

Algorithm	SGS	Reference	Iterations 1,000	5,000
GA – activity list	s	Hartmann (1998)	2.59	0.89
SA – activity list	s	Bouleimen, Lecocq (1998)	5.73	1.86
sampling – LFT	p	Kolisch (1996a)	2.92	2.32
sampling – WCS	p	Kolisch (1996b)	2.94	2.34
sampling – adaptive	s/p	Schirmer (1998)	3.28	2.55
sampling – adaptive	s/p	Kolisch, Drexl (1996)	3.95	3.33
GA – random key	mod. p	Leon, Ramamoorthy (1995)	5.33	3.76
sampling – LFT	s	Kolisch (1996a)	4.78	4.10

Table 7:9 Average deviations from best solution — $J = 120$

Algorithm	SGS	Reference	Iterations 1,000	5,000
GA – activity list	s	Hartmann (1998)	12.68	11.89
SA – activity list	s	Bouleimen, Lecocq (1998)	12.75	11.90
sampling – adaptive	s/p	Schirmer (1998)	13.02	12.62
sampling – adaptive	s/p	Kolisch, Drexl (1996)	13.51	13.06
sampling – WCS	p	Kolisch (1996b)	13.66	13.21
sampling – LFT	p	Kolisch (1996a)	13.59	13.23
TS – sched. scheme	related	Baar et al. (1997)	13.80	13.48
GA – random key	mod. p	Leon, Ramamoorthy (1995)	14.33	13.49
sampling – LFT	s	Kolisch (1996a)	13.96	13.53

Table 7:10 Average deviations from critical path based lower bound — $J = 60$

activity list representation. The use of one metaheuristic paradigm itself does not necessarily lead to consistently good solutions. This can be seen by the results of the two GA's of Hartmann (1998) and Leon and Ramamoorthy (1995). The activity list based GA excells the problem space based GA.

Analyzing the priority rule based sampling procedures, we observe a strong influence of the SGS when used together with the LFT rule. While the serial SGS leads to better results on the j30 instance set, the parallel one is superior on the j60 and j120 instance sets, respectively. The two rules WCS and

Algorithm	SGS	Reference	Iterations 1,000	5,000
GA – activity list	s	Hartmann (1998)	39.37	36.74
SA – activity list	s	Bouleimen, Lecocq (1998)	42.81	37.68
sampling – LFT	p	Kolisch (1996a)	39.60	38.75
sampling – WCS	p	Kolisch (1996b)	39.65	38.77
sampling – adaptive	s/p	Schirmer (1998)	40.08	39.08
sampling – adaptive	s/p	Kolisch, Drexl (1996)	41.37	40.45
GA – random key	mod. p	Leon, Ramamoorthy (1995)	42.91	40.69
sampling – LFT	s	Kolisch (1996a)	42.84	41.84

Table 7:11 Average deviations from critical path based lower bound — $J = 120$

LFT give almost identical results when employed within the parallel SGS. The adaptive sampling strategies do not consistently dominate the simple sampling procedures. Compared to each other, we observe that the adaptive sampling approach of Schirmer (1998) outperforms the one of Kolisch and Drexl (1996) on the j60 and j120 instance sets while the latter yields better results on the j30 set.

We finally remark that the metaheuristic algorithms which make use of the activity list representation can be assumed to be the fastest approaches. This is due to the fact that the underlying serial SGS for activity lists (cf. Section 7.2) does neither compute the eligible set nor does it select an activity on the basis of priority values.

7.7 OUTLOOK ON RESEARCH OPPORTUNITIES

Our computational results indicate that the best heuristics currently available are metaheuristic strategies which make use of activity lists. Further investigations (cf. Hartmann and Kolisch 1998) are headed towards a deeper insight of the functioning of different heuristics subject to different problem characteristics as given in Kolisch et al. (1995) and Kolisch and Sprecher (1996).

Although priority rule based methods do not give the best results, they are important for several reasons. First, they are indispensable when solving large problem instances in a short amount of time. Second, good priority rule based methods are needed to determine the initial solution(s) for metaheuristic procedures. Hence, further efforts in this area are still justified.

Summarizing, recent years have brought a considerable progress in designing more efficient heuristics for the RCPSP, and we believe that this will remain a fruitful field of research in the future. In addition to the development of even better heuristics, extending the current approaches to more general project scheduling problems are of special interest.

Acknowledgement. We are indebted to Tonius Baar, Peter Brucker, and Sigrid Knust (University of Osnabrück), Kamel Bouleimen and Henri Lecocq (University of Liège), Jorge Leon and Balakrishnan Ramamoorthy (Texas A&M University) as well as Andreas Schirmer and Sven Riesenberg (University of Kiel) for their help in this research. Furthermore, we would like to thank Andreas Drexl for his continuous support. This research has been supported by the Deutsche Forschungsgemeinschaft and the Studienstiftung des deutschen Volkes.

References

ALVAREZ–VALDÉS, R. AND J.M. TAMARIT 1989a. Heuristic algorithms for resource–constrained project scheduling: A review and an empirical analysis. *Advances in project scheduling.* 113–134, Słowiński, R. and J. Węglarz (eds.). Elsevier, Amsterdam.

ALVAREZ-VALDÉS, R., AND J.M. TAMARIT 1989b. Algoritmos heuristicos deterministas y aleatorios en secuenciacón de proyectos con recursos limitados. *Qüestiió* **13**, 173–191.

BAAR, T., BRUCKER, P., AND S. KNUST 1997. Tabu–search algorithms for the resource–constrained project scheduling problem. Technical Report, Universität Osnabrück.

BEAN, J. 1994. Genetic algorithms and random keys for sequencing and optimization. *ORSA Journal on Computing*, **6**, 2, 154–160.

BELL, C., AND J. HAN 1991. A new heuristic solution method in resource–constrained project scheduling. *Naval Research Logistics* **38**, 315–331.

BIXBY, N., AND E. BOYED 1996. *Using the CPLEX callable library.* CPLEX Optimization Inc., Houston.

BŁAŻEWICZ, J., LENSTRA, J., AND A. RINNOOY KAN 1983. Scheduling subject to resource constraints: Classification and complexity. *Discrete Applied Mathematics* **5**, 11–24.

BOCTOR, F. 1990. Some efficient multi–heuristic procedures for resource–constrained project scheduling. *European Journal of Operational Research* **49**, 3–13.

BOULEIMEN, K., AND H. LECOCQ 1998. A new efficient simulated annealing algorithm for the resource–constrained project scheduling problem. Technical Report, Service de Robotique et Automatisation, Université de Liège.

BRUCKER, P., AND S. KNUST 1998 Solving large–sized resource–constrained project scheduling problems. *Handbook on recent advances in project scheduling*. Chapter 18, J. Weglarz (ed.). Kluwer, Dordrecht.

BRUCKER, P., KNUST, S., SCHOO, A., AND O. THIELE 1998. A branch & bound algorithm for the resource–constrained project scheduling problem. *European Journal of Operational Research*, forthcoming.

CHO, J.-H., AND Y.–D. KIM 1997. A simulated annealing algorithm for resource constrained project scheduling problems. *Journal of the Operational Research Society* **48**, 735–744.

CHRISTOFIDES, N., ALVAREZ-VALDÉS, R., AND J.M. TAMARIT 1987. Project scheduling with resource constraints: A branch and bound approach. *European Journal of Operational Research* **29**, 262–273.

COOPER, D. 1976. Heuristics for scheduling resource–constrained projects: An experimental investigation. *Management Science* **22**, 11, 1186–1194.

COOPER, D. 1977. A note on serial and parallel heuristics for resource–constrained project scheduling. *Foundations of Control Engineering* **2**, 4, 131–133.

DAVIES, E. 1973. An experimental investigation of resource allocation in mulitactivity projects. *Operational Research Quarterly* **24**, 587–591.

DAVIS, E., AND J. PATTERSON 1975. A comparison of heuristic and optimum solutions in resource–constrained project scheduling. *Management Science* **21**, 944–955.

DELLA CROCE, F. 1995. Generalized pairwise interchanges and machine scheduling. *European Journal of Operational Research* **83**, 310–319.

DEMEULEMEESTER, E., AND W. HERROELEN 1997a. New benchmark results for the resource–constrained project scheduling problem. *Management Science* **43**, 11, 1485–1492.

DORNDORF, U., AND E. PESCH 1994. Evolution based learning in a job shop scheduling environment. *Computers & Operations Research* **22**, 1, 25–40.

DREXL, A. 1991. Scheduling of project networks by job assignment. *Management Science* **37**, 12, 1590–1602.

ELMAGHRABY, S. 1977. *Activity networks: Project planning and control by network models*. Wiley, New York.

ELSAYED, E. 1982. Algorithms for project scheduling with resource constraints. *International Journal of Production Research* **20**, 1, 95–103.

GLOVER, F. 1989a. Tabu search - Part I. *ORSA Journal on Computing* **1**, 190–206.

GLOVER, F. 1989b. Tabu search - Part II. *ORSA Journal on Computing* **2**, 4–32.

HARTMANN, S. 1998. A competitive genetic algorithm for resource–constrained project scheduling. *Naval Research Logistics*, forthcoming.

HARTMANN, S., AND R. KOLISCH 1998. An experimental investigation of state–of–the–art heuristics for the resource–constrained project scheduling problem. Technical Report, Universität Kiel, 1998. In preparation.

HERROELEN, W., DEMEULEMEESTER, E., AND B. DE REYCK 1998. A classification scheme for project scheduling. *Handbook on recent advances in project scheduling.* Chapter 1, J. Weglarz (ed.). Kluwer, Dordrecht.

HOLLAND, H. 1975. *Adaptation in natural and artificial systems.* University of Michigan Press, Ann Arbor.

KELLEY, J. 1963. The critical–path method: Resources planning and scheduling. 347–365. *Industrial scheduling,* Muth, J. and G. Thompson (eds.). Prentice–Hall, New Jersey.

KIM, Y.-D. 1995. A backward approach in list scheduling algorithms for multi–machine tardiness problems. *Computers & Operations Research* **22**, 3, 307–319.

KIRKPATRICK, S., GELATT JR., C. D., AND M.P. VECCHI 1983. Optimization by simulated annealing. *Science* **220**, 671–680.

KOHLMORGEN, U., SCHMECK, H., AND K. HAASE 1998. Experiences with fine–grained parallel genetic algorithms. *Annals of Operations Research,* forthcoming.

KOLISCH, R. 1995. *Project scheduling under resource constraints — Efficient heuristics for several problem classes.* Physica, Heidelberg.

KOLISCH, R. 1996a. Serial and parallel resource–constrained project scheduling methods revisited: Theory and computation. *European Journal of Operational Research* **90**, 320–333.

KOLISCH, R. 1996b. Efficient priority rules for the resource–constrained project scheduling problem. *Journal of Operations Management* **14**, 3, 179–192.

KOLISCH, R., AND A. DREXL 1996. Adaptive search for solving hard project scheduling problems. *Naval Research Logistics* **43**, 23–40.

KOLISCH, R., AND R. PADMAN 1997. An integrated survey of project scheduling. Technical Report 463, Manuskripte aus den Instituten für Betriebswirtschaftslehre der Universität Kiel.

KOLISCH, R., SCHWINDT, C., AND A. SPRECHER 1998. Benchmark instances for project scheduling problems. *Handbook on recent advances in project scheduling.* Chapter 9, J. Weglarz (ed.). Kluwer, Dordrecht.

KOLISCH, R., AND A. SPRECHER 1996. PSPLIB — a project scheduling problem library. *European Journal of Operational Research* **96**, 205–216.

KOLISCH, R., SPRECHER, A., AND A. DREXL 1995. Characterization and generation of a general class of resource–constrained project scheduling problems. *Management Science* **41**, 10, 1693–1703.

LAWRENCE, S. 1985. Resource constrained project scheduling – A computational comparison of heuristic scheduling techniques. Technical Report,

Graduate School of Industrial Administration, Carnegie–Mellon University, Pittsburgh.

LEE, J.-K., AND Y.-D. KIM 1996. Search heuristics for resource constrained project scheduling. *Journal of the Operational Research Society* **47**, 678–689.

LEON, V., AND B. RAMAMOORTHY 1995. Strength and adaptability of problem–space based neighborhoods for resource–constrained scheduling. *OR Spektrum* **17**, 2/3, 173–182.

LI, R.-Y., AND J. WILLIS 1992. An iterative scheduling technique for resource–constrained project scheduling. *European Journal of Operational Research* **56**, 370–379.

MAUSSER, H., AND S. LAWRENCE 1998. Exploiting block structure to improve resource–constrained project schedules. *Metaheuristics*, forthcoming.

MINGOZZI, A., MANIEZZO, V., RICCIARDELLI, S., AND L. BIANCO 1998. An exact algorithm for project scheduling with resource constraints based on a new mathematical formulation. *Management Science*, forthcoming.

NAPHADE, K., WU, S., AND R. STORER 1997. Problem space search algorithms for resource–constrained project scheduling. *Annals of Operations Research* **70**, 307–326.

OĞUZ, O., AND H. BALA 1994. A comparative study of computational procedures for the resource constrained project scheduling problem. *European Journal of Operational Research* **72**, 406–416.

ÖZDAMAR, L., AND G. ULUSOY 1994. A local constraint based analysis approach to project scheduling under general resource constraints. *European Journal of Operational Research* **79**, 287–298.

ÖZDAMAR, L., AND G. ULUSOY 1996a. A note on an iterative forward/backward scheduling technique with reference to a procedure by Li and Willis. *European Journal of Operational Research* **89**, 400–407.

ÖZDAMAR, L., AND G. ULUSOY 1996b. An iterative local constraint based analysis for solving the resource constrained project scheduling problem. *Journal of Operations Management* **14**, 3, 193–208.

PATTERSON, J. 1973. Alternate methods of project scheduing with limited resources. *Naval Research Logistics Quarterly* **23**, 767–784.

PATTERSON, J. 1976. Project scheduling: The effects of problem structure on heuristic performance. *Naval Research Logistics Quarterly* **20**, 95–123.

PINSON, E., PRINS, C., AND F. RULLIER 1994. Using tabu search for solving the resource–constrained project scheduling problem. *Proceedings of the 4. International Workshop on Project Management and Scheduling*, 102–106, Leuven.

POLLACK–JOHNSON, B. 1995. Hybrid structures and improving forecasting and scheduling in project management. *Journal of Operations Management* **12**, 101–117.

PRITSKER, A., WATTERS, L., AND P. WOLFE 1969. Multiproject scheduling with limited resources: A zero–one programming approach. *Management Science* **16**, 93–107.

SAMPSON, S., AND E. WEISS 1993. Local search techniques for the generalized resource constrained project scheduling problem. *Naval Research Logistics* **40**, 665–675.

SCHIRMER, A., AND S. RIESENBERG 1997. Parameterized heuristics for project scheduling — Biased random sampling methods. Technical Report 456, Manuskripte aus den Instituten für Betriebswirtschaftslehre der Universität Kiel.

SCHIRMER, A. 1998. Case-based reasoning and improved adaptive search for project scheduling. Technical Report 472, Manuskripte aus den Instituten für Betriebswirtschaftslehre der Universität Kiel.

SCHUTTEN, J. 1996. List scheduling revisited. *Operations Research Letters* **18**, 167–170.

SHAFFER, L., RITTER, J., AND W. MEYER 1965. *The critical–path method.* McGraw Hill, New York.

SPRECHER, A. 1996. Solving the RCPSP efficiently at modest memory requirements. Technical Report 425, Manuskripte aus den Instituten für Betriebswirtschaftslehre der Universität Kiel.

SPRECHER, A., KOLISCH, R., AND A. DREXL 1995. Semi–active, active, and non–delay schedules for the resource–constrained project scheduling problem. *European Journal of Operational Research* **80**, 94–102.

STINSON, J., DAVIS, E., AND B. KHUMAWALA 1978. Multiple resource–constrained scheduling using branch and bound. *AIIE Transactions* **10**, 252–259.

STORER, R., WU, S., AND R. VACCARI 1992. New search spaces for sequencing problems with application to job shop scheduling. *Management Science* **38**, 10, 1495–1509.

THESEN, A. 1976. Heuristic scheduling of activities under resource and precedence restrictions. *Management Science* **23**, 4, 412–422.

THOMAS, P., AND S. SALHI 1997. An investigation into the relationship of heuristic performance with network–resource characteristics. *Journal of the Operational Research Society* **48**, 1, 34–43.

ULUSOY, G., AND L. ÖZDAMAR 1989. Heuristic performance and network/resource characteristics in resource–constrained project scheduling. *Journal of the Operational Research Society* **40**, 12, 1145–1152.

ULUSOY, G., AND ÖZDAMAR, L. 1994. A constraint–based perspective in resource constrained project scheduling. *International Journal of Production Research* **32**, 3, 693–705.

VALLS, V., PÉREZ, M., AND M. QUINTANILLA 1992. Heuristic performance in large resource–constrained projects. Technical Report 92–2, Departament D'Estadistica I Invecigacio Operativa, Universitat de Valencia.

WHITEHOUSE, G., AND J. BROWN 1979. GENRES: An extension of Brooks algorithm for project scheduling with resource constraints. *Computers & Industrial Engineering* **3**, 261–268.

8 A HEURISTIC PROCEDURE FOR THE MULTI-MODE PROJECT SCHEDULING PROBLEM BASED ON BENDERS' DECOMPOSITION

Vittorio Maniezzo
Aristide Mingozzi

University of Bologne

8.1. Introduction

In this paper we consider the problem of determining the starting times of the activities of a project with the objective to minimize the total project duration (makespan). The activities cannot be interrupted once put in progress and are subject to constraints that impose a partial precedence relation among them and a limit to their resource usage. Moreover, the activities can be executed in different *modes*, involving different durations and resource requirements. This problem is known as Multi-Mode Resource-Constrained Project Scheduling Problem (MRCPSP), and it is NP-hard being a generalization of the Resource-Constrained Project Scheduling Problem (RCPSP), which is also NP-hard.

Several exact and heuristic techniques have been proposed for the MRCPSP. Talbot (1982) was the first to propose an exact enumeration scheme, followed by Patterson et al. (1989). These early methods were able to solve instances up to 15 activities. Sprecher and Drexl (1996) proposed new dominance criteria, which enabled their branch and bound algorithm to solve problems up to 20 activities.

With reference to heuristic approaches, Talbot (1982) and Sprecher and Drexl (1996) proposed the truncated versions of their tree searches. Different heuristic strategies have also been investigated, such as biased random sampling (Drexl, Grünewald, 1993), local search (Kolisch, Drexl, 1994), constraint-based approaches (Özdamar, Ulusoy, 1994) and a mix of solution strategies (Boctor, 1994). Slowinski et al. (1994) developed a decision support systems, which helps the user to identify strategies to choose the activities to put in progress in case of resource conflicts.

This paper proposes a new mathematical formulation for the MRCPSP, that is a generalization of the formulation proposed by Mingozzi et al.(1994) for the RCPSP. The new formulation is used to derive two new lower bounds, and a new heuristic algorithm based on Benders' decomposition.

8.2. Problem description

The MRCPSP problem can be formulated as follows. A set X={1, ... , n} of activities (jobs) and two sets RR and NR of resources are given. The set RR is the set of *renewable resources*, resources whose quantities are available on a period-by-period basis and are renewed from period to period, while NR is the set of *nonrenewable resources*, i.e., the set of resources which have a limited overall availability over all project duration. In each period of the planning horizon t, $t = 1, \dots, T_{max}$, K_r units of renewable resource r are available, $\forall r \in$ RR, while there is a global availability of W_s units of nonrenewable resource s, $\forall s \in$ NR. We assume that time period t corresponds to the time interval $[t, t+1)$.

Every activity i is associated with a set M_i of possible *modes* it can be executed in. When activity i is executed in mode m, it has a processing time (duration) d_{im}, it requires a constant amount k_{imr} of renewable resource r, $r \in$ RR, during each time interval of its execution and w_{ims} of nonrenewable resource s, $s \in$ NR. All quantities, d_{im}, k_{imr}, w_{ims}, K_r and W_s, are assumed to be non-negative integers, no pre-emption is allowed and setup times are included in the processing times.

With each activity j is associated a set $\Gamma_j^{-1} \subseteq X \backslash \{j\}$ of immediate predecessors: activities that must be completed before starting the execution of j. The precedence constraints can be represented by an acyclic digraph G=(X,H) where $H=\{(i,j): i \in \Gamma_j^{-1}, j \in X\}$. We assume, without loss of generality, that activities are topologically ordered, i.e., each predecessor of activity j has a smaller activity number than j.

Activities 1 and n are used to represent the beginning and the end of the whole project: activity 1 must be completed before starting activities $X \backslash \{1\}$ and activity n can start after the completion of activities $X \backslash \{n\}$; let $X' = X \backslash \{1,n\}$ and $n' = |X'|$. Both activities 1 and n have only one execution mode ($|M_1| = 1$, $|M_n| = 1$). Furthermore, we assume $d_1 = d_n = 0$, $k_{11r} = k_{n1r} = 0$, $\forall r \in$ RR, and $w_{11s} = w_{n1s} = 0$, $\forall s \in$ NR.

The cost of a feasible solution is given by the project completion time (*project makespan*). The objective is to assign a mode to each activity and to find a schedule that minimizes the project completion time, satisfying precedence and resource constraints.

Note that an obvious upper bound to the project makespan is given by the sum of the maximum possible duration of each activity: $T_{max} = \sum_{i=2}^{n-1} \max_{m \in M_i} \{d_{im}\}$.

Moreover, let $\hat{d}_i = \min_{m \in M_i} \{d_{im}\}$ and let $c_{ij} = \hat{d}_i$ be the cost associated with arc $(i,j) \in H$. A time window $[es_i, ls_i]$ of earliest and latest start times for each activity $i \in X$ can be computed by performing a forward and backward recursion on the graph G, by setting $es_1 = 0$ and $ls_n = T_{max}$ (Elmaghraby, 1977).

8.2.1. Simple lower bounds

It is possible to associate to the MRCPSP a single-mode RCPSP instance, where activity durations are set to $\hat{d}_i$, $\forall i \in X$, nonrenewable resource constraints are relaxed and renewable resource consumptions are set to $\hat{k}_{ir} = \min_{m \in M_i} [k_{imr}]$, $\forall i \in X$, $\forall r \in RR$.
The cost of an optimal solution to the accordingly defined RCPSP problem is a lower bound to the MRCPSP optimal cost, and such is obviously also any lower bound for the RCPSP problem. Simple lower bounds for MRCPSP can thus be obtained as follows.

Lower bound BND0

BND0 is the value of the longest path in G from vertex 1 to vertex n, computed using activity durations $\{\hat{d}_i\}$. This bound has been originally proposed by Talbot (1982) and successively used also by Sprecher and Drexl (1996).

Lower bound BND1

Lower bound BND1 makes use of bound LB3 proposed by Mingozzi et al. (1994) for the RCPSP. It is computed as follows.

Consider the intersection graph $\tilde{G} = (X', E)$, where $(i\,j) \in E$ if and only if the pair of activities i and j satisfies the two following conditions:

i) the precedence graph G does not contain any path from vertex i to vertex j if $i<j$, or from vertex j to vertex i if $j<i$;

ii) the two activities i and j satisfy renewable resource constraints: $\hat{k}_{ir} + \hat{k}_{jr} \leq K_r$, $\forall r \in RR$.

Let $\hat{d}_i$ be the weight associated to node i of $\tilde{G}$. Lower bound BND1 is given by the solution of a weighted node packing problem (WNP) on graph $\tilde{G}$, which consists of finding an independent set of $\tilde{G}$ of maximum total weight. Mingozzi et al. (1994) describe a heuristic algorithm for solving problem WNP. This solution produces a lower bound, BND1, not smaller than BND0.

8.2.2. Mathematical formulation

In this section we present a 0-1 integer programming formulation of the MRCPSP.

Let $B_i = \{(i,m) : m \in M_i\}$ and $B = \bigcup_{i \in X'} B_i$. For a given subset $F \subseteq B$, we denote with $X(F)=\{i : (i,m) \in F\}$ the different activities of F.

A subset $F \subset B$ is called a *feasible subset* if it satisfies the three following conditions:

A) $|X(F)| = |F|$

B) *resource constraints*:

$$\sum_{(i,m) \in F} k_{imr} \leq K_r \qquad \forall r \in RR \tag{1}$$

$$\sum_{(i,m) \in F} w_{ims} \leq W_s \qquad \forall s \in NR \tag{2}$$

C) *precedence constraints*:
no path exists in G from vertex i to vertex j, for every pair $i, j \in X(F)$ with $i < j$.

Any feasible MRCPSP solution of cost t^* can be represented by a sequence $\mathbf{S}=(F_{l_1}, F_{l_2}, \ldots, F_{l_{t^*}})$ where each F_{l_t} of $\mathbf{S}$ represents the feasible subset of activities in progress at time t. Each F_{l_t} of $\mathbf{S}$ satisfies conditions A, B and C above.

Let $\mathbf{F} = \{1, 2, \ldots, \hat{f}\}$ be the index set of all feasible subsets of B, and let $\mathbf{F}_{im} \subset \mathbf{F}$ be the index set of all feasible subsets containing the pair $(i,m) \in B$.

Let $y_{l\,t}$ be a (0-1) variable equal to 1 if and only if all activities of the feasible subset F_l are being executed at time period t.

Define a (0-1) variable x_{it} equal to 1 if and only if activity i terminates its execution at time period t and a (0-1) binary variable ζ_{im} equal to 1 if and only if job i is executed in mode m.

The mathematical formulation P of MRCPSP is as follows:

$$\text{(P)} \quad z_P = \min \sum_{t=es_n}^{ls_n} t \cdot x_{nt} \tag{3}$$

$$\text{s.t.} \quad \sum_{l \in F_{im}} \left(\sum_{t=es_i}^{ls_i} y_{lt} \right) = d_{im}\, \zeta_{im} \qquad i \in X', m \in M_i \tag{4}$$

$$\sum_{l \in F} y_{l\,t} \le 1 \qquad t = 1,\ldots,T_{max} \qquad (5)$$

$$\sum_{m \in M_i} \zeta_{im} = 1 \qquad i \in X \qquad (6)$$

$$x_{it} \ge \sum_{l \in F_i} y_{l\,t} - \sum_{l \in F_i} y_{l\,t+1} \qquad i \in X',\ t = es_i,\ldots,ls_i \qquad (7)$$

$$\sum_{t=es_j}^{ls_j} t \cdot x_{jt} - \sum_{t=es_i}^{ls_i} t \cdot x_{it} \ge \sum_{m \in M_i} d_{jm}\zeta_{jm} \qquad (i,j) \in H \qquad (8)$$

$$\sum_{i \in X} \sum_{m \in M_i} w_{ims}\zeta_{im} \le W_s \qquad s \in NR \qquad (9)$$

$$x_{it} \in \{0,1\} \qquad i \in X,\ t = 1,\ldots,T_{max} \qquad (10)$$

$$y_{l\,t} \in \{0,1\} \qquad l \in F,\ t = 1,\ldots,T_{max} \qquad (11)$$

$$\zeta_{im} \in \{0,1\} \qquad i \in X,\ m \in M_i. \qquad (12)$$

Constraints (4) force the solution to have in progress feasible subsets contained in F_{im} for exactly d_{im} time periods if activity i is executed in mode m. Constraints (5) ensure that at each time period t at most one feasible subset is in progress. Constraints (6) force each activity to be executed in only one mode. Constraints (7) ensure that the execution of activity i terminates at time period t if the activity i is contained in the feasible subset in execution at time period t but is not contained in the feasible subset in execution at time period t+1. Finally, constraints (8) and (9) are the precedence and the nonrenewable resource constraints, respectively.

Note that, since a feasible subset F_l is executed for a number of time periods equal to $\sum_t y_{l\,t}$, the objective function (3) can be rewritten as

$$z_P = min \sum_{l \in F} \sum_{t=1}^{T_{max}} y_{l\,t} \qquad (3')$$

8.3. New lower bounds

Different relaxations of problem (P) give rise to different lower bounds to MRCPSP. For example, lower bound BND0 can be obtained removing constraints (4), (5), (7), (9) and (11). In the following we will present other relaxations that lead to formulations from which bounds stronger than BND0 can be derived.

8.3.1. Lower bound BND2

Lower bound BND2 is obtained from problem (P) by:

- removing precedence and non-preemption constraints (5), (7) and (8);
- defining $h_l = \sum_t y_{l\,t}$ to be the total execution time of feasible subset l.

The resulting relaxed problem, called RP1, is as follows.

$$\text{(RP1)} \qquad z_P = \min \sum_{l \in F} h_l \qquad \text{(3')}$$

$$\text{s.t.} \quad \sum_{l \in F_{im}} h_l = d_{im}\,\zeta_{im} \qquad i \in X',\ m \in M_i \qquad (4)$$

$$\sum_{m \in M_i} \zeta_{im} = 1 \qquad i \in X \qquad (6)$$

$$\sum_{i \in X} \sum_{m \in M_i} w_{ims}\zeta_{im} \le W_s \qquad s \in NR \qquad (9)$$

$$\zeta_{im} \in \{0,1\} \qquad i \in X,\ m \in M_i \qquad (12)$$

$$h_l \ge 0 \qquad l \in F \qquad (13)$$

Lower bound BND2 is given by the optimal solution cost of the LP-relaxation of problem (RP1).

8.3.2. Lower bounds BND3 and BND4

Let RP2 be a relaxation of RP1 obtained by relaxing equations (4) by replacing "=" with "≥", to give a system of inequalities. This relaxation allows to reduce the set F of all feasible subsets used by RP2 by removing redundant subsets. This does not change the optimal solution cost of RP2.

Consider two different feasible subsets $F_{l'}$ and $F_{l''}$ such that $F_{l'} \subset F_{l''}$. It is easy to show that the cost of the optimal solution to RP2 does not change if $F_{l'}$ is removed from F. Let (h^*, ζ^*) be an optimal RP2 solution of cost z^*_{RP2} such that $h^*_{\ell} > 0$. Consider a new solution (h', ζ'), where $h'_{\ell} = h^*_{\ell}$, $\ell \in \mathcal{F} \setminus \{\ell', \ell''\}$, $h'_{\ell'} = 0$ and $h'_{\ell''} = h^*_{\ell'} + h^*_{\ell''}$. Since $F_{\ell'} \subset F_{\ell''}$, we have $\sum_{\ell \in \mathcal{F}_{im}} h'_{\ell} \geq \sum_{\ell \in \mathcal{F}_{im}} h^*_{\ell} \geq d_{im} \zeta^*_{im}$, $i \in X$, $m \in M_i$, which shows that (h', ζ') is a feasible RP2 solution. Moreover, from the definition of h', we have $\sum_{\ell \in \mathcal{F}} h'_{\ell} = \sum_{\ell \in \mathcal{F}} h^*_{\ell}$, which shows that the cost of the new solution is equal to z^*_{RP2}.

Therefore, we can remove from $\mathcal{F}$ any feasible subset ℓ' such that $F_{\ell'} \subset F_{\ell''}$, for some $\ell'' \in \mathcal{F}$, without changing the optimal cost of RP2.

Let $\mathcal{M} \subseteq \mathcal{F}$ be the set of *maximal feasible subsets*, defined as

$$\mathcal{M} = \{ \ell \in \mathcal{F} : F_{\ell} \not\subset F_{\ell'}, \ \forall \ \ell' \in \mathcal{F} \setminus \{\ell\} \}$$

We denote with $\mathcal{M}_{im} \subset \mathcal{M}$ the index set of all maximal feasible subsets containing job i executed in mode m. The relaxed problem RP2 is the following.

$$\text{(RP2)} \quad z_P = \min \sum_{\ell \in \mathcal{M}} h_{\ell} \tag{3''}$$

$$\text{s.t.} \quad \sum_{\ell \in \mathcal{M}_{im}} h_{\ell} > d_{im} \zeta_{im} \qquad i \in X', \ m \in M_i \tag{4'}$$

$$\sum_{m \in M_i} \zeta_{im} = 1 \qquad i \in X \tag{6}$$

$$\sum_{i \in X} \sum_{m \in M_i} w_{ims} \zeta_{im} \leq W_s \qquad s \in NR \tag{9}$$

$$\zeta_{im} \in \{0,1\} \qquad i \in X, \ m \in M_i \tag{12}$$

$$h_{\ell} \geq 0 \qquad \ell \in \mathcal{M} \tag{14}$$

Lower bound BND3 is given by the cost of the optimal solution of the LP-relaxation of problem (RP2), while lower bound BND4 is given by the cost of the optimal solution cost of the LP-relaxation of problem (RP2) after further removing constraints (9).

It can be shown that BND2 ≥ BND3 ≥ BND4 ≥ BND1 ≥ BND0.

8.3.3 Experimental comparison of the lower bounds

We implemented all lower bounds so far described in order to obtain an experimental evaluation of their strength, utilizing a PC 486 dx 66 MHz machine equipped with 16 Mb of RAM. The problem testbed we used is due to Kolisch et al. (1995) and consists of two sets of problems with the following characteristics: all problems of the two sets have a feasible solution, two renewable resources ($|RR|=2$), two nonrenewable resources ($|NR|=2$) and three possible modes for each nondummy activity ($|M_i|=3$, $i \in X'$). The first set of instances is composed of 536 problems with 10 activities, the second set is composed of 554 problems with 20 activities each.

Table 8:1 presents the results obtained. The rows GAP%avg and GAP%max show the average and maximum percentage distance from optimality of the five lower bounds, respectively.

Table 8:1. Lower bounds comparison

		BND2	BND3	BND4	BND1	BND0
10 jobs	GAP%avg	3.20	3.68	7.67	10.47	10.87
	GAP%max	16.66	19.04	52.17	57.35	61.76
20 jobs	GAP%avg	4.01	4.52	11.10	16.07	17.24
	GAP%max	26.66	26.66	68.96	70.41	74.07

Table 8:1 shows that all new lower bounds (BND2, BND3, BND4 and BND1) are tighter than BND0, both on the 10 and on the 20 activity projects. Moreover, the average results induce a tight ordering, where BND2 is the best lower bound, followed by BND3, BND4, BND1 and then BND0.

The better results obtained by the tighter bounds are paid by the higher CPU time required to compute them. Specifically, on the 20 activities problems, lower bounds BND2, BND3 and BND4 require an average CPU time of 4.23 and 1.66 and 1.55 seconds, respectively. BND1 and BND0 are computed in average in 0.04 and 0.02 seconds on those instances, respectively.

The higher computing time for BND2 and BND3 is due to the dimensions of problems (RP1) and (RP2) which need to be solved. The number of variables of (RP1), corresponding to all feasible subsets $\mathcal{F}$, and of (RP2), corresponding to all maximal feasible subsets $\mathcal{M}$, for the 20 activities problem instances is presented in table 8:2.

Table 8:2. Number of variables of problems RP1 and RP2

			10 jobs	20 jobs
Problem RP1	| F |	Avg	633	15905
		Max	2991	88114
Problem RP2	| M |	Avg	288	5362
		Max	972	24252

8.4. A new heuristic algorithm

In this section we describe a new heuristic algorithm for MRCPSP, called HBEND, based on a decomposition of problem RP2 which follows Benders' approach (Benders, 1962).

Problem RP2 is decomposed into a master problem MP and a subproblem SP. MP defines a RCPSP instance, by assigning modes to activities, in order to satisfy nonrenewable resource constraints. SP computes a valid lower bound to the completion time of the resulting RCPSP instance. HBEND solves problems MP and SP iteratively. At each iteration, a heuristic method is used for finding a feasible solution of the RCPSP instance defined by MP. A valid Benders' cut is obtained from the solution of SP and added to MP in the following iteration.

Consider problem RP2 and let $\zeta=\zeta'$, where ζ' is any (0-1) vector that satisfies constraints (6), (9) and (12). The LP problem resulting from RP2 replacing ζ with ζ' is as follows.

$$\text{(SP)}\quad z_{SP}(\zeta') = \min \sum_{\ell \in M} h_\ell \tag{15}$$

$$\text{s.t.} \quad \sum_{\ell \in M_{im}} h_\ell \geq d_{im}\, \zeta'_{im} \qquad i \in X',\ m \in M_i \tag{4'}$$

$$h_\ell \geq 0 \qquad \ell \in M \tag{14}$$

The dual of SP is

$$\text{(DSP)}\quad z_{DSP}(\zeta') = \max \sum_{i \in X'} \sum_{m \in M_i} u_{im}\, d_{im}\, \zeta'_{im} \tag{16}$$

$$\text{s.t.} \quad \sum_{(i,m) \in F_\ell} u_{im} \leq 1, \qquad \ell \in M \tag{17}$$

$$u_{im} \geq 0 \qquad i \in X', m \in M_i \quad (18)$$

Problem RP2 can be written as

$$z_{RP2} = \min_{\zeta} \; z_{SP}(\zeta)$$

s.t. constraints (6), (9) and (12)

but also as

$$z_{RP2} = \min_{\zeta} \; z_{DSP}(\zeta)$$

s.t. constraints (6), (9) and (12)

or, equivalently, as

$$\text{(MP)} \quad z_{RP2} = \min \; z$$

$$\text{s.t.} \quad z \geq \sum_{i \in X'} \sum_{m \in M_i} u^t_{im} \, d_{im} \, \zeta_{im} \qquad \mathbf{u}^t \in \mathbf{U} \quad (16)$$

$$\sum_{m \in M_i} \zeta_{im} = 1 \qquad i \in X' \quad (6)$$

$$\sum_{i \in X} \sum_{m \in M_i} w_{ims} \zeta_{im} \leq W_s \qquad s \in NR \quad (9)$$

$$\zeta_{im} \in \{0,1\} \qquad i \in X, m \in M_i \quad (12)$$

where $\mathbf{U}$ is the set of all extreme points of the constraint set of DSP.

Problem MP is a mixed integer program, having one continuous variable, z. This problem is impractical to solve, even for moderate size MRCPSP instances, since the number of constraints (16) is as large as $|\mathbf{U}|$. Moreover, an exact solution of MP would provide a valid lower bound, but not an optimal MRCPSP solution.

By means of the Benders' method, MP can be solved iteratively, generating constraints (16) only when they are needed. At iteration t, let $\mathbf{U}(t)$ be the set of known extreme points of DSP, and let MP^t be problem MP when $\mathbf{U}$ is replaced by $\mathbf{U}(t)$. Any feasible solution of MP^t corresponds to an assignment of modes to the activities; therefore, it defines a RCPSP instance whose optimal solution cost provides an upper bound to the MRCPSP.

At each iteration of procedure HBEND, we propose to solve the resulting RCPSP using a heuristic method, since an exact method would be too time-consuming. In order to improve the quality of the MRCPSP upper bound, we store the K best RCPSP instances and solve them to optimality, at the end of HBEND, using the branch and bound method described by Mingozzi et al. (1994) for the RCPSP.

The heuristic procedure HBEND is as follows.

Step 0: Initialize $\mathbf{u}^0$, set t=0, $\mathbf{U}(0) = \varnothing$ and $z_{UB}=\infty$.

Step 1: If t = NITER then STOP; otherwise solve problem MP^t by means of a heuristic procedure (see section 8.4.1).
Let ζ^t be the solution of MP^t of cost z^t.

Step 2: Assign to each job i the mode $\mu_i = \sum_{m \in M_i} m\zeta_{im}^t$.

a) Solve, by means of a heuristic, the resulting RCPSP instance. In our computational results, we applied the heuristic rules described by Patterson (1973). We denote with z_H the cost of the best solution obtained by the different heuristics.
Set $z_{UB} = \min(z_{UB}, z_H)$.

b) Solve in a heuristic way (see section 8.4.2) the following subproblem:

$$(S) \quad \max \quad z_S = \sum_{i \in X'} u_{i\mu_i}^{t+1} d_{i\mu_i}$$

$$\text{s.t.} \quad \sum_{(i,\mu_i) \in F_\ell} u_{i\mu_i}^{t+1} \leq 1 \qquad \ell \in M_{i\mu_i},\ i \in X'$$

$$u_{i\mu_i}^{t+1} \geq 0 \qquad i \in X'$$

Let z_S^* be the cost of the heuristic solution of (S). Set $u_{im}^{t+1} = 0$, $m \in M_i \setminus \{\mu_i\}$, $i \in X'$.

Step 3: If $z^t \geq z_S^*$ then STOP; otherwise set $t = t+1$ and go to Step 1.

Figure 8:1 presents the trace of the evolution of z^t, of the MRCPSP upper bound z_{UB}, and of the RCPSP upper bound z_H during a run on a 20 activities problem instance. 700 Benders cuts were generated, until z^t and z_{UB} coincided.

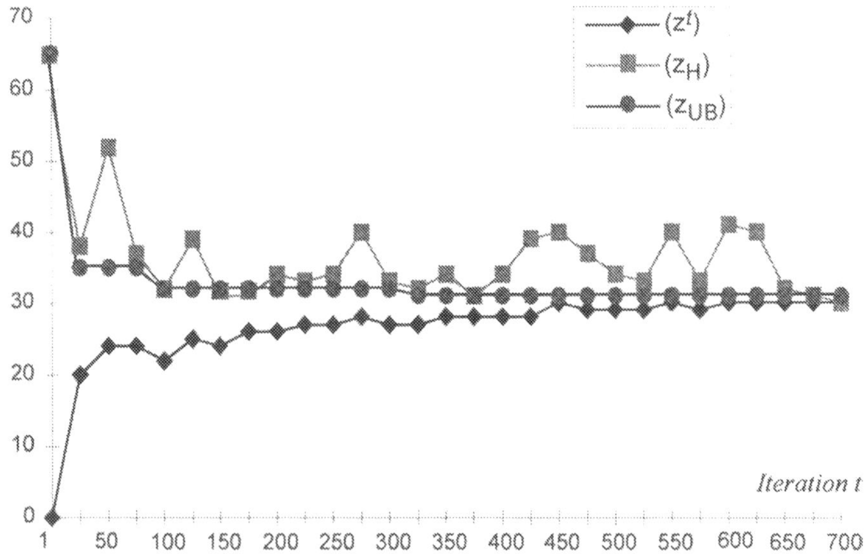

Figure 8:1. Upper and lower bounds evolution

8.4.1. Solving the master problem MP^t

Finding an optimal solution of M^t may be too time-consuming, due to the number of Benders' cuts which can be as large as $|\mathbf{U}(t)|$. We use therefore the following heuristic method.

Let z^{t-1} be the solution found for problem M^{t-1} at iteration t-1. At iteration t ($t \geq 1$) we start by setting $\zeta^t = \zeta^{t-1}$, being obvious that ζ^t is a feasible solution of M^t.

The solution ζ^t is iteratively improved by assigning to an activity i a feasible mode that minimizes the cost function when every other activity $j \in X' \setminus \{i\}$ is assigned to the mode $\mu^t_j = \sum_{m \in Mj} m \zeta^t_{jm}$. This iterative procedure is repeated until no further improvement is achieved.

This recursive way of solving M^t requires a feasible solution of M^0. Note that a feasible solution of M^0, when NR=1, can be simply obtained by setting $\zeta^0_{im^*_i}=1$

where m^*_i is such that $w_{im^*_i 1} = \min_{m \in M_i} w_{im1}$.

The case NR>1 is more complex. In this case, we find ζ^0 as a feasible solution of the following integer program:

$$\text{(ILP)} \qquad \min \sum_{i \in X'} \sum_{m \in M_i} d_{im} \zeta_{im}$$

s.t. constraints (6), (9) and (12).

A truncated branch and bound algorithm has been used to solve ILP.

8.4.2. Solving subproblem SP

Any feasible solution of problem S, at iteration t, provides a valid lower bound to the RCPSP defined by the solution of M^t. Therefore, a feasible solution $\mathbf{u}^t$ of S can be obtained by any of the bounding procedures described in Mingozzi et al. (1994). In our computational results we used bound LB3, already summarized in section 8.2.1.

Let us consider the RCPSP instance, where each activity i is executed in mode μ_i, as given by the solution of MP^t. Define the intersection graph $\tilde{G}$=(X′,E), where $\hat{k}_{ir} = k_{i\mu_i r}$, $r \in$RR, and $\hat{d}_i = d_{i\mu_i}$, see section 8.2.1. Let $\overline{X} \subset X'$ be the maximum weighted independent set of $\tilde{G}$. A feasible solution of S is obtained by setting $u^{t+1}_{i\mu_i}=1$, $\forall i \in \overline{X}$, and $u^{t+1}_{i\mu_i}=0$, $\forall i \in X' \setminus \{\overline{X}\}$.

5. Computational results

Algorithm HBEND has been implemented in C and run on a PC 486 dx 66 MHz, 16Mb machine. We tested the algorithm on the two sets of test problems described in section 8.3.3, and on a third set of 550 problems with 30 activities each, due to Kolisch et al. (1995). Also the problems of this third set involve two renewable resources (|RR|=2), two nonrenewable resources (|NR|=2) and three possible modes for each nondummy activity ($|M_i|=3$, $i \in X'$). Optimal solutions are known for the first two sets of problems, but not for the third one.

We now want to compare the results obtained by HBEND with those produced, on the same sets of problems, by two other heuristics presented in the literature. The first one, hereafter called HKD, is the algorithm proposed by Kolisch and Drexl (1994), which was run on a PC 386 dx 40MHz. The second one, hereafter HSD, has been proposed by Sprecher and Drexl (1996) and run on a PC 486 dx 66MHz. Only the results on the 10 activities projects are known for HKD, while, for HSD, we have complete results both on the 10 and on the 20 activities problem instances. For the third set of problems we have the solution produced by HSD, but not the CPU time needed to produce them.

The three algorithms operated under different terminating conditions: HBEND is executed until a predefined number of Benders cuts is generated, HKD is run for a given number of iterations while HSD is run for a given CPU time interval.

Table 8:3. Distance from optimality and CPU times on the 10 activities problems

	HKD		HBEND				HSD
Num. Iterations	100	500					
Num. Benders cuts			100	200	500	1000	
CPU interval (sec.)							10
GAP% avg.	4.08	1.75	6.07	1.34	1.34	1.34	0
GAP% max	47.50	47.50	58.10	27.27	27.27	27.27	0
CPU time avg.	0.29	1.22	0.79	1.60	3.96	12.48	0.14
CPU time max	0.44	1.92	1.30	2.62	7.35	23.31	2.31

Tables 8:3 and 8:4 present the results obtained on the 10 and on the 20 activities projects, respectively. In particular, table 8:3 shows the average and the maximum percentage distance from optimality (GAP% avg and GAP% max, respectively), and the average and maximum CPU time, in seconds (CPU time avg and CPU time max, respectively). The terminating conditions were: 100 and 500 iterations for HKD, 100 to 1000 Benders cuts for HBEND, and 10 seconds CPU seconds for HSD. In the case of HSD, the CPU times (avg and max) are shorter than the 10 seconds allowed, because the algorithm was prematurely stopped as soon as it found an optimal solution, which happened on all problems of table 8:3.

Table 8:4 shows the results of HBEND and of HSD on the second set of test problems. In the case of HBEND we report the results obtained introducing 670 cuts, because the introduction of a greater number of cuts experimentally does not produce any improvement on the quality of the solutions found.

Table 8:4. Distance from optimality and CPU times on the 20 activities problems

	HBEND				HSD	
Num. Benders cuts	100	500	670	1000		
CPU interval (sec.)					10	60
GAP% avg.	8.08	4.24	2.13	2.13	6.05	2.38
GAP% max	80.11	50.20	16.66	16.66	78.95	51.61
CPU time avg.	6.32	32.11	43.03	80.27	10.00	60.00
CPU time max	8.41	43.10	56.47	141.33	10.00	60.00

Both tables 8:3 and 8:4 show that HBEND requires a number of cuts before achieving good results, and that a threshold on their number exists (200 in the case of 10 activities, 670 in the case of 20 activities), after which adding more cuts does not produce any improvement.

On the 10 activities projects (see table 8:3), HSD performs better than HBEND, which in turns produces better solutions than HKD. When the number of activities per project increases (see table 8:4), HBEND outperforms also HSD both in terms of solution quality and of speed of convergence.

A further comparison of procedures HBEND and HSD was made on the third set of problems, with 30 activities. The results of HSD have been made publicly available on the Internet at the URL ftp://ftp.bwl.uni-kiel.de/pub/operations-research/progen/. As we do not know the CPU time required by HSD for producing the results, we ran HBEND with 1000, 1500 and 2000 cuts and compared the quality of the obtained solutions against that produced by HSD. The result of the comparison is reported in figures 8:2, 8:3 and 8:4. In each of these figures, we report the number of instances for which the difference between the solution generated by HSD and that generated by HBEND (i.e., HSD – HBEND) is equal to the abscissa value; negative values correspond therefore to problems for which HBEND produced worse solutions than HSD.

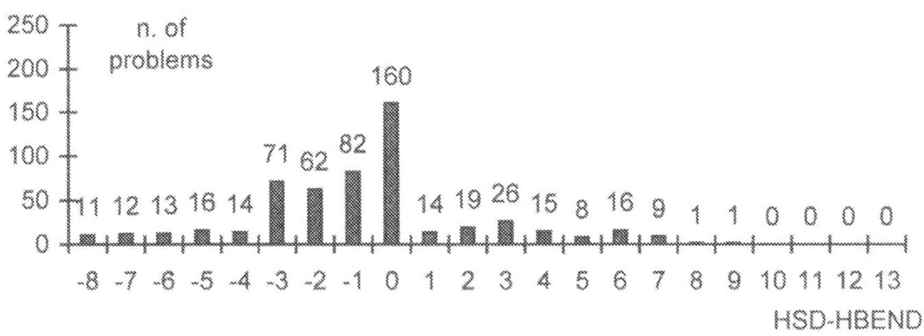

Figure 8:2. HBEND vs. HSD (1000 Benders cuts)

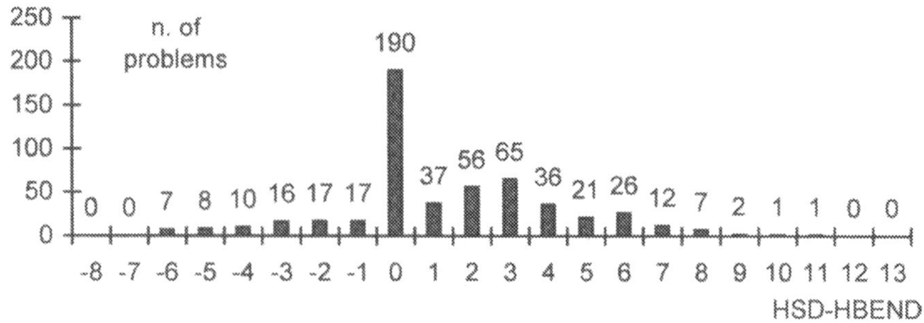

Figure 8:3. HBEND vs. HSD (1500 Benders cuts)

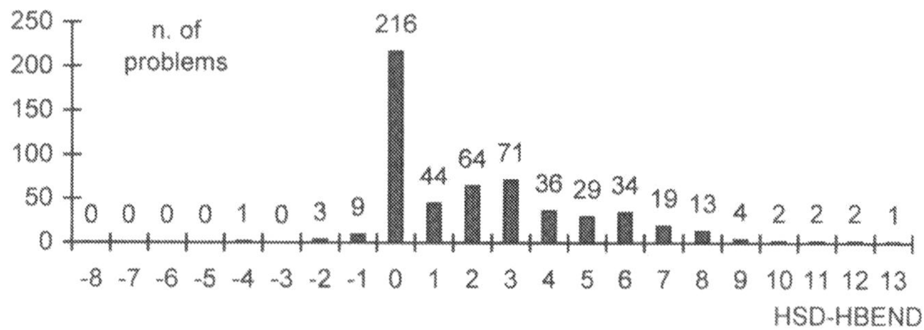

Figure 8:4. HBEND vs. HSD (2000 Benders cuts)

Figures 8:2 to 8:4 show that, by increasing the number of allowed Benders cuts, the comparative effectiveness of the HBEND heuristic improves substantially, moving from a situation where the majority of the problems were solved better with HSD to one where the reverse is true.

The percentage gain, obtained as the percentage ratio between the difference HSD-HBEND and the HSD value over all instances of a set, underlines this result. Using 1000 cuts the percentage gain has been of –2.4, and the average CPU time needed by HBEND to produce its solutions is of 260.34 sec.; with 1500 cuts HBEND already performs better than HSD having a gain of 4.0 (average CPU time 390.28 sec), while with 2000 cuts the gain rises to 6.7 (average CPU time 574.28 sec).

8.6. Conclusions

This paper focuses on a resource-constrained scheduling problem where the activities have different execution modes and resources can be renewable and nonrenewable (MRCPSP). By means of a new mathematical formulation, we derived new lower bounds for MRCPSP and identified a Benders type decomposition strategy which lead to the design of a new heuristic for MRCPSP, called HBEND.

The lower bounds have been obtained by means of different relaxations of the new mathematical formulation. All of them dominate the longest path bound, the most widely used bound in current tree search exact algorithms for MRCPSP.

The new heuristic procedure has been computationally evaluated on standard problem testsets against two state-of-the-art heuristics proposed in the literature. The results show that HBEND requires a number of Benders cuts increasing with the problem dimension before providing good results, but when a sufficient number of cuts is granted, the quality of the solution obtained compares favorably against that of the solution obtained by competing heuristics in comparable CPU time.

References

BENDERS J.F. 1962. Partitioning Procedures for Solving Mixed-Variables Programming Problems, *Numerische Mathematik*, **4**, 280 – 322.

DREXL A., GRUNEWALD J. 1993. Nonpreemptive multi-mode resource-constrained project scheduling, *IIE Transaction*, **25** (5), 74 – 81.

ELMAGHRABY S.E. 1977. *Activity Networks: Project Planning and Control by network Models*, J.Wiley, New York.

KOLISCH R., DREXL A. 1994. Local Search for Nonpreemptive Multi-Mode Resource Constrained Project Scheduling, Research Report No. 360, Institut für Betriebswirtschaftslehre, Christian-Albrechts-Universität zu Kiel.

KOLISCH R., SPRECHER A., DREXL A. 1995. Characterization and Generation of a General Class of Resource-Constrained Project Scheduling Problems, *Management Science*, **41** (11), 1693 – 1703.

MINGOZZI A., MANIEZZO V., RICCIARDELLI S., BIANCO L. 1994. An Exact Algorithm for the Resource Constrained Project Scheduling Problem Based on a New Mathematical Formulation (to appear on *Management Science*, **8**, 1998).

PATTERSON J.H. 1973. Alternate Methods of Project Scheduling with Limited Resources, *Naval Research Logistics Quarterly*, **20**, 767 – 784.

PATTERSON J.H., SLOWINSKY R., TALBOT F.B., WEGLARZ J. 1989. An algorithm for a general class of precedence and resource constrained scheduling problems, in Slowinsky R., Weglarz J. (Eds), *Advances in project scheduling*, Elsevier, Amsterdam, 3 – 28.

SLOWINSKI R., SONIEWICKI B., WEGLARZ J. 1994. DSS for multiobjective project scheduling subject to multiple-category resource constraints, *European Journal of Operational Research*, **79**, 220 – 229.

SPRECHER A., DREXL A. 1996. Solving Multi-Mode Resource-Constrained Project Scheduling Problems by a Simple, General and Powerful Sequencing Algorithm. Part I: Theory, Research Report No. 385, Institut für Betriebswirtschaftslehre, Christian-Albrechts-Universität zu Kiel.

SPRECHER A., DREXL A. 1996. Solving Multi-Mode Resource-Constrained Project Scheduling Problems by a Simple, General and Powerful Sequencing Algorithm. Part II: Computation,

Research Report No. 386, Institut für Betriebswirtschaftslehre, Christian-Albrechts-Universität zu Kiel.

SPRECHER A. 1994. Resource-Constrained Project Scheduling. Exact Methods for the Multi-Mode Case, *Lecture Notes in Economics and Mathematical Systems*, n. 409, Springer-Verlag, Berlin.

TALBOT B. 1982. Resource-constrained project scheduling with time-resource tradeoffs: The nonpreemptive case, *Management Science*, **28**, 1197 – 1210.

9 BENCHMARK INSTANCES FOR PROJECT SCHEDULING PROBLEMS

Rainer Kolisch
Christoph Schwindt
Arno Sprecher

Christian–Albrechts–Universität zu Kiel
Universität Karlsruhe
Christian–Albrechts–Universität zu Kiel

9.1 INTRODUCTION

With the development of project scheduling models and methods arose the need for data instances in order to benchmark the solution procedures. Generally, benchmark instances can be distinguished by their origin into real world problems and artificial problems. The analysis of algorithmic performance on real world problem instances is of a high practical relevance, but at the same time it is only an analysis of individual cases. Consequently, general conclusions about the algorithms cannot be drawn. A solution procedure which shows very good performance on one real world instance might produce poor results on another. In order to allow a systematic evaluation of the performance of algorithms, characteristics of the projects have to be identified. The characteristics can then serve as the parameters for the systematic generation of artificial instances. The variation of the levels of these problem parameters in a full factorial design study allows to produce a set of well–balanced instances (cf. Montgomery 1976).

The majority of experimental investigations on project scheduling methods have used artificially generated problem instances. Exceptions are, e.g., Valadares Tavares (1986) and Tsubakitani & Deckro (1990). In order to free researchers from the demanding task of generating well–balanced benchmark instance sets, we propose a library with benchmark instances for different types of project scheduling problems such as duration and levelling problems with single and multiple modes, renewable and nonrenewable resources, and minimum and maximum time lags. The library is fully accessible under the world wide web (WWW). Latest benchmark results (lower bounds, upper bounds, and objective function values of optimal solutions) are available, too. The benchmark results are updated whenever improved results are submitted to the library via e–mail.

We proceed as follows: In Section 9.2 we use a generic model in order to describe different types of project scheduling problems for which benchmark instances are provided in the library. In Section 9.3 we provide a brief summary of benchmark sets and problem generators which have been proposed in the literature. Section 9.4 presents the instance library. The functionality is outlined and the available benchmark sets are characterized. The chapter ends with a summary and an outlook in Section 9.5.

9.2 MODELS

The project scheduling problems PSP to be dealt with throughout this paper can be stated as follows: A set $\mathcal{V} = \{0, 1, \ldots, n, n+1\}$ of activities has to be processed. Fictitious activity 0 and $n+1$ correspond to the "project start" and to the "project end", respectively. The activities use renewable resources and consume nonrenewable resources. The sets of renewable and nonrenewable resources are denoted by $\mathcal{K}^\rho$ and $\mathcal{K}^\nu$, respectively. For each renewable resource $k \in \mathcal{K}^\rho$, R_k^ρ units are available at any point in time. For each nonrenewable resource $k \in \mathcal{K}^\nu$, R_k^ν units are available in total.

Each activity $j \in \mathcal{V}$ has to be performed in one out of its several alternative execution modes $m_j \in \mathcal{M}_j$.

If performed in mode m_j, activity j uses $r_{kjm_j}^\rho$ units of the renewable resources $k \in \mathcal{K}^\rho$ while being processed and consumes $r_{kjm_j}^\nu$ units of the nonrenewable resources $k \in \mathcal{K}^\nu$. As for the resource requirements, the durations

p_{jm_j} of activities $j \in \mathcal{V}$ depend on the execution mode $m_j \in \mathcal{M}_j$. While being processed, activities must not be interrupted.

Between the starts of two activities minimum and maximum time lags may be prescribed. In general, minimum and maximum time lags depend on the execution modes of the two activities. Let $i, j \in \mathcal{V}$ be two different activities executed in modes m_i and m_j, respectively. A minimum time lag $d^{min}_{im_ijm_j} \geq 0$ means that activity j can start $d^{min}_{im_ijm_j}$ units of time after the start of activity i at the earliest. A maximum time lag $d^{max}_{im_ijm_j} \geq 0$ implies that activity j has to be started $d^{max}_{im_ijm_j}$ units of time after the start of activity i at the latest. A minimum time lag $d^{min}_{im_ijm_j} = p_{im_i}$ is called precedence constraint between activities i and j, whereas general minimum and maximum time lags are referred to as time constraints. In the sequel, we will distinguish between *precedence–constrained problems* and *time–constrained problems.* In precedence–constrained project scheduling problems no maximum time lags have to be observed and any minimum time lag corresponds to a precedence constraint.

In the following, all input data are supposed to be integer–valued. Activities $j \in \mathcal{V}$ as well as minimum and maximum time lags can be represented by an activity–on–node project network $\vec{G} = \langle \mathcal{V}, \mathcal{E}; \boldsymbol{d} \rangle$ with node set $\mathcal{V}$, arc set $\mathcal{E}$, and arc weights $\boldsymbol{d}$. If between two activities i and j a minimum time lag is given, we introduce an arc $\langle i, j \rangle$ from node i to node j weighted by the matrix $(d^{min}_{im_ijm_j})_{m_i \in \mathcal{M}_i, m_j \in \mathcal{M}_j}$. Maximum time lags between activities i and j are represented by (backward) arcs $\langle j, i \rangle$ with weight $(-d^{max}_{im_ijm_j})_{m_i \in \mathcal{M}_i, m_j \in \mathcal{M}_j}$. In general, network $\vec{G}$ may be cyclic. There is no feasible solution to the (PSP) instance under consideration if $\vec{G}$ contains cycles of positive length.

For each activity $j \in \mathcal{V}$ we have to determine a starting time $S_j \geq 0$ and an execution mode $m_j \in \mathcal{M}_j$ such that usage of the renewable resources by activities performed simultaneously does not exceed the resource capacities at any point in time, the consumption of nonrenewable resources is restricted to the respective availabilities, the minimum and maximum time lags are met, and a given objective function in the activities' starting times and mode assignments is minimized.

A schedule can be specified by a vector $\boldsymbol{S}$ of starting times

$$\boldsymbol{S} = (S_0, \ldots, S_{n+1})$$

and a vector $\boldsymbol{x}$ of binary-valued mode assignments

$$\boldsymbol{x} = (x_{01}, x_{11}, \ldots, x_{1|\mathcal{M}_1|}, \ldots, x_{n1}, \ldots, x_{n|\mathcal{M}_n|}, x_{n+1,1})$$

A binary decision variable x_{jm_j} equals 1 iff activity j is performed in mode m_j. Given a schedule $(\boldsymbol{S}, \boldsymbol{x})$ and a time instant t, the set of activities which are in process at time t is

$$\mathcal{V}(\boldsymbol{S},\boldsymbol{x},t) \quad = \quad \{j \in \mathcal{V} \,|\, S_j \le t < S_j + \sum_{m_j \in \mathcal{M}_j} x_{jm_j} p_{jm_j}\}.$$

The corresponding usage of renewable resource $k \in \mathcal{K}^\rho$ at time t, is then given by

$$r_k^\rho(\boldsymbol{S},\boldsymbol{x},t) = \sum_{j \in \mathcal{V}(\boldsymbol{S},\boldsymbol{x},t)} r_{kjm_j}^\rho$$

and the consumption of resource $k \in \mathcal{K}^\nu$ is determined by

$$r_k^\nu(\boldsymbol{x}) \quad = \quad \sum_{j \in \mathcal{V}} \sum_{m_j \in \mathcal{M}_j} r_{kjm_j}^\nu x_{jm_j}$$

Defining $M := \sum_{j \in \mathcal{V}} |\mathcal{M}_j|$ to be the total number of execution modes we

$\min f(\boldsymbol{S},\boldsymbol{x})$		(1)
s.t.		
$S_0 = 0$		(2)
$S_j - S_i \ge \sum_{m_i \in \mathcal{M}_i} \sum_{m_j \in \mathcal{M}_j} x_{im_i} x_{jm_j} d_{im_ijm_j}$	$(\langle i,j \rangle \in \mathcal{E})$	(3)
$\sum_{j \in \mathcal{M}_j} x_{jm_j} = 1$	$(j \in \mathcal{V})$	(4)
$r_k^\rho(\boldsymbol{S},\boldsymbol{x},t) \le R_k^\rho$	$(k \in \mathcal{K}^\rho; t \ge 0)$	(5)
$r_k^\nu(\boldsymbol{x}) \le R_k^\nu$	$(k \in \mathcal{K}^\nu)$	(6)
$\boldsymbol{S} \ge 0$		(7)
$\boldsymbol{x} \in \{0,1\}^M$		(8)

Table 9:1 The PSP Model

obtain the representation of the PSP in Table 9:1. (1) defines the objective function $f(\cdot,\cdot)$ as a function of the starting times and the mode assignments. (2) fixes the project start at time 0. (3) describes the minimum and maximum time lags that have to be taken into account. (4) secures that each activity is assigned exactly one execution mode. (5) and (6) ensure that the renewable and the nonrenewable resource constraints are met. (7) and (8) define the domain of the decision variables.

9.2.1 Project Duration Problems

The PSP with objective $\min f(\boldsymbol{S}, \boldsymbol{x}) := S_{n+1}$ is called multi–mode resource–constrained project scheduling problem with minimum and maximum time lags (MRCPSP/max). A MRCPSP/max with only a single processing alternative per activity is referred to as resource–constrained project scheduling problem with minimum and maximum time lags (RCPSP/max). If all the time constraints of MRCPSP/max and RCPSP/max are precedence constraints then we obtain the multi–mode resource–constrained project scheduling problem (MR-CPSP) and the resource–constrained project scheduling problem (RCPSP), respectively.

Several exact and heuristic procedures have been proposed for the RCPSP. The most competitive exact algorithms are from Brucker et al. (1998), Demeulemeester & Herroelen (1997), Mingozzi et al. (1998), and Sprecher (1996). An extensive survey of heuristic algorithms for the RCPSP is given by Kolisch and Hartmann (1998) in Chapter 7 of this book. The best performing exact algorithm for the MRCPSP is the one of Sprecher & Drexl (1996a, 1996b). Different exact algorithms have been compared by Hartmann & Drexl (1997). The best heuristic procedure for solving the MRCPSP is currently the genetic algorithm of Hartmann (1997). A survey of optimal and heuristic methods for the MRCPSP is given in Kolisch & Padman (1997) .

Heilmann (1998) devises a branch–and–bound algorithm for solving the MRCPSP/max where resource conflicts and the determination of appropriate modes are treated alternately. Starting with fictitious minimal modes, a feasible solution has been reached if each activity has been assigned to a real mode and no resource conflict occurs in the current schedule.

Exact algorithms for the RCPSP/max can be found in Bartusch et al. (1988), De Reyck & Herroelen (1996), and Schwindt (1997). All these branch–and–bound algorithms rely on the relaxation of the resource constraints which are subsequently ensured by the stepwise introduction of additional precedence constraints. The most advanced priority–rule based heuristics for RCPSP/max have been proposed by Franck & Neumann (1997).

9.2.2 Resource Levelling Problems

Resource levelling problems correspond to PSP with unlimited resource capacities, a maximum time lag $\overline{d} := d^{max}_{0,1,n+1,1}$ between the start and the completion of the project (deadline), and an objective function $f(\boldsymbol{S}, \boldsymbol{x})$ in the jumps of the resource profiles $r_k^\rho(\boldsymbol{S}, \boldsymbol{x}, t)$ of renewable resources $k \in \mathcal{K}^\rho$. Following Zimmermann (1997) it can be distinguished between problems for which the maximal resource requirements over time are to be minimized (resource investment problems), problems for which the overloads w.r.t. a given reference profile have to be kept as small as possible (minimal deviation problems), and problems for which the variation in the resource profiles have to be minimized (minimal variation problems).

To the best of our knowledge, the literature on multi–mode resource levelling problems is completely void. In the following, we will briefly sketch out two approaches to single–mode resource–levelling problems with minimum and maximum time lags.

A branch–and–bound algorithm for the resource investment problem with minimum and maximum time lags (RIP/max) and objective function

$$f(\boldsymbol{S}, \boldsymbol{x}, t) = \sum_{k \in \mathcal{K}^\rho} c_k \max_t r_k(\boldsymbol{S}, \boldsymbol{x}, t)$$

has been proposed by Nübel (1998). Starting with the vector $\boldsymbol{ES}$ of earliest (time–feasible) starting times, activities are right–shifted such that at least one peak of the resource profile of a renewable resource is decreased. By enumeration of alternatives to delay activities, an optimal solution can be determined.

Priority–rule based methods for various types of resource–levelling problems with minimum and maximum time lags can be found in Neumann & Zimmermann (1997). The activities of the project are scheduled consecutively extending the current partial schedule. Depending on the type of the resource–levelling problem under consideration, the starting time (completion time) of an eligible activity is set to the completion (start) of an activity in the partial schedule.

9.3 DATA SETS AND INSTANCE GENERATORS

9.3.1 Data Sets

So far, three widely used instance sets of project scheduling have been proposed: the instance sets of Patterson (1984), of Alvarez–Valdés & Tamarit (1989), of Kolisch et al. (1995) and of Kolisch & Sprecher (1996). An overview of the use of the instance sets can be found in Kolisch & Sprecher (1996) and in Kolisch & Padman (1997).

Patterson (1984) collected 110 instances for the RCPSP from different sources – 83 instances from Davis (1969), 11 instances from Patterson & Huber (1974), 10 instances from Talbot & Patterson (1978), and 6 more instances generated by Davis (1969). The benchmark set consists of projects with n=5 to 49 activities. A detailed analysis of the data can be found in Kolisch (1996). While the great benefit of the Patterson problems was the availability of optimal makespan values and the wide usage of the test set for both exact and heuristic algorithm evaluation, a major disadvantage was the fact that the instances were not generated based on well-defined problem parameters.

Alvarez-Valdés & Tamarit (1989) generated 144 instances for the RCPSP. The benchmark set consists of projects with n=25, 49, and 101 activities. Five different problem parameters have been used to generate the instances in a full factorial design.

Kolisch et al. (1995) generated 480 instances for the RCPSP and 536 instances for the MRCPSP with their project generator ProGen. The RCPSP benchmark set consists of projects with n=30 activities and the MRCPSP benchmark set consists of projects with n=10 activities. Later, Kolisch & Sprecher (1996) extended these instance sets, and made it available via ftp.

9.3.2 *Instance Generators*

Kolisch et al. (1995) were the first to propose a full-size instance generator, named ProGen, for a general class of project scheduling problems. The instances are represented by activity-on-node networks. ProGen allows to generate problem instances in accordance with a full factorial design of a wide range of well-defined problem parameters. A new problem parameter, called resource strength, measures the availability of a specific resource as a convex combination of a lower bound and an upper bound of the resource availability. The resource strength can be applied to different categories and types of resources. Schwindt (1996) extended ProGen to ProGen/max in order to incorporate minimum and maximum time lags. Further problem parameters, such as the estimator of the network restrictiveness, suggested by Thesen (1977), and the project's slack factor have been taken into account. The estimator of the network restrictiveness – for activity-on-node networks – measures the number of precedence-feasible activity sequences as a fraction of the number of all activity sequences. The slack factor of a project corresponds to the ratio of the project deadline and the earliest time-feasible completion time of the project. Drexl et al. (1997) extended ProGen to ProGen/πx in order to incorporate new modelling extensions such as partially renewable resources, changeover times, mode identity and set of mode identity. A survey of these modelling extensions is given by Drexl et al. (1998) in Chapter 18 of this book.

Beside ProGen and its extensions, generators have been developed by Agrawal et al. (1996) and Demeulemeester et al. (1993).

Agrawal et al. (1996) developed an instance generator for activity-on-arc networks which employs the problem parameter complexity index as introduced by Bein et al. (1992). The complexity index measures how far a given network is from a series–parallel network. Costs, duration, and resource requirements associated with the activities are generated randomly from uniform distributions.

Demeulemeester et al. (1993) proposed a generator for strongly random activity–on–arc networks for a given number of arcs and nodes.

9.4 A PROJECT SCHEDULING LIBRARY

In what follows we outline a project scheduling problem library which has been first proposed in Kolisch & Sprecher (1996). The current state is much more elaborated. It contains new benchmark sets for the RCPSP and MR-CPSP and comprises new problems of the type RCPSP/max, MRCPSP/max, RIP/max, and RLP/max. Finally, the whole library is now web–based which eases its use. The library is accessible under the address *http://www.bwl.uni-kiel.de/Prod/psplib/index.html.* Since the library is self contained and underlies a dynamic development we will not give details here. Rather, we present the general functionality and structural principles. All information contained in the library can be classified into three broad categories. First, general information about the library, second, benchmark instances, and third, access to instance generators. We will refer to the latter two categories in the Subsections 9.4.1 and 9.4.2. General information about the library include the scope of the library and its development which is given in Kolisch & Sprecher (1996) and which will not be detailed here. Also, technical information about downloading instance sets and submitting improved solutions are given in the library under the category general information.

9.4.1 Benchmark Instances

The benchmark instances are the core of the library. Currently, benchmark sets for the six different type of project scheduling models which have been given in Section 9.2 are available. For each type of model there is at least one instance set. Each instance set is characterized by the levels of the fixed and systematically varied problem parameters. Details of the parameter levels are provided in the library. All instances which belong to one set are packed in a file with UNIX–based tgz–format. Up to three additional files can come with the tgz–file. The opt–file contains the optimal objective function values, the hrs–file gives the best objective function values which have been obtained by heuristic algorithms, and finally, the lb–file comprises lower and upper bounds where the upper bounds are the same values as given in the hrs–file.

Tables 9:2 to 9:9 present all instance sets available in the library which were generated with ProGen or Progen/max. For each instance set, the levels of the major parameters n, the number of activities, $|\mathcal{K}^\rho|$, the number of renewable resource types, $|\mathcal{K}^\nu|$, the number of nonrenewable resource types, NC, the network complexity, RT, the network restrictiveness, RF^ρ and RF^ν, the resource factor for renewable and nonrenable resources, RS^ρ and RS^ν, the resource strength for renewable and nonrenewable resources, as well as $\overline{d}/ES_{n+1}$, the project's slack factor are given. The parameters have been briefly described in Section 9.2. Details about the problem parameters are provided in Kolisch et al. (1995), Kolisch & Sprecher (1996), and Schwindt (1996). Postscript files of these papers are also available from the library. In addition to the ProGen– and ProGen/max–instance sets, the library contains the instance sets of Patterson (1984), Alvarez–Valdés & Tamarit (1989), and Boctor (1993). The former two instance sets are described in Section 9.3.

9.4.2 Access to Instance Generators

Researchers might not always find the particular test instances needed in the library because different types of project scheduling models are needed or instances with specific problem parameter levels are required. In this case the project scheduling generators ProGen and ProGen/max as outlined in Subsection 9.3.2 can be used in order to generate indiviudal benchmark instances. The generators can be employed in two different ways. First, each program itself can be downloaded and second, each program can be used remotely through the MatSe–System. The system provides an electronic marketplace that supports researchers to create Web-accessible computational services. It combines software that was originally developed for Web-based statistical computing (cf. Guenther et al. 1997) with a language for large scale programming on the Internet (cf. Becker 1996) .

9.5 SUMMARY AND OUTLOOK

A dynamic extension of the library is planned for the future. First, new benchmark solutions for currently available instance sets will be updated according to the progress made by researchers. Second, larger instance sets for problem types which are already represented in the library will be added. Third, paralleling the emergence of new modelling concepts as reported by Drexl et al. (1998) in Chapter 18, we will add instance sets for new types of project scheduling models. This will comprise project scheduling problems with non–regular performance measures such as the maximization of the net present value (cf. Herroelen et al. 1997 and Kolisch & Padman 1997). Finally, we plan to enhance the library by adding real world problems.

Instance Set	No. of instances	Fixed Levels n	Fixed Levels $\|\mathcal{K}^\rho\|$	Variable Levels NC	Variable Levels RF^ρ	Variable Levels RS^ρ
j30	480	30	4	1.5	0.25	0.2
j60	480	60	4	1.8	0.50	0.5
j90	480	90	4	2.1	0.75	0.7
					1.00	1.0

Table 9:2 Fixed and Varied Parameter Levels RCPSP — Instance Sets j30, j60, and j90

Instance Set	No. of instances	Fixed Levels n	Fixed Levels $\|\mathcal{K}^\rho\|$	Variable Levels NC	Variable Levels RF	Variable Levels RS
j120	600	120	4	1.5	0.25	0.1
			4	1.8	0.50	0.2
			4	2.1	0.75	0.3
					1.00	0.4
						0.5

Table 9:3 Fixed and Varied Parameter Levels RCPSP — Instance Set j120

Set	No. of Inst.	Fixed Levels n	$\|\mathcal{M}_j\|$	$\|\mathcal{K}^\rho\|$	$\|\mathcal{K}^\nu\|$	NC	Varied Levels RF^ρ	RS^ρ	RF^ν	RS^ν
j10	536	10	3	2	2	1.8	0.50	0.2	0.50	0.2
							1.00	0.5	1.00	0.5
								0.7		0.7
								1.0		1.0

Table 9:4 Fixed and Varied Parameter Levels MRCPSP — Instance Set j10

Set	No. of Inst.	Fixed Levels n	$\|\mathcal{M}_j\|$	$\|\mathcal{K}^\rho\|$	$\|\mathcal{K}^\nu\|$	NC	Varied Levels RF^ρ	RS^ρ	RF^ν	RS^ν
j20	554	20	3	2	2	1.8	0.50	0.25	0.50	0.25
j30	640	30	3	2	2	1.8	1.00	0.50	1.00	0.50
								0.75		0.75
								1.00		1.00

Table 9:5 Fixed and Varied Parameter Levels MRCPSP — Instance Sets j20 and j30

Set	No. of Inst.	Fixed Levels n	$\|\mathcal{K}^\rho\|$	Varied Levels RT	RF^ρ	RS^ρ
j10	270	10	5	0.25	0.50	0.00
j20	270	20	5	0.50	0.75	0.25
j30	270	30	5	0.75	1.00	0.50

Table 9:6 Fixed and Varied Parameter Levels RCPSP/max

Set	No. of Inst.	Fixed Levels n	RF^ρ	RF^{nu}	$\|\mathcal{K}^\rho\|$	$\|\mathcal{K}^\nu\|$	Varied Levels $\|\mathcal{M}_j\|$	RT	$RS^\rho = RS^\nu$
j10	270	10	1.0	1.0	5	2	2	0.25	0.00
j20	270	20	1.0	1.0	5	2	3	0.50	0.25
j30	270	30	1.0	1.0	5	2	5	0.75	0.50

Table 9:7 Fixed and Varied Parameter Levels MRCPSP/max

Set	No. of Inst.	Fixed Levels n	Fixed Levels RF^ρ	Varied Levels RT	Varied Levels $\|\mathcal{K}^\rho\|$	Varied Levels $\overline{d}/ES_{n+1}$
j10	270	10	1.0	0.25	1	1.00
j20	270	20	1.0	0.50	3	1.25
j30	270	30	1.0	0.75	5	1.50

Table 9:8 Fixed and Varied Parameter Levels RIP/max

Set	No. of Inst.	Fixed Levels n	Fixed Levels RF^ρ	Varied Levels RT	Varied Levels $\|\mathcal{K}^\rho\|$	Varied Levels $\overline{d}/ES_{n+1}$
j10	270	10	1.0	0.25	1	1.00
j20	270	20	1.0	0.50	3	1.25
j30	270	30	1.0	0.75	5	1.50

Table 9:9 Fixed and Varied Parameter Levels RLP/max

References

AGRAWAL, M.K., ELMAGHRABY, S.E., AND W. HERROELEN 1996. DAGEN: A generator of testsets for project activity nets. *European Journal of Operational Research*, **90**, 376–382.

ALVAREZ–VALDÉS, R. AND J.M. TAMARIT 1989. Heuristic algorithms for resource–constrained project scheduling: A review and an empirical analysis. *Advances in project scheduling*. 113–134, Słowiński, R. and J. Węglarz (eds.). Elsevier, Amsterdam.

BARTUSCH, M., MÖHRING, R.H., AND F.J. RADERMACHER 1988. Scheduling project networks with resource constraints and time windows. *Annals of Operations Research*, **16**, 201–240.

BECKER, P. 1996. An embeddable and extendable language for large-scale programming on the Internet. *Proceedings of the 16–th International Conference on Distributed Computing Systems (ICDCS'96)*.

BEIN, W.W., KAMBUROWSKI, J., AND M.F.M. STALLMANN 1992. Optimal reduction of two–terminal directed graphs. *SIAM Journal on Computing*, **221**, 6, 1112–1129.

BOCTOR, F. 1993. Heuristics for scheduling projects with resource restrictions and several resource duration modes. *International Journal of Production Research*, **31**, 11, 2547–2558.

BRUCKER, P., KNUST, S., SCHOO, A., AND O. THIELE 1998. A branch & bound algorithm for the resource–constrained project scheduling problem. *European Journal of Operational Research*, forthcoming.

DAVIS, E.W. 1969. An exact algorithm for the multiple constrained project scheduling problem. *PhD thesis*, Yale University.

DEMEULEMEESTER, E., DODIN, B., AND W. HERROELEN 1993. A random activity network generator. *Operations Research*, **41**, 972–980.

DEMEULEMEESTER, E. AND W. HERROELEN 1997. New benchmark results for the resource–constrained project scheduling problem. *Management Science*, **43**, 11, 1485–1492.

DE REYCK, R. AND W. HERROELEN 1996. A Branch–and Bound Procedure for the Resource–Constrained Project Scheduling Problem with Generalized Precedence Relations. *Technical Report 9613, Operations Management Group*, Department of Applied Economics, Katholieke Universiteit Leuven.

DREXL, A., JURETZKA, J., SALEWSKI, F., AND A. SCHIRMER 1998. New modelling concepts and their impact on resource–constrained project scheduling. *Handbook on recent advances in project scheduling*. Chapter 18, Weglarz, J. (ed.). Kluwer, Dordrecht.

DREXL, A., NISSEN, R., PATTERSON, J.H., AND F. SALEWSKI 1997. ProGen/πx — An instance generator for resource–constrained project scheduling problems with partially renewable resources and further extensions *Technical Report, Institut für Betriebswirtschaftslehre*, Universität Kiel.

FRANCK, B. AND K. NEUMANN 1997. Resource–constrained project scheduling with time windows — structural questions and priority–rule methods. *Technical Report 492, Institut für Wirtschaftstheorie und Operations Research*, Universität Karlsruhe.

GÜNTHER, O., MÜLLER, R., SCHMIDT, P., BHARGAVA, H.K., AND R. KRISHNAN 1997. MMM: A Web–based system for sharing statistical computing modules. *IEEE Internet Computing*, **3**, 1, 59–68.

HARTMANN, S. 1997 Project scheduling with multiple modes: A genetic algorithm. *Technical Report 435, Manuskripte aus den Instituten für Betriebswirtschaftslehre der Universität Kiel.*

HARTMANN, S. AND A. DREXL 1997. Project scheduling with multiple modes: A comparison of exact algorithms. *Technical Report 430, Manuskripte aus den Instituten für Betriebswirtschaftslehre der Universität Kiel.*

HERROELEN, W., DEMEULEMEESTER, E., AND B. DE REYCK 1997. Project network models with discounted cash flows: A guided tour through recent developments. *European Journal of Operational Research*, **100**, 1, 97–121.

HEILMANN, R. 1998. A branch–and–bound procedure for the MRCPSP/max. *Technical Report 512, Institut für Wirtschaftstheorie und Operations Research*, Universität Karlsruhe.

KOLISCH, R. AND S. HARTMANN 1998. Heuristic algorithms for the resource constrained project scheduling problem: Classification and Computational Analysis. *Handbook on recent advances in project scheduling.* Chapter 7, Weglarz, J. (ed.). Kluwer, Dordrecht.

KOLISCH, R. AND R. PADMAN 1997. An integrated survey of project scheduling. *Technical Report 463, Manuskripte aus den Instituten für Betriebswirtschaftslehre der Universität Kiel.*

KOLISCH, R. 1996. Efficient priority rules for the resource–constrained project scheduling problem. *Journal of Operations Management*, **14**, 3, 179–192.

KOLISCH, R. AND A. SPRECHER 1996. PSPLIB – a project scheduling problem library. *European Journal of Operational Research*, **96**, 205–216.

KOLISCH, R., SPRECHER, A., AND A. DREXL 1995. Characterization and generation of a general class of resource–constrained project scheduling problems. *Management Science*, **41**, 10, 1693–1703.

MINGOZZI, A., MANIEZZO, V., RICCIARDELLI, S., AND L. BIANCO 1998. An exact algorithm for project scheduling with resource constraints based on a new mathematical formulation. *Management Science*, forthcoming.

MONTGOMERY, D.C. 1976. *Design and analysis of experiments.* Wiley, New York.

NEUMANN, K. AND J. ZIMMERMANN 1997. Resource–levelling for projects with schedule–dependent time windows, *Technical Report 508, Institut für Wirtschaftstheorie und Operations Research*, Universität Karlsruhe.

NÜBEL, H. 1998 A branch–and–bound procedure for the resource investment problem with generalized precedence constraints. *Technical Report 516, Institut für Wirtschaftstheorie und Operations Research*, Universität Karlsruhe.

PATTERSON, J.H. 1984. A comparison of exact approaches for solving the multiple constrained resource, project scheduling problem. *Management Science*, **30**, 7, 854–867.

PATTERSON, J.H. AND W.D. HUBER 1974. A horizon–varying, zero–one approach to project scheduling. *Management Science*, **20**, 990–998.

SCHWINDT, C. 1997. Verfahren zur Lösung des ressourcenbeschränkten Projektdauerminimierungsproblems mit planungsabhängigen Zeitfenstern. *Doctoral Dissertation*, Universität Karlsruhe.

SCHWINDT, C. 1996. Generation of resource–constrained project scheduling problems with minimal and maximal time lags. *Technical Report 489, Institut für Wirtschaftstheorie und Operations Research*, Universität Karlsruhe.

SPRECHER, A. 1996. Scheduling resource constrained projects competitively at modest memory requirements. *Technical Report 425, Manuskripte aus den Instituten für Betriebswirtschaftslehre der Universität Kiel.*

SPRECHER, A. AND A. DREXL 1996a. Solving multi–mode resource–constrained project scheduling by a simple, general, and powerful sequencing algorithm. Part I: Theory. *Technical Report 385, Manuskripte aus den Instituten für Betriebswirtschaftslehre der Universität Kiel.*

SPRECHER, A. AND A. DREXL 1996b. Solving multi–mode resource–constrained project scheduling problems by a simple, general, and powerful sequencing algorithm. Part II: Computation. *Technical Report 386, Manuskripte aus den Instituten für Betriebswirtschaftslehre der Universität Kiel.*

TALBOT, F.B. AND J.H. PATTERSON 1978. An efficient integer programming algorithm with network cuts for solving resource–constrained scheduling problems. *Management Science*, **24**, 11, 1163–1174.

THESEN, A. 1977. Measures of the restrictiveness of project networks. *Networks*, **7**, 193–208.

TSUBAKITANI, S. AND R.F. DECKRO 1990. A heuristic for multi–project scheduling with limited resources in the housing industry. *European Journal of Operational Research*, **49**, 80–91.

VALADARES TAVARES, L. 1986. Multicriteria scheduling of a railway renewal program. *European Journal of Operational Research*, **25**, 395–405.

ZIMMERMANN, J. 1997. Heuristics for resource–levelling problems in project scheduling with minimum and maximum time lags, *Technical Report 491,*

Institut für Wirtschaftstheorie und Operations Research, Universität Karlsruhe.

10 A SURVEY OF INTERVAL CAPACITY CONSISTENCY TESTS FOR TIME- AND RESOURCE-CONSTRAINED SCHEDULING

Ulrich Dorndorf
Toàn Phan Huy
Erwin Pesch

University of Bonn, Faculty of Economics, BWL 3
Adenauerallee 24-42, D-53113 Bonn, Germany
udorndorf@acm.org

10.1 Overview

Interval capacity consistency tests consider the resource capacities available and required within certain time intervals. The goal of the tests is to draw conclusions that allow to rule out inadmissible activity start times or sequences. The tests can be effectively used to reduce the search space of difficult time- and resource-constrained scheduling problems. They have successfully been applied in algorithms for solving idealised problems such as the classical job shop scheduling problem (JSP) or the resource-constrained project scheduling problem (RCPSP) as well as for solving industrial scheduling problems. For instance, it seems fair to say that the advances in modern branch and bound algorithms for the JSP that have been made in the last decade can to a large extent be attributed to the effect of interval consistency tests, some of which are also known under the names of immediate selection, edge finding, and energetic reasoning. The tests can also serve to derive tight lower bounds for makespan minimisation problems.

This chapter presents interval consistency tests for disjunctive and general resource-constrained (cumulative) scheduling within a unified framework, using numerous examples for illustration. We review the state of the art and derive some new results for disjunctive scheduling.

10.2 Introduction

Time and resource constrained scheduling is concerned with the task of scheduling a number of activities subject to temporal constraints between activities, such as precedence or synchronisation, and constraints on the availability of several shared resources. An activity i is characterised by its processing time p_i and resource requirements r_{ik}: It requires r_{ik} units of a renewable resource k for each of p_i time units, and it releases the resource units again upon completion. Once begun, an activity must be processed in an uninterrupted fashion. A resource k is available in constant amount R_k.

By imposing an upper bound on the latest completion time and a lower bound on the earliest start of all activities, activity start time windows can be derived in a straightforward way. The start time window of activity i is the interval between the earliest start time est_i and the latest start time lst_i of i. The *domain* δ_i is the set of all possible start time assignments st_i of i. It is bounded by the start time window; because some values in the start time window may be excluded, e.g. by temporal constraints of the type $st_i \neq t_x$, for some time t_x, we can state in general that $\delta_i \subseteq [est_i, lst_i]$.

The purpose of this chapter is to present a class of logical tests called *interval capacity consistency* tests which are based on resource constraints. These tests allow to reduce activity domains by ruling out infeasible start time assignments. They can be applied in scheduling algorithms such as list scheduling or branch and bound procedures, or in constraint propagation based scheduling systems. The benefit of the tests is that they can reduce the search space and direct an algorithm towards good solutions. Here, we are only interested in the tests themselves and do not address scheduling algorithms in which they can be embedded. Since the tests only eliminate solutions incompatible with the capacity constraints, they are independent of the overall objective function to be optimised. The assumptions that we have made so far are rather general and cover $\mathcal{NP}$-hard models such as the classical job shop problem (JSP) (Błażewicz et al. 1996a,b) and the resource-constrained project scheduling problem with simple (RCPSP) or generalised precedence constraints (RCPSP/max) (Brucker et al. 1998, Domschke and Drexl 1991). In models with generalised precedence constraints time varying resource supply can easily be reflected by introducing dummy activities (Bartusch et al. 1988).

In the following sections we assume that all activity domains have been made consistent with the temporal constraints. Efficient algorithms for this are well known; a detailed description of this process in a constraint propagation based scheduling system is for instance given by Nuijten (1994). We are interested in further reducing the domains by applying interval consistency tests.

In the literature, activity domains are often approximated by start time windows, and this approximation is then referred to as activity release times and due dates, or heads and tails. The domain reduction process may then be called adjustment of heads and tails or time bound adjustment. Specific interval consistency tests have become known under the names immediate selection, edge finding, and energetic reasoning.

The remainder of this chapter is organised as follows. Section 10.3 introduces some additional formal notation and a graphical representation used for the examples throughout the chapter. Section 10.4 discusses the concept of interval consistency and shows the general idea how to deduce domain reductions. Section 10.5 then goes into more detail and takes a bottom up approach to derive the tests for the case of unit resource capacities and requirements (*disjunctive* scheduling). Section 10.6 generalises the results for arbitrary resource capacities and requirements (*cumulative* scheduling).

10.3 Notation

Let us first briefly introduce some additional notation. We denote the set of all activities as $\mathcal{V}$ and the subset of all activities to be processed by resource k as $\mathcal{V}_k \subseteq \mathcal{V}$, with $\mathcal{V}_k = \{i \in \mathcal{V} | r_{ik} > 0\}$. We will often consider subsets $\mathcal{A} \subseteq \mathcal{V}_k$ of activities. To deduce domain reductions for an activity i and activities in a set $\mathcal{A}$ we often try to show that i must start before or finish after all activities in $\mathcal{A}$. This is denoted by $i \to \mathcal{A}$ if i must start first, and $\mathcal{A} \to i$ if i must finish last. We also use the notation $\mathcal{A} \to \mathcal{A}'$, with $\mathcal{A}, \mathcal{A}' \subseteq \mathcal{V}_k$ to express that all activities in set $\mathcal{A}$ must start before all activities in set $\mathcal{A}'$. Table 10.1 summarises the notation used throughout this chapter.

For illustration and motivation of the tests, we use examples in the style of Figure 10.1, which shows two activities that must be processed by the same resource with capacity $R_k = 1$. The style is similar to the one used by Nuijten (1994). Consider

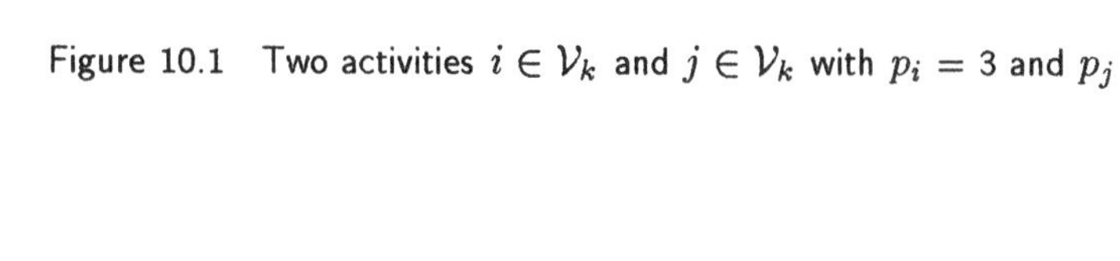

Figure 10.1 Two activities $i \in \mathcal{V}_k$ and $j \in \mathcal{V}_k$ with $p_i = 3$ and $p_j = 4$

activity i where several points on the time scale have been annotated for illustration. The figure shows the time between the earliest start of i, est_i, and its latest completion, lct_i, as a horizontal line segment. For convenience, the processing time of i, p_i, is depicted as a hollow bar beginning at est_i with rounded right end at $ect_i = est_i + p_i$; the length of this bar is, of course, equal to lct_i - lst_i. Admissible start times, i.e. the values in δ_i, are shown as black circles. Times in the interval $[lst_i + 1, lct_i)$ at which i may be in process, but at which it cannot start, are marked with tick marks. You can think of scheduling an activity as positioning the processing time bar at one of the admissible start times. Activity j in Figure 10.1 appears in the usual style without annotations. Initially, possible start times of j are in the interval $[2, 8]$. The x appearing under j's scale at time 2 indicates that we have — by applying a test described below — deduced that j cannot start at time 2.

Table 10.1 Summary of notation

Symbol	*Description*
δ_i	domain of activity i: the set of possible start times of i
est_i	earliest start time of activity i
ect_i	earliest completion time of activity i
lst_i	latest start time of activity i
lct_i	latest completion time of activity i
p_i	processing time of activity i
$p_i(t_1, t_2)$	smallest amount of time during which activity i must be processed in time interval $[t_1, t_2)$
r_{ik}	amount of resource k required by activity i
st_i	actual start time of activity i
t	time period
$w_{ik}(t_1, t_2)$	smallest amount of work of activity i that must fall into the time interval $[t_1, t_2)$: $r_{ik}p_i(t_1, t_2)$
$\mathcal{A}$	set of activities to be processed by resource k: $\mathcal{A} \subseteq \mathcal{V}_k$
$Ect(\mathcal{A})$	preemptive bound for the earliest completion time of activity set $\mathcal{A}$
$\mathcal{H}$	set of hypothetical constraints
$Lst(\mathcal{A})$	preemptive bound for the latest start time of activity set $\mathcal{A}$
$P(\mathcal{A})$	total processing time of activities in set $\mathcal{A}$: $\sum_{i \in \mathcal{A}} p_i$
$P(\mathcal{A}, t_1, t_2)$	interval processing time of $\mathcal{A}$: $\sum_{i \in \mathcal{A}} p_i(t_1, t_2)$
R_k	constant capacity of resource k
$\mathcal{V}$	set of all activities
$\mathcal{V}_k$	set of activities requiring resource k: $\{i \in \mathcal{V} \mid r_{ik} > 0\}$
$\mathcal{V}_k(t_1, t_2)$	set of activities requiring resource k that must be completely or partially processed within $[t_1, t_2)$: $\{i \in \mathcal{V}_k \mid p_i(t_1, t_2) > 0\}$
$W(\mathcal{A})$	total work of activities in set $\mathcal{A}$: $\sum_{i \in \mathcal{A}} w_{ik}$
$W(\mathcal{A}, t_1, t_2)$	interval work of activities in set $\mathcal{A}$: $\sum_{i \in \mathcal{A}} w_{ik}(t_1, t_2)$

10.4 Interval consistency

Before looking at the individual consistency tests, this section introduces a general framework for the tests.

An activity i requires an amount of work $w_{ik} = r_{ik}p_i$ from resource k. We say a time interval is capacity consistent if the amount of work available from resource k within the interval is greater than the interval work required by all activities. Let us consider the work of an activity that must fall into a time interval $[t_1, t_2)$. Figure 10.2 shows the four possible relations between an activity and a given time interval where the activity must be completely or partially processed within the interval. The activity can be (1) completely contained within the interval, (2) completely overlap the interval when started as early (left-shifted) or as late (right-shifted) as possible, (3) have a minimum processing time within the interval that is realised when started as early as possible, or (4) have a minimum processing within the interval that is realised when started as late as possible. The smallest amount of time, or *interval processing time*,

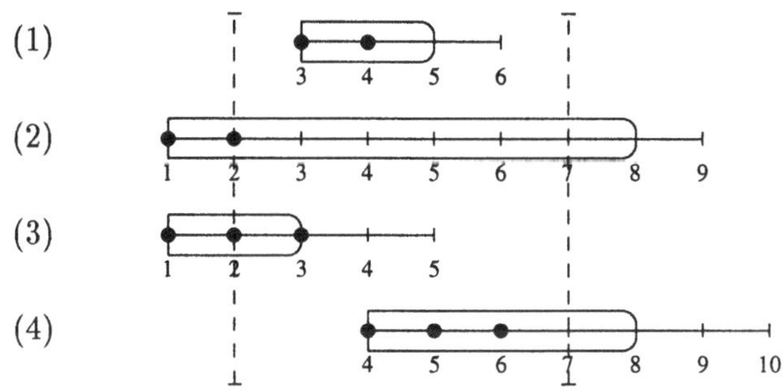

Figure 10.2 Types of intersections between activities and a time interval

during which an activity i must be executed in $[t_1, t_2)$ is

$$p_i(t_1,t_2) = \max\{0, \min\{p_i, t_2 - t_1, ect_i - t_1, t_2 - lst_i\}\}, \tag{1}$$

and the corresponding *interval work* is $w_{ik}(t_1,t_2) = r_{ik}p_i(t_1,t_2)$. The interval processing time for a set $\mathcal{A} \subseteq \mathcal{V}_k$ is $P(\mathcal{A},t_1,t_2) = \sum_{i\in\mathcal{A}} p_i(t_1,t_2)$; the corresponding work within the interval is $W(\mathcal{A},t_1,t_2) = \sum_{i\in\mathcal{A}} w_{ik}(t_1,t_2)$. Inversely, we denote the set of all activities that must be processed completely or partially within $[t_1,t_2)$ as $\mathcal{V}_k(t_1,t_2) = \{i \in \mathcal{V}_k | p_i(t_1,t_2) > 0\}$.

Using the interval work definition we can now state an interval consistency constraint in its general form:

$$R_k \cdot (t_1 - t_2) \geq W(\mathcal{V}_k, t_1, t_2). \tag{2}$$

Clearly, a scheduling problem can only have a feasible solution if this constraint holds for all resources k and time intervals $[t_1,t_2)$. However, this is only a necessary condition and does not guarantee that a feasible schedule exists.

The basic idea behind all interval consistency tests described in this chapter is as follows: We consider a set of additional, hypothetical constraints $\mathcal{H}$ and try to show that if $\mathcal{H}$ is satisfied then Constraint 2 is violated for some resource and time interval; in this case we can conclude: $\neg\mathcal{H}$. This leads to two main questions which we will try to answer in the following sections:

1. How should we choose $\mathcal{H}$ so that the conclusion $\neg\mathcal{H}$ leads to useful domain reductions?

2. For which intervals $[t_1,t_2)$ should we test Constraint 2?

Tests are usually applied in an iterative fashion for as long as some domain can be reduced.

The notion of interval capacity consistency as defined here has to the best of our knowledge first been suggested by Lopez (1991) (see also Lopez et al. 1992) under the name *energetic reasoning*. Schwindt (1997) has independently developed a concept of interval work. He and, independently, Baptiste et al. (1998) were the first to answer Question 2. The area of the rectangles defined by activity processing time and resource requirements can be interpreted as work or energy, and we use the terms interchangeably.

Although our focus is primarily on the use of interval consistency tests for deducing domain reductions, we remark that Constraint 2 can, of course, also serve to derive lower bounds for makespan minimisation problems in the following way: Impose a hypothetical upper bound UB on the makespan; if this leads to a violation of Constraint 2 then $UB + 1$ is a lower bound. This approach is for example used by Nuijten (1994), Pesch and Tetzlaff (1996), Klein and Scholl (1997), Schwindt (1997), Heilmann and Schwindt (1997), and has been called destructive improvement due to the principle of refuting hypothetical constraints. Test values for UB are usually chosen through a dichotomising search. This type of bound argument can, equivalently, be applied whenever any of the tests described in the following sections causes an activity domain to become empty.

Finally, let us point out that resource capacity constraints in the form of Constraint 2, but mostly limited to intervals defined by earliest start and latest completion times of activities, have often been used in constraint logic based scheduling; see e.g. the description of solving a famous bridge scheduling problem (an RCPSP/max instance) in Van Hentenryck (1989) or the implementation of the cumulative constraint in CHIP (Aggoun and Beldiceanu 1993).

10.5 Consistency tests for disjunctive scheduling

The idea behind all tests described in the following subsections is to consider subsets $\mathcal{A} \subseteq \mathcal{V}_k$ of activities to be processed by the same resource k. Within these subsets, all possible activity sequences with a particular property are examined, e.g. the property that the sequence does not start with an activity $i \in \mathcal{A}$. If all such sequences are infeasible, then we can draw the conclusion that the sequence must *not* have this property and deduce that i must be first in $\mathcal{A}$.

The consistency tests are presented in order from strongest to weakest condition. While a stronger condition allows a stronger conclusion, it is at the same time more likely to be inapplicable. After developing the individual tests we generalise the results and show how they relate to the concept of interval consistency.

All tests in this section are derived for disjunctive scheduling problems where all activities $i \in \mathcal{A} \subseteq \mathcal{V}_k$ are mutually exclusive in the sense that they exclusively occupy resource k throughout their processing time. This is of course always the case if all activity sizes and resource capacities are equal to one, as in many machine scheduling problems, for example in the JSP. However, even if $r_{ik} > 1$ for some $i \in \mathcal{V}_k$, and consequently $R_k > 1$ the tests may still be used for subsets of "disjunctive" activities. For instance, it is likely that difficult RCPSP instances contain such disjunctive scheduling elements (Kolisch et al. 1995). Hence, disjunctive consistency tests have been suggested for the makespan minimisation RCPSP (Brucker et al. 1996b, Klein and Scholl 1997, Baptiste et al. 1998) and RCPSP/max (Schwindt 1997).

10.5.1 Input/output test

Figure 10.3 shows an example with a set $\mathcal{A} = \{i, j, k\}$ of three activities to be processed by the same resource. We can deduce that i must be scheduled first in the following way: Suppose i does not start first. Then all three operations must be processed in the interval $[2, 9)$. This means that a total processing time of $8 = 3 + 2 + 3$ must be scheduled in $7 = 9 - 2$ available time units, which is a contradiction. Thus we can conclude that i must start first; we can then deduce that start times of i greater than 1 can be removed from δ_i. Note that this conclusion cannot be drawn by separately considering any two of the three activities. Carlier and Pinson (1989) have formalised

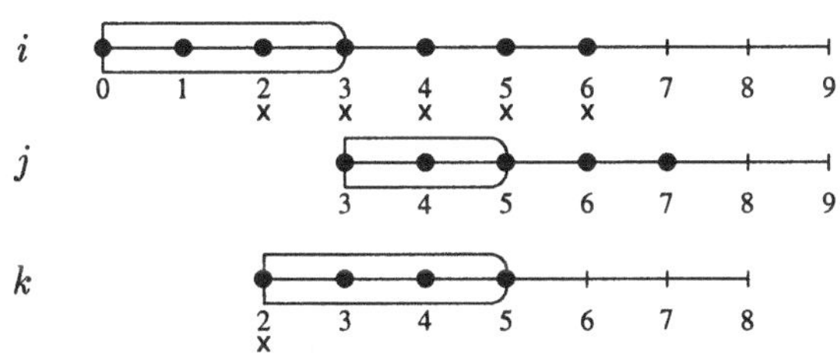

Figure 10.3 An example for the input test

the observation made in the example and have derived conditions under which it can be concluded that an operation $i \in \mathcal{A}$ must be scheduled first or last in $\mathcal{A}$. If i is scheduled before or after $\mathcal{A} \setminus \{i\}$ we may also think of i as the input or output of $\mathcal{A} \setminus \{i\}$, hence the name of the conditions. We use the shorthand notation $P(\mathcal{A}) = \sum_{i \in \mathcal{A}} p_i$ for the total processing time of $\mathcal{A}$.

Theorem 1 (Input/output). *Let $i \in \mathcal{A} \subseteq V_k$. If*

$$\max_{u \in \mathcal{A} \setminus \{i\}, v \in \mathcal{A}, u \neq v} (lct_v - est_u) < P(\mathcal{A}) \tag{3}$$

then i must precede all activities in $\mathcal{A} \setminus \{i\}$ (input condition). Likewise, if

$$\max_{u \in \mathcal{A}, v \in \mathcal{A} \setminus \{i\}, u \neq v} (lct_v - est_u) < P(\mathcal{A}) \tag{4}$$

then i must succeed all operations in $\mathcal{A} \setminus \{i\}$ (output condition).

Proof. If i does not precede $\mathcal{A} \setminus \{i\}$, then all activities in $\mathcal{A}$ must be scheduled within $\max_{u \in \mathcal{A} \setminus \{i\}, v \in \mathcal{A}, u \neq v} (lct_v - est_u)$ time units. If Condition 3 holds this is not possible. The second part can be shown symmetrically. □

If the input or output condition holds we face the question how to adjust activity domains. Assume we have concluded that $\mathcal{A} \setminus \{i\} \rightarrow i$. Clearly, i can only start after the minimum completion time t^* of all activities in $\mathcal{A} \setminus \{i\}$. Unfortunately, finding t^* is an $\mathcal{NP}$-hard problem: It is equivalent to solving the one-machine makespan minimisation problem with release time and due dates (Carlier 1982). Therefore we resort to approximating t^*. A simple and obvious approximation is the maximal earliest completion time in $\mathcal{A} \setminus \{i\}$. However, we can do better by considering the preemptive relaxation of the one-machine problem (preemptive bound). For this problem, an optimal solution known as Jackson's Preemptive Schedule (JPS) can be efficiently obtained by scheduling the activities in $\mathcal{A} \setminus \{i\}$ "from left to right" according to the "earliest due date" priority dispatching rule (Jackson 1956): Whenever the resource is free, schedule the activity i with minimal lct_i; if an activity j with $lct_j < lct_i$ becomes available while i is in process then interrupt i and start j. We denote the completion time of JPS for $\mathcal{A} \setminus \{i\}$ by $Ect(\mathcal{A} \setminus \{i\})$. Clearly, $Ect(\mathcal{A} \setminus \{i\})$ is a lower bound on the earliest start of i, and the same holds true for all subsets $\mathcal{A}' \subseteq \mathcal{A} \setminus \{i\}$. However, Carlier (1982) has shown that

$$Ect(\mathcal{A} \setminus \{i\}) = \max_{\mathcal{A}' \subseteq \mathcal{A} \setminus \{i\}} \{\min_{u \in \mathcal{A}'} est_u + P(\mathcal{A}')\}. \tag{5}$$

This implies that $Ect(\mathcal{A}') \leq Ect(\mathcal{A} \setminus \{i\})$, if $\mathcal{A}' \subset \mathcal{A} \setminus \{i\}$. We can now state a domain reduction rule to be applied if the output condition holds:

$$\delta_i := \delta_i \setminus \{t | t < Ect(\mathcal{A} \setminus \{i\})\} \tag{6}$$

A symmetric rule applies if the input condition holds. Using $Lst(\mathcal{A} \setminus \{i\})$ as the preemptive bound for the latest start time of $\mathcal{A} \setminus \{i\}$, obtained by preemptively scheduling the activities in $\mathcal{A} \setminus \{i\}$ "from right to left" as late as possible according to the "maximum latest start" priority dispatching rule, the domain reduction rule is:

$$\delta_i := \delta_i \setminus \{t | t > Lst(\mathcal{A} \setminus \{i\}) - p_i\} \tag{7}$$

In addition to these domain reductions we can adjust lct_j to be less than or equal to lst_i (output) or est_j to be greater than or equal to ect_i (input) for all $j \in \mathcal{A} \setminus \{i\}$. Alternatively, we can wait until a subsequent application of this test for a set $\mathcal{A} = \{i, j\}$ automatically yields this reduction.

Before returning to the initial example, let us point out that — like all consistency tests — the input/output test consists of (a) conditions and (b) corresponding domain reduction rules.

For the example in Figure 10.3 the maximum of the expression on the left side of the input condition is $9 - 2$, and $P(\mathcal{A}) = 8$; since $9 - 2 < 8$, we can deduce $i \rightarrow \{j, k\}$. With $Lst(\{j, k\}) = 4$ the domain of i becomes $\delta_i := [0, 6] \setminus \{t | t > 4 - 3\} = [0, 1]$. Note the effect of using the preemptive bound: By using $Lst(\{j, k\})$ we have obtained a stronger domain reduction for i than we would have by considering lst_j and lst_k separately, which would have left the value 2 in δ_i. Finally, the domain of k becomes $\delta_k := [2, 5] \setminus \{t | t < ect_i\} = [3, 5]$. As pointed out above, this reduction in δ_k could also be achieved through a further application of the input test for $\mathcal{A} = \{i, k\}$.

The input condition also applies in the example in Figure 10.1. For activity i and $\mathcal{A} = \{i, j\}$, we obtain $8 - 2 < 7$ and deduce $i \rightarrow j$. The domain of i remains unmodified, and the domain of j reduces to $\delta_j := \delta_j \setminus \{t | t < 3\} = [3, 8]$.

In branch and bound procedures that branch over disjunctive edges, the test may be employed to immediately select the orientation of edges, a process often called immediate selection, as first suggested by Carlier and Pinson (1989), or edge finding, a term introduced by Applegate and Cook (1991). The input/output test was first described by Carlier and Pinson in the context of a branch and bound algorithm for the JSP; the tests that they actually implemented in their initial algorithm were limited to two-element sets $\mathcal{A}$ and one additional heuristically determined $\mathcal{A}$ and $i \in \mathcal{A}$ for each resource. Using these tests, they were for the first time able to optimally solve a notoriously difficult 10×10 JSP instance (Fisher and Thompson 1963) that — despite many attempts — had defied solution for over 25 years.

Efficient algorithms that have later been developed for testing the input/ output conditions for all $\mathcal{A}$ and i and performing the corresponding domain reductions based on the preemptive bounds usually use an ordering of activities according to earliest start and latest completion times. The challenging part is to test the input/output conditions and calculate preemptive bounds at the same time. Carlier and Pinson (1990), Martin and Shmoys (1996), and Nuijten (1994) have designed $O(|\mathcal{V}_k|^2)$ algorithms for testing all subsets $\mathcal{A} \subseteq \mathcal{V}_k$. The algorithm of Nuijten has the interesting property that it can be generalised for cumulative scheduling. $O(|\mathcal{V}_k| \log |\mathcal{V}_k|)$ algorithms for testing all subsets have been described by Brucker et al. (1996a) and Carlier and Pinson (1994). Caseau and Laburthe (1994, 1995, 1996) describe an algorithm based on the concept of "task intervals" for checking all sets $\mathcal{A}$ with effort $O(|\mathcal{V}_k|^3)$. The advantage of their approach is that the consistency conditions can be evaluated incrementally within a search procedure. When used within a branch-and-bound algorithm this means that the effective time complexity for performing the tests at each node of the search tree is usually lower than $O(|\mathcal{V}_k|^3)$ because it is not necessary to test all $\mathcal{A}$; although the worst case complexity for performing the tests at a node is still $O(|\mathcal{V}_k|^3)$, the average complexity is lower. This contrasts with the usual approach of applying the full test at each node of a branch-and-bound tree.

Finally, we would like to mention that to our knowledge all algorithms discussed above do not test the input/output conditions in the form of Theorem 1, where we have required in the maximum expressions that $u \neq v$, but rather allow for $u = v$, thus actually testing a weaker condition. Although the extension may seem trivial it does lead to additional deductions in certain cases. However, it is not always obvious how to include it in existing algorithms without increasing their time complexity.

10.5.2 Input-or-output test

From the input/output condition it can be deduced that an operation $i \in \mathcal{A} \subseteq \mathcal{V}_k$ must be scheduled first or last in $\mathcal{A}$. The weaker input-or-output condition can be used to show that a precedence relation $i \rightarrow j$ must exist between a pair of activities i and j from set $\mathcal{A}$.

Figure 10.4 shows an example with a set $\mathcal{A} = \{i, j, k, l\}$ of four activities to be processed by the same resource. The input/output condition does not allow to draw any conclusions about the order in which the activities must be scheduled. However, we can deduce that i must precede j: Suppose i is not scheduled first *and* j is not scheduled last. Then all four activities with a total processing time of $7 = 3+2+1+1$ must be scheduled within the interval $[2, 8]$, which is a contradiction. Hence we can conclude that it is impossible that at the same time i is not first *and* j is not last. If either i must be first or j must be last, then i must precede j, and we can remove the start time 3 from δ_j. This observation leads to the following theorem.

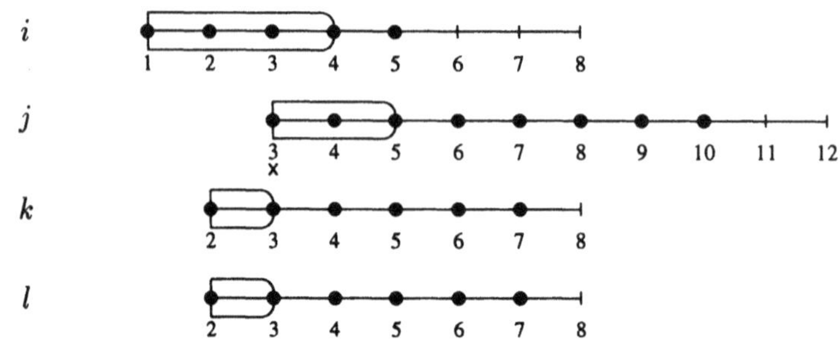

Figure 10.4 An example for the input-or-output test

Theorem 2 (Input-or-output). *Let $i, j \in \mathcal{A} \subseteq \mathcal{V}_k$. If*

$$\max_{u \in \mathcal{A}\setminus\{i\}, v \in \mathcal{A}\setminus\{j\}, u \neq v} (lct_v - est_u) < P(\mathcal{A}) \tag{8}$$

then i must be scheduled first or j must be scheduled last in $\mathcal{A}$. If $i \neq j$ then i must precede j.

Proof. Suppose neither i is scheduled first nor j is scheduled last. All activities in $\mathcal{A}$ must then be scheduled within $\max_{u \in \mathcal{A}\setminus\{i\}, v \in \mathcal{A}\setminus\{j\}, u \neq v}(lct_v - est_u)$ time units. If Condition 8 holds, this is impossible and we can conclude that either i must be first or j must be last in $\mathcal{A}$. In both cases i must precede j if $i \neq j$. □

Comparison to the very similar input/output conditions shows in what sense the input-or-output condition is weaker. If this condition holds and $i \neq j$ which means that $i \rightarrow j$, we can reduce the domains of i and j in the following way:

$$\delta_i := \delta_i \setminus \{t | t > lst_j - p_i\} \tag{9}$$

$$\delta_j := \delta_j \setminus \{t | t < ect_i\} \tag{10}$$

If the condition holds for $i = j$, the new domain of i after reduction is:

$$\delta_i := \delta_i \setminus \{t | Lst(\mathcal{A} \setminus \{i\}) - p_i < t < Ect(\mathcal{A} \setminus \{i\})\} \tag{11}$$

While the domain reduction rules 9 and 10 always lead to a domain reduction at the domain bounds if any, Rule 11 may remove values within the domain but leaves the bounds untouched.

For the example in Figure 10.4 we obtain $8 - 2 < 7$ and deduce $i \to j$. By applying domain reduction rule 10 we can remove the value 3 from δ_j. Figure 10.5 shows another example where the input-or-output condition can deduce that a single activity must either start first or last; in terms of Theorem 2 this is the case where $i = j$. We obtain $5-3 < 4$ and conclude that i must start before or after $\{k, l\}$. Domain reduction rule 11 allows to remove the values $[2, 4]$ from δ_i. As a final example, note that the

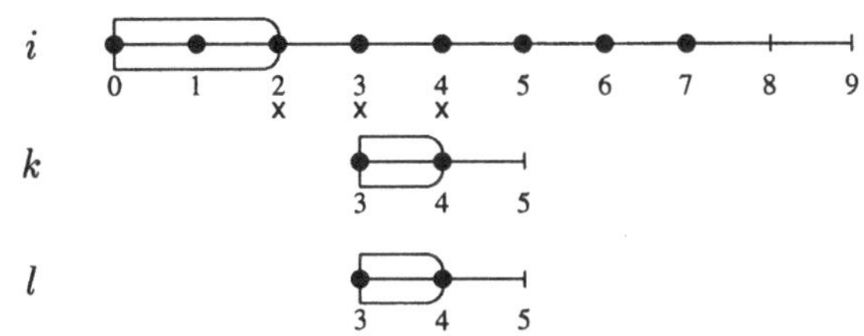

Figure 10.5 Input-or-output condition example: i must be first or last

result $i \to \{j, k\}$ that we have obtained with the input/output condition for the example in Figure 10.3 can also be deduced in two steps with the input-or-output condition, resulting in the conclusions $i \to j$ and $i \to k$. However, the corresponding reduction in δ_i is weaker, leaving the value 2 in δ_i.

To our knowledge, the input-or-output test in its general form has not been discussed in the literature. A similar condition for the special case where $i = j$ has been described by Carlier and Pinson (1990); see also Błażewicz et al. (1997, 1998). Stronger conditions based on considering all sets $\mathcal{A}$ of cardinality r, hence called r-set conditions, have been discussed by Brucker et al. (1996a). They describe an $O(|\mathcal{V}_k|^2)$ 3-set algorithm that checks all activity sets of cardinality three and detects all pairwise ordering relations derivable from triples. The algorithm thus implements the input-or-output test for $|\mathcal{A}| = 3$. Judging from the implementation within their branch-and-bound procedure for the JSP, the efficiency of the 3-set tests is comparable to that of the input/output tests. It is unclear whether a low polynomial time-complexity r-set algorithm could be developed for $r > 3$.

The development of an algorithm with low polynomial time complexity for testing the input-or-output conditions is an open issue. Based on experience with other consistency tests, we conjecture that in order to be of practical value such an algorithm must at most have time complexity $O(|\mathcal{V}_k|^2)$. There is an obvious $O(|\mathcal{V}_k|^4)$ algorithm using task or activity intervals, and we have designed an $O(|\mathcal{V}_k|^3)$ algorithm.

10.5.3 Input/output negation test

By further relaxing the condition to be tested, we can still draw additional conclusions in situations where the input-or-output condition and the stronger input/output conditions do not hold. Figure 10.6 shows an example with a set $\mathcal{A} = \{i, j, k\}$ of three activities to be processed by the same resource. Although we cannot conclude that activity i must be last or must precede j or k, we can deduce that i must *not be first*, and therefore remove the value 2 from δ_i. By generalising the observations made in

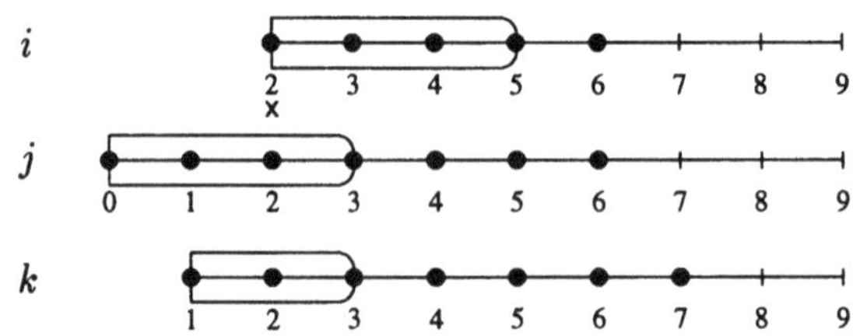

Figure 10.6 An example for the input negation test

the example, we arrive at the following theorem.

Theorem 3 (Input/output negation). *Let $i \in \mathcal{A} \subseteq V_k$. If*

$$\max_{v \in \mathcal{A} \setminus \{i\}} (lct_v - est_i) < P(\mathcal{A}) \tag{12}$$

then i must not start first in $\mathcal{A} \setminus \{i\}$ (input negation). If

$$\max_{u \in \mathcal{A} \setminus \{i\}} (lct_i - est_u) < P(\mathcal{A}) \tag{13}$$

then i must not end last in $\mathcal{A} \setminus \{i\}$ (output negation).

Proof. If i precedes $\mathcal{A} \setminus \{i\}$, all activities in $\mathcal{A}$ must be scheduled in the interval $[est_i, \max_{v \in \mathcal{A} \setminus \{i\}} lct_v)$. If Condition 12 holds, this is not possible. The second part can be shown symmetrically. □

Again, it is easy to see in which sense these conditions are weaker than in the preceding tests. Domain reduction rules can be based on the observation that i must succeed (input negation) or precede (output negation) at least one other activity in $\mathcal{A}$:

$$\delta_i := \delta_i \setminus \{t | t < \min_{u \in \mathcal{A} \setminus \{i\}} ect_u\} \tag{14}$$

$$\delta_i := \delta_i \setminus \{t | t > \max_{u \in \mathcal{A} \setminus \{i\}} lst_u - p_i\} \tag{15}$$

For the example in Figure 10.6 the input negation condition yields $9 - 2 < 8$ and we conclude $i \not\rightarrow \{j, k\}$. According to domain reduction rule 14 we can therefore remove all values less than 3 from δ_i.

Conclusions similar to those obtained in the examples for the input/output and input-or-output test could also have been produced through successive application of the input/output negation test. Since the condition to be tested for the input/output negation conclusion is weaker than the preceding conditions, it will of course hold whenever the stronger conditions apply. Consider again the example in Figure 10.3. Here, the input/output negation conditions allows to conclude $j \not\rightarrow \{i,k\} \Leftrightarrow i \rightarrow j \vee k \rightarrow j$ and $k \not\rightarrow \{i,j\} \Leftrightarrow i \rightarrow k \vee j \rightarrow k$, which implies $i \rightarrow \{j,k\}$. However, this implication is not automatically deduced by the input/output negation condition. This demonstrates that input/output negation conditions alone do not deduce all interesting domain reductions. A similar effect can be seen in the example in Figure 10.4. Here, the input/output negation conditions can be used to deduce $\{j,k,l\} \not\rightarrow i$ and $j \not\rightarrow \{i,k,l\}$, but this does not allow to remove the value 3 from δ_j as in the input-or-output test.

The input/output negation test has first been suggested by Carlier and Pinson (1989). Most authors working on consistency tests have considered the test in some form. However, an algorithm that tests all interesting $\mathcal{A}$ and i with effort $O(|\mathcal{V}_k|^2)$ has only recently been developed by Baptiste and Le Pape (1996). Other researchers have often applied the tests in an incomplete way, testing only some $\mathcal{A}$ and i (Carlier and Pinson 1989, 1990, Nuijten 1994, Baptiste and Le Pape 1995). Caseau and Laburthe (1994, 1995) have integrated the tests in their task interval algorithm which tests input/output conditions and the negation conditions with effort $O(|\mathcal{V}_k|^3)$.

10.5.4 Summary and generalisation

All disjunctive interval consistency conditions that we have discussed can be regarded as special cases of the following theorem.

Theorem 4 (Sequence consistency). *Let $\mathcal{A}', \mathcal{A}'' \subset \mathcal{A} \subseteq \mathcal{V}_k$. If*

$$\max_{u \in \mathcal{A} \setminus \mathcal{A}', v \in \mathcal{A} \setminus \mathcal{A}'', u \neq v} (lct_v - est_u) < P(\mathcal{A}) \tag{16}$$

then an activity in $\mathcal{A}'$ must start first or an activity in $\mathcal{A}''$ must end last in $\mathcal{A}$.

Proof. If none of the activities in $\mathcal{A}''$ succeeds $\mathcal{A} \setminus \mathcal{A}''$ *and* none of the activities in $\mathcal{A}'$ precedes $\mathcal{A} \setminus \mathcal{A}'$, then $\mathcal{A}$ must be processed within $\max_{u \in \mathcal{A} \setminus \mathcal{A}', v \in \mathcal{A} \setminus \mathcal{A}'', u \neq v}(lct_v - est_u)$ units of time. If Condition 16 holds this is a contradiction. □

The results of the preceding sections are summarised in Table 10.2. For each condition, the table shows the values of $\mathcal{A} \setminus \mathcal{A}'$ and $\mathcal{A} \setminus \mathcal{A}''$ that, when used in Theorem 4, yield the condition. The conclusions of Theorem 4 have been reformulated to match the tests presented above. Note that the conclusion is always the negation of the hypothesis $\mathcal{H}$ falsified by the test.

10.5.5 Relation to general interval consistency

We will now relate the Sequence Consistency Theorem to the general interval consistency constraint 2. For disjunctive scheduling, Constraint 2 reduces to $t_2 - t_1 \geq$

Table 10.2 Summary of disjunctive interval consistency tests, $\mathcal{A}', \mathcal{A}'' \subset \mathcal{A} \subseteq \mathcal{V}_k$

Test	$\mathcal{A} \setminus \mathcal{A}''$	$\mathcal{A} \setminus \mathcal{A}'$	*Conclusion* $(\neg\mathcal{H})$
input	$\mathcal{A}$	$\mathcal{A} \setminus \{i\}$	$i \rightarrow \mathcal{A} \setminus \{i\}$
output	$\mathcal{A} \setminus \{i\}$	$\mathcal{A}$	$\mathcal{A} \setminus \{i\} \rightarrow i$
input-or-output	$\mathcal{A} \setminus \{j\}$	$\mathcal{A} \setminus \{i\}$	$i \rightarrow \mathcal{A} \setminus \{i\} \vee \mathcal{A} \setminus \{j\} \rightarrow j$
input negation	$\mathcal{A} \setminus \{i\}$	$\{i\}$	$i \not\rightarrow \mathcal{A} \setminus \{i\}$
output negation	$\{i\}$	$\mathcal{A} \setminus \{i\}$	$\mathcal{A} \setminus \{i\} \not\rightarrow i$

$P(\mathcal{V}_k, t_1, t_2)$. The following theorem shows how we can efficiently test violations of this constraint.

Theorem 5. *If, for some time interval* $[t_1, t_2)$,

$$t_2 - t_1 < P(\mathcal{V}_k, t_1, t_2), \tag{17}$$

then

$$\max_{i,j \in \mathcal{V}_k(t_1,t_2), i \neq j} (lct_j - est_i) < P(\mathcal{V}_k(t_1, t_2)). \tag{18}$$

Proof. From Equation 1 we know that $0 < t_2 - t_1 < P(\mathcal{V}_k, t_1, t_2)$ implies that $|\mathcal{V}_k(t_1, t_2)| \geq 2$. We consider two activities $i, j \in \mathcal{V}_k(t_1, t_2), i \neq j$, and start to transform Condition 17 into Condition 18 by rewriting the left hand side of 17:

$$lst_j + t_2 - lst_j - t_1 < P(\mathcal{V}_k, t_1, t_2),$$

By observing that $t_2 - lst_j \geq p_j(t_1, t_2) \geq 0$, according to Equation 1, we can approximate the left side. We rewrite the right side and obtain:

$$lst_j + p_j(t_1, t_2) - t_1 < p_i(t_1, t_2) + p_j(t_1, t_2) + \sum_{l \in \mathcal{V}_k \setminus \{i,j\}} p_l(t_1, t_2)$$

Again, we know from Equation 1 that $ect_i - t_1 \geq p_i(t_1, t_2) \geq 0$. This approximation leads to:

$$lst_j - t_1 < ect_i - t_1 + \sum_{l \in \mathcal{V}_k \setminus \{i,j\}} p_l(t_1, t_2)$$

Next, we approximate the sum on the right hand side, once again using Equation 1 which tells us that $p_k \geq p_k(t_1, t_2) \geq 0$, and obtain:

$$lst_j - ect_i < \sum_{l \in \mathcal{V}_k(t_1,t_2) \setminus \{i,j\}} p_l$$

By adding $p_i + p_j$ on both sides we arrive at:

$$lct_j - est_i < P(\mathcal{V}_k(t_1, t_2))$$

As it is always possible to choose i and j in such a way that the maximum difference $lct_j - est_i$ is realised, Condition 18 must hold. □

The theorem tells us two interesting things. First, it states that if an interval capacity constraint is violated for some arbitrary time interval $[t_1, t_2)$, then there will also be a violation for an interval defined by the earliest start and latest completion time of two activities in $\mathcal{V}_k(t_1, t_2)$. When checking for violations this allows us to restrict our attention to intervals defined by earliest start and latest completion times — called task or activity intervals (Caseau and Laburthe 1994) — instead of considering all possible time intervals. Inconsistencies can thus be detected by testing $O(|\mathcal{V}_k|^2)$ intervals. This answers the initial question (posed in Section 10.4) what intervals we should test. Second, the theorem states that, as long as we test all activity intervals, there is nothing to be gained from considering *interval* processing time instead of *simple* processing time. If interval processing time has an effect on the test for a given set $\mathcal{A}$ then we can obtain the same effect by considering a different set $\mathcal{A}'$. In summary, this means that an algorithm which tests Condition 18 for all activity intervals will detect all violations according to the more general concept of Condition 17 which is the negation of the disjunctive version of the general interval consistency constraint 2.

It is worth emphasising that this statement is independent of the particular set of hypothetical constraints $\mathcal{H}$ to be tested. This can be seen as follows: For any set of constraints, it is always possible to first add and propagate the constraints, and then test the interval consistency constraints. The particular form of the sequence consistency tests is simply an accelerated version of this "add and propagate, then test" process. For illustration, consider again the example shown in Figure 10.3, where the conclusion $i \rightarrow \{j, k\}$ could also have been obtained in the following way: (1) Add $\mathcal{H} = \{i \not\rightarrow \{j, k\}\}$, (2) update the domain of i based on $\mathcal{H}$, which yields $\delta_i := \delta_i \backslash \{t | t < \min_{u \in \{j,k\}} ect_u\} = [5, 6]$, and (3) test the interval consistency constraint 2 for the activity interval defined by $\{i, j, k\}$ which has the left time bound 2 and the right time bound 9. Because $9 - 2 \not\geq 8$ this test fails and we conclude $\neg\mathcal{H} \Leftrightarrow i \rightarrow \{j, k\}$.

For disjunctive scheduling, Theorem 5 improves the results obtained by Schwindt (1997) and Baptiste et al. (1998) for the cumulative case discussed in Section 10.6.

10.5.6 2-consistency and 2-bound-consistency

This section relates the consistency tests to the general concept of 2-consistency and 2-bound-consistency that is commonly used in the constraint satisfaction and constraint propagation literature (Kumar 1992, Tsang 1993, Nuijten 1994, Van Hentenryck 1989). We derive a 2-consistency test and show that the sequencing tests can be used to achieve 2-bound-consistency.

We use the following informal consistency definitions: Activity domains are called 2-consistent if, for any pair $i, j \in \mathcal{V}$, and for any value $x \in \delta_i$ there is some value $y \in \delta_j$ such that $st_i = x$ and $st_j = y$ is permitted by the constraints of the scheduling problem. The weaker definition of bound consistency looks at domain bounds: Activity domains are called 2-bound-consistent, or 2-b-consistent for short, if, for any pair $i, j \in \mathcal{V}$, and for every value $x \in \{\min \delta_i, \max \delta_i\}$ there is a value $y \in \delta_j$ such that $st_i = x$ and $st_j = y$ is permitted. Clearly, 2-consistency implies 2-b-consistency.

The concept of bound consistency is of interest because, as we have seen, many consistency tests are based on domain bound considerations. In addition, the propagation of temporal constraints depends on domain bounds. Any change in domain bounds is therefore likely to trigger further domain reductions. Finally, if domains are approximated by start time windows — and this is often done for reasons of implementation efficiency — bound-consistency is the only reasonable concept of consistency.

Figure 10.7 shows an example, taken from Nuijten (1994), with a pair of activities $i, j \in \mathcal{V}_k$ where any 2-inconsistent value is marked. For example, j cannot start at time 2 since this does neither leave enough room for i to be processed before j nor after j. In general, i cannot start in the open interval $(lst_j - p_i, ect_j)$. Note that the interval can be empty if $ect_j \leq lst_j - p_i$. The observation is summarised in the following theorem due to Nuijten (1994).

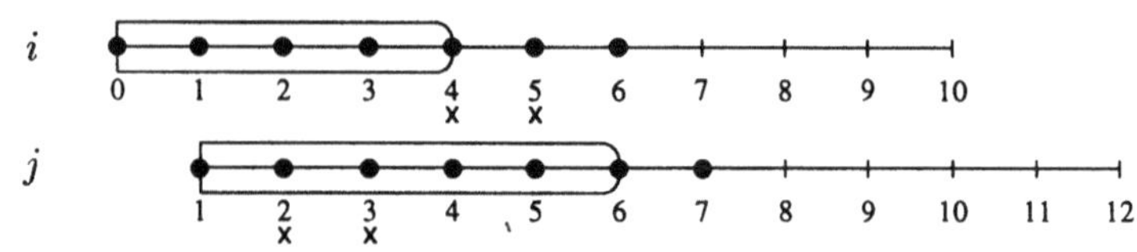

Figure 10.7 2-consistency

Theorem 6 (2-consistency). *Let $i, j \in \mathcal{V}_k, i \neq j$. δ_i and δ_j are 2-consistent if and only if*

$$\delta_i \cap (lst_j - p_i, ect_j) = \emptyset. \tag{19}$$

Proof. If j is started at time $t \in \delta_j$ then i is blocked during the open interval $(t - p_i, t + p_j)$. The left bound of the interval is maximal for $t = lst_j$, and the right bound is minimal for $t = est_j$. Thus the minimal interval during which i cannot start is $(lst_j - p_i, ect_j)$. All other possible start times of j leave possible start times for i. □

An $O(|\mathcal{V}_k|^2)$ algorithm for achieving 2-consistency is described by Nuijten (1994).

The following result shows that the sequence consistency tests based on Theorem 4 can be used to ensure 2-b-consistency.

Theorem 7. *The input/output, input-or-output, and input/output negation tests all achieve 2-b-consistency.*

Proof. For $\mathcal{A} = \{i, j\}$ all the tests simplify to:

$$lct_j - est_i < p_i + p_j \Rightarrow j \rightarrow i \Rightarrow \left\{ \begin{array}{lcl} \delta_i & := & \delta_i \setminus \{t | t < ect_j\} \\ \delta_j & := & \delta_j \setminus \{t | t > lst_i - p_j\} \end{array} \right.$$

To achieve 2-b-consistency any 2-inconsistent value must be removed from the domain bounds. According to Equation 19, the left domain bound can only be 2-inconsistent if

$$lst_j - p_i < est_i \Leftrightarrow lct_j - est_i < p_i + p_j.$$

In this case, the condition of the tests is satisfied and any inconsistent values are removed by the first domain reduction rule above. The proof for the right domain bound is symmetrical. □

10.5.7 Sequence consistency does not imply k-consistency

Trivial though it may be, it is worth emphasising that the consistency tests only check necessary, but not sufficient conditions for the existence of a feasible schedule and that the domain reduction rules employed are heuristic. While we could show that the sequence consistency tests always achieve 2-b-consistency, this means that they in general do not achieve k-b-consistency for $k > 2$. The example in Figure 10.8 illustrates these two points.

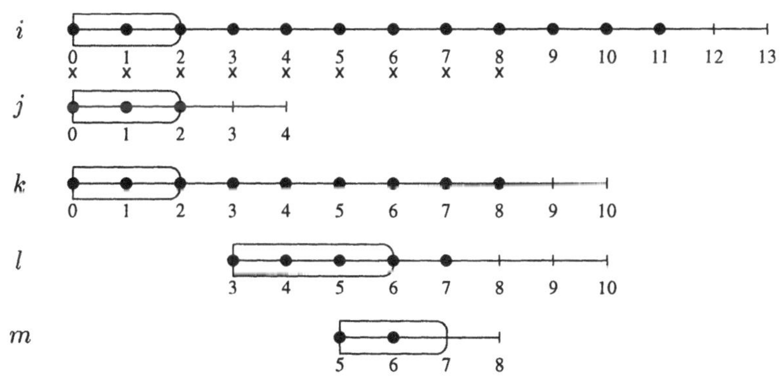

Figure 10.8 When sequence consistency tests fail

In the example, $\mathcal{A} = \{i, j, k, l, m\}$. The output condition allows to conclude $\{j, k, l, m\} \rightarrow i$, since $10 - 0 < 11$. The preemptive bound $Ect(\mathcal{A} \setminus \{i\})$ for the earliest completion time of $\{j, k, l, m\}$ is 9. According to domain reduction rule 6 this leaves the value 9 as the left bound of δ_i. However, manual inspection shows that the earliest completion time of $\{j, k, l, m\}$ is actually 10. Thus, the input/output test leaves an inconsistent value at the left bound of δ_i. This demonstrates that the domain reduction rule based on the preemptive bound is heuristic.

Now modify the example by reducing lct_i to 11. The input/output test still yields the same result, and none of the other sequence consistency tests leads to an inconsistency (by producing an empty domain). Again, manual inspection shows that there is no feasible schedule for $\mathcal{A}$.

10.5.8 Shaving

In the tests based on the Sequence Consistency Theorem 4 we have tried to refute hypothetical constraints on the sequence in which activities in a set $\mathcal{A} \subseteq \mathcal{V}_k$ execute. Now, we take a purely time-oriented approach and consider hypothetical constraints on individual activity start times. If we can falsify such a constraint, then we can reduce the corresponding activity domain in an obvious way. The process of reducing activity domains based on this kind of reasoning has been called *shaving* (Martin and Shmoys 1996, Caseau and Laburthe 1996).

For example, we can test a hypothetical constraint of the type $st_i > t_x$ for some $t_x \in \delta_i$. If this leads to a contradiction, then we can conclude that st_i must be less than or equal to t_x and remove all values greater t_x from δ_i. A contradiction may be caused by a direct violation of the interval capacity constraint 2 or after propagating the hypothetical constraint by repeatedly applying other consistency tests. Values of t_x can for example be chosen by a dichotomising search over δ_i.

A shaving approach for disjunctive scheduling has been proposed by Carlier and Pinson (1994) for solving the JSP. Martin and Shmoys (1996) have, independently, applied the technique within a time-oriented branch-and-bound algorithm for the JSP. Using a shaving technique, Caseau and Laburthe (1996) were able to obtain a proof of optimality for the famous 10×10 JSP due to Fisher and Thompson (1963) with only 7 backtracks.

10.6 Consistency tests for cumulative scheduling

While disjunctive scheduling or sequencing is concerned with unit resource requirements and capacities, cumulative scheduling considers the general case of arbitrary requirements and capacities.

10.6.1 Activity interval consistency

The disjunctive sequence consistency tests developed in the preceding section can be generalised for cumulative scheduling in a straightforward way by considering interval work instead of processing times. This relation was first pointed out by Nuijten (1994) (see also Nuijten and Aarts 1996). The following rule extends Theorem 4 for cumulative scheduling. As the time intervals considered are activity or task intervals defined by activity sets, we have chosen the name activity interval consistency.

Theorem 8 (Activity interval consistency). *Let $\mathcal{A}', \mathcal{A}'' \subset \mathcal{A} \subseteq \mathcal{V}_k$.If*

$$R_k \cdot \max_{u \in \mathcal{A} \setminus \mathcal{A}', v \in \mathcal{A} \setminus \mathcal{A}''} (lct_v - est_u) < W(\mathcal{A}) \qquad (20)$$

then an activity in $\mathcal{A}'$ must start first or an activity in $\mathcal{A}''$ must end last.

Proof. Similar to proof of Theorem 4. □

In contrast to sequencing, we can no longer assume that $i \neq j$ because an activity that starts first may now also end last. Comparison of Condition 20 to the general interval capacity constraint 2 shows that the condition only considers activity or task intervals and has a weaker right hand side. In the disjunctive case we were able to show that it was sufficient to consider activity intervals and that there was nothing to be gained from using interval work instead of set based work on the right side. However, it turns out that this is not the case for cumulative scheduling, so that the condition can actually be strengthened. The reason for presenting the condition in the above form is that this extension of the disjunctive case allows to generalise algorithms originally designed for sequencing. We will discuss a sharper form in Section 10.6.2.

The theorem can be used to derive tests in analogy to the sequencing tests by using suitable values for $\mathcal{A}'$ and $\mathcal{A}''$, as shown in Table 10.2. Note that the meaning of conclusions such as $\mathcal{A} \setminus \{i\} \rightarrow i$ or $i \rightarrow \mathcal{A} \setminus \{i\}$ is that i must end after (start before) all activities in $\mathcal{A} \setminus \{i\}$; in contrast to the disjunctive case this, however, does not imply that it must also start after (end before) $\mathcal{A} \setminus \{i\}$.

Useful domain reductions can be deduced for the cumulative version of the input-or-output test with $i = j$. For $\mathcal{A}' = \mathcal{A}'' = \{i\}$, i.e. for testing the hypothetical constraints $\mathcal{H} = \{\, i \not\rightarrow \mathcal{A} \setminus \{i\}, \mathcal{A} \setminus \{i\} \not\rightarrow i\}$, Theorem 8 yields:

$$R_k \cdot \max_{u \in \mathcal{A}\setminus\{i\}, v \in \mathcal{A}\setminus\{i\}} (lct_v - est_u) < W(\mathcal{A}) \Rightarrow i \rightarrow \mathcal{A} \setminus \{i\} \vee \mathcal{A} \setminus \{i\} \rightarrow i \qquad (21)$$

Clearly, the excess amount of work that cannot be processed in the interval defined by $\min_{u \in \mathcal{A}\setminus\{i\}} est_u$ and $\max_{v \in \mathcal{A}\setminus\{i\}} lct_v$ is the difference of the total work required by $\mathcal{A}$ and the capacity available within the interval. Since only activity i can move partially or completely out of the interval, we can conclude that the amount of processing time of i to be moved outside to the left and/or right, denoted by rest$(\mathcal{A}, i)$, is:

$$\text{rest}(\mathcal{A}, i) = \lceil (W(\mathcal{A}) - R_k \cdot \max_{u \in \mathcal{A}\setminus\{i\}, v \in \mathcal{A}\setminus\{i\}} (lct_v - est_u))/r_{ik} \rceil . \qquad (22)$$

This observation allows to deduce domain reductions if the minimum amount of processing time that is always outside of the interval, regardless of the chosen start time, is less than the required amount:

$$p_i - \max_{u \in \mathcal{A}\setminus\{i\}, v \in \mathcal{A}\setminus\{i\}} (lct_v - est_u) < \text{rest}(\mathcal{A}, i). \qquad (23)$$

If Condition 23 holds, then the part of i that must be outside of the interval must either be completely on the left or be completely on the right side of the interval. This leads to the following domain reduction rule that can be applied if Conditions 21 and 23 hold:

$$\delta_i := \delta_i \setminus \{t | \min_{u \in \mathcal{A}\setminus\{i\}} est_u - \text{rest}(\mathcal{A}, i) < t < \max_{v \in \mathcal{A}\setminus\{i\}} lct_v + \text{rest}(\mathcal{A}, i) - p_i\} \quad (24)$$

This rule can actually be sharpened as follows: If the left or right bound reduction may be applied for $\mathcal{A} \setminus \{i\}$ then it can also be applied for all subsets $\mathcal{A}' \subseteq \mathcal{A} \setminus \{i\}$; this is not shown here. Note that the sharpened form of the rule is equivalent to domain reduction rule 11. We refer to the conditions and this domain reduction rule as the cumulative input-or-output test.

Figure 10.9 illustrates the test. It shows an example, taken from Nuijten (1994), with four activities to be processed by the same resource k with capacity $R_k = 2$.

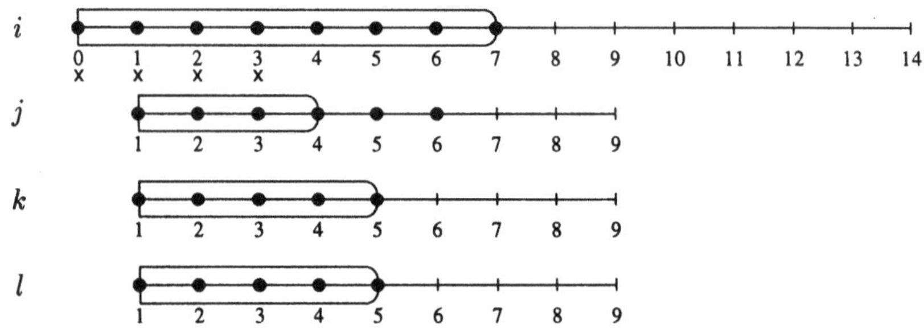

Figure 10.9 Four activities requiring 1 unit of a resource with capacity 2

Inspection shows that if activity i is started before time 4, then it is impossible to schedule all of the other activities j, k, l within their time window. This is detected by the input-or-output test in the following way: Because $2 \cdot (9 - 1) < 18$ we conclude that $i \rightarrow \{j, k, l\} \vee \{j, k, l\} \rightarrow i$. The amount of processing time of i that must take place outside of the interval $[1, 9)$ is $\text{rest}(\{i, j, k, l\}, i) = \lceil (18 - 2 \cdot (9 - 1))/1 \rceil = 2$. Because $7 - 8 < 2$, Condition 23 is satisfied and we apply the domain reduction rule $\delta_i := \delta_i \setminus \{t | 1 - 2 < t < 9 + 2 - 7\} = [4, 7]$, as shown in Figure 10.9.

It is interesting to consider a slight modification of the example: For $p_i = 6$ the reduction rule yields $\delta_i := \delta_i \setminus \{t | 0 < t < 4\} = \{0, 4, \ldots, 7\}$; the value 0 is thus left in δ_i and the domain bounds remain untouched.

The test presented here is similar to the three cumulative tests described by Nuijten (1994). Nuijten uses the tests to extend the disjunctive sequence consistency algorithm for the cumulative case. The time complexity of the resulting algorithm is $O(|\{r_{ik}\}| \cdot |\mathcal{V}_k|^3)$, where $|\{r_{ik}\}|$ is the number of distinct resource capacity requirements. As mentioned before, the tests could be strengthened by using interval work instead of simple work and by considering additional time intervals other than activity intervals. This is explained in the following section.

10.6.2 General interval consistency (energetic reasoning)

Figure 10.10 shows an example, similar to an example used by Baptiste et al. (1998), with five activities that require one unit of a resource with capacity 2. We can conclude that activity i must start after time 6. This can be deduced by testing the hypothetical constraint $\mathcal{H} = \{lct_i \leq 10\}$ against the general interval consistency constraint 2 for the interval $[1, 9)$. If i is constrained to finish at time 10 or before, then the total amount of interval work to be processed within $[1, 9)$ is $2 \cdot 4 + 3 \cdot 3 = 17$ units, whereas only $2 \cdot (9 - 1) = 16$ units are available. We can thus conclude $\neg\mathcal{H}$ and remove values less than or equal to 6 from δ_i. We emphasise that $\mathcal{H}$ can only be refuted by testing

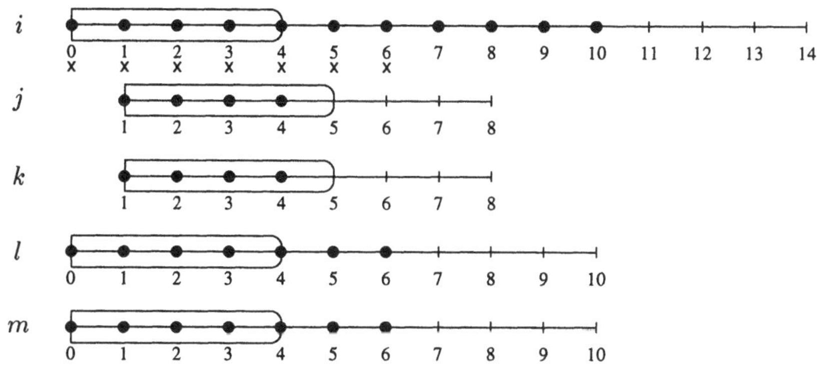

Figure 10.10 Five activities requiring one unit of a resource with capacity 2

the interval $[1, 9)$, while the interval consistency constraint is satisfied for all other intervals, including all activity or task intervals.

The example leads us back to the initial questions (1) how we should choose the hypothesis $\mathcal{H}$ and (2) for what time intervals the capacity constraint should be tested. Let us first address the second question.

The question has recently been answered by Schwindt (1997) and, independently, by Baptiste et al. (1998). They consider the difference between available and required resource capacity for a given time interval, i.e. the difference between the left and right side of the interval consistency constraint 2, as defined by the following slack function:

$$\text{slack}(\mathcal{V}_k, t_1, t_2) = R_k \cdot (t_2 - t_1) - W(\mathcal{V}_k, t_1, t_2). \qquad (25)$$

By studying the possible extrema of the slack function for a given $\mathcal{V}_k$, the set of intervals $[t_1, t_2)$ can be characterised for which the slack function can take a local or global minimum and may thus violate an interval consistency constraint. Schwindt and Baptiste et al. have shown that the number of such minimum-slack intervals is of order of magnitude $O(|\mathcal{V}_k|^2)$ and have given a characterisation of the intervals (the one in Schwindt (1997) is slightly tighter). Thus, as we know from the initial example, the set of minimum-slack intervals is larger than the set of activity intervals but still of order $O(|\mathcal{V}_k|^2)$. Since an intuitive description of the minimum-slack intervals is hard to give, and because the proof is lengthy, we do not describe the set of minimum-slack intervals in more detail.

Baptiste et al. have developed an $O(|\mathcal{V}_k|^2)$ algorithm for testing the interval consistency constraint for all minimum-slack intervals, and an $O(|\mathcal{V}_k|^2 \log |\mathcal{V}_k|)$ algorithm has been described by Schwindt, who has used the interval consistency condition for deriving lower bounds for RCPSP/max instances.

In order to reduce activity domains, Baptiste et al. suggest to use hypothetical constraints of the type $st_i \leq t_x$, similar to the initial example. The time complexity of an algorithm that computes all domain reductions which can be obtained on the minimum-slack intervals follows from the fact that there are $|\mathcal{V}_k|$ activities to be tested and $O(|\mathcal{V}_k|^2)$ minimal-slack intervals. The development of a quadratic algorithm to compute all domain reductions is an open issue.

10.6.3 Unit-interval consistency

An important special case of the general interval consistency constraint 2 is obtained when we consider time intervals of width one. The following theorem describes a test for checking unit-intervals, based on the slack function defined in Equation 25.

Theorem 9 (Unit-width interval test). *Let $i \in \mathcal{V}_k$. If, for some time t,*

$$slack(\mathcal{V}_k \setminus \{i\}, t, t+1) < r_{ik} \tag{26}$$

then the domain of i can be reduced in the following way:

$$\delta_i := \delta_i \setminus \{t' | t - p_i < t' < t+1\}. \tag{27}$$

Proof. Obvious. □

The test is interesting because it is easy to design efficient algorithms for it. There is an obvious $O(T|\mathcal{V}_k|)$ algorithm for performing all reductions deducible by the test, where T is the time horizon of the problem. The test can also be efficiently implemented through tables reflecting remaining or used capacity over time, using either vector or support point representations of the capacity profile; whenever there is a change in the table for time t — caused by an update of some activity domain — the domain bounds of all activities $i \in \mathcal{V}_k$, for which $t \in \delta_i$, can be tested and updated with effort of at most $O(|\mathcal{V}_k|)$.

Tests equivalent or similar to the unit-interval consistency test have for instance been described in Le Pape (1994, 1995), Nuijten (1994), Caseau and Laburthe (1996), Klein and Scholl (1997).

For disjunctive scheduling, the unit-interval test is subsumed in the 2-consistency test.

10.6.4 Fully and partially elastic relaxations

This section describes two relaxations of the scheduling problem that have been suggested by Baptiste et al. (1998). The relaxations describe necessary conditions for the existence of a feasible schedule. They are based upon the idea of trying to answer the question whether there exists an integer function $es_k(t, i)$, for elastic schedule, that describes the number of work units assigned to all activities throughout their time windows so that for every activity the total number of units assigned equals the required work. The capacity assignment defined by $es_k(t, i)$ is elastic in the sense that it does not necessarily take the strict rectangular shape required in the original problem.

The fully elastic relaxation is the decision problem of deciding whether a function $es_k(t, i)$ exists such that the following constraints hold:

$$es_k(t,i) = 0, \quad \text{for all } i \in \mathcal{V}_k \text{ and } t \notin \delta_i \tag{28}$$

$$\sum_t es_k(t,i) = p_i r_{ik}, \quad \text{for all } i \in \mathcal{V}_k \tag{29}$$

$$\sum_{i \in \mathcal{V}_k} es_k(t,i) \leq R_k, \quad \text{for all } t. \tag{30}$$

A tighter relaxation can be obtained by adding the two following constraints.

$$\sum_{t'<t} es_k(t',i) \leq R_k \cdot (t - \min \delta_i), \quad \text{for all } i \text{ and } t \in \delta_i \tag{31}$$

$$\sum_{t \leq t'} es_k(t',i) \leq R_k \cdot (\max \delta_i - t), \quad \text{for all } i \text{ and } t \in \delta_i. \tag{32}$$

The resulting decision problem is called partially elastic relaxation; the way in which assigned work may float within the activity time window is more restricted than in the fully elastic case.

The partially and fully elastic relaxations can be used to deduce activity domain reductions in the usual way. If, after adding hypothetical constraint $\mathcal{H}$, it can be shown that no function $es_k(t, i)$ exists that satisfies Constraints 28–32, then $\neg\mathcal{H}$ must hold. Baptiste et al. describe an $O(|\mathcal{V}_k|^2)$ domain reduction algorithm based upon the fully elastic relaxation and an $O(\log|\{r_{ik}\}| \cdot |\mathcal{V}_k|^2)$ algorithm using the partially elastic relaxation, where $|\{r_{ik}\}|$ is the number of distinct resource capacity requirements.

The partially elastic relaxation is strictly weaker than the general interval consistency constraint (Baptiste et al. 1998).

10.7 Summary

We have presented a general, unifying framework for understanding interval capacity consistency tests. Within this framework, we have surveyed and extended previous results that have been obtained in the areas of Operations Research and Artificial Intelligence. We have related the concept of energetic reasoning to sequence consistency tests known under the names of immediate selection or edge finding.

The interval consistency tests described in this chapter have been applied frequently and with great success for solving disjunctive scheduling problems. Fewer and so far less conclusive results have been reported for the application of the tests for cumulative scheduling. Many of the tests that we have described are available in general purpose scheduling software libraries such as ILOG Scheduler (Le Pape 1994, 1995, Nuijten and Le Pape 1998), CHIP (Aggoun and Beldiceanu 1993), or CLAIRE Schedule (Le Pape and Baptiste 1996).

Acknowledgments

This research has been supported by the Deutsche Forschungsgemeinschaft under grant Pe 514/7 and by the Friedrich Ebert Stiftung.

References

AGGOUN, A. AND N. BELDICEANU. 1993. Extending CHIP in Order to Solve Complex Scheduling and Placement Problems. *Mathematical and Computer Modelling* **17**, 57–73.

APPLEGATE, D. AND W. COOK. 1991. A Computational Study of the Job-Shop Scheduling Problem. *ORSA Journal on Computing* **3**, 149–156.

BAPTISTE, P. AND C. LE PAPE. 1995. A Theoretical and Experimental Comparison of Constraint Propagation Techniques for Disjunctive Scheduling. In *Proceedings of the* 14th *International Joint Conference on Artificial Intelligence*. Montreal.

BAPTISTE, P. AND C. LE PAPE. 1996. Edge-finding Constraint Propagation Algorithms for Disjunctive and Cumulative Scheduling. In *Proceedings of the* 15th *Workshop of the U.K. Planning Special Interest Group*. Liverpool, UK.

BAPTISTE, P., C. LE PAPE AND W. P. NUIJTEN. 1998. Satisfiability Tests and Time-Bound Adjustments for Cumulative Scheduling Problems. Research report, Université de Technologie de Compiègne.

BARTUSCH, M., R. MÖHRING AND F. RADERMACHER. 1988. Scheduling Project Networks with Resource Constraints and Time windows. *Annals of Operations Research* **16**, 201–240.

BŁAŻEWICZ, J., W. DOMSCHKE AND E. PESCH. 1996a. The Job Shop Scheduling Problem: Conventional and New Solution Techniques. *European Journal of Operational Research* **93**, 1–33.

BŁAŻEWICZ, J., K. H. ECKER, E. PESCH, G. SCHMIDT AND J. WĘGLARZ. 1996b. *Scheduling Computer and Manufacturing Processes*. Springer, Berlin.

BŁAŻEWICZ, J., E. PESCH AND M. STERNA. 1997. Application of Modified Disjunctive Graphs for Job Shop Scheduling Problems. Working paper.

BŁAŻEWICZ, J., E. PESCH AND M. STERNA. 1998. A Branch and Bound Algorithm for the Job shop Scheduling Problem. In *Beyond Manufacturing Resource Planning (MRP II)*, A. Drexl and A. Kimms, eds. Springer, Berlin. To appear.

BRUCKER, P., A. DREXL, R. MÖHRING, K. NEUMANN AND E. PESCH. 1998. Resource-Constrained Project Scheduling: Notation, Classification, Models, and Methods. Working paper.

BRUCKER, P., B. JURISCH AND A. KRÄMER. 1996a. The Job-Shop Problem and Immediate Selection. *Annals of Operations Research* **50**, 73–114.

BRUCKER, P., A. SCHOO AND O. THIELE. 1996b. A Branch and Bound Algorithm for the Resource-Constrained Project Scheduling Problem. Osnabrücker Schriften zur Mathematik 178, Universität Osnabrück.

CARLIER, J. 1982. The One-Machine Sequencing Problem. *European Journal of Operational Research* **11**, 42–47.

CARLIER, J. AND E. PINSON. 1989. An Algorithm for Solving the Job-Shop Problem. *Management Science* **35**, 164–176.

CARLIER, J. AND E. PINSON. 1990. A Practical Use of Jackson's Preemptive Schedule for the Job Shop Problem. *Annals of Operations Research* **26**, 269–287.

CARLIER, J. AND E. PINSON. 1994. Adjustments of Heads and Tails for the Job-Shop Problem. *European Journal of Operational Research* **78**, 146–161.

CASEAU, Y. AND F. LABURTHE. 1994. Improved CLP Scheduling with Task Intervals. In *Proceedings of the* 11th *International Conference on Logic Programming*, P. van Hentenryck, ed. MIT-Press.

CASEAU, Y. AND F. LABURTHE. 1995. Disjunctive Scheduling with Task Intervals. Tech. Rep. 95-25, Laboratoire d'Informatique de l'Ecole Normale Supérieure Paris.

CASEAU, Y. AND F. LABURTHE. 1996. Cumulative Scheduling with Task Intervals. In *Proceedings of the Joint International Conference on Logic Programming*. MIT-Press.

DOMSCHKE, W. AND A. DREXL. 1991. Kapazitätsplanung in Netzwerken: Ein Überblick über neuere Modelle und Verfahren. *OR Spektrum* **13**, 63–76.

FISHER, H. AND G. THOMPSON. 1963. Probabilistic Learning Combinations of Local Job-Shop Scheduling Rules. In *Industrial Scheduling*, J. Muth and G. Thompson, eds. Prentice-Hall, Englewood Cliffs, NF.

HEILMANN, R. AND C. SCHWINDT. 1997. Lower Bounds for RCPSP/max. Tech. Rep. WIOR-511, Institut für Wirtschaftstheorie und Operations Research, Universität Karlsruhe.

JACKSON, J. 1956. An Extension of Johnson's Results on Job Lot Scheduling. *Naval Research Logistics Quarterly* **3**, 201–203.

KLEIN, R. AND A. SCHOLL. 1997. Computing Lower Bounds by Destructive Improvement — an Application to Resource-Constrained Project Scheduling. Schriften zur Quantitativen Betriebswirtschaftslehre 4/97, Technische Hochschule Darmstadt.

KOLISCH, R., A. SPRECHER AND A. DREXL. 1995. Characterization and Generation of a General Class of Resource-constrained Project Scheduling Problems. *Management Science* **41**, 1693–1703.

KUMAR, V. 1992. Algorithms for Constraint-Satisfaction Problems: A Survey. *A.I. Magazine* **13**, 32–44.

LE PAPE, C. 1994. Implementation of Resource Constraints in ILOG SCHEDULE: A Library for the Development of Constraint-Based Scheduling Systems. *Intelligent Systems Engineering* **3**, 55–66.

LE PAPE, C. 1995. Three Mechanisms for Managing Resource Constraints in a Library for Constraint-Based Scheduling. In *Proceedings of the INRIA/IEEE Conference on Emerging Technologies and Factory Automation*. Paris.

LE PAPE, C. AND P. BAPTISTE. 1996. A Constraint Programming Library for Preemptive and Non-Preemptive Scheduling. In *Proceedings of the* 12th *European Conference on Aritificial Intelligence*.

LOPEZ, P. 1991. *Aproche énergétique pour l'ordonnancement de tâches sous contraintes te temps et de ressources*. Ph.D. thesis, Université Paul Sabatier, Toulouse. Cited after Lopez et al. 1992.

LOPEZ, P., J. ERSCHLER AND P. ESQUIROL. 1992. Ordonnancement de tâches sous contraintes: une approche énergétique. *RAIRO Automatique, Productique, Informatique Industrielle* **26**, 453–481.

MARTIN, P. AND D. B. SHMOYS. 1996. A New Approach to Computing Optimal Schedules for the Job-Shop Scheduling Problem. In *Proceedings of the* 5th *International IPCO Conference*.

NUIJTEN, W. P. 1994. *Time and Resource Constrained Scheduling: A Constraint*

Satisfaction Approach. Ph.D. thesis, Eindhoven University of Technology.

NUIJTEN, W. P. AND E. AARTS. 1996. A Computational Study of Constraint Satisfaction for Multiple Capacitated Job-Shop Scheduling. *European Journal of Operational Research* **90**, 269–284.

NUIJTEN, W. P. AND C. LE PAPE. 1998. Constraint-based Job Shop Scheduling with ILOG SCHEDULER. *Journal of Heuristics* **3**, 271–286.

PESCH, E. AND U. TETZLAFF. 1996. Constraint Propagation Based Scheduling of Job Shops. *INFORMS Journal on Computing* **8**, 144–157.

SCHWINDT, C. 1997. *Verfahren zur Lösung des Ressourcenbeschränkten Pro-jektdauerminimierungsproblems mit Planungsabhängigen Zeitfenstern*. Ph.D. thesis, Universität Fridericiana zu Karlsruhe.

TSANG, E. 1993. *Foundations of Constraint Satisfaction*. Academic Press, London.

VAN HENTENRYCK, P. 1989. *Constraint Satisfaction in Logic Programming*. Logic Programming Series. MIT Press, Cambridge, MA.

11 THE EVOLUTION OF SOFTWARE QUALITY IN PROJECT SCHEDULING

Concepción Maroto
Pilar Tormos
Antonio Lova

Departamento de Estadística e Investigación Operativa
Universidad Politécnica de Valencia (Spain)

11.1. Introduction

Since the end of the 50's when CPM –Critical Path Method– and PERT –Program Evaluation and Review Technique- were developed, the evolution of the use in practice of these and other methods of project scheduling has been closely linked to the evolution of the software and hardware which has made it possible from the start. We can divide its four decades of history into three principle periods. The first until the 70's, during which large machines, mainframes, were necessary. Therefore, project scheduling in that period was quite costly, reserving it in practice for large projects and big companies. The PC's are the protagonists of the second period, taking place in approximately the decade of the 80's. In 1981 the IBM PC came out and in 1983 Harvard Project Manager software appeared. Later, tens of packages for the PC flooded the market, making it possible for any company to have access to software at a low cost and with few hardware requirements (Maroto and Tormos, 1994; Burke, 1997). Finally, the third period corresponds to the decade of the 90's, in which software and hardware have made high quality and low price compatible. The improved quality of some programs has been achieved in successive versions by correcting flaws and deficiencies detected in previous ones, parallel to the improvements in the capabilities of the hardware and software in general. The programs are becoming more and more interactive and user-friendly, offering greater ease in connecting with other tools and more and better capabilities in resource-constrained project scheduling.

This evolution of the software has had an effect on the management of information in companies by moving project management computing from the data processing department to the desk of the project manager. In addition, it has created other conditions which strengthen the current use of project management techniques. In this sense, it is important to note that many companies have evolved from being structured in an hierarchical manner to having a project based structure. In this new context, project management software is a tool which should allow both good scheduling of activities and resources as well as communication between resources, managers and decision-makers.

In general, software quality factors can be grouped in three main categories with reference to their operational characteristics, adaptation to new interfaces and their ability to adapt to changes. Thus, in the first group we may find correction, reliability, efficiency, integrity and user friendliness. Within the second we have portability and rejectability and in the last, facility in maintenance and testing as well as flexibility. From the viewpoint of a professional software user all factors are not equally important and some are difficult to measure (Maroto and Tormos, 1994).

Various authors have evaluated the quality of project management software. We can divide the works into two groups, the first including the general software capabilities (Assad and Wasil, 1986; Wasil and Assad, 1988; De Wit and Herroelen, 1990 and Maroto and Tormos, 1994) and the second efficiency in the area of resource-constrained project scheduling (Johnson, 1992; Kolisch and Hempel, 1996; Maroto et al. 1993, 1994, 1996). On the one hand, these works demonstrate the great improvement in capabilities of software in recent years and, on the other hand, point out the need to establish a rigorous methodology for evaluation in the specific area of resource-constrained project scheduling, which is definitively one of the principle keys to the quality of project management software today. There are many packages which allow us to plan, schedule and control project activities, but not so many which assign resources well when these have limited availabilities, which is the most frequent situation in practice. This is true because in the first case it is easy to obtain the optimal solution and, in general, it is only possible to find „good solutions" through the use of heuristic methods when we want to schedule projects with resource constraints, due to the combinatorial nature of the problem.

This chapter deals with the current quality and recent evolution of project management software. In particular, we evaluate the general capabilities and their efficiency in the key area of the resource allocation, both in single project scheduling as well as multi-project scheduling, through a representative sample of systems. In this evaluation we have taken into account both the parameters most commonly used in published studies, as well as those obtained in a survey of Spanish companies. Likewise, the software has been analyzed with respect to the optimal solution –only possible in small projects-, and to the solutions obtained with the best heuristic methods. The evaluation methodology is based on an extensive design of experiment, where we control all relevant project characteristics which make the scheduling tools more or less efficient. In addition, this evaluation includes an algorithm developed by the authors which can be applied to the schedules obtained by the software and improve their performance.

11.2. Project Management Software Tools in Planning, Scheduling and Controling

A recent Project Management Software Survey (PM Network,1996) included 63 packages, whose price varied from 40$ to 50,000$. We could expect their capabilities to vary in accordance with their price. Although the majority run with Windows, there are many packages which are available for other platforms, such as DOS, OS/2, Macintosh, Unix and Windows NT. In recent years many of the older DOS products were moved to Windows, with varying success. Some authors have pointed out that Primavera Project Planner is one of the better adaptations. On the other extreme we have Artemis which does not represent a good adaptation of the windows interface (Levine, 1995). We confirm this with our evaluation.

The sample we have included in our evaluation has four of the best-known packages from the popular-priced group of Project Management Software - CA-Superproject v.4.0, Microsoft Project v.4.0, Project Scheduler 6 v.1.5 and Time Line v.6-, as well as two other higher-priced products - Primavera Project Planner v.2 and Artemis Schedule Publisher v.4.2- all of them are PC-based project management software.

Regarding Time Analysis tools all of the packages include critical path, as well as the precedence constraints: Finish-Start, Start-Start and Finish-Finish. Almost all of them allow Start-Finish links, except Project Scheduler 6 and CA-Superproject. In addition, in all the packages analyzed time lag can be associated to the different types of links. Regarding the duration of activities, 3 or more units of measure appear, being able to include up to years, months, weeks, days, hours and minutes, as with CA-Superproject and Time Line.With regard to activity duration type all of the software has at least one fixed duration and others dependent on resources (resource-driven and/or effort-driven). Finally, only CA-Superproject and Project Scheduler 6 allow PERT probability analysis.

In Resource Analysis we can divide the software into two categories in function with the number of capabilities. Among the most complete are CA-Superproject, Primavera Project Planner and Artemis, while Project Scheduler 6, Time Line and Microsoft Project, especially the latter, offer fewer capabilities in this field. The first have labour, material and other types of resources. These three packages and Time Line have variable resource availability, all packages have resource leveling, but only some offer resource smoothing, leveling by data range and activity splitting during leveling. In multi-project scheduling, only half of the packages include single and multiple approach, the first option allowing the user to include all the projects in a big one, while independence is maintained in the multi-project approach. Refering to the criteria of the different packages when resolving resource-constrained project scheduling, due to proprietary rights, the details in this sense are not in general very explicit. We have used the default options in the evaluation carried out on single and multiple projects which appears in sections 4 and 5.

All software systems permit the calculation of project costs, but none carry out time/cost trade-off analysis. Finally, in project tracking and controling there are few differences between the project management software analyzed. All the packages can track by achievement, cost and time, they include actual start/finish dates, resources used and remaining duration, costs and resources. Also, they have earned value and Cost/Schedule Control System Criteria.

Regarding the recent evolution in software it is important to note its adaptation to new technologies, making for improvements in interactivity, charting and connectivity with other software. Therefore it is currently more user-friendly than at the beginning of the 90's. Likewise, the accelerated rate with which new versions come out is remarkable. For example, CA-Superproject has come out with at least 4 versions from the end of 1988 until the beginning of 1995 -Superproject Expert v. 1.1 (1988), CA-Superproject for DOS v. 2.0 (1991), CA-Superproject for Microsoft Windows v.3 (1992) and CA-Superproject for Microsoft Windows v.4 (1995)- in which not only has the graphic interface been changed, but also other capabilities such as the inclusions of a programming language (CA-Realizer), more possibilities in the consideration of "time lag" associated with precedence constraints, better capabilities in exporting-importing data (Dynamic Data Exchange, Open Data Base Connectivity -ODBC-), etc. For example, ODBC support allows the use of external programs to update project data or to store multiproject data into a single relational database. Moreover, in each of the three first versions cited CA-Superproject has changed the calculating method in resource leveling, improving the schedules in successive versions, as the authors have confirmed in a previous work (Maroto et al., 1993) and the inclusion of resource smoothing, which provides partial leveling without project delay in Superproject Expert v. 1.1, with respect to previous versions. In the same way, it is important to point out that almost all the packages have incorporated a complete system of help and tutorials to aid learning how to use them.

11.3. Methodology to evaluate Software Quality in Resource-Constrained Project Scheduling

11.3.1. Literature review

De Wit and Herroelen (1990) pointed out that the jargon used by the user and reference manuals of professional project management software basically neglects the standardised terminology that is internationally accepted among teachers and researchers in the field. This continues to be the case, which is why it is important to keep in mind that the software sometimes calls „resource smoothing" what we know as the resource leveling problem and uses the term „resource leveling" for the resource-constrained project scheduling problem. The first studies which evaluated the efficiency of the software packages with regard to this latter problem used a single test project (Suarez, 1987; De Wit and Herroelen, 1990; Maroto and Tormos, 1994). The results already pointed out important differences between software packages. Nevertheless, all of the authors admit that the results cannot be generalized due to the use of a single test project.

Johnson (1992) evaluated 7 commercial packages using the 110 problems of Patterson. These projects have between 7 and 51 activities per project and between 1 and 3 resources. This has been the benchmark set chiefly used by the researchers in this field since 1982. Nonetheless, various authors in recent years have pointed out the need to generate nets with different degrees of „difficulty" and in fact indicate that the projects of the Patterson set are „easy" nets. From this comes the project generator PROGEN (Kolisch, Sprecher and Drexl, 1995). In 1996 Agrawal, Elmaghraby and Herroelen present another generator of test sets for project activity nets, DAGEN. This generator includes a Complexity Index, which according to the authors is an important parameter that adds one more dimension to the measure of a net's „difficulty". In our opinion these studies are a milestone in the methodology of the evaluation of project scheduling methods, both with software as well as with algorithms, in the sense that it will be necessary to use a group of perfectly defined instances and controlled parameters. Nevertheless, it seems that both generators are especially designed to generate nets to predict how easy or difficult it will be to solve the problem that interests us. However, generating test projects with controlled characteristics can have other objectives, such as establishing the best heuristic method for a certain type of project or predicting the increase in project delay which will allow us to solve the resource conflicts and whether or not this delay is near the optimal solution, both being of great practical interest (Boctor, 1990; Álvarez-Valdes and Tamarit, 1989; Maroto et al, 1996). With these objectives in mind the authors have defined two new parameters which characterize projects. One is *Nodes per Level* which measures the grade of parallelism of the activities of the net and the other is a *new definition* of resource *overload.* For this reason we have generated the test projects with a tool designed for this end (Vives, 1995).

Finally, in the studies published in this field, there is a lack of data for two important aspects; on the one hand, evaluations which use data from real companies' projects and, on the other, evaluations of the capabilities in multi-project scheduling, which according to our data are of great interest to practitioners.

11.3.2. Characteristics of real projects: a company survey

The parameter values from real projects were obtained by means of a survey carried out between October '96 and January '97 in the Valencian Region -Spain- (Lova, 1997). The company sectors of the survey are made up of small and medium-sized companies from the area of construction, textile, computers and information systems and public administrations. The latter were included because they manage many important projects. The construction sector was selected because it is a „classic sector" in the use of project management, textile because it is an important sector in the Valencian region and computers and information systems because it is a sector in which the characteristics of the projects they carry out would probably be different from the previous sectors. 1000 surveys were sent out of which we received 202, which is around 20%.

The first part of the survey is to determine the general characteristics of the companies. In the second part the questions refer to the characteristics of their activities, resources and projects –single and multiple projects-. Finally, the third part is used to determine to what extent project scheduling techniques are being used. To characterize the size of the companies surveyed we have used the number of workers and the annual sales turnover. Specifically, in total 46% of the companies which responded to the survey have between 10 and 50 workers, 23% have less than 10, 13% between 50 and 100 and only 17% have more than 100 workers. By sectors, it is important to note that in the construction sector the proportion of companies with the number of workers between 10 and 50 is 65% and in the textile sector it is 55%. In this region the companies from the computers and information systems sector have a lower number of workers, given that 2 out of 3 have less than 10 workers. In the case of the public administrations, given that we sent the survey to all of the larger ones, more than half of those that answered had more than 100 workers. This data confirms that the sample, obtained through a stratified sampling of all the companies in the Valencian Region pertaining to the cited sectors, corresponds to small and medium-sized companies. By annual sales volume 37% presented values less than 2 million dollars, 28% between 2 and 4 million dollars and the rest with higher values. In accordance with the number of workers, lower sales turnover appears in the computer and information system sector and higher in public administrations.

Refering to the number of activities of the projects we found that its value depends on the sector studied. Nevertheless, we can see the predominance of projects with less than 50 activities (84%) and about 95% of the projects have less than 100 activities. Larger sized projects were practically non-existent in this geographical area characterized by small and medium sized companies. Regarding the duration of activities, on average the minimum and maximum values were 6 and 73, respectively. Greater durations appeared in construction and public administrations and smaller values appeared with the textile sector, while the computers and information system sector had intermediate values. With regard to the parameters related to the net structure, the survey points out than on average half of the projects have a grade of parallelism between 1 and 5 *Nodes per Level* and one quarter have between 5 and 10, this being a new parameter defined by the authors and different from „network complexity".

One of the main results of the survey is to quantify how many companies work with single projects and how many with multiple projects. The results are conclusive; 84% of the companies which responded to the survey indicated that they work with multiple projects, while only 16% worked with single projects. Although the different sectors vary a great deal, one third share less than 3 resources and approximately 85% share less than 10 resources. This data strikes a contrast with that which appears in PM Network (1996) where it indicates that 50% of the project management community work on single projects. Both sources seem to indicate that the evaluation of software should be carried out in both areas, for single projects as well as for multiple projects. The units of resource requirements per activity vary on average between 2 and 8, and there were no significant differences between sectors.

11.3.3. Design of experiment to evaluate software quality in resource allocation

Project scheduling software has high capabilities in the area of resource allocation to activities: different resource types, activity durations which depend on the resources, the possibility to split activities in the leveling process, leveling by priority, variable resource availability, the possibility to solve resource conflicts in an interval of dates selected by the user, etc. In theory a higher number of capabilities means higher quality software. Nevertheless, does the software carry out all of these possibilities in an efficient manner? Not always. For example, the capability „resource smoothing" does not exactly correspond to the well known „resource leveling problem" (the smoothing of resource requirements by rescheduling activities within their available slack subject to a fixed project duration). Thus, in CA-Superproject the option of resource smoothing does the same as resource leveling, with the only difference being that in the first case the project duration does not increase, with respect to the initial critical path. Therefore, it does not usually solve the resource conflicts that the project had and moreover can give schedules where the sum of the square deviations of the resource requeriments from the average is greater than the initial schedule. This fact brings to light that the real quality of the software needs to be compared through variables indicative of the same and in conditions which allow us to generalize the obtained results, as is the case with an adequate design of experiment.

We consider the software to be more efficient and therefore to be of higher quality when the project duration is lower and we are able to schedule activities with the available resources. With the single project environment the response variable that we have analyzed is the increase in project duration with regard to the optimal solution. In larger projects with more than 50 activities we only can measure the efficiency through of the increase in project duration with regard to the critical path. We also have analyzed this latter response variable. In multi-project approach, where it is not possible to obtain the optimal solution, we have measured the efficiency through the classic response variable *„mean project delay"*, which measures the average increase in project completion time with respect to its initial critical path.

The software sample used in the evaluation includes 4 packages pertaining to what we could call medium-range software (CA-Superproject v.4.0, Microsoft Project v.4.0, Project Scheduler 6 v.1.5 and Time Line v.6) and 2 of the high-range with respect to price (Artemis Schedule Publisher v.4.2 and Primavera Project Planner v.2). These 6 systems are evaluated both in the scheduling of projects with constrained resources in single projects as well as in multi-projects with shared resources. In the first case its efficiency is measured against the optimal solution obtained by the Branch and Bound procedure from Demeulemeester (1992) and against the heuristic algorithms Minimum Latest Finish Time (MINLFT), First Come First Served (FCFS) and a new algorithm based on resource slack concept, which is applied to a feasible schedule and obtains a better one. This method we will call Time and Resource Criticality (TRC) (Tormos, 1996). TRC-LFT is this algorithm applied to the schedule obtained with the LFT rule and TRC-CASPJ indicates this same algorithm applied to the schedule obtained with CA-Superproject. In the case of scheduling multi-projects, the algorithms against which we have evaluated the relative efficiency of the software are Shortest Activity from Shortest Project (SASP), Maximum Total Work Content (MAXTWK) and FCFS.

Table 11:1. Parameter values of projects from literature review, survey and design of experiment: single projects

Project Characteristics	*Parameter Values*			
	Authors (year)		*Survey*	*Design of Experiment*
Number of Activities	Davis and Patterson (1975) Cooper (1976) Christofides et al.(1987) Álvarez-Valdés and Tamarit (1989) Ulusoy and Özdamar (1989) Boctor (1990) Kolisch and Drexl (1996)	20-27 60 25 27,51,103 7-33 5-30;38- 111 30	84%≤ 50 97%≤ 100	30, 50
Durations of Activities	Davis and Patterson (1975) Cooper (1976) Christofides et al.(1987) Álvarez-Valdés and Tamarit (1989) Ulusoy and Özdamar (1989) Kolisch and Drexl (1996)	1-9 1-10 1-9 1-10 1-9 1-10	6-73	1-10
Network Complexity (Arcs/Nodes)	Davis and Patterson (1975) Christofides et at.(1987) Ulusoy and Özdamar (1989) Kolisch and Drexl (1996)	1-3 1-3 1-3 1-3	60 %≤ 3	3
Nodes per Level (NL)	Maroto et al. (1996) Tormos et al. (1996)	2-4, 5-7 2-4, 5-7	50 %≤ 5 75 %≤ 10	2-4, 5-7
Number of Resources	Davis and Patterson (1975) Cooper (1976) Christofides et al.(1987) Álvarez-Valdés and Tamarit (1989) Ulusoy and Özdamar (1989) Boctor (1990) Kolisch and Drexl (1996)	3 5 3 6 3 3 4	34%≤ 3 75%≤ 10	4
Resource Requirements	Cooper (1976) Álvarez-Valdés and Tamarit (1989) Christofides et al. (1987)	1-10 1-10 0-6	2-8	1-6
Resource Factor (RF)	Davis and Patterson (1975) Cooper (1976) Álvarez-Valdés and Tamarit (1989) Ulusoy and Özdamar (1989) Kolisch and Drexl (1996)	1/3,2/3,3/3 2/5, 5/5 3/6,6/6 1/3,2/3,3/3 1/4,...4/4	44%≤ 3 92%≤ 5	1/4, 2/4, 3/4, 4/4
Resource Overload	Tormos (1996) Tormos et al. (1996)	0.6, 0.9 0.6, 0.9	-	0.6 0.9

To design the computational experience we have analyzed the survey together with the literature review, so that on the one hand the results can be applicable to a large number of companies and on the other hand the results can be comparable to those found in the literature. Tables 11:1 and 11:2 contain the values of the parameters which characterize the projects that have been used to evaluate the algorithms and software in resource-constrained project scheduling, obtained in the survey made of companies and which are those that we have used in our design of experiments. In general terms we can say that the parameters appearing in the literature reflect the characteristics of part of the real projects, even if these in practice have more variability.

Obviously it is necessary to do two designs of experiments, one for single projects and the other for multiple projects. In the first case and observing table 11:1, the activity durations of the test projects have been randomly generated between 1 and 10. We have done the same with resource requirements which vary between 1 and 6. Regarding the parameter network complexity we have set it at 3, due to the fact that in previous studies it has not proved to be a significant parameter in explaining the efficiency of the different methods. For the parameter indicative of network structure we have proposed *Nodes per Level*, which measures the grade of parallelism of the project activities. The instances have been obtained with a generator especially designed to control this parameter, given that it cannot be controlled with PROGEN or with DAGEN. The generator used allows us to specify the minimum and maximum number of nodes for each level of the network. Thus, a project with values of Nodes per Level of 5-7 is refering to a project ordered in levels generated in such a way that the number of nodes (activities) of each level is a random number between 5 and 7. As we will see in the next section this factor decisively influences the response variable with which we measure the efficiency and therefore the software quality.

In short, in the design for single projects we have considered four factors: Number of Activities, Nodes per Level, Resource Factor and Resource Overload. The Number of Activities is a factor with two levels, 30 and 50 activities. Nodes per Level also has two levels, one of 2 to 4 nodes per level and the other of 5 to 7. Resource Factor is analyzed at 4 levels, in which activities have assigned 1 of 4 resources (1/4), 2 of 4 (2/4), 3 of 4 (3/4) and 4 of 4 (4/4). Finally, the Overload Factor is also a new parameter defined by the authors, which establishes the relationship between the total demand of a resource and the availability of the same in the time that it is assigned. This overload parameter is a factor that characterizes the resources and by extension the project, given that its corresponding value will be that of the resource with the greatest resource overload (Tormos, 1996). The aforementioned factors and their respective levels define 2x2x4x2=32 treatments, of which three replications have been carried out. Therefore, 96 projects have been solved with each one of the six software packages and the exact and heuristic methods.

The factors that we have considered in the design of experiment to analyze efficiency in multiple projects are: Number of Projects, Number of Activities per Project, Resource Factor and Resource Overload, all of them at two levels. We have carried out two replications for each one of the 2x2x2x2= 16 treatments. In this case, as in the single projects we have used intances with activities which have randomly generated duration between 1 and 10, resource requirements between 1 and 6, and an arcs/nodes relationship of 3 and 5 shared resources. Regarding the values of the different levels of the factors we have: the Number of Projects which make up the multiple project is of 4 and 8, the number of activities of each single project is of 30 and 60, the number of resources assigned to an activity is 2 of 5 and 4 of 5. In total, we have solved 32 multiprojects integrated by 192 single projects. Finally, the overload parameter is the one which measures the grade of resource conflict and is defined by Kurtulus and Davis (1982).

Table 11:2. Parameter values of projects from literature review, survey and design of experiment: multi-projects

Project Characteristics	Parameter Values			
	Authors (year)		Survey	Design of Experiment
Number of Projects	Fendley (1968) Kurtulus and Narula (1985) Lawrence and Morton (1993)	3, 5 3 5	28%≤ 5 56%≤10	4 8
Number of Activities per Project	Kurtulus and Davis (1982) Kurtulus and Narula (1985) Dumond and Mabert (1988) Lawrence and Morton (1993)	34-63 24-66 6-49 25-50	84%≤ 50 97%≤100	30 60
Durations of Activities	Kurtulus and Narula (1985) Lawrence and Morton (1993)	1-9 1-10	6-73	1-10
Network Complexity (Arcs/Nodes)	Lova (1997) Maroto et al.(1997)	3 3	60%≤3	1-3
Nodes per Level (NL)	Lova (1997) Maroto et al. (1997)	3-5 3-5	50%≤ 5 75%≤ 10	3-5
Number of Shared Resources	Dumond and Mabert (1988) Lawrence and Morton (1993)	3 5	34%≤ 3 75%≤10	5
Resource Factor (RF)	Dumond and Mabert (1988) Lawrence and Morton (1993)	1-3 1-3	44%≤ 3 92%≤5	2/5 4/5
Resource Requirements	Kurtulus and Davis (1982) Kurtulus and Narula (1985) Dumond and Mabert (1988)	2-6 1-9 0-8	2-8	1-6
Resource Overload	Kurtulus and Davis (1982) Kurtulus and Narula (1985) Lova (1997) Maroto et al. (1997)	0.6-1.6 0.6-1.6 0.8, 1.2 0.8, 1.2	-	0.8, 1.2

The two aforementioned designs allow us to correctly evaluate software quality, the performance of which is measured as the makespan in single projects and mean project delay in multiple projects, given that they consider as factors the principle project characteristics which affect the performance of the response variable. ANOVA is the appropriate tool to carry out this analysis, even though it is a particular ANOVA which considers fixed and random factors (the project factor is random). Then, we have analyzed the results through a mixed ANOVA with crossed and nested factors and fixed and random effects with a significance level of 5%.

11.4. Efficiency in scheduling single projects

Given the wide range of software capabilities in project scheduling under resource constraints described in section 2, we have evaluated its efficiency in the specific case of the resource-constrained project scheduling problem (often denoted RCPSP) which is the scheduling of project activities subject to finish-start precedence constraints with zero time lag and constant renewable resource constraints in order to minimize the project duration. Activities have a single execution mode with a fixed integer duration, preemption is not allowed and renewable resource requirements are constant throughout the duration of an activity (Herroelen, Demeulemeester and De Reyck, 1998).

As Agrawal, Elmaghraby and Herroelen (1996) say to answer the question of „goodness of performance", one must have on hand two elements: measures of relative efficacy and a field of experimentation (standard set of test problems). The latter has been defined in the previous section with the designs of experiments. Refering to the first element and taking into account the classification scheme for project scheduling from Herroelen, Demeulemeester and De Reyck (1998) we use a regular performance measure, in particular the *makespan*. That is to say, the objective is to minimize the project makespan. Then the problem is denoted as $m,1|cpm|C_{max}$.

Instead of analyzing the makespan absolute values, we analyze it expressed in the percentage increase with respect to the optimal project duration. Obviously the results of the statistic analysis will be the same. Thus, this is a measurement of the absolute efficiency, given that we compared the performance of the professional software with respect to the best possible solution. This optimal solution has been obtained with the Branch and Bound procedure from Demeulemeester (1992).

Therefore, the response variable considered has been the increase in project duration by resolving the resource conflicts with the different project scheduling software and heuristic algorithms in comparison to the optimal solution, which is expressed in percentage and denoted by OS_INC as follows:

$$(1)\quad OS\_INC = \frac{Project\ Duration - Optimal\ Solution}{Optimal\ Solution} * 100$$

The performance of professional software packages appears in figure 11:1. The most noteworthy result is the significantly worse performance of one of the higher priced software packages: Artemis Schedule Publisher. On average for all the 92 instances used, Artemis gives schedules which are 14.6% worse than the optimal solution. The results of Primavera Project Planner are also surprising. This higher priced software package with greater options in resource allocation gives, with its standard option, project durations similar to those from the medium range software analysed such as Microsoft Project and Project Scheduler 6. Primavera Project Planner has been evaluated not only in its standard option, but also in two of the many possibilities it offers, specifically in assigning resources using LFT and free slack rules. The former has an efficiency similar to the standard option and the latter performs sufficiently worse, as can be seen in figure 11:1. CA-Superproject and Time Line represent the most efficient software with reference to resource constrained project scheduling and give project durations which are on average similar to those obtained with the classic rule LFT. The priority rule Time and Resource Criticality, defined by the authors, applied to the schedules of CA-Superproject and LFT improves their results as appears in figure 11:1.

In general, we can say that the software analyzed can be divided into three groups regarding the quality of the schedules that they produce when the project resources have limited availabilities. The first group is formed by those which present a higher performance and which is similar to the classic method LFT: CA-Superproject and Time Line. The second group is formed by the medium quality software which is similar to that obtained with the rule First Come First Served (FCFS): Project Scheduler 6, Microsoft Project and Primavera Project Planner, both in its default option as well as the LFT option. In the third group we place the worst ones whose performance is much worse than the one obtained with the rule for randomly assigning resources: Artemis and Primavera with the free slack option. Finally, it is important to note that the most efficient software also can be improved as is shown by the fact that the priority rule TRC is significantly better than Time Line (figure 11:1).

All of the remaining factors considered in the design of experiment have resulted significant, except for number of activities. Therefore, the increase in project duration with respect to the optimal solution does not depend on the number of activities -for the levels that we are considering which are 30 and 50 activities-, but rather the values of the following parameters: Resource Factor, Nodes per Level and Overload. In addition, some interactions resulted significant, such as Software x Nodes per Level, Software x Overload and Software x Nodes per Level x Overload, meaning that the influence of some factors depends on the value of the others. We will explain the results in detail below.

The influence of Resource Factor is shown in the significantly higher increase in project duration with respect to the optimal solution when resolving conflicts of more than one resource. Thus, when a resource is implicated the duration increases 3.7% with respect to the optimal solution, while this value is between 6.1 and 7.8%

in resource constrained project scheduling with 2 to 4 resources considered, observing no significant differences between the three levels- Resource Factor 0.5, 0.75 and 1-, even if the lowest value of the three appears with RF=1, as in the evaluations carried out by other authors (Álvarez-Valdés and Tamarit, 1989).

Refering to Overload factor, its well-known significant influence in the response variable analyzed stands out. The projects with low overload increase their duration 3.7% with respect to the optimal solution, while those with higher overload have an increase of 8.9% on average. A similar result appears with the factor Nodes per Level.

The projects which have 2 to 4 nodes per level obtain schedules that are on average 4% worse than the optimal, while those with higher parallelism (from 5 to 7 activities) present a value of 8.6%. The interactions were significant in the ANOVA and we should therefore interpret the results in the light of this. For example, the effect of increasing the number of nodes per level is much greater in projects with greater resource conflicts, which we have measured with overload factor. In general, our analysis allows us to state that projects in which a duration higher than the optimal can be expected- between 12 and 17.6%- are those which present higher values in the factors Overload and Nodes per Level together with the assigning of more than one resource. All of the other combinations of factors –Resource Factor, Nodes per Level and Overload- presented increases between 1.7 and 6%.

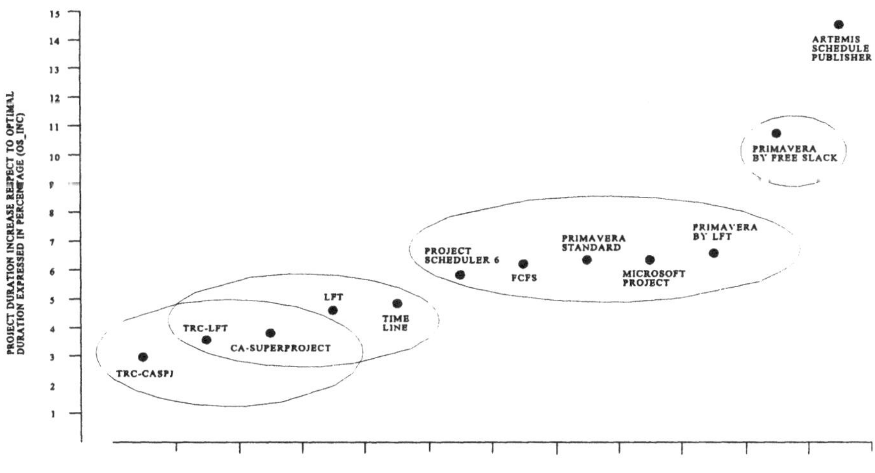

Figure 11:1. Performance in Resource-Constrained Project Scheduling: Single Projects. The methods which appear within an ellipse do not present a significantly different performance according to the Newman-Keuls test

Special mention should be made of the interactions between Software by Nodes per Level and Software by Overload, which shed a new light on the previous results. Although the order of efficiency is maintained with respect to the one indicated in figure 11:1, when projects are more sequential the performance of Project Scheduler and Primavera is not significantly different from that of CA-Superproject and Time

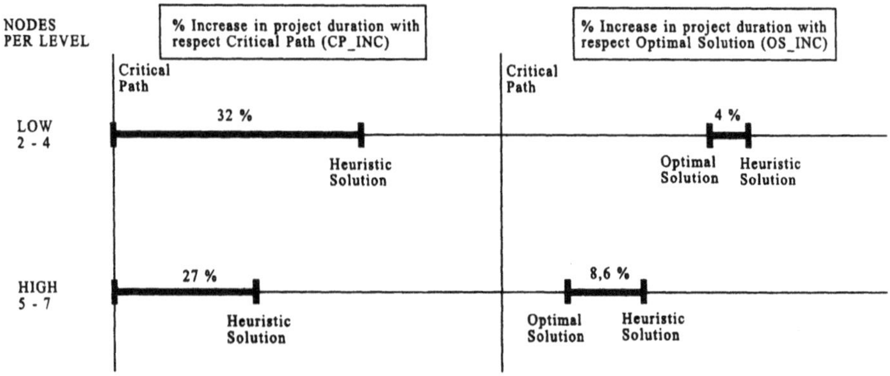

Figure 11:2.The influence of Nodes per Level parameter in the variables which measure the software and algorithm performance

Line, although on average its output is worse. On the other hand, the anomalous

$$(2)\quad CP\_INC = \frac{Project\ Duration - Initial\ Critical\ Path}{Initial\ Critical\ Path} * 100$$

performance of Artemis stands out, which with sequential projects gives increments of 10% and with those with a higher level of parallelism of 18.7%. Regarding the interaction Software by Overload, it is apparent that the differences between the different packages are basically due to the greater efficiency in scheduling with constrained resources when the level of resource conflicts is high. With low levels of overload the order reflected in figure 11:1 is maintained, but the differences between the average values are not significant in the ANOVA carried out.

In general, when the number of project activities is higher than 50 activities it is not possible to obtain the optimal solution, thus being a NP-hard problem. In this case, it is only possible to evaluate the relative efficiency of the different software packages. A variable which allows us to measure this relative efficiency is the percentage increase in the project duration with respect to the initial critical path. This variable is represented by CP_INC and evaluated as follows:

The interpretation of the results obtained through the ANOVA carried out with this response variable is similar for all factors, except for Nodes per Level. Thus, the performance of the software is similar to that represented in figure 11:1 in which CA-Superproject and Time Line appear as the most efficient, Artemis as the worst, and the rest in an intermediate position. The first ones offer an increase in project duration of 26.6% and 27.5% respectively. Project Scheduler 6, Microsoft Project and Primavera have values around 29.5% and Artemis of 40%. Regarding the overload, it can be said that the projects with few resource constraints increase their duration 13.2% on average and those with higher overload have an increase of 46%.

The nodes per level affects the increase in project duration with respect to the initial critical path in the opposite way in which it influences the increase in project duration with respect to the optimal solution. The more sequential the project the lower the difference between the optimal solution and the heuristic solution offered by the software, but the increase in project duration is greater compared to that offered by CPM. In contrast, the greater the number of nodes per level of a project the less its duration will increase with respect to the critical path, but the difference between the optimal and heuristic solutions will be greater, as can be seen in figure 11:2. If we do not include Artemis in this analysis the differences between the optimal and the heuristic solutions set out in figure 11:2 would be reduced in 2 points. We believe that this result is very interesting from the practitioners′ point of view as it allows them to predict, according to the characteristics of the project network, the magnitude of the advantages or disadvantages of applying an exact or heuristic method.

11.5. Professional software performance in multiproject scheduling

In this section we continue to evaluate the capabilities of the professional software in resource-constrained project scheduling as in the previous section, but now these resources are shared by various projects. From a conceptual point of view, and following the classification scheme defined by Herroelen, Demeulemeester and De Reyck (1998), we could say that it is the same problem, except for the variable which is going to measure the relative efficiency of the different software packages. In fact, if we could find the optimal solution to this problem this differential treatment would make no sense. Nevertheless, scheduling multi-projects with shared resources is quite relevant in practice given that the heuristic solution is the only one possible and the implicit or explicit consideration of each project from multi-project is a relevant factor in scheduling method efficiency.

We can distinguish two possible focuses to the problem: single and multiple approaches. There are programs such as CA-Superproject and Microsoft Project which allow the user to integrate various projects in a bigger one or maintain the independence of the same, although they share resources. Results can vary according to the option chosen. The measurement of efficiency which we could call classic of the single project approach is the increase in the duration of the multiproject. The classic measurement of the multi-project approach is the *mean project delay*. This latter is the one we are going to use in our evaluation and it is calculated through the summation for all the projects of the increase in finish time with respect to the initial duration given by the critical path and dividing it by the number of projects, as appears in the expression (3).

$$(3)\quad MEAN\ PROJECT\ DELAY = \frac{\sum_{i=1}^{M}(Finish\ Time_i - Critical\ Path_i)}{M}$$

To analyze the results obtained upon evaluating this response variable in the 32 test multiprojects, for the 6 software systems and the 3 algorithms specified in the design of experiment, we have carried out a mixed ANOVA with crossed and nested factors and fixed and random effects, due to the nature of the multi-project factor. The level of significance used is 5%. Figure 11:3 reflects the average value of the results obtained for each one of the methods evaluated. All of the software packages have been evaluated in their default options and Primavera has also been evaluated with the option of giving priority to projects. This priority has been randomly assigned to projects. The methods which in figure 11:3 do not appear within the same ellipse present a performance significantly different through the ANOVA carried out. In the same way, the systems which appear within the same ellipse offer a similar performance. Comparing these results with those obtained for single projects, we can see a different performance and with greater statistically contrasted differences between the packages which make up the sample. The software which obtains the best schedules is Primavera with the option of giving random priority to the projects. In this case its performance is similar to that of the rule MAXTWK – Maximum Total Work Content- and this presents a slightly better performance than SASP –Shortest Activity from Shortest Project-. Both algorithms implicitly give priority to the projects, the first to the one that has more work scheduled and the second to the shortest project. The fact that these three methods are the most efficient demonstrates the need and importance of considering the projects independently in this context.

Another relevant result is the quality of the solutions obtained by Primavera in its default option, which on average are better than those of the other packages, although the differences with CA-Superproject are not significant, both systems having an efficiency similar to the one offered by the FCFS rule as appears in figure 11:3. Time Line, Project Scheduler 6, Artemis and Microsoft Project present a significantly worse performance, in that order, both compared to Primavera and CA-Superproject, as well as between themselves as can be seen in figure 11:3. Artemis continues to be among the worst in quality when solving the problem in which we are interested, but in this case better than Microsoft Project. The order of magnitude of these differences is such that, for example, the Mean Project Delay that Microsoft gives is 36% worse than Primavera's. In absolute terms, while Microsoft Project gives average increases of 254 days, this data in Primavera, in its default option, is 187. These differences can have a considerable economic impact on companies.

Regarding the influence of the remaining factors contemplated in the design of experiment the results agree with those predicted a priori. The number of activities of the projects is a significant factor, obtaining a Mean Project Delay of 153 days when the single projects of the multiple projects have 30 activities, while the value reaches 237 days for those formed by projects with 60 activities. The Resource Factor which has also turned out to be significant offers a somewhat inferior value when resolving conflicts of 4 of the 5 resources (175 days), against an average obtained when the resources are 2 of 5 (215 days). This result is similar to that obtained in the evaluation carried out for single projects. The overload also has an important and significant effect on the increase in Mean Project Delay due to the resource constraints. The higher the overload of the multi-project, the greater the average increase in finish time of the projects that make it up. Thus, the multi-projects with a value of 0.8 in overload present average increases of 148 days, while this value is 242 for multi-projects with 1.2 of overload.

The only factor considered in the design of experiment for multi-projects which did not turn out to be significant is the number of projects which make up the multi-project. This result is not unusual if we take into account the variable with which we are measuring the performance and therefore the quality of the software packages. That is to say, the number of projects which share resources does not effect the Mean

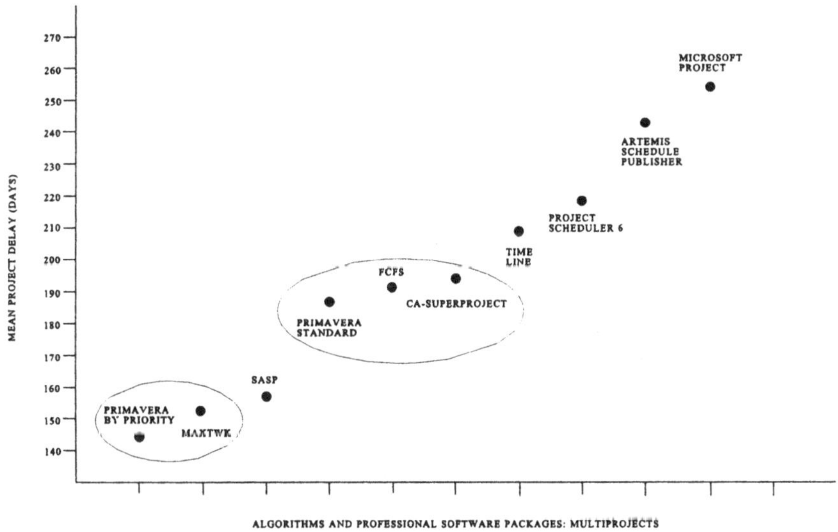

Figure 11:3. Performance in resource-constrained project scheduling: multiprojects. The methods which appear within an ellipse do not present a significantly different performance according to the Newman-Keuls test

Project Delay, for the levels that we are considering, which are 4 and 8 projects. If we had measured the performance in function with other response variables such as the average increase in project duration and the idle resources, the number of projects factor would have turned out to be significant and/or its interaction with the different solving methods would have been significant (Lova, 1997).

Finally, we want to make a special point of how complex it is to carry out a complete evaluation of the software capabilities in resource-constrained project scheduling in general and in particular when the resources are shared by various projects, due to the diversity of possible measurements of performance in function of the objective that we established and also for the high number of factors –and levels of each of those- that can have an influence. Therefore, it is essential that the conditions are precisely established when evaluating the performance of software packages, in order to determine the level of generality of the obtained results.

11.6. Discussion of results, conclusions and future perspectives

From the end of the 80's until today, the evolution of Project Management Software has been characterized by the incorporation of new technologies, translated into more interactive systems with higher charting and connectivity capabilities. Thus, this software is more userfriendly and open. In addition, new tools have been incorporated which are useful for planning, scheduling and controling, such as PERT Probabilility Analysis, more general precedence constraints and time lag, more and better resource allocation methods, earned value and so on.

Our evaluation has demonstrated that the area in which there are more differences regarding capabilities and quality of the same is resource-constrained project scheduling. Some systems only allow the consideration of renewable resources, type labour, while others also include non-renewable resources. All the packages analyzed resolved the resource-constrained project scheduling problem, and in some cases are able to partially solve the resource conflicts without increasing the project duration given by the critical path or without passing the duration indicated by the user. CA-Superproject, Primavera Project Planner and Time Line solve the preemptive and generalized resource-constrained project scheduling problem, while Artemis Schedule Publisher only the generalized case and Microsoft Project and Project Scheduler neither of the two problems. Finally, regarding the criteria for assigning the resources to different activities in the resolution of the aforementioned problems, there is great variability. All of them allow the user to give priority to activities and at least another default option, although most of the time they do not specify which one. Primavera is the most complete in this sense, given that it allows the user to choose the first, second, etc. criteria from a list of 43 possibilities.

To correctly evaluate the quality of the schedules obtained with the different software packages we have created two designs of experiments, one in the scheduling of constrained resources in single projects and the other for multi-projects with shared resources. In both we have generated instances in which all the relevant project characteristics with respect to the aforementioned problem have been controlled. The results point out significant differences in performance of the different packages. Thus, in single projects CA-Superproject and Time Line are the most efficient because they offer the schedules with the shortest duration, even better than more expensive software, such as Artemis Schedule Publisher and Primavera Project Planner. It is also important to note that the most efficient systems have a performance similar to that of the classical methods such as LFT, although these results can be improved by applying algorithms based on priority rules like the one included in this work –TRC rule-.

The evaluation carried out of the resource allocation in multi-projects with shared resources demonstrates their great complexity. The majority of the software packages analyzed permit single and multi-project approaches. The results show that the performance of the software in this case differs from that obtained in single projects. The highest quality software is Primavera Project Planner, a program with which we have evaluated the usefulness of the priorities of the projects as criteria for assigning resources, with an efficiency comparable to that of the classical algorithms MAXTWK and SASP. In its standard option, Primavera Project Planner offers an efficiency similar to CA-Superproject, while the rest of the programs perform worse and in the following order: Time Line, Project Scheduler 6, Artemis and Microsoft Project. These four present performances significantly different amongst themselves.

Both in single and multiple projects our analysis has demonstrated that certain project characteristics, such as nodes per level (structure factor) and overload decisively affect both the absolute as well as the relative quality of the professional software. Thus, when projects have few resource conflicts the differences in performance of the different packages are not significant. We found similar results when the structure of the projects is very sequential, that is to say it has a low number of activities for each graph level of the project. In contrast, in projects with many nodes per level and/or high overload the differences in software performance are more acute and in addition to being significant in the statistical analysis they are relevant in absolute value and therefore with possible important economic repercussions from the practitioners' point of view. It is important for them to know that the efficiency of software is different in single and multiple projects. Not always are the best packages for scheduling single projects with constrained resources the best in the multi-project approach. Thus, the work environment in this sense should be taken into account when choosing software. In addition, the usefulness of project priorities when scheduling multi-projects should also be taken into account, which even used randomly can improve scheduling as well as offering greater flexibility to the user, as in activity priorities.

Despite the many capabilities in the field of project scheduling, the software does not allow us to solve problems like time/cost trade-off, multi-mode resource-constrained project scheduling and maximizing the net present value, while only some of them include the option of solving the resource levelling problem. Its incorporation in the professional software would mean an improvement in quality. This could also be increased by including the optimal solution in the cases where it was possible. Along these lines, parallel computing could increase its area of application in the future, as well as the application of the heuristic techniques, like genetic algorithms. Finally, another important improvement would be the increase in flexibility of the Project Management Software allowing the user to choose one or various objectives when scheduling projects.

Notes

1. Supported by GV-2216/94, Programa de Proyectos de Investigación Científica y Desarrollo Tecnológico, Generalitat Valenciana.

2. Acknowledgement: We are grateful to Erik Demeulemeester of the Katholieke Universiteit Leuven for providing his B&B procedure.

References

AGRAWAL, M.K., S.E. ELMAGHRABY AND W.S. HERROELEN. 1996. DAGEN: A generator of test sets for project activity nets. *European Journal of Operational Research*. 90, 376-382.

ÁLVAREZ-VALDÉS, R. AND J.M. TAMARIT. 1989. Heuristics Algorithms for a Resource-Constrained project scheduling: A review and an empirical analysis in *Advances in Project Scheduling*, R. Slowinski and J. Weglarz (editors). Elsevier, Amsterdam, 113-134.

ASSAD, A.A. AND E.A. WASIL. 1986. Project Management using a microcomputer. *Computers and Operations Research*. 12/2-3, 231- 260.

BOCTOR, F.F. 1990. Some efficient multi-heuristic procedures for resource constrained project scheduling. *European Journal of Operational Research*. 49, 3-13.

BURKE, R. 1997. Project Management. Planning and Control. Wiley & Sons.

CHRISTOFIDES, N., R. ALVAREZ-VALDES AND J.M. TAMARIT. 1987. Project scheduling with resource constraints: a branch and bound approach. *European Journal of Operational Research*. 29, 262-273.

COOPER, D.F. 1976. Heuristics for scheduling resource constrained scheduling projects: an experimental investigation. *Management Science*. 22, 1186-1194.

DAVIS, E.W. AND J.H. PATTERSON.1975. A comparison of heuristics and optimum solutions in resource constrained project scheduling. *Management Science*. 21, 944-955.

DAVIS, E.W. AND R.D. MARTIN. 1985. Project management software for the personal computer: an evaluation. *Industrial Management*. 1-21.

DE WIT, J. AND W.HERROELEN.1990. An evaluation of microcomputer-based software packages for project management. *European Journal of Operational Research*. 49, 102-139.

DEMEULEMEESTER, E. AND W. HERROELEN. 1992. A branch and bound procedure for the multiple resource constrained project scheduling problem. *Management Science*. 38, 1803-1818.

DUMOND, J. AND V.A. MABERT.1988. Evaluating project scheduling and due date assignment procedures: An experimental analysis. *Management Science*. 34, 101-118.

FENDLEY, L.G. 1968. Towards the Development of a Complete Multiproject Scheduling System. *Journal of Industrial Engineering*. 19,10, 505-515.

HERROELEN, W., E.DEMEULEMEESTER AND B. DE REYCK. 1998. A classification scheme for project scheduling in *Handbook on Recent Advances in Project Scheduling*. Chapter 1. Kluwer.

JOHNSON, R.V. 1992. Resource constrained scheduling capabilities of commercial project management software. *Project Management Journal*. 22, 39-43.

KOLISCH, R., A. SPRECHER AND A.DREXL. 1995. Characterization and Generation of General Class of Resource-Constrained Project Scheduling Problems. *Management Science*. 41/10, 1693-1703.

KOLISCH, R. AND A.DREXL.1996. Adaptative Search for Solving Hard Project Scheduling Problems. *Naval Research Logistics*. 43, 23-40.

KOLISCH, R. AND K. HEMPEL. 1996. Finite Scheduling Capabilities of Commercial Project Management Systems. Working Paper n° 397. Institut für Betriebswirtschaftslehre, Lehrstuhl für Produktion und Logistik, Kiel, Germany.

KURTULUS, I.S. AND E.W.DAVIS. 1982. Multi-Project scheduling: Categorization of heuristic rules performance. *Management Science*. 28, 161-172.

KURTULUS, I.S. AND S.C. NARULA. 1985. Multi-Project Scheduling: Analysis of Project Performance. *IIE Transactions*. 17/1,58-66.

LAWRENCE, S.R. AND T.E.MORTON. 1993. Resource-constrained multi-project scheduling with tardy costs: Comparing myopic, bottleneck, and resource pricing heuristics. *European Journal of Operational Research*. 64, 168-187.

LEVINE, A. H. 1995. Innovators? Or Tripping Over Their Egos? Winners and Losers in the PM Software Industry. *IEEE Engineering Management Review*. Winter 50-62.

LOVA, A. 1997. Programación de multiproyectos con recursos limitados: un enfoque multicriterio. Unpublished Ph. D. Dissertation. Departamento de Estadística e Investigación Operativa. Universidad Politécnica de Valencia. Spain.

MAROTO, C., C.CAPILLA, P.TORMOS AND C.ESTEVE.1993. Constrained Resource Scheduling in Project Management Software. Paper presented at Decision Making:Towards the 21st Century. Madrid, June 2-5.

MAROTO, C. AND P.TORMOS. 1994. Project Management: An Evaluation of Software Quality. *International Transactions in Operational Research.* 1,2, 209-221.

MAROTO,C., S.BALASCH, P.TORMOS, C. CAPILLA, AND F.CRESPO.1994. Project Management Software: Capabilities and efficiency in resource constrained Scheduling. Paper presented at the Fourth International Workshop on Project Management and Scheduling. Leuven, Belgium. July 12-15.

MAROTO, C., P.TORMOS, A.LOVA AND F.CRESPO. 1996. Professional Software for Resource Constrained Scheduling: Heuristic and Exact Methods. Paper presented at the Fifth International Workshop on Project Management and Scheduling. Poznan, Poland, April 11-13.

MAROTO, C., A.LOVA, J.ALCARAZ AND P.TORMOS.1997. Resource allocation in single and multiple projects: algorithms and professional software. Paper presented at the XV EURO and XXXIV INFORMS Joint International Meeting: OR/MS for the New Millennium. Barcelona, Spain, July 14-17.

PM Network. 1996. The software selection project: Tips from a pro, September,28-40.

SUÁREZ, L.F. 1987. Resource Allocation: A Comparative Study. *Project Management Journal.*

TORMOS, P., C. MAROTO AND A. LOVA. 1996. New Project Management Method based on priority rules: Focusing on resources. Paper presented at the XIV World Conference on Operations Research, Vancouver, Canada, July 8-12.

TORMOS, P. 1996. *Programación de proyectos con recursos limitados: Técnicas heurísticas basadas en reglas de prioridad.* Unpublished Ph. D. Dissertation. Departamento de Estadística e Investigación Operativa, Universidad Politécnica de Valencia, Spain.

TORMOS, P., C.MAROTO, AND A.LOVA. 1997. The resource constrained project scheduling problem: New criteria based on resources. Paper presented at the XV EURO and XXXIV INFORMS Joint International Meeting: OR/MS for the New Millennium. Barcelona, Spain, July 14-17.

ULUSOY G. AND L.ÖZDAMAR. 1989. Heuristic Performance and Network/Resource Characteristics in Resource-constrained Project Scheduling. *Journal of Operational Research Society.* 40/12, 1145-1152.

VIVES CASINO, E.1995. *Modelización de grafos para Gestión de Proyectos*, Proyecto Fin de Carrera. Facultad de Informática, Universidad Politécnica de Valencia, Spain. Directora: C. Maroto.

WASIL, E.A. AND A.A. ASSAD. 1988. Project Management on the PC: Software, Applications, and Trends. *Interfaces.* 18/2, 75-84.

12 METHODS FOR RESOURCE–CONSTRAINED PROJECT SCHEDULING WITH REGULAR AND NONREGULAR OBJECTIVE FUNCTIONS AND SCHEDULE–DEPENDENT TIME WINDOWS

Klaus Neumann
Jürgen Zimmermann

Universität Karlsruhe, Germany

12.1 INTRODUCTION

Projects with *minimum and maximum time lags* between the start of different activities (i.e. there are *schedule–dependent time windows*) play an important role in practice. Unfortunately, maximum time lags have been discussed in literature very rarely and only recently. However, maximum time lags (in addition to minimum ones) are needed to model a variety of practical requirements such as deadlines for subprojects or individual activities, simultaneous or non-delay execution of several activities, time–varying resource requirements, time windows for resources, scheduling of make–to–order production, and resource allocation in process–determined industry (cf. Neumann and Schwindt 1995, 1997).

In addition to the *temporal constraints* given by minimum and maximum time lags, *resource constraints* have to be observed in general taking into account the limited availability of renewable resources. Hence, the resource–

constrained project scheduling problems considered in this chapter consist of minimizing or maximizing some objective function subject to temporal and resource constraints.

In practice, the objective function to be minimized mostly represents the *project duration*. The corresponding project scheduling problem is denoted by *RCPSP/max* (*R*esource–*C*onstrained *P*roject *S*cheduling *P*roblem with *max*imum time lags in addition to minimum ones). The project duration is a *regular objective function*, i.e. it is nondecreasing in the completion times of activities (in the case of minimizing the objective function). Sometimes, some measure of the variation of resource utilization is to be minimized, for example, if expensive machinery should be utilized highly and evenly over time. In the latter case, we speak of the ***resource levelling problem*** (*RLP/max*). Finally, the net present value of the project, which is to be maximized, may represent the objective function: ***net present value problem*** (*NPVP/max*). The objective functions of RLP/max and NPVP/max are *nonregular*.

In this chapter, we study all three types of project scheduling problems: RCPSP/max, RLP/max, and NPVP/max. Section 12.2 summarizes some basic concepts from project scheduling. Section 12.3 is devoted to RCPSP/max. First, we discuss some properties of RCPSP/max. After that, branch–and–bound procedures, truncated branch–and–bound methods, and priority–rule methods for (approximately) solving RCPSP/max are presented. An experimental performance analysis compares these methods. In Section 12.4 we are concerned with RLP/max and NPVP/max. Again, structural questions, exact and heuristic solution methods, and an experimental performance analysis are dealt with.

12.2 BASIC CONCEPTS FROM PROJECT SCHEDULING

Assume that the project in question consists of n activities $1, \ldots, n$ each of which is to be carried out without interruption. In addition, we introduce the fictitious activities 0 and $n+1$, which represent the beginning and completion of the project, respectively. Let $p_i \in \mathbb{Z}_{\geq 0}$ be the *duration* (or *processing time*) and $S_i \in \mathbb{Z}_{\geq 0}$ be the *start time* of activity i where $p_0 = p_{n+1} = 0$ and

$$S_0 = 0 \tag{1}$$

that is, the project always begins at time 0. Then S_{n+1} represents the *project duration* (or *makespan*).

We now assign the nodes $0, 1, \ldots, n+1$ of a directed graph to the activities of the project (*activity–on–node representation*) and identify activity i with node i ($i = 0, 1, \ldots, n+1$). If there is a prescribed minimum time lag $d_{ij}^{min} \in \mathbb{Z}_{\geq 0}$ between the start of two different activities i and j, i.e. $S_j - S_i \geq d_{ij}^{min}$, we introduce an arc $\langle i,j\rangle$ with weight $b_{ij} := d_{ij}^{min}$. If there is a maximum time lag $d_{ij}^{max} \in \mathbb{Z}_{\geq 0}$ between the start of activities i and j, i.e. $S_j - S_i \leq d_{ij}^{max}$, we introduce a backward arc $\langle j,i\rangle$ with weight $b_{ji} := -d_{ij}^{max}$ (see Fig. 12:1).

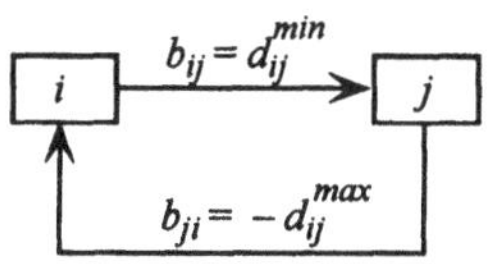

Figure 12:1 Representation of minimum and maximum time lags

The resulting weighted directed graph with node set $V = \{0, 1, \ldots, n+1\}$, arc set E, and weights b_{ij}, where

$$S_j - S_i \geq b_{ij} \qquad (\langle i,j\rangle \in E) \tag{2}$$

represents the *project network N* (*activity–on–node network*). Nodes 0 and $n+1$ are also called the ***beginning*** and ***end node***, respectively, of network N. In general, N contains cycles due to maximum time lags. By an appropriate specification of the minimum and maximum time lags (for example, in the case where several activities have to be begun at the same point in time), it can be ensured that the network assigned to a given project is uniquely determined. For the latter statement and more details of the construction and properties of activity–on–node networks we refer to Neumann and Schwindt (1997) and Franck and Neumann (1997).

A sequence of integer start times $S = (S_0, S_1, \ldots, S_{n+1})$ is called a *schedule*. We speak of a *time–feasible schedule* S if S satisfies the temporal constraints (1) and (2). We now state the basic

Theorem 1. There exists a time–feasible schedule for an activity–on–node network N exactly if N does not contain a cycle of positive length.

For a proof we refer to Bartusch et al. (1988). In what follows, we assume that project network N contains only cycles with nonpositive length. Moreover, each schedule is assumed to be time–feasible.

Several heuristic procedures for solving project scheduling problems require a *strict order* $\prec$ in the node set V of project network N. To define such an order, let d_{ij} be the length of a longest path from node i to node j in N. If there is no path from i to j, we set $d_{ij} := -\infty$. In case that $d_{ij} > 0$, $d_{ij}^{min} := d_{ij}$ represents a minimum time lag between the start of activities i and j, possibly induced by several individual time lags between activities on a longest path from i to j. If $d_{ij} < 0$, then $d_{ji}^{max} := -d_{ij}$ is a maximum time lag between the start of activities j and i. If $d_{ij} = 0$, the time lag may be either minimum or maximum. For $i, j \in V$, $i \neq j$, we then define $i \prec j$ exactly if either (a) $d_{ij} > 0$ or (b) $d_{ij} = 0$ and $d_{ji} < 0$.

Now assume that the renewable resources $1, \ldots, K$ are required for carrying out the project. Let $R_k > 0$ be the capacity of resource k available and let $r_{ik} \geq 0$ be the amount of resource k used by activity i ($k = 1, \ldots, K; i \in V$) where $0 \leq r_{ik} \leq R_k$ and $r_{0k} = r_{n+1,k} = 0$. For simplicity, R_k and r_{ik} are assumed to be constant.

Since most solution procedures schedule the activities successively, it is expedient to consider ***partial schedules*** $S' = (S_i)_{i \in V'}$ with $V' \subseteq V$. Given a partial schedule S', let for $t \in \mathbb{Z}_{\geq 0}$

$$\mathcal{A}^{S'}(t) := \{i \in V' \mid t - p_i < S_i \leq t\} \tag{3}$$

be the set of activities in progress at time t (or in time interval $[t, t+1[$ or time period $t+1$, respectively), also called the ***active set***. Moreover, let

$$r_k^{S'}(t) := \sum_{i \in \mathcal{A}^{S'}(t)} r_{ik} \tag{4}$$

be the amount of resource k used at time t given partial schedule S'. The step functions $r_k^{S'}$ are also called ***resource profiles***. The requirement that the usage of resources must not exceed the capacity of resources available leads to the constraints

$$r_k^S(t) \leq R_k \quad (k = 1, \ldots, K;\ t \in \mathbb{Z}_{\geq 0}) \tag{5}$$

for given schedule $S = (S_i)_{i \in V}$.

12.3 MINIMIZATION OF PROJECT DURATION: RCPSP/MAX

12.3.1 Formulation and properties of RCPSP/max

The problem of minimizing the project duration subject to temporal and resource constraints can be formulated as follows:

$$\left.\begin{array}{ll}
\text{Minimize} & S_{n+1} \\
\text{subject to} & S_j - S_i \geq b_{ij} \quad (\langle i,j \rangle \in E) \\
& S_0 = 0 \\
& S_i \in \mathbb{Z}_{\geq 0} \quad (i \in V) \\
& r_k^S(t) \leq R_k \quad (k = 1, \ldots, K;\ t = 0, \ldots, \overline{d} - 1)
\end{array}\right\} \text{RCPSP/max}$$

where

$$\overline{d} := \sum_{i \in V} \max(p_i, \max_{\langle i,j \rangle \in E} b_{ij}) \tag{6}$$

is an upper bound on the minimum project duration.

Whereas a feasible solution to RCPSP (i.e. without maximum time lags) can be found in polynomial time, testing whether there is a feasible solution to RCPSP/max is strongly $\mathcal{NP}$–complete (see Bartusch et al. 1988). If we omit the integer constraints $S_i \in \mathbb{Z}_{\geq 0}$ $(i \in V)$, the feasible region of RCPSP/max is generally disconnected and represents the union of convex polyhedra, whose number increases exponentially in the number of activities (cf. Nübel and Schwindt 1997 and Schwindt 1998).

A feasible (optimal) solution to RCPSP/max is also called a *feasible (optimal) schedule* for the underlying project network N. A *cycle structure* of N is a strong component which contains at least two nodes. For each cycle structure C of N treated as a separate subproject and started at time 0, a project scheduling problem corresponding to that one for project network N can be formulated. Analogously, the concept of a *feasible (optimal) subschedule* for the subproject corresponding to cycle structure C is defined. The next theorem (cf. Neumann and Zhan 1995 and Franck and Neumann 1997) is important for decomposition methods for solving RCPSP/max.

Theorem 2. There is a feasible schedule for project network N exactly if, for each cycle structure C of N, there is a feasible subschedule for C.

12.3.2 Branch–and–bound and truncated branch–and–bound procedures

To solve RCPSP/max exact methods of the branch–and–bound type and heuristic procedures (truncated branch–and–bound methods as well as priority–rule methods) have been developed. The basic idea of the branch–and–bound procedures is as follows. The *resource relaxation* of RCPSP/max, i.e. RCPSP/max without resource constraints (5), can be solved within polynomial time. Starting with an optimal solution ES to the resource relaxation, where $ES = (ES_i)_{i \in V}$ and $ES_i = d_{0i}$, so–called *resource conflicts*, i.e. points in time, where the resource usage of the activities in execution exceeds at least one resource capacity, are determined successively. To resolve a resource conflict, one or several activities are delayed, which means that we introduce additional precedence constraints. If there is a resource conflict at time t for resource k, it holds that $\sum_{i \in \mathcal{A}^S(t)} r_{ik} > R_k$. Each set $\mathcal{M} \subseteq \mathcal{A}^S(t)$ with $\sum_{i \in \mathcal{M}} r_{ik} > R_k$ for at least one resource k is referred to as a *forbidden set.*

The generation of additional precedence constraints such that all resource constraints are resolved represents a solution to the so–called *sequencing problem.* The resource relaxation with additional precedence constraints is termed *scheduling problem.* If there exists a feasible solution to RCPSP/max, there is always a sequencing solution such that each optimal solution to the corresponding scheduling problem represents an optimal solution to RCPSP/max. Thus, we have to enumerate the sequencing solutions in an appropriate way.

To obtain sequencing solutions, Bartusch et al. (1988) and De Reyck and Herroelen (1996) introduce precedence constraints of the type

$$S_j \geq S_i + p_i$$

until all resource conflicts are resolved. Schwindt (1998) introduces so-called *disjunctive precedence constraints*

$$\min_{j \in \mathcal{M}_2} S_j \geq \min_{i \in \mathcal{M}_1} (S_i + p_i)$$

to solve the underlying resource conflicts at time t, where $\mathcal{M}_1 \cap \mathcal{M}_2 = \emptyset$ and $\mathcal{M}_1 \cup \mathcal{M}_2 = \mathcal{A}^S(t)$. Furthermore, $\mathcal{M}_1$ is not a forbidden set and $\mathcal{M}_1 \cup \{j\}$ is a forbidden set for each $j \in \mathcal{M}_2$. Set $\mathcal{M}_2$ is referred to as a *minimal delaying alternative* (cf. De Reyck and Herroelen 1996). The precedence constraints introduced by Bartusch et al. (1988) and De Reyck and Herroelen (1996) correspond to disjunctive precedence constraints between single–element sets.

To enumerate sequencing solutions, Schwindt (1998) successively considers minimal delaying alternatives w.r.t. forbidden sets $\mathcal{A}^S(t)$. Each node ν of the enumeration tree corresponds to the introduction of an additional disjunctive precedence constraint induced by a minimal delaying alternative. The corresponding scheduling problem is specified by this disjunctive precedence constraint and the disjunctive precedence constraints introduced by the ancestors of ν as well as by the original temporal constraints (1) and (2). That means that each leaf of the enumeration tree represents a set of disjunctive precedence constraints which solve all resource conflicts incurred by schedule ES.

The concept of disjunctive precedence constraints markedly reduces the number of enumeration nodes which have to be considered (cf. Schwindt 1998). On the other hand, the introduction of disjunctive precedence constraints leads to feasible regions of the resulting scheduling problems which are no longer convex but represent a union of convex polyhedra. Such a scheduling problem represents a special linear reverse convex program (cf. Horst and Pardalos 1995) which can be solved by a pseudopolynomial fixed–point algorithm (cf. Schwindt 1998).

To speed up the branch–and–bound procedure, so–called *preprocessing* algorithms can be used (cf. De Reyck and Herroelen 1996, Heilmann and Schwindt 1997, and Schwindt 1998). Based on the examination of resource conflicts induced by two activities as well as the "workload" of time intervals, additional precedence and temporal constraints can be added to the underlying project network. These additional constraints must be met by each optimal schedule and do not represent alternative manners of resolving resource conflicts. Let us consider an example with two activities i and j, $p_i = p_j = 4$, $r_{ik} = r_{jk} = 3$, $R_k = 5$, and a minimum time lag $d_{ij}^{min} = 2$. Since activities i and j cannot be performed simultaneously and activity i has to be started two time units before activity j, the precedence constraint $S_j \geq S_i + p_i$ can be added. The preprocessing procedure can be used in the root of the enumeration tree as well as in each enumeration node to dynamically restrict the feasible region of the respective sequencing problem (*immediate selection*, cf. Brucker et al. 1996).

The branch–and–bound procedure by Schwindt (1998) uses two tight lower bounds: (i) The workload–based bound *LBW* results from the resource requirements and the resource supply for time intervals $[t_1, t_2[$ with $t_1, t_2 \in [0, \overline{d}]$ and $t_1 < t_2$. (ii) The preprocessing bound *LBP* is based on destructive improvement (cf. Klein and Scholl 1997). *LBP* corresponds to the maximum hypothetic upper bound which cannot be contradicted by preprocessing algorithms.

Algorithm 3 summarizes the branch–and–bound algorithm of Schwindt (1998).

Algorithm 3 B & B for RCPSP/max

Step 1. $S^* := nil$, $upperBound := \overline{d}$.
$\mathcal{D} := \emptyset$ ($\mathcal{D}$ is a stack).
Preprocessing.
Compute the longest path lengths d_{ij} $(i, j \in V)$.
If $ES := (0, d_{01}, \ldots, d_{0,n+1})$ is feasible, then
    $S^* := ES$ and terminate;
else $Push(ES, \mathcal{D})$ and compute $LBP(ES)$.

Step 2. While $\mathcal{D} \neq \emptyset$
$S := Pop(\mathcal{D})$ (depth–first search)
Compute $LBW(S)$
If $LBW(S) < upperBound$, then
    If S is feasible, then $S^* := S$ and $upperBound := S^*_{n+1}$
        If $upperBound \leq LBP(ES)$, then terminate;
    else identify first resource conflict at time t
        For all minimal delaying alternatives w.r.t. $\mathcal{A}^S(t)$
            Determine disjunctive precedence constraint and schedule S^+ by solving the corresponding scheduling problem
            If $S^+_{n+1} < upperBound$, then $Push(S^+, \mathcal{D})$. □

In Schwindt (1998) several *truncated branch–and–bound algorithms* based on the aforementioned exact procedure have been discussed. The *filtered beam search technique* restricts the number of children of an enumeration node to the beam width B. First, a simple lower bound LBF is determined for all minimal delaying alternatives arising from the current enumeration node. Then, the delaying alternatives are sorted according to nondecreasing values of LBF. For the first F alternatives, where $F \geq B$ is a prescribed filter width, tighter lower bounds LBB are determined. The B alternatives with smallest LBB are selected for further enumeration. Algorithm 3 modified in that way is denoted as **Algorithm 3FB** in the following.

The *decomposition approach* by Schwindt (1998) exploits Theorem 2. First, an optimal subschedule S^C is determined for each cycle structure C of project network N using Algorithm 3. Then, the nodes i of each cycle structure C are sorted according to nondecreasing values of S_i^C providing the node sequence $i_1, i_2, \ldots, i_\sigma$. The arcs of C are eliminated. Instead, for every two successive nodes i_μ and $i_{\mu+1}$ of C, the arc $\langle i_\mu, i_{\mu+1}\rangle$ with weight $S_{i_{\mu+1}}^C - S_{i_\mu}^C$ is added ($\mu = 1, \ldots, \sigma - 1$). Moreover, we introduce the arc $\langle i_\sigma, i_1\rangle$ with weight $-S_{i_\sigma}^C$. The resulting network is referred to as the decomposition network N'. Next, a feasible solution to the RCPSP/max instance induced by network N' is determined by Algorithm 3 where we replace "*upperBound*" in Step 2 by "*upperBound*$/(1+\epsilon)$" with $\epsilon > 0$. As a consequence, Algorithm 3 becomes an ϵ–approximate algorithm which will be called **Algorithm 3ϵ** in what follows. Algorithm 4 briefly sketches the described decomposition approach.

Algorithm 4 Decomposition for RCPSP/max

Step 1. Find all cycle structures of N.

$N' := N$.

Step 2. For each cycle structure C

Compute optimal subschedule S^C using Algorithm 3
Eliminate all arcs of C (update N')
Order the nodes i of C according to nondecreasing S_i^C
Let $i_1, \ldots, i_\sigma$ be the resulting node sequence
For $\mu = 1, 2, \ldots, \sigma - 1$ add $\langle i_\mu, i_{\mu+1}\rangle$ with $b_{i_\mu i_{\mu+1}} = S_{i_{\mu+1}}^C - S_{i_\mu}^C$ and $\langle i_\sigma, i_1\rangle$ with $b_{i_\sigma i_1} = -S_{i_\sigma}^C$ to N'.

Step 3. Compute feasible schedule S for the decomposition network N' by Algorithm 3ϵ. □

12.3.3 Priority–rule methods

Heuristic procedures which schedule the activities according to some priority rule generally provide feasible solutions much faster than truncated branch–and–bound methods. Such priority–rule methods for RCPSP/max have been proposed by Zhan (1994), Neumann and Zhan (1995), Brinkmann and Neumann (1996), and Franck and Neumann (1997).

In what follows, we sketch two different kinds of priority–rule methods (for details we refer to Franck and Neumann 1997). The simple *sequential* or *direct method* processes the activities one after another without considering the cycle structures separately. The *contraction method* represents a decomposition procedure which exploits Theorem 2. First, a feasible subschedule is determined for each cycle structure. Second, each cycle structure is replaced by a single node and the resulting acyclic "contracted" network is treated by the direct method.

To construct a feasible schedule (whether for the whole project network in the direct method or for a cycle structure or the contracted network in the contraction method), a *serial* or a *parallel schedule generation scheme* can be used. Later we will briefly sketch the serial scheme. For more details and for the parallel scheme we refer to Franck and Neumann (1997).

Before applying a generation scheme, the longest path lengths d_{ij} from nodes i to j $(i, j \in V)$ have to be computed and node set V has to be ordered strictly. Moreover, the upper bound $\overline{d}$ on the project duration has to be determined by (6). Then $ES_i = d_{0i}$ and $LS_i = \overline{d} - d_{i,n+1}$ are the *earliest* and *latest start times*, respectively, of activities $i \in V$.

Next, we specify the direct and contraction method for project network N with node set V.

Algorithm 5 Direct for RCPSP/max

Step 1. Find all cycle structures of N.

Step 2. Compute the longest path lengths d_{ij} $(i, j \in V)$, the upper bound $\overline{d}$ on the project duration, and the earliest and latest start times ES_i and LS_i $(i \in V)$. Order the node set V strictly.

Step 3. Determine a feasible schedule using a generation scheme, where activities of one and the same cycle structure are scheduled directly one after the other. □

Algorithm 6 Contract for RCPSP/max

Step 1. Find all cycle structures of N.

Step 2. Expand each cycle structure C of N to obtain a subproject network with beginning and end nodes and node set V^C. For each C, compute the longest path lengths d_{ij}^C $(i, j \in V^C)$, the maximum duration $\overline{d}^C$ of the subproject corresponding to C, and the earliest and latest start times ES_i^C and LS_i^C within the subproject $(i \in V^C)$. Order the node set V^C strictly.

Step 3. For each cycle structure C, compute a feasible subschedule $(S_i^C)_{i \in V^C}$ using some generation scheme. If for some cycle structure, a feasible subschedule cannot be found, then terminate.

Step 4. Replace each cycle structure C by a single node or activity i_C, respectively, with appropriate duration and (time–dependent) resource usage. Construct the acyclic contracted project network N' with node set V'.

Step 5. For N', compute the longest path lengths d'_{ij} $(i, j \in V')$, the upper bound $\overline{d}'$ on the project duration, and the earliest and latest start times ES'_i and LS'_i $(i \in V')$. Order the node set V' strictly.

Step 6. Compute a feasible schedule $(S'_i)_{i \in V'}$ for N' using some generation scheme. Determine a feasible schedule $(S_i)_{i \in V}$ for N as follows:

$$S_i := \begin{cases} S'_{i_C} + S_i^C, & \text{if } i \in V^C \text{ for some cycle structure } C \\ S'_i, & \text{otherwise} \end{cases}$$

□

The schedule generation schemes use several subsets of set V. The *complete set* $\mathcal{C}$ contains all activities that have been scheduled and are thus completed. The *eligible set* $\mathcal{E}$ contains the activities that have not been scheduled yet but all of whose immediate predecessors (with respect to strict order $\prec$) have been scheduled. The set of immediate predecessors of node i with respect to $\prec$ is denoted by $Pred_i^{\prec}$. The activities which neither belong to $\mathcal{C}$ nor to $\mathcal{E}$ form the *unscheduled set* $\mathcal{U}$.

The activity to be scheduled next is always an activity $j \in \mathcal{E}$ with highest priority $\pi(j)$, i.e. $\pi(j) = \text{ext}_{i \in \mathcal{E}} \pi(i)$, provided there is enough resource capacity available. A large number of different *priority rules* have been examined, for example,

$\pi(i) = LS_i, \ \text{ext}_{i \in \mathcal{E}}\, \pi(i) = \min_{i \in \mathcal{E}} LS_i$ (*LST rule*: smallest "latest start time" first)

$\pi(i) = |\{j|j \prec i\}|$, $\operatorname{ext}_{i\in\mathcal{E}} \pi(i) = \min_{i\in\mathcal{E}} |\{j|j \prec i\}|$ (*MSO rule*: "minimum" number of activities before i w.r.t. "strict order" $\prec$ first)

$\pi(i) = p_i \sum_{k=1}^{K} r_{ik}$, $\operatorname{ext}_{i\in\mathcal{E}} \pi(i) = \max_{i\in\mathcal{E}} p_i \sum_{k=1}^{K} r_{ik}$ (*GRD rule*: "greatest resource demand" first)

$\pi(i) = TF_i$, $\operatorname{ext}_{i\in\mathcal{E}} \pi(i) = \min_{i\in\mathcal{E}} TF_i$ (*MST rule*: "minimum slack time" first)

where

$$TF_i := LS_i - ES_i \tag{7}$$

is the *total float* or *slack time* of activity i. For additional priority rules we refer to Franck and Neumann (1997).

The serial schedule generation scheme runs as follows:

Algorithm 7 Serial Generation Scheme

> $\mathcal{U} := V$, $\mathcal{C} := \emptyset$.
> While $\mathcal{U} \neq \emptyset$
> > $\mathcal{E} := \{i \in \mathcal{U} | h \in \mathcal{C}$ for all $h \in Pred_i^{\prec}\}$
> > $j := \min\{i \in \mathcal{E} | \pi(i) = \operatorname{ext}_{h\in\mathcal{E}}\pi(h)\}$
> > Compute earliest start time $t \geq ES_j$ for activity j such that the resource constraints are satisfied
> > If $t \leq LS_j$, then *schedule* activity j at time t;
> > else do *backward scheduling*. □

Scheduling activity j at time t means that we put $S_j := t$. Moreover, the sets $\mathcal{C}$ and $\mathcal{U}$, the remaining resource capacities, and the earliest and latest start times ES_i and LS_i for $i \in \mathcal{U}$ are updated.

The *backward scheduling* process is, in principle, as follows. If for the earliest resource–feasible start time t of activity j we have $t > LS_j$, there must be an activity h already scheduled such that $LS_j = S_h + d_{hj}^{max}$. Then the scheduling of all activities between start times S_h and t is nullified and performed again. To do so, we assign an appropriate new start time $S_h^{new} > S_h$ to activity h which induces a new latest start time LS_j^{new} of activity j such that $LS_j^{new} \geq t$. Note that the backward scheduling also requires an updating of $\mathcal{C}$, $\mathcal{U}$, and ES_i and LS_i for $i \in \mathcal{U}$. If the backward scheduling results in exceeding the upper bound $\overline{d}$ on the project duration, the algorithm has to be determined because no feasible schedule can be found.

12.3.4 Experimental performance analysis

We discuss the performance of different truncated branch–and–bound and priority–rule methods. For the experimental tests, we have used an IBM–compatible PC with a Pentium 200 processor running MS NT 4.0. All procedures have been coded in ANSI C.

We make use of three test sets A, B, and C, which have been generated by the problem generator ProGen/max (cf. Schwindt 1996). ProGen/max randomly generates problem instances w.r.t. 20 control parameters. There are control parameters for the network structure, the activities, and the resource constraints.

Parameters for the network structure are, for example, the number n of activities, the number nc of cycle structures, the number nnc of nodes per cycle structure, and the ***restrictiveness*** P. P is a $[0,1]$–normalized measure of the degree to which precedence constraints restrict the number of feasible activity sequences. Since the calculation of P is $\mathcal{NP}$–hard, we use the ***approximation of the restrictiveness***

$$RT := \frac{2\sum_{i,j\in V} \rho_{ij} - 6(|V|-1)}{(|V|-2)(|V|-3)}$$

(cf. Thesen 1977 and Schwindt 1996), where $\rho_{ij} := 1$ if there is a path from i to j, and $\rho_{ij} := 0$ otherwise. We have $RT = 0$ for a parallel network and $RT = 1$ for a series network.

Parameters for an activity $i \in V$ are the resource usage r_{ik} $(k = 1, \ldots, K)$ and the activity duration p_i. Parameters for the resource constraints are the number K of different resources, the resource factor, and the resource strength. The ***resource factor*** RF denotes the average fraction of the K resources used per activity. The ***resource strength*** RS is a $[0,1]$–normalized parameter which measures the scarcity of resources and is defined by

$$RS := \frac{R_k - R_k^{min}}{R_k^{max} - R_k^{min}} \qquad \text{for all } k = 1, \ldots, K$$

where $R_k^{min} := \max_{i\in V} r_{ik}$ and $R_k^{max} := \max_{t=0,\ldots,\bar{d}-1} \sum_{i\in\mathcal{A}^{ES}(t)} r_{ik}$. Note that for each $k = 1, \ldots, K$, R_k has to be specified such that $R_k = RS\,(R_k^{max} - R_k^{min}) + R_k^{min}$.

Table 12:1 shows the relevant parameter values for the three test sets A, B, and C used. Test set A with 27 parameter combinations and 270 problem instances corresponds to test set A in Neumann and Zimmermann (1997) and will be used in Section 12.4.4. Test set B with 54 parameter combinations and 1080 problem instances, where 1059 instances possess a feasible solution, as well as test set C with 12 parameter combinations and 120 problem instances, where 119 instances possess a feasible solution, correspond to test sets B and C in Franck and Neumann (1997) and test sets A and C in Schwindt (1998), respectively.

Table 12:1 Control parameters of test sets A, B, and C

Parameters	A	B	C
n	$10, 15, 20$	100	500
nc	$\{1, \ldots, 4\}$	$\{2, \ldots, 5\}, \{6, \ldots, 9\}$	$\{10, \ldots, 21\}$
nnc	$\{2, \ldots, 5\}$	$\{2 \ldots, 15\}$	$\{2, \ldots, 21\}$
RT	$0.35, 0.5, 0.65$	$0.35, 0.5, 0.65$	$0.25, 0.5$
r_{ik}	$\{1, \ldots, 5\}$	$\{1, \ldots, 5\}$	$\{1, \ldots, 10\}$
p_i	$\{1, \ldots, 10\}$	$\{1, \ldots, 15\}$	$\{1, \ldots, 10\}$
K	$1, 3, 5$	5	5
RF	1	$0.5, 0.75, 1$	$0.5, 0.75, 1$
RS	0.5	$0.2, 0.5, 0.7$	$0.25, 0.5$

In Schwindt (1998) an improved version of Algorithm 3 has been compared with the branch–and–bound procedure by De Reyck and Herroelen (1996). For test set B, the percentage p_{feas} of the 1059 solvable problem instances which are solved to feasibility and the percentage p_{opt} of the 1059 instances solved to optimality have been determined. Table 12:2 shows p_{feas} and p_{opt} for both algorithms and running times of 1 sec, 10 sec, and 100 sec. We see that the algorithm of Schwindt solves more problem instances of test set B to optimality within 10 sec than the algorithm of De Reyck and Herroelen in 100 sec.

Table 12:2 Comparison between exact procedures for RCPSP/max

	p_{feas}	p_{opt}
De Reyck & Herroelen (1 s)	83%	51%
De Reyck & Herroelen (10 s)	94%	56%
De Reyck & Herroelen (100 s)	95%	57%
Schwindt (1 s)	62%	54%
Schwindt (10 s)	81%	59%
Schwindt (100 s)	94%	61%

For test set C, we have used Algorithm 3FB with $F = 1$, Algorithm 4 with $\epsilon = 0.05$ and an imposed time limit of 200 sec, and Algorithms 5 and 6 with serial generation scheme and priority rule LST. Table 12:3 shows p_{feas}, p_{opt}, Δ_{LBP}, and the average running time per instance solved. Δ_{LBP} is the average relative deviation of the project duration for the best solution computed from the lower bound LBP. We see that Algorithms 4 and 6 solve nearly all problem instances to feasibility whereas Algorithms 3FB and 5 solve more problem instances to optimality. An additional result is that Algorithm 4 solves more than 80 % of all problem instances to ϵ-optimality (with respect to the decomposition network N').

Table 12:3 Truncated branch–and–bound and priority–rule methods

	p_{feas}	p_{opt}	Δ_{LBP}	CPU [sec]
Alg. 3FB	80%	62%	0.1%	13.7
Alg. 4	100%	6%	5%	50.7
Alg. 5	53%	33%	0.3%	0.9
Alg. 6	98%	4%	6%	1.9

12.4 NONREGULAR OBJECTIVE FUNCTIONS: RLP/MAX AND NPVP/MAX

12.4.1 *Formulation and properties of RLP/max and NPVP/max*

In addition to the constraints of RCPSP/max, we assume that there is a *prescribed maximum project duration* $\overline{d} \in \mathbb{Z}_{\geq 0}$ with $\overline{d} \geq d_{0,n+1}$ where $d_{0,n+1}$ is the minimum project duration without resource constraints. That is, we have the additional restriction

$$S_{n+1} \leq \overline{d} \tag{8}$$

Note that $\overline{d}$ is no longer defined by (6) from now on.

Recall that $r_k^{S'}(t)$ is the amount of resource k used at time t given partial schedule $S' = (S_i)_{i \in V'}$ with $V' \subseteq V$ (see (4)). The $K \times \overline{d}$ matrix $\boldsymbol{R}^{S'}$ with elements $r_k^{S'}(t)$ ($k = 1, \ldots, K$; $t = 0, 1, \ldots, \overline{d} - 1$) is called the *resource utilization matrix*. The objective function f of RLP/max represents some measure of the variation of resource utilization, i.e. f is a function of $\boldsymbol{R}^S$ for schedule $S = (S_i)_{i \in V}$. Several different types of objective functions f will be discussed later on. The resource levelling problem can then be formulated as follows:

$$\left.\begin{array}{ll}\text{Minimize} & f(\boldsymbol{R}^S) \\ \text{subject to} & S_j - S_i \geq b_{ij} \quad (\langle i,j\rangle \in E) \\ & S_0 = 0 \\ & S_{n+1} \leq \overline{d} \\ & S_i \in \mathbb{Z}_{\geq 0} \quad (i \in V) \\ & r_k^S(t) \leq R_k \quad (k = 1,\dots,K;\ t = 0,\dots,\overline{d}-1)\end{array}\right\} \text{RLP/max}$$

In contrast to RLP/max, the objective function f of NPVP/max is independent of the resource utilization matrix and emphasizes the financial aspects of a project by maximizing the *net present value* of the project. For the basic concepts of the theory of cash flows and net present values of projects we refer to Russell (1970) and Herroelen et al. (1997).

For each activity $i \in V$, a cash flow c_i^F, which may be negative, is assumed to be given, which arises at the completion time $S_i + p_i$ of i. Given an interest rate per period I, the discount factor $\beta = 1/(1+I)$ denotes the present value of a cash flow of 1 received a time period hence. Thus, the net present value of a project depends on the start times of its activities and is given by

$$f(S) := \sum_{i \in V} c_i^F \, \beta^{S_i + p_i} \tag{N}$$

Obviously, function f is in general neither nonincreasing nor nondecreasing in the completion times of the activities and thus nonregular.

The net present value problem can then be formulated as follows:

$$\left.\begin{array}{ll}\text{Maximize} & \sum\limits_{i=1}^{n} c_i^F \, \beta^{S_i + p_i} \\ \text{subject to} & S_j - S_i \geq b_{ij} \quad (\langle i,j\rangle \in E) \\ & S_0 = 0 \\ & S_{n+1} \leq \overline{d} \\ & S_i \in \mathbb{Z}_{\geq 0} \quad (i \in V) \\ & r_k^S(t) \leq R_k \quad (k = 1,\dots,K;\ t = 0,\dots,\overline{d}-1)\end{array}\right\} \text{NPVP/max}$$

Testing whether there is a feasible solution to RLP/max or NPVP/max is strongly $\mathcal{NP}$-complete (see Neumann and Zimmermann 1997).

Now we briefly discuss several objective functions f for RLP/max all of them are nonregular. To interpret these functions as resource cost functions, let $c_k > 0$ be the cost per unit of resource k. The first type A of objective function says that the sum of the maximum resource costs per period is to be minimized. This is the objective function of the so–called *resource investment problem*, which is used in practice if expensive resources have to be purchased.

$$f(\boldsymbol{R}^S) \quad = \quad \sum_{k=1}^{K} c_k \max_{t=0,1,\ldots,\bar{d}-1} r_k^S(t) \tag{A}$$

The second type B of objective function measures the *deviation* of the consumption of resource k *from* a given *resource supply* or *target value for resource usage* $Y_k \geq 0$. Y_k can also be replaced by the *average resource utilization*

$$\bar{r}_k := \sum_{i \in V} r_{ik} p_i / \bar{d}$$

The first objective function of type B is

$$f(\boldsymbol{R}^S) \quad = \quad \sum_{k=1}^{K} c_k \sum_{t=0}^{\bar{d}-1} \left[r_k^S(t) - Y_k\right]^+ \tag{B1}$$

If we replace $[\ldots]^+$ in (B1) by $|\ldots|$ or $[\ldots]^2$, we obtain objective functions (B2) or (B3), respectively.

The third type C of objective function refers to the *variation of resource utilization over time.* It is used if the resources represent different kinds of manpower where changing the size of work force from period to period should be smoothed.

$$f(\boldsymbol{R}^S) \quad = \quad \sum_{k=1}^{K} c_k \sum_{t=0}^{\bar{d}} \left[r_k^S(t) - r_k^S(t-1)\right]^+ \tag{C1}$$

$$\text{where } r_k^S(-1) := r_k^S(\bar{d}) := 0$$

If we replace $[\ldots]^+$ in (C1) by $|\ldots|$ or $[\ldots]^2$, we obtain objective functions (C2) or (C3), respectively.

Given a resource cost function f, a partial schedule $S' = (S_i)_{i \in V'}$, $V' \subset V$, and the corresponding resource utilization matrix $\boldsymbol{R}^{S'}$, the additional cost arising when activity $j \in V \setminus V'$ is scheduled at time S_j is

$$f^p(\boldsymbol{R}^{S'}, j, S_j) := f(\tilde{\boldsymbol{R}}^{S'}) - f(\boldsymbol{R}^{S'})$$

$$\text{where } \tilde{r}_k^{S'}(t) := \begin{cases} r_k^{S'}(t) + r_{jk}, & \text{for} \quad S_j \leq t < S_j + p_j \\ r_k^{S'}(t), & \text{otherwise} \end{cases}$$

$f^p(\boldsymbol{R}^{S'}, j, S_j)$ can be viewed as a *penalty cost* incurred when activity j is scheduled at time S_j.

Analogously, for the net present value function f, we define the additional cash flow

$$f^a(j, S_j) := c_j^F \beta^{S_j+p_j}$$

arising when activity $j \in V \setminus V'$ is scheduled at time S_j.

Next we briefly discuss some properties of RLP/max and NPVP/max which will be exploited in Section 12.4.2, where we will sketch heuristics for both problems. For details we refer to Neumann and Zimmermann (1997).

The heuristics schedule the activities successively where all critical activities i (i.e. $TF_i = LS_i - ES_i = 0$) are scheduled immediately. On the other hand, for a scheduled activity i it holds that $S_i = ES_i = LS_i$, i.e. activity i is critical. Note that at the beginning of the algorithm, we have $ES_i = d_{0i}$ and $LS_i = \overline{d} - d_{i,n+1}$ for all $i \in V$. Suppose that a partial schedule $S' = (S_i)_{i \in V'}$ has been determined and we want to schedule an activity $j \in V \setminus V'$. Then the earliest start time of j is

$$ES_j^{S'} = \max(d_{0j}, \max_{i \in V'}(S_i + d_{ij})) \tag{9}$$

because of temporal constraint (2). The latest start time of activity j is

$$LS_j^{S'} = \min(\overline{d} - d_{j,n+1}, \min_{i \in V'}(S_i - d_{ji})) \tag{10}$$

To find the start time S_j of activity j, some noncritical activities $h \in V \setminus V' \setminus \{j\}$ and their resource usage have to be taken into account as well, which we now discuss.

Obviously, each activity $i \in V$ has to be carried out during the so–called *base time interval* $[LS_i, ES_i + p_i[$ independently of its start time $S_i \in [ES_i, LS_i]$. Activity $i \in V$ is called *near–critical* if $0 < TF_i < p_i$. It is easy to see that $[LS_i, ES_i + p_i[\neq \emptyset$ exactly if activity i is critical or near–critical. Consider a partial schedule $S' = (S_i)_{i \in V'}$, $V' \subset V$. If we want to schedule an activity $j \in V \setminus V'$, we have to observe the set of activities different from j that have to be in progress at time t, i.e. set

$$\mathcal{A}_j^{S'}(t) := \{i \in V \setminus \{j\} | LS_i^{S'} \leq t < ES_i^{S'} + p_i\} \text{ for } t = 0, 1, \ldots, \overline{d} - 1$$

The least amount of resource k used at time t for activities $i \in V \setminus \{j\}$ is

$$r_{kj}^{S'}(t) := \sum_{i \in \mathcal{A}_j^{S'}(t)} r_{ik} \tag{11}$$

Let $\boldsymbol{R}$ be the $K \times \overline{d}$ resource utilization matrix with elements $r_{kj}^{S'}(t)$ ($k = 1, \ldots, K$; $t = 0, 1, \ldots, \overline{d} - 1$). Then the heuristics for RLP/max (or NPVP/max) schedule activity j at a point in time S_j^+ which is a minimizer (or maximizer) of function $f^p(\boldsymbol{R}, j, \cdot)$ (or $f^a(j, \cdot)$) on set

$$\mathcal{S}_j := \{ES_j, ES_j + 1, \ldots, LS_j\}$$

For NPVP/max, we have

$$S_j^+ := \begin{cases} ES_j & \text{for } c_j^F \geq 0 \\ LS_j & \text{for } c_j^F < 0 \end{cases} \tag{12}$$

Next is a result which ensures that for RLP/max, the time complexity of the heuristics is polynomial as well: There is at least one minimizer S_j^+ of function $f^p(\boldsymbol{R}, j, \cdot)$ on $\mathcal{S}_j$ with $S_j^+ \in \mathcal{D}_j(\boldsymbol{R})$ where the cardinality of the so-called *decision set* $\mathcal{D}_j(\boldsymbol{R})$ for activity j is linear in n.

Let $\mathcal{D}(\boldsymbol{R})$ be the set of *jump discontinuities* of the resource profiles, that is, the set of points in time $t = 0$ and $t \in \{1, \ldots, \bar{d}-1\}$ such that $r_k^{S'}(t) \neq r_k^{S'}(t-1)$ for some $k \in \{1, \ldots, K\}$. Since at each jump point $t \in \mathcal{D}(\boldsymbol{R})$, at least one activity is begun or is terminated, we have $|\mathcal{D}(\boldsymbol{R})| \leq 2n$.

Let j be again the activity to be scheduled. Since for objective function (A) and two immediately successive elements $t_1 < t_2$ of set

$$\mathcal{D}_j^1(\boldsymbol{R}) := (\mathcal{S}_j \cap \mathcal{D}(\boldsymbol{R})) \cup \{ES_j, LS_j\}$$

it holds that

$$f^p(\boldsymbol{R}, j, t_1) \leq f^p(\boldsymbol{R}, j, t) \text{ for } t_1 \leq t < t_2$$

$\mathcal{D}_j^1(\boldsymbol{R})$ is the decision set for function (A).

Similarly, it can be shown (see Neumann and Zimmermann 1997) that

$$\mathcal{D}_j^2(\boldsymbol{R}) := \mathcal{D}_j^1(\boldsymbol{R}) \cup \{t \in \mathcal{S}_j | t + p_j \in \mathcal{D}(\boldsymbol{R})\}$$

is the decision set for objective functions (B1), (B2), (B3), (C1), and (C2), and

$$\mathcal{D}_j^3(\boldsymbol{R}) := \mathcal{D}_j^2(\boldsymbol{R}) \cup \{t \in \mathcal{S}_j | t - 1 \in \mathcal{D}_j^2(\boldsymbol{R})\}$$

is the decision set for objective function (C3).

12.4.2 Priority–rule heuristics

For resource levelling problems with special objective functions, without resource constraints (5), and without maximum time lags, pseudopolynomial priority–rule heuristics have been proposed by Burgess and Killebrew (1962), Ahuja (1976), Harris (1978, 1990), Takamoto et al. (1995), and Savin et al. (1997). None of these studies contains a performance analysis and only small examples with up to 20 activities have been considered. For resource levelling problems with general objective functions, without resource constraints (5), and with maximum time lags in addition to minimum ones, Brinkmann and Neumann (1996) have devised pseudopolynomial heuristics, which have been tested for problem instances with up to 100 activities and several resources.

Polynomial priority–rule methods for the general RLP/max with resource constraints (5) have been offered for the first time by Neumann and Zimmermann (1997). The experimental performance analysis in Section 12.4.3 will show that these heuristics can be successfully used for problem instances with up to 500 activities and several resources.

For the net present value problem with resource constraints (5) and without maximum time lags, priority–rule heuristics have been presented, for example, by Russell (1986) and Padman et al. (1997). Both studies contain a performance analysis where problem instances with 1000 or more activities and several resources are considered. Moreover, Yang et al. (1995) have devised a simulated annealing approach and have compared that approach with priority–rule methods using stochastic scheduling rules (i.e. random sampling), for problem instances with 20 activities and several resources.

In what follows, we specify a priority–rule method for RLP/max without resource constraints (5). This heuristic exploits the properties of RLP/max discussed in Section 12.4.1. For details we refer to Neumann and Zimmermann (1997).

Algorithm 8 Heuristic for RLP/max without resource constraints

Step 1. $\mathcal{U} := V$. Initialize $\boldsymbol{R}$.

Step 2. Compute the longest path lengths d_{ij} $(i, j \in V)$, the earliest and latest start times ES_i and LS_i $(i \in V)$, and the total floats TF_i $(i \in V)$.

Step 3. For all $i \in V$ with $TF_i = 0$ set $S_i := ES_i$ and $\mathcal{U} := \mathcal{U} \setminus \{i\}$.

Step 4. While $\mathcal{U} \neq \emptyset$

$j := \min\{i \in \mathcal{U} | \pi(i) = \text{ext}_{h \in \mathcal{U}} \pi(h)\}$
Update $\boldsymbol{R}$
Compute $\mathcal{D}_j(\boldsymbol{R})$ and the greatest minimizer S_j^+ of $f^p(\boldsymbol{R}, j, \cdot)$ on $\mathcal{D}_j(\boldsymbol{R})$
$\mathcal{U} := \mathcal{U} \setminus \{j\}$. Update ES_i, LS_i, and TF_i for $i \in \mathcal{U}$
For all $i \in \mathcal{U}$ with $TF_i = 0$ set $S_i := ES_i$ and $\mathcal{U} := \mathcal{U} \setminus \{i\}$. □

In Algorithm 8, $\mathcal{U}$ is again the unscheduled set. Note that decision set $\mathcal{D}_j(\boldsymbol{R})$ depends on the type of objective function as discussed in Section 12.4.1. For π, several different priority rules can be used (see Section 12.3.3). For example, rule GRD–MST means that the activity to be scheduled next is always that activity with greatest resource demand (main priority rule) where ties are broken on the basis of nondecreasing total float values.

Algorithm 8 can also be used to solve NPVP/max without resource constraints approximately. In that case, S_j^+ is given by (12) and the so–called *GCF rule* ("greatest cash flow" first) can be used, i.e. $\text{ext}_{h\in\mathcal{U}}\pi(h) = \max_{h\in\mathcal{U}} |c_h^F|$.

Next we show how to generalize Algorithm 8 to the case where there are resource constraints (5). To solve an instance of RLP/max approximately, we consider the corresponding instance of RCPSP/max, that is, we omit constraint (8) and replace $f(\boldsymbol{R}^S)$ by S_{n+1}. To solve the latter problem approximately, we apply Algorithm 6 with serial generation scheme and priority rule LST which, on the average, provides a feasible solution with "small" S_{n+1} for more than 98% of all instances (see Sections 12.3.3 and 12.3.4). If Algorithm 6 does not provide a feasible schedule $\hat{S} = (\hat{S}_i)_{i\in V}$ with $\hat{S}_{n+1} \leq \overline{d}$ (for $\overline{d} = 1.1\, d_{0,n+1}$, this has happened for 3% of all instances of test set C in Section 12.3.4), we apply Algorithm 8 without any alteration. Otherwise, the start times $\hat{S}_i$ $(i \in V)$ are used as lower bounds on the start times to be determined for the RLP/max instance. That means, we set $ES_i := \hat{S}_i$ $(i \in V)$ at the beginning of Algorithm 8. After performing Algorithm 8, we examine whether the resulting schedule S is feasible for problem RLP/max or not. Algorithm 9a summarizes this heuristic for RLP/max. For a detailed description we refer to Neumann and Zimmermann (1997).

Algorithm 9 a Heuristic for RLP/max

Step 1. Initialization.
(see Steps 1 and 2 of Algorithm 8)

Step 2.
Apply Algorithm 6 to corresponding RCPSP/max instance.
If the resulting schedule $\hat{S}$ is feasible and $\hat{S}_{n+1} \leq \overline{d}$, then
For all $i \in V$ set $ES_i := \hat{S}_i$.

Step 3. Scheduling the activities.
(see Steps 3 and 4 of Algorithm 8)

Step 4.
If $r_k(t) \leq R_k$ for all $t \in \mathcal{D}(\boldsymbol{R})$ and all $k = 1, \ldots, K$, then schedule S obtained at the end of Step 3 is feasible;
else terminate without a feasible solution. □

In Neumann and Zimmermann (1997) a second generalization of Algorithm 8 for RLP/max has been proposed. When scheduling an activity j, only start times S_j are considered for which j can be performed without exceeding the resource capacities R_k $(k = 1, \ldots, K)$. If there is no such start time, the scheduling of all activities which can be performed simultaneously with j is nullified and performed again in a different order. This procedure will be referred to as **Algorithm 9b** in the following.

12.4.3 Exact procedures

For the net present value problem without resource constraints (5) and without maximum time lags, polynomial exact procedures based upon linear programming or recursive search have been presented by Grinold (1972) and Herroelen et al. (1996). The latter paper contains an experimental performance analysis with up to 30 activities. For the resource–unconstrained net present value problem with maximum time lags, which can also be solved in polynomial time, De Reyck (1998) has offered an exact procedure, which is an extension of the procedure by Herroelen et al. (1996), as well as a comprehensive experimental performance analysis for problem instances with up to 100 activities. For the resource–constrained net present value problem without maximum time lags, integer programming and branch–and–bound procedures have been devised, for instance, by Yang et al. (1992) and Icmeli and Erengüç (1995). The two papers contain a performance analysis for problem instances with up to 21 and 51 activities, respectively. For the resource–constrained net present value problem with maximum time lags, De Reyck (1998) has presented a branch–bound–procedure and a comprehensive performance analysis for problem instances with up to 50 activities and 5 resources.

For resource levelling problems with special objective functions, without resource constraints (5), and without maximum time lags, exact algorithms based upon enumeration, integer programming, or dynamic programming techniques have been devised by Ahuja (1976), Easa (1989), Bandelloni et al. (1994), and Younis and Saad (1996). These studies, however, do not contain any experimental performance analysis, and only small examples with at most 9 activities and a single resource have been discussed.

A branch–and–bound method for resource levelling problems with maximum time lags in addition to minimum ones has been proposed for the first time by Zimmermann and Engelhardt (1998). For RLP/max without resource constraints (5), this branch–and–bound procedure is as follows. Starting with the partial schedule $S^0 = (S_0)$ with $S_0 = 0$, at each level of the enumeration tree an activity $i \in \mathcal{U}$ is scheduled. Thus, each node of the enumeration tree corresponds to a partial schedule S'. Given a partial schedule $S' = (S_i)_{i \in V'}$,

$V' \subset V$, the earliest and latest start times $ES_j^{S'}$ and $LS_j^{S'}$ of $j \in V \setminus V'$ can be computed by (9) and (10), respectively. Hence, for the node of the enumeration tree corresponding to partial schedule S', we choose an activity $j \in V \setminus V'$ with minimum $TF_j^{S'} = LS_j^{S'} - ES_j^{S'}$, and for each $t \in \{ES_j^{S'}, \ldots, LS_j^{S'}\}$, we generate a child S^t by setting $S_j^t := t$. Algorithm 10 summarizes this branch-and-bound algorithm.

Algorithm 10 B & B for RLP/max without resource constraints

Step 1. Compute the longest path lengths d_{ij} $(i, j \in V)$.

$S^* := (0, d_{01}, \ldots, d_{0,n+1})$, $upperBound := f(\boldsymbol{R}^{S^*})$.

$S^0 := (S_0 = 0)$, $\mathcal{D} := \emptyset$ ($\mathcal{D}$ is a stack).

$Push(S^0, \mathcal{D})$.

Step 2. While $\mathcal{D} \neq \emptyset$

$S' := Pop(\mathcal{D})$ (depth–first search)

If $LB(S') < upperBound$, then

If $V' = V$, then $S^* := S'$ and $upperBound := f(\boldsymbol{R}^{S^*})$;
else choose activity $j \in \mathcal{U}$ with minimum $TF_j^{S'}$

For $t = ES_j^{S'}$ to $LS_j^{S'}$

Generate a partial schedule S^t with $S_i^t := S_i'$ for all $i \in V'$ and $S_j^t := t$

If $LB(S^t) < upperBound$, then $Push(S^t, \mathcal{D})$. □

In Algorithm 10, $LB(S')$ denotes a lower bound on the minimum objective function value for all feasible schedules which represent expansions of partial schedule S'. For a so called *r–monotonous* objective function f (that is, $f(\boldsymbol{R}^{S_1}) \leq f(\boldsymbol{R}^{S_2})$ if $r_k^{S_1}(t) \leq r_k^{S_2}(t)$ for all $k = 1, \ldots, K$ and $t = 0, 1, \ldots, \bar{d}-1$), a partial schedule $S' = (S_i)_{i \in V'}$, $V' \subset V$, and each $j \in V \setminus V'$, $LB(S') := f(\boldsymbol{R})$ can be used. Recall that the elements $r_{kj}^{S'}(t)$ of matrix $\boldsymbol{R}$ are given by (11). For example, objective functions (A), (B1), as well as (B2) and (B3) with $Y_k = 0$ are r–monotonous. For a generalization of Algorithm 10 to RLP/max with resource constraints (5), we refer to Zimmermann and Engelhardt (1998), which also contains a performance analysis.

12.4.4 *Experimental performance analysis*

We briefly sketch a performance analysis for Algorithms 8, 9a, and 9b, where we use the test sets A and C from Section 12.3.4. For more details we refer to

Neumann and Zimmermann (1997). In practice, there may be some leeway for the prescribed maximum project duration $\overline{d}$. We have chosen three different values for $\overline{d}$: $\overline{d}_{1.0} = d_{0,n+1}$, $\overline{d}_{1.1} = 1.1\, d_{0,n+1}$, $\overline{d}_{1.25} = 1.25\, d_{0,n+1}$.

For test set A, Algorithm 8 with priority rules MSO–LST, GRD–MST, and MST–GRD (i.e. three heuristics), as well as objective functions (A) and (B3) with $c_k = 1$ and $Y_k = 0$ $(k = 1, \ldots, K)$, we have determined Δ_{opt}. Δ_{opt} is the relative deviation of the objective function value for the best solution computed from the minimum objective function value. The latter value has been determined by Algorithm 10. Table 12:4 shows the average relative deviation Δ_{opt} depending on $\overline{d}$. The last row gives the average relative deviations if we always choose the best solution obtained by any of the three heuristics. A more detailed analysis shows that the results become better and better as the number K of different resources increases. For example, for five resources, the average relative deviation Δ_{opt} obtained by any of the three heuristics is less than 3% for objective function (B3) independently of $\overline{d}$.

Table 12:4 Δ_{opt} for test set A

	(A)			(B3)		
	$\overline{d}_{1.0}$	$\overline{d}_{1.1}$	$\overline{d}_{1.25}$	$\overline{d}_{1.0}$	$\overline{d}_{1.1}$	$\overline{d}_{1.25}$
Alg.8 MSO–LST	14%	29%	32%	8%	12%	13%
Alg.8 GRD–MST	15%	32%	44%	4%	9%	15%
Alg.8 MST–GRD	14%	29%	40%	5%	8%	12%
Best	6%	15%	27%	2%	5%	8%

The minimum objective function values for the instances of test set C are unknown. Therefore, we compare Algorithm 8 with the heuristic by Brinkmann and Neumann (1996) denoted by **Algorithm Bri & Neu**. To do so, we have measured the performance of a heuristic H by the average relative deviation Δ_{best} of the objective function value provided by H from the smallest objective function value provided by any heuristic considered (cf. Brinkmann and Neumann 1996). Note that Δ_{best} for heuristic H only compares H with other heuristics and does not represent an "absolute" measure of accuracy.

Table 12:5 shows Δ_{best} for $\overline{d} = \overline{d}_{1.1}$, Algorithm 8 with priority rules MSO–LST, GRD–MST, and MST–GRD as well as objective functions (A) and (B3) with $Y_k = 0$. We see that Algorithm 8 with priority rules MSO–LST outperforms the remaining heuristics for objective functions of type (A). For objective functions of type (B3) it turns out to be best to choose Algorithm 8 with priority rule MST–GRD. The average running time of Algorithm 8 for a problem instance of test set C is smaller than 2 sec.

Table 12:5 Δ_{best} for test set C and $\overline{d}_{1.1}$

	(A)	(B3)
Alg.8 MSO–LST	3%	12%
Alg.8 GRD–MST	18%	4%
Alg.8 MST–GRD	11%	0.2%
Alg. Bri & Neu	21%	23%

Finally, we briefly discuss the performance of Algorithms 9a and 9b, where we use the priority rule MST–GRD. We consider objective function (B3) with $Y_k = 0$ and a maximum project duration $\overline{d}_{1.1}$.

Table 12:6 gives the average relative deviation Δ_{opt} and the percentage p_{feas} of problem instances solved to feasibility obtained by the respective heuristic. Moreover, the last column gives the values if we choose the best solution obtained by any of the two heuristics for each problem instance. It turns out to be best to apply both heuristics and choose the "better" schedule.

Table 12:6 Δ_{opt} and p_{feas} for test set A and $\overline{d}_{1.1}$

	Alg. 9a	Alg.9b	Best
Δ_{opt}	8%	9%	6%
p_{feas}	72%	90%	95%

12.5 CONCLUSIONS

Resource–constrained project scheduling problems with minimum and maximum time lags and different kinds of regular and nonregular objective functions have been considered. The regular objective function of RCPSP/max is the project duration. The nonregular objective function of the resource levelling problem RLP/max is some measure of the variation of resource utilization. The nonregular objective function of NPVP/max is the net present value of the project.

For RCPSP/max, a new branch–and–bound procedure based upon introducing disjunctive precedence constraints has been discussed. Moreover, truncated branch–and–bound algorithms and priority–rule methods have been studied. An experimental performance analysis has shown that the new branch–and–

bound procedure is markedly superior to the exact algorithms known so far. The heuristics approximately solve problem instances with 500 activities and 5 resources, where the running time per instance amounts to a few seconds. The truncated branch–and–bound techniques provide more accurate solutions whereas the priority–rule methods run faster.

For RLP/max without resource constraints, a branch–and–bound method has been offered for the first time. Moreover, polynomial priority–rule heuristics have been devised for the first time, which again solve problem instances with 500 activities and 5 resources within a computing time of a few seconds approximately. These heuristics can be used for NPVP/max as well and can be generalized to the case where resource constraints are present.

Acknowledgement

This research has been supported in part by the Deutsche Forschungsgemeinschaft (Grant Ne 137/4).

References

AHUJA, H.N. 1976. *Construction Performance Control by Networks.* John Wiley and Sons, New York.

BANDELLONI, M., M. TUCCI, AND R. RINALDI. 1994. Optimal Resource Leveling Using Non–Serial Dynamic Programming. *European Journal of Operational Research* **78**, 162–177.

BARTUSCH, M., R.H. MÖHRING, AND F.J. RADERMACHER. 1988. Scheduling Project Networks with Resource Constraints and Time Windows. *Annals of Operations Research* **16**, 201–240.

BRINKMANN, K. AND K. NEUMANN. 1996. Heuristic Procedures for Resource–Constrained Project Scheduling with Minimal and Maximal Time Lags: The Resource–Levelling and Minimum Project Duration Problems. *Journal of Decision Systems* **5**, 129–155.

BRUCKER, P., A. SCHOO, AND O. THIELE. 1996. A Branch & Bound Algorithm for the Resource–Constrained Project Scheduling Problem. *Osnabrücker Schriften zur Mathematik 178,* University of Osnabrück.

BURGESS, A.R. AND J.B. KILLEBREW. 1962. Variation in Activity Level on a Cyclical Arrow Diagram. *Journal of Industrial Engineering* **13**, 76–83.

DE REYCK, B. 1998. Scheduling Projects with Generalized Precedence Relations: Exact and Heuristic Procedures. *Ph. D. Dissertation*, Katholieke Universiteit Leuven.

DE REYCK, B. AND W. HERROELEN. 1996. A Branch–and–Bound Procedure for the Resource–Constrained Project Scheduling Problem with Generalized

Precedence Relations. *Research Report 9613,* Department of Applied Economics, Katholieke Universiteit Leuven.

EASA, S.M. 1989. Resource Leveling in Construction by Optimization. *Journal of Construction Engineering and Management* **115,** 302–316.

FRANCK, B. AND K. NEUMANN. 1997. Resource–Constrained Project Scheduling with Time Windows: Structural Questions and Priority–Rule Methods. *Technical Report WIOR-492,* University of Karlsruhe.

GRINOLD, R.C. 1972. The Payment Scheduling Problem. *Naval Research Logistics Quarterly* **19**, 123–136.

HARRIS, R.B. 1978. *Precedence and Arrow Networking Techniques for Construction.* John Wiley and Sons, New York.

HARRIS, R.B. 1990. Packing Method for Resource Leveling (Pack). *Journal of Construction Engineering and Management* **116,** 39–43.

HEILMANN, R. AND C. SCHWINDT. 1997. Lower Bounds for RCPSP/max. *Technical Report WIOR-511,* University of Karlsruhe.

HERROELEN, W.S., E.L. DEMEULEMEESTER, AND P. VAN DOMMELEN. 1996. An Optimal Recursive Search Procedure for the Deterministic Unconstrained Max–NPV Project Scheduling Problem. *Research Report 9603,* Department of Applied Economics, Katholieke Universiteit Leuven.

HERROELEN, W.S., P. VAN DOMMELEN, AND E.L. DEMEULEMEESTER. 1997. Project Network Models with Discounted Cash Flows: A Guided Tour through Recent Developments. *European Journal of Operational Research* **100,** 97–121.

HORST, R. AND P.M. PARDALOS. 1995. *Handbook of Global Optimization.* Kluwer Academic Publishers, Dordrecht.

ICMELI, O. AND S.S. ERENGÜÇ. 1996. A Branch and Bound Procedure for the Resource–Constrained Project Scheduling Problem with Discounted Cash Flows. *Management Science* **42,** 1395–1408.

KLEIN, R. AND A. SCHOLL. 1997. Computing Lower Bounds by Destructive Improvement: An Application to Resource–Constrained Project Scheduling. *Schriften zur quantitativen BWL 4/97,* University of Darmstadt.

NEUMANN, K. AND C. SCHWINDT. 1995. Projects with Minimal and Maximal Time Lags: Construction of Activity–on–Node Networks and Applications. *Technical Report WIOR-447,* University of Karlsruhe.

NEUMANN, K. AND C. SCHWINDT. 1997. Activity–on–Node Networks with Minimal and Maximal Time Lags and their Application to Make–to–Order Production. *OR Spektrum* **19,** 205–217.

NEUMANN, K. AND J. ZHAN. 1995. Heuristics for the Minimum Project–Duration Problem with Minimal and Maximal Time–Lags under Fixed Resource Constraints. *Journal of Intelligent Manufacturing* **6,** 145–154.

NEUMANN, K. AND J. ZIMMERMANN. 1997. Resource Levelling for Projects with Schedule-Dependent Time Windows. *Technical Report WIOR-508,* University of Karlsruhe.

NÜBEL, H. AND C. SCHWINDT. 1997. A Classification of Shifts, Schedules and Objective Functions in Project Scheduling. *Technical Report WIOR-509,* University of Karlsruhe.

PADMAN, R., D.E. SMITH-DANIELS, AND V.L. SMITH-DANIELS. 1997. Heuristic Scheduling of Resource-Constrained Projects with Cash Flows: An Optimization Approach. *Naval Research Logistics Quarterly* **44,** 365-381.

RUSSELL, A.H. 1970. Cash Flows in Networks. *Management Science* **16,** 357-373.

RUSSELL, R.A. 1986. A Comparison of Heuristics for Scheduling Projects with Cash Flows and Resource Restrictions. *Management Science* **32,** 1291-1300.

SAVIN, D., S. ALKASS, AND P. FAZIO. 1997. A Procedure for Calculating the Weight-Matrix of a Neuronal Network for Resource Leveling. *Advances in Engineering Software* **28,** 277-283.

SCHWINDT, C. 1996. ProGen/max: Generation of Resource-Constrained Scheduling Problems with Minimal and Maximal Time Lags. *Technical Report WIOR-489,* University of Karlsruhe.

SCHWINDT, C. 1998. Verfahren zur Lösung des ressourcenbeschränkten Projektdauerminimierungsproblems mit planungsabhängigen Zeitfenstern. Shaker, Aachen.

TAKAMOTO, M., N. YAMADA, Y. KOBAYASHI, AND H. NONAKA. 1995. Zero-One Quadratic Programming Algorithm for Resource Leveling of Manufacturing Process Schedules. *Systems and Computers in Japan* **26,** 68-76.

THESEN, A. 1977. Measures of the Restrictiveness of Project Networks. *Networks* **7,** 193-208.

YANG, K.K., F.B. TALBOT, AND J.H. PATTERSON. 1992. Scheduling a Project to Maximize its Net Present Value: An Integer Programming Approach. *European Journal of Operational Research* **64,** 188-198.

YANG, K.K., L.C. TAY, AND C.C. SUM. 1995. A Comparison of Stochastic Scheduling Rules for Maximizing Project Net Present Value. *European Journal of Operational Research* **85,** 327-339.

YOUNIS, M.A. AND B. SAAD. 1996. Optimal Resource Leveling of Multi-Resource Projects. *Computers and Industrial Engineering* **31,** 1-4.

ZHAN, J. 1994. Heuristics for Scheduling Resource-Constrained Projects in MPM Networks. *European Journal of Operational Research* **76,** 192-205.

ZIMMERMANN, J. AND H. ENGELHARDT. 1998. Lower Bounds and Exact Algorithms for Resource Levelling Problems. *Technical Report WIOR-517,* University of Karlsruhe.

13 PROJECT SCHEDULING UNDER DISCRETE AND CONTINUOUS RESOURCES

J. Józefowska
M. Mika
R. Różycki
G. Waligóra
J. Węglarz

Institute of Computing Science, Poznań University of Technology

13.1. Problem formulation

We consider a project consisting of n precedence- and resource-constrained, nonpreemptable activities which require renewable resources of two types: discrete and continuous ones. We assume m discrete resources are available and vector $\mathbf{r}_i = [r_{i1}, r_{i2}, \ldots, r_{im}]$, $i = 1,\ldots, n$, determines the (fixed) discrete resource requirements of activity i. The total number of units of the discrete resource j, $j = 1,\ldots, m$, is limited by R_j. The single continuous, renewable resource can be allotted to activities in (arbitrary) amounts from the interval [0, 1]. The amount (unknown in advance) of the continuous resource $u_i(t)$, allotted to activity i at time t determines the processing rate of activity i as it is described by the following equation:

$$\dot{x}_i(t) = \frac{dx_i(t)}{dt} = f_i[u_i(t)], \; x_i(0) = 0, \; x_i(C_i) = \tilde{x}_i \tag{1}$$

where $x_i(t)$ is the state of activity i at time t,

f_i is a continuous, nondecreasing function, where $f_i(0)=0$,

$u_i(t)$ is the amount of the continuous resource allotted to activity i at time t,

C_i is the completion time (unknown in advance) of activity i,

$\widetilde{x}_i$ is the processing demand (final state) of activity i.

State $x_i(t)$ of activity i at time t is an objective measure of work related to the processing of activity i up to time t. It may denote, for example, the number of man-hours already spent on processing of activity i, the number of standard instructions in processing of computer program i, and so on. All activities are available at the start of the process.

The problem is to find a feasible assignment of discrete resources and, simultaneously, a continuous resource allocation which minimize the schedule length, i.e. the project duration, $M = \max\{C_i, i = 1, 2,\ldots, n\}$. The continuous resource allocation is defined by a piece-wise continuous, non-negative vector function $\mathbf{u}^*(t) = [u_1^*(t) u_2^*(t),\ldots,u_n^*(t)]$ whose values $\mathbf{u}^* = [u_1^*, u_2^*,\ldots,u_n^*]$ are (continuous) resource allocations corresponding to M^* — the minimal value of M.

Problems of this type where a set of identical parallel machines was the single discrete resource was defined in Józefowska and Węglarz (1998) and examined for nonpreemptable, independent jobs. Extension of the problem to the case of precedence-constrained jobs was presented in Józefowska et. al. (1998b). In this chapter we consider the project scheduling problem with multiple arbitrary discrete resources and a single continuous resource, all renewable ones. We adapt three local search metaheuristics to solve this problem. Although the same metaheuristics have been already applied in the case of machine scheduling problems, the extended problem required that the solution representations and neighbourhood generation mechanisms were significantly modified leading to new algorithms.

13.2. Basic notions

Let us divide a feasible schedule (i.e. a solution of a discrete-continuous project scheduling problem) into $p \le n$ intervals of lengths M_k, $k = 1,\ldots, p$, defined by the completion times of the consecutive activities.

Let Z_k denote the combination of activities corresponding to the k-th interval. Thus, a *feasible sequence* S of combinations Z_k, $k = 1,2,\ldots, p$, is associated with each feasible schedule. The feasibility of such a sequence requires that:

1. the number of units of the discrete resource j, $j = 1,\ldots, m$, assigned to all activities in combination Z_k, $k = 1,\ldots, p$, does not exceed R_j, i.e. $\sum_{i\in Z_k} r_{ij} \le R_j$, $j = 1,\ldots, m$, $k = 1,\ldots, p$,
2. each activity appears in at least one combination,
3. precedence constraints between activities are satisfied,

4. nonpreemptability of each activity is guaranteed.

The last condition requires that each activity appeared in exactly one or in consecutive combinations in S.

Example 1

Consider the following problem of scheduling 9 activities with one discrete resource ($R = 4$, $r_i = 2$, $i = 1,\ldots, 9$) and precedence constraints given by the following activity-on-node graph (Fig. 13:1a).

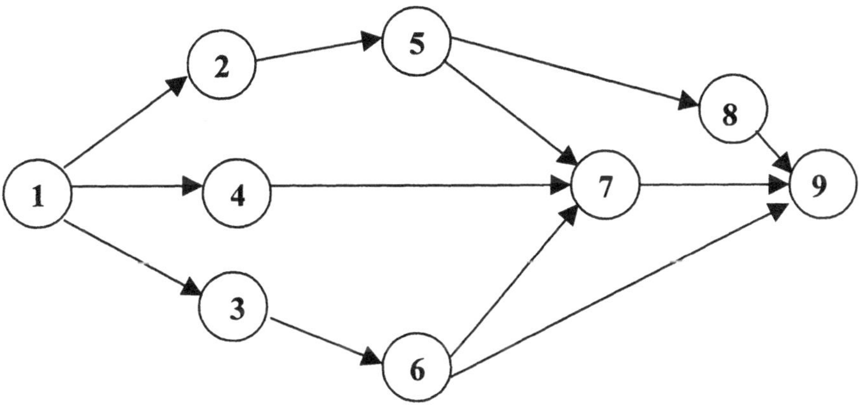

Fig. 13:1a. Graph of precedence constraints

An example of a feasible sequence of combinations for this problem has the form: $S = \{1\}, \{2,3\}, \{4,3\}, \{4,5\}, \{6,5\}, \{6,8\}, \{7,8\}, \{9\}$ (see Fig. 13:1b).

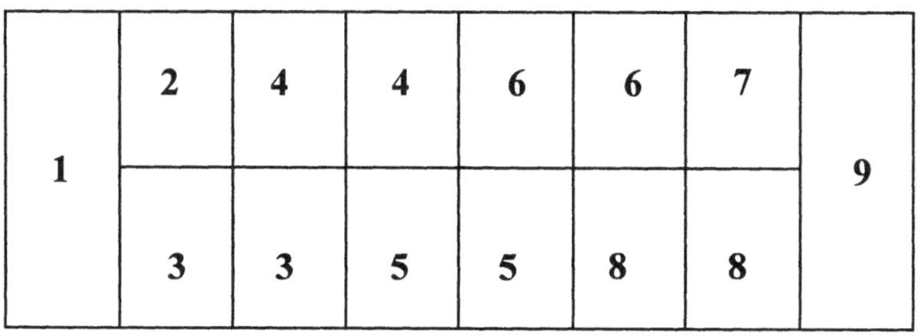

Fig. 13:1b. The idea of a feasible sequence

Further, for given S, the processing demand of each activity can be divided into parts $\widetilde{x}_{ik} \geq 0$ (unknown in advance) corresponding to particular time intervals (combinations). An illustration of such a division is shown in Fig. 13:1c.

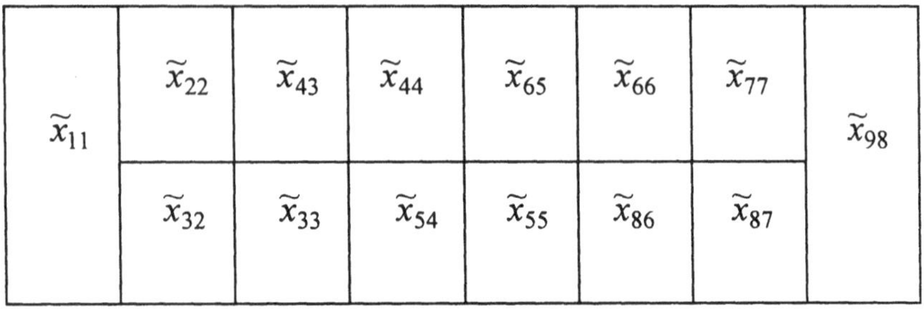

Fig. 13:1c. A division of processing demands of activities

It has been proved in Józefowska and Węglarz (1998) that for convex processing rate functions f_i, $i = 1, 2,\dots, n$, a makespan optimal schedule is obtained by allotting the total amount of the continuous resource to one activity at a time only. Concluding, in an optimal schedule activities are ordered in a sequence fulfilling the precedence constraints and the total amount of the continuous resource is allotted to each activity. Of course, if we assume $r_{ij} \le R_j$, $i = 1, 2,\dots, n$, $j = 1, 2,\dots, m$, the discrete resource constraints are fulfilled, because only one activity is performed at a time. For concave processing rate functions in a makespan optimal schedule as many activities as possible are performed in parallel (see Józefowska and Węglarz (1998)). In this case for a given feasible sequence S one can find an optimal division of processing demands of activities, $\widetilde{x}_i$, $i = 1,2,\dots, n$, among combinations in S, i.e. a division which leads to a minimum length schedule from among all feasible schedules generated by S. To this end a nonlinear programming problem can be formulated in which the sum of the minimum-length intervals (i.e. parts of a feasible schedule) generated by consecutive combinations in S, as functions of the $\{\widetilde{x}_{ik}\}_{k \in Z_k}$, where $\widetilde{x}_{ik}$ is a part of activity i processed in combination Z_k, is minimized subject to the constraints that each activity has to be completed. For concave processing rates f_i, $i = 1,\dots, n$, it is sufficient to consider feasible schedules in which the resource allocations among activities remain constant in each interval k, $k = 1, 2,\dots, p$. In the sequel we will assume that f_i, $i = 1, 2,\dots, n$, are concave.

Let $M_k^*(\widetilde{\mathbf{x}}_k)$ be the minimum length of the part of the schedule generated by $Z_k \in S$, as a function of $\widetilde{\mathbf{x}}_k = \{\widetilde{x}_{ik}\}_{i \in Z_k}$. Let K_i be the set of all indices of Z_k's such that $i \in Z_k$. The following mathematical programming problem is obtained to find an optimal demand division of activities for a given feasible sequence S.

Problem P

Minimize
$$M\left(\{\widetilde{\mathbf{x}}_k\}_{k=1}^{p}\right) = \sum_{k=1}^{p} M_k^*(\widetilde{\mathbf{x}}_k) \qquad (2)$$

subject to

$$\sum_{k \in K_i} \tilde{x}_{ik} = \tilde{x}_i, \qquad i = 1,2,\dots, n \tag{3}$$

$$\tilde{x}_{ik} \geq 0, \qquad i = 1,2,\dots, n; \quad k \in K_i \tag{4}$$

where $M_k^*(\tilde{\mathbf{x}}_k)$ is calculated as a unique positive root of the equation:

$$\sum_{i \in Z_k} f_i^{-1}\left(\tilde{x}_{ik} / M_k\right) = 1$$

which can be solved analytically for some important cases (see Węglarz (1982)).

Of course, constraints (3) correspond to the condition of fulfilling processing demands of all activities. It was proved in Węglarz (1982) that the objective function (2) is always a convex function. In consequence, our problem is to minimize a convex function subject to linear constraints.

After finding an optimal division $\tilde{x}_{ik}$, $i = 1, 2,\dots, n$, $k \in K_i$ of $\tilde{x}_i$'s the corresponding continuous resource allocation for combination Z_k is given as

$$u_{ik}^* = f_i^{-1}\left(\tilde{x}_{ik}^* / M_k^*\right), \; i \in Z_k .$$

Thus, an optimal schedule for a given feasible sequence can be found by solving Problem P. In consequence, a globally optimal schedule can be found by solving the continuous resource allocation problem optimally for all feasible sequences and choosing a schedule with the minimum length. In this sense the process of finding an optimal schedule can be viewed as a search process over the space of all feasible sequences for a given problem instance. Unfortunately, in general, the number of all feasible sequences grows exponentially with the number of activities. Therefore it is justified to apply local search metaheuristics such as Simulated Annealing (SA), Tabu Search (TS) or Genetic Algorithms (GA) operating on a space of all feasible sequences. In the following sections three algorithms are presented being adaptations of the above mentioned metaheuristics for discrete-continuous project scheduling problems with concave functions f_i, $i = 1, 2,\dots, n$. Below, the common features of all the three algorithms are summarised.

13.2.1. Search space

We have stated above that a way to find an optimal schedule is to calculate optimal resource allocations for all the feasible sequences and choose the best one of all the schedules generated. However, since the number of combinations in a feasible sequence p is less than or equal to n, it is easy to observe that in order to find an optimal schedule, it is sufficient to look through n-element feasible sequences only.

Thus we define a *feasible solution* as an n-element feasible sequence. The set of all such feasible solutions constitutes the search space for all the three heuristics.

We may assume without loss of generality that activities are numbered in such a way that if $i < j$, then i precedes j. This can be always accomplished using the level numbering of nodes in a directed acyclic graph.

13.2.2. Objective function

For each feasible solution the objective function is calculated by solving Problem P for the corresponding feasible sequence.

13.2.3. Stop criterion

The stop criterion has been defined as a number of feasible solutions visited.

In the following sections three basic metaheuristics (SA, TS and GA) to solve the discrete-continuous scheduling problem are presented.

13.3. Tabu Search

Tabu Search, like Simulated Annealing, is a local search technique based on neighbourhood search with overcoming local optimality but unlike Simulated Annealing or Genetic Algorithms it works in a deterministic way trying to model human memory processes. Memory is implemented by the implicit recording of previously seen solutions using simple but effective data structures. Tabu Search was originally developed by Glover (1989) and Glover (1990) and a comprehensive account of the basic concepts and recent developments is given by Glover and Laguna (1993).

13.3.1. Solution representation

A feasible solution has been defined as an n-element feasible sequence and in the implementation of the tabu search algorithm it has a natural representation of an n-element sequence of combinations of activities.

13.3.2. Starting solution

In order to ensure the feasibility of a starting solution (i.e. to observe precedence constraints and discrete resource constraints) it is generated by placing each activity in a different combination in such a way that the precedence constraints are not violated. Taking into account that activities are ordered as described above the simplest way to obtain such a solution is to consider the sequence: $\{1\}, \{2\}, \ldots, \{n\}$ as the starting solution.

13.3.3. Neighbourhood generation mechanism

There are two mechanisms of generating the neighbours of a current solution: *replacement* and *exchange*. Replacement is the main method, exchange is used in special cases only.

13.3.3.1. Replacement. Consider a given feasible solution consisting of at most n^2 positions. A neighbour is generated by replacing an activity in a chosen position by another activity. An activity may be replaced only in the first or in the last combination in that it occurs (nonpreemtability) provided that these combinations are different (each activity must appear in at least one combination).

It is easy to see that in order to maintain the nonpreemtability, an activity in combination Z_k may be replaced only by activities from combinations Z_{k-1} or Z_{k+1}.

Assume that activity i in combination Z_k is to be replaced by activity j from combination Z_{k-1} (Z_{k+1}). It may be done only if the following conditions are simultaneously satisfied:

- activity j is not in combination Z_k
- there is no successor (predecessor) of activity j in combination Z_k (apart from activity i)
- replacing activity i by activity j in combination Z_k does not violate the constraint of any discrete resource.

A special case of replacement is what could be called *insertion*. It works exactly as described above with the exception that activity i is so-called *empty activity* (no processing demand, no resource requirements, no either predecessors or successors). As a result of replacing an empty activity, activity j is, in fact, added to combination Z_k and the number of activities in Z_k increases by 1. An empty activity must not replace any other activity, it may only be replaced.

Example 2

Let $n = 4$. Assume there are no precedence constraints and no discrete resource requirements. Let the current solution be as follows: {1}, {1,2}, {2,3}, {4}.

Applying all possible replacements results in the following set of neighbouring solutions:

{2}, {1,2}, {2,3}, {4},
{1,2}, {1,2}, {2,3}, {4},
{1}, {3,2}, {2,3}, {4},
{1}, {1,3}, {2,3}, {4},
{1}, {1,2,3}, {2,3}, {4},
{1}, {1,2}, {1,3}, {4},
{1}, {1,2}, {2,3,1}, {4},
{1}, {1,2}, {4,3}, {4},
{1}, {1,2}, {2,3,4}, {4},
{1}, {1,2}, {2,3}, {4,2},
{1}, {1,2}, {2,3}, {4,3}.

Let us continue the above example and assume that there are precedence constraints: 1→3, 1→4, and there is one discrete resource available in 10 units. If the vector of discrete resource requirements of activities is: **r** = [3, 5, 1, 7] then the above set of neighbouring solution will not contain the following elements:

{1}, {1,3}, {2,3}, {4},	/precedence constraints
{1}, {1,2,3}, {2,3}, {4},	/precedence constraints
{1}, {1,2}, {1,3}, {4},	/precedence constraints
{1}, {1,2}, {2,3,1}, {4},	/precedence constraints
{1}, {1,2}, {2,3,4}, {4},	/discrete resource constraints
{1}, {1,2}, {2,3}, {4,2}	/discrete resource constraints.

The above example illustrates that under certain precedence and discrete resource constraints some of the potential neighbours may not be generated.

13.3.3.2. Exchange. Exchange is specially designed for the case when an activity with a large discrete resource requirement cannot be processed in parallel with any other activity and blocks the mechanism of replacement.

Example 3

If the current solution is: {1,2}, {2}, {3}, {4} and a discrete resource requirement of activity 3 is equal to the total available number of units of that resource then activity 4 will never be processed in parallel with activity 1 or 2 (activity 3 blocks the replacement mechanism). In such a case two activities may be exchanged in consecutive combinations provided that each of these activities occurs in exactly one combination (otherwise the nonpreemtability would be violated). For example, in the above example exchanging activities 3 and 4 gives a new solution: {1,2}, {2}, {4}, {3}.

Although exchange was originally designed for cases described above, it is useful to perform it every time it is possible to check if it leads to a better solution.

Assume that activity i in Z_k and activity j in Z_{k+1} are to be exchanged (i.e. Z_k is the only combination with activity i and Z_{k+1} is the only one with activity j). It may be done only if the following conditions are simultaneously satisfied:

- there is no successor of activity i in combination Z_{k+1}
- there is no predecessor of activity j in combination Z_k
- exchanging activity i with activity j does not violate any discrete resource constraints either in Z_k or in Z_{k+1}.

A neighbourhood of a current solution consists of feasible solutions obtained by performing all possible replacements and exchanges.

13.3.4. Tabu list

The tabu list is managed by the *Reverse Elimination Method* - REM (Glover 1990). As far as the replacement mechanism is concerned, a move is represented by a 3-tuple: (number of a combination, activity replaced, activity introduced). A *Residual*

Cancellation Sequence - RCS is a minimal set of moves leading back from a current solution to a solution visited in the past. After a move (k,i,j) is performed, each non-empty RCS is modified in such a way that if move (k,i,j) belongs to the RCS then it is removed, otherwise the reverse move: (k,j,i) is added. A new RCS is created with the move (k,j,i) in order to eliminate the possibility of reversing the performed move in the next iteration. Every RCS which consists of one move only defines a tabu replacement in the next iteration.

After the number of RCS's equals to the tabu list length the largest RCS is removed from the list. It is obvious that the probability of going back to the most distant from among solutions already visited is least of all. The tabu list length is set at 7. A few preliminary experiments showed that for the considered problems this value effectively prevents cycles.

For more details on the Reverse Elimination Method for problems with discrete and continuous resources see Józefowska et.al. (1998a).

As far as the exchange mechanism is concerned, it is easy to see that it may be represented by a pair of moves of the form described above. E.g. if activity i in Z_k is exchanged with activity j in Z_{k+1}, it can be seen as if moves (k,i,j) and $(k+1,j,i)$ were performed simultaneously. All non-empty RCS's are modified in the way described above as if the moves were performed sequentially and a new RCS with moves (k,j,i) and $(k+1,i,j)$ is created to eliminate the possibility of reversing the performed exchange in the next iteration. Every RCS which consists of two moves being a pair representing an exchange defines a tabu exchange in the next iteration.

Thus, in order to define the status of a replacement (tabu or not) one has to examine all one-element RCS's whereas to define the status of an exchange one must examine all two-element ones.

Example 4

Let the current solution be of the form: {1,2}, {2,3}, {2,4}, {4,5}, {5}.

Under the sequence of moves presented below the Reverse Elimination Method works as follows:

current solution	tabu list
1. {1,2}, {2,3}, {2,4}, {4,5}, {5}	
replacement (4,4,2)	
2. {1,2}, {2,3}, {2,4}, {2,5}, {5}	1. (4,2,4)
tabu replacements: (4,2,4)	
tabu exchanges: none	
exchange (1,1,3) (2,3,1)	
3. {3,2}, {2,1}, {2,4}, {2,5}, {5}	1. (4,2,4) (1,3,1) (2,1,3)
	2. (1,3,1) (2,1,3)
tabu replacements: none	
tabu exchanges: (1,3,1) (2,1,3)	

replacement (5,5,2)

4. {3,2}, {2,1}, {2,4}, {2,5}, {2}

tabu replacements: (5,2,5)
tabu exchanges: none

1. (4,2,4) (1,3,1) (2,1,3) (5,2,5)
2. (1,3,1) (2,1,3) (5,2,5)
3. (5,2,5)

exchange (1,3,1) (2,1,3)

5. {1,2}, {2,3}, {2,4}, {2,5}, {2}

tabu replacements: (5,2,5)
tabu exchanges: (1,1,3) (2,3,1)

1. (4,2,4) (5,2,5)
2. (5,2,5)
3. (5,2,5) (1,1,3) (2,3,1)
4. (1,1,3) (2,3,1)

replacement (4,2,4)

6. {1,2}, {2,3}, {2,4}, {4,5}, {2}

tabu replacements: (5,2,5) and (4,4,2)
tabu exchanges: none

1. (5,2,5)
2. (5,2,5) (4,4,2)
3. (5,2,5) (1,1,3) (2,3,1) (4,4,2)
4. (1,1,3) (2,3,1) (4,4,2)
5. (4,4,2)

13.4. Simulated Annealing

Simulated Annealing is a metaheuristic that belongs to the class of local search algorithms which are known as threshold algorithms as presented by Aarts, Korst and van Laarhoven (1997). In its original form this algorithm is based on the Monte Carlo method Metropolis et. al. (1953) which is used to simulate a physical annealing process. This method was applied to the combinatorial optimization problems for the first time in the 1980's by Kirkpatrick, Gelatt Jr. and Vecchi (1983) and by Černy (1985). Later, this algorithm was popularized by the works of van Laarhoven, Aarts {1987) and Aarts and Korst (1989). To apply the SA algorithm to a particular combinatorial problem one must specify a set of parameters, which can be divided into two categories. The first category of parameters is a general one that is usually called a cooling scheme and the second one contains problem specific parameters. Below we will describe how these parameters are used in our implementation of SA to the presented problem.

13.4.1. Cooling scheme

The cooling scheme described by Aarts and Korst (1989) is used except the stop criterion, common for all the presented metaheuristics.

13.4.1.1. The initial value T_0 of the control parameter is calculated at the start of the algorithm from the following equation:

$$T_0 = \frac{\overline{\Delta f}^{(+)}}{ln\left(\frac{m_2}{m_2 \cdot \chi - m_1(1-\chi)}\right)}$$

where χ_0 is the initial acceptance ratio (the proportion between transitions accepted and all the transitions generated for T_0); m_1 is the number of cost-decreasing transitions and m_2 is the number of cost-increasing transitions from among m_0 trial transitions generated to determine the initial value of the control parameter. Of course, $m_1 + m_2 = m_0$. $\overline{\Delta f}^{(+)}$ is the average difference in cost over the m_2 cost-increasing transitions. We assume that $\chi_0 = 0.97$ and $m_0 = 50$.

13.4.1.2. The decrement of the control parameter T_k depends on the average value of the control parameter and the standard deviation σ_{T_k} for the kth Markov chain. The next value T_{k+1} is calculated according to the following formula:

$$T_{k+1} = \frac{T_k}{1 + \frac{T_k \cdot ln(1+\delta)}{3 \cdot \sigma_{T_k}}}, \quad k = 0, 1, \ldots .$$

The amount by which the value of T_k is decreased by the above decrement function is determined by the value of the distance parameter δ. We assume that $\delta = 0.5$.

13.4.1.3. The length of the Markov chains determines the number of transitions performed for each value of the control parameter. We assumed that the length of Markov chains is fixed and depends on the size of the neighbourhood according to the following equation $L_k = L = n \cdot (n / 2)$, $k = 0, 1, \ldots$, where n is the number of activities.

13.4.2. Solution representation

A feasible solution is represented by two n-element sequences. The first one is a permutation of activities and defines an order in which activities are started. This sequence is called a *Sequence of Activity Starts* (*SAS*). The *SAS* is constructed in the way that precedence constraints are not violated. The second one is also a permutation of activities that does not violate the precedence constraints and denotes the order in which activities should be completed if not determined by other rules. It is called a *Sequence of Activity Completions* (*SAC*). We have to stress that *SAC* does not define the actual order in which activities will be finished. It is used only when introducing a new activity would violate the discrete resources constraints in a combination and in order to maintain feasibility of the sequence some activities must be finished before this new one is started. We introduce one additional n-element sequence which is used while generating a neighbour of a current solution. This sequence is called an *Auxiliary Sequence* (*AS*) and consists of binary values only. A feasible solution is constructed according to the following rules:

- a feasible solution consists of exactly n combinations of activities;
- each combination consists of at least one and at most n activities;
- there is exactly one activity in each combination that does not appear in the previous one; it means that only one activity is started in each combination i.e. the first activity from *SAS* is started in the first combination Z_1, the second one is started in Z_2 and so on;
- activities are finished when it is forced either by the precedence constraints or by the discrete resource constraints;
- when an activity is finished in order to release the discrete resources for the next activity the value of 1 is set in *AS* in the position that corresponds to the position of this activity in *SAC* otherwise the value of 0 is set in this position;
- each activity must appear in all combinations between the combination in which it is started and the combination in which it is finished because of the nonpreemptability of activities.

Below we will show how to obtain a feasible solution from a given pair of sequences (*SAS* and *SAC*) applying the above listed rules.

Example 5

To illustrate the idea let us assume that we have $n = 9$ activities with precedence constraints given by the graph in Fig. 13:1a with only one discrete resource available in $R = 5$ units. Each activity i, $i = 1,\ldots,n$ requires for its processing r_i units of the resource where the vector of resource requirements $\mathbf{r} = [1,2,3,1,3,3,2,4,1]$.

Let us assume that *SAS* and *SAC* take the following forms:

SAS	1	2	3	4	5	6	7	8	9

SAC	1	2	3	4	5	6	7	8	9

The corresponding feasible solution looks as follows: {1}, {2,3}, {3,4}, {4,5}, {4,5}, {6}, {7}, {8}, {9} (see below).

Feasible solution	1	2	2	4	4	6	7	8	9
			3	3	5				

It represents among other the following feasible sequence $S = \{1\}, \{2,3\}, \{3,4\}, \{5,4\}, \{6\}, \{7\}, \{8\}, \{9\}$.

The *AS* obtained during the transformation of *SAS* and *SAC* to the feasible solution has the following form:

0	1	1	0	1	0	1	0	0

13.4.3. Starting solution

The initial *SAS* and *SAC* are obtained by setting all activities in an ascending order. The feasible solution presented in Example 5 is an example of a starting solution.

13.4.4. Neighbourhood generation mechanism

A neighbour of a current solution is generated according to the following algorithm.

1. Randomly choose an integer $k \in \langle 1, K \rangle$, where K is the sum of n plus the number of the positions in *AS* which are set at one.
2. If $k \le n$ then $j := k$ and obtain the new solution by a small perturbation in *SAS* otherwise let j be the position of the $(k - n)$-th non-zero element of *AS* and obtain the new solution by a small perturbation in *SAC*. In steps 3 through 4 we will describe how the small perturbation is generated.
3. Randomly choose an integer $l \in (p, s)$, where p is the position of the nearest predecessor of activity SAS_j (SAC_j) from the left hand side of j in *SAS* (*SAC*) and s is the position of the nearest successor of activity SAS_j (SAC_j) from the right hand side of j in *SAS* (*SAC*).
4. Remove activity SAS_j (SAC_j) from position j and put it into position l. Move all the remaining activities between positions l and $j \pm 1$ by one position towards position j.

13.5. Genetic Algorithm

The theoretical basis of genetic algorithms were laid down by J. von Neumann and J. Holland almost thirty years ago. A comprehensive account of the basic concept and related topics (Evolutionary Programming) is given in Michalewicz (1992).

Although Genetic Algorithm is considered to be a local search optimization method (as Tabu Search and Simulated Annealing), it has some distinctive features. In each iteration (*generation*) GA operates on a group (so-called *population*) of the potential solutions (*individuals*) of the considered problem. There is not an evident definition of a solution neighbourhood in GA. The process of looking over new solutions imitates the evolution process by using so-called *recombination operators*.

13.5.1. Solution representation

In a GA suited to solve complex combinatorial problems the way in which a feasible solution is represented, is strongly related to the genetic operators used. The simpler the representation of a solution, the more complicated the genetic operators maintaining the feasibility of the whole population. In our implementation, we decided to simplify the genetic operators as much as possible. Thus, a sophisticated decoding procedure had to be prepared for resulting representation of a feasible solution.

A single feasible solution is represented by an individual consisting of two *chromosomes*. The first chromosome defines the order of activities. It is a vector of n unique numbers, each one representing a single activity. The second chromosome is a vector of $n-1$ arbitrary numbers from the set $\{-n, -n+1, \ldots, -2, -1, 1, 2, \ldots, n-1, n\}$.

The decoding procedure fulfils the following assumptions:

- each individual represents a feasible solution,
- combination Z_k ($k = 1, \ldots, n$) contains a single activity from the first chromosome which has not been scheduled yet (namely activity i occurring in position k of this chromosome) and activities from *transitive vector* V_k.

Transitive vector V_k of activities is constructed iteratively in the following three steps:

1. (creating step) All the activities from combination Z_{k-1}, which are not precedence related with activity i are added to transitive vector V_k preserving the order defined by the first chromosome ($V_0 = []$),
2. (control step) If the total number of units of the discrete resources needed by activity i and all the activities from transitive vector V_k does not exceed the discrete resources constraints the construction of the transitive vector is finished. Otherwise, to ensure feasibility of the solution, the transitive vector must be reduced. It is realised in the third step,
3. (reducing step) Let us assume that l is the number of activities in transitive vector V_k after performing step 2. The following formula gives the position o of the activity which is removed from the transitive vector:

$$o = ((|q| - 1) \textbf{ modulo } l) + 1$$

where q is the value in the $(k-1)$-th position of the second chromosome. All activities in positions $o+1, \ldots, l$ in vector V_k are shifted one position to the left (i.e. $l := l - 1$) Then the second step is repeated.

Of course, for a given order of activities defined by the first chromosome, there may appear some (maybe all) positions in the second chromosome which will not be used by the decoding procedure. If a positive value of the second chromosome is not used by the decoding procedure, then its value is multiplied by (-1) and becomes negative. Similarly if a negative value occurs in a position used by the decoding procedure its value is multiplied by (-1) and becomes positive. A positive value in the second chromosome means that the corresponding position has been the significant one for the decoding procedure.

Example 6

Let us consider the precedence constraints from Example 1 and $n = 9$, $m = 1$, $R = 10$, $\mathbf{r} = [2,5,4,2,4,3,5,1,3]$ ($i = 1,\dots, n$). The decoding of an individual is presented below.

	the first chromosome									the second chromosome							
individual	1	2	3	4	5	6	7	8	9	-2	-3	-4	-9	-2	-1	-1	-2

feasible solution	{**1**}	{**2**}	{2,**3**}	{2,**4**}	{4,**5**}	{4,5,**6**}	{**7**}	{7,**8**}	{**9**}
transitive vectors:		[]	[2]	[2,~~3~~]	[4]	[4,5]	[]	[7]	[]

removed in the reducing step (↑ [2,~~3~~])

It is worth mentioning, that a single solution may be represented by multiple individuals. For example, the above feasible solution may be represented also by the following individual:

individual	1	2	3	4	5	6	7	8	9	-2	-3	**-2**	-9	-2	-1	-1	-2

13.5.2. Creating an initial population

An initial population is a starting point for a multidirectional evolution search process. The considerable variety of individuals in this population usually improves the effectiveness of the Genetic Algorithm. The commonly used method of creating an initial population — the random creating — is used in our implementation. In fact, only the second chromosome is generated fully randomly, whereas a more sophisticated procedure is used for the first chromosome. We assume that the activities are ordered according to the precedence constraints, as described in Section *Search Space*. Then the first chromosomes of the new generated individuals are obtained by a feasible (not violating the precedence constraints) exchange of two activities in the original order.

13.5.3. Fitness of an individual

The fitness of an individual depends on the minimum schedule length for the relevant feasible sequence. The shorter the schedule length, the higher the fitness of the individual.

13.5.4. Genetic operators

The selection operator is responsible for selecting relatively good individuals, in the sense of their fitness, which are promising material for constructing new individuals representing new feasible solutions. As a selection strategy, the linear ranking with elitism was selected. It is easy to implement, and does not need any scaling method, which is necessary in the case of the roulette wheel for minimization problems.

After the selection step the recombination operators generate individuals representing feasible solutions on the basis of randomly selected parent individuals. New offspring individuals replace their parents in a population. In such a way a new generation is completed.

There are three recombination operators especially suited for the assumed representation. The first one, the *crossover,* operates on the pairs of the parent individuals while *mutation* and *exchange* operate on a single parent individual.

The crossover can be applied to the second chromosomes of the parent individuals only. It simply cuts the second chromosomes of both the parents in two randomly chosen cut points and exchanges their central segments. Of course, the crossover leads to new solutions only if the central segments contain at least one positive value.

parent individual 1	1	2	3	4	5	6	7	8	9	-2	-3	-4	-4	-2	3	-1	-2
												cut points					
parent individual 2	1	3	2	5	4	6	7	9	8	-2	-3	2	-4	-2	3	-1	-2
offspring individual 1	1	2	3	4	5	6	7	8	9	-2	-3	**2**	**-4**	-2	3	-1	-2
offspring individual 2	1	3	2	5	4	6	7	9	8	-2	-3	**-4**	**-4**	-2	3	-1	-2

The second operator, *mutation*, operates on the second chromosome only. A position with a positive value is randomly chosen in a parent individual. Then this value is replaced by another one from the set $\{1, 2,..., n\}$.

parent individual	1	2	3	4	5	6	7	8	9	-2	-3	2	-4	-2	3	-1	-2
												↓					
offspring indiviidual	1	2	3	4	5	6	7	8	9	-2	-3	**6**	-4	-2	3	-1	-2

The third operator, called *exchange*, operates on the first chromosome only. This operator exchanges (if it does not violate the precedence constraints) randomly chosen activity with its immediate predecessor or successor in the first chromosome. The direction of the exchange is also random.

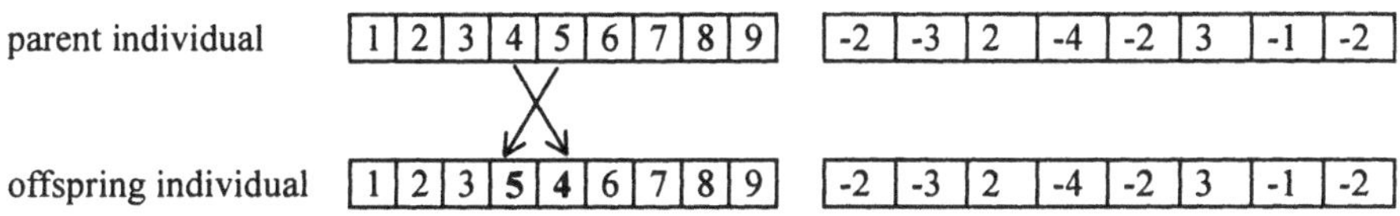

13.6. Computational experiments

A number of computational experiments have been carried out to compare the performance of the considered metaheuristics on a randomly generated set of instances by a comparable computational effort.

All the heuristics have been implemented in C++ and have run on SGI PowerChallenge XL with 12 RISC 8000 processors. As we have mentioned before, for each solution visited in the solution space, the corresponding schedule has been found by solving the suitable convex mathematical programming problem. In this step specially adapted solver CFSQP 2.3 – C Code for Solving (Large Scale) Constrained Nonlinear (Minimax) Optimization Problems, Generating Iterates Satisfying All Inequality Constraints (see Lawrence, Zhou and Tits, 1995) was applied. The solver stopped when the absolute difference in consecutive values of the objective function was less than or equal to 10^{-3}.

Three heuristics have been compared: TS, GA and SA. In order to ensure a comparable computational effort devoted to each heuristic, the stop criterion has been defined as a number of solutions visited. This number has been set at 1000. It is clear that neither Tabu Search nor Genetic Algorithm can stop exactly after the required number of solutions visited, so they stop after visiting at least the required number of solutions. In fact, Tabu Search performed the smallest number of iterations in which the total number of visited solutions exceeds 1000 and the Genetic Algorithm generated the smallest number of generations in which the total number of individuals exceeded 1000.

The experiment has been performed for several values of n starting with $n = 5$ and no considerable difference in the results has been observed for different n. Moreover, for $n \geq 10$, the computational time for a single instance ran into hours. Thus we decided not to continue the experiment for larger n. Typical results for n = 10 are shown in Table 13:1.

Four groups of problems, 30 instances each, with $R = 2, 5, 10$ and 15, respectively, have been generated. The values of r_i, have been generated randomly from the interval [0, $R/2$]. The experiments have been carried out for processing rates of activities of the form $f_i = u_i^{1/\alpha_i}$. Values of α_i from the set {1,2} have been generated randomly with equal probability. Processing demands have been generated from the interval [1,100] with a uniform distribution. The average density of a precedence graph was 0.5.

Table 13:1. Results of the computational experiment

R	*Algorithm*	#	*Average deviation [%]*	*Maximum deviation [%]*
	TS	13	1.3463	12.4451
2	GA	23	0.2513	2.9953
	SA	30	0	0
	TS	10	1.4525	12.5193
5	GA	16	0.6318	4.9142
	SA	30	0	0
	TS	10	1.5564	12.4451
10	GA	23	0.1406	2.0287
	SA	30	0	0
	TS	10	1.7217	11.4951
15	GA	20	0.4741	3.7331
	SA	28	0.0100	0.1644

The comparison among heuristics is presented in Table 13:1 on the basis of three values: the number of instances for which the relevant heuristic found the best solution of all heuristics, the average relative deviation from the best solution found and the maximum relative deviation from that value.

The results presented in Table 13:1 show that SA has found the best solution of all the three heuristics for all instances generated. The average relative deviation of the solutions found by both the remaining heuristics from the solution found by SA is less than 2%, while the maximum deviation is over 10% for TS. GA has produced slightly better results than TS, especially the number of instances for which the best solution has been found is larger for GA. These results differ from the results of earlier experiments carried out for the machine scheduling problems, where TS usually outperformed both the other heuristics.

To the end let us stress that the high computational effort required follows from the fact that for every feasible solution visited by a heuristic an optimal resource allocation has been calculated by solving Problem P. Solving the nonlinear mathematical programming problem was very time consuming. Thus it is important to develop and test efficient heuristics for allocating the continuous resource among activities in a given solution (feasible sequence).

Acknowledgement

This research has been supported by a KBN grant No. 8T11C 022 10.

References

AARTS E.H.L., J.H.M. KORST. 1989. Simulated Annealing and Boltzmann Machines: A Stochastic Approach to Combinatorial Optimization and Neural Computing. John Wiley, Chichester.

AARTS E.H.L., J.H.M.KORST, P.J.M. VAN LAARHOVEN. 1997. Simulated Annealing. In: AARTS E.H.L. AND J.K. LENSTRA ed. Local Search in Combinatorial Optimization. John Wiley, Chichester, 91-120.

ČERNY, V. 1985. Thermodynamical approach to the traveling salesman problem: an efficient simulation algorithm. Journal of Optimization Theory And Applications **45**, 41-51.

GLOVER F. 1989. Tabu Search - part 1. ORSA J. Computing. **1**, 190-206.

GLOVER F. 1990. Tabu Search - part 2. ORSA J. Computing **2**, 4-32.

GLOVER F., M . LAGUNA. 1993. Tabu Search. In: REEVES C.R., ed. Modern Heuristic Techniques for Combinatorial Problems. Blackwell Scientific Publications, Oxford.

JÓZEFOWSKA J., M. MIKA, R. RÓŻYCKI , G. WALIGÓRA, J . WĘGLARZ. 1998a. Local search metaheuristics for discrete-continuous scheduling problems. European Journal of Operational Research **107**.

JÓZEFOWSKA J., M. MIKA, R. RÓŻYCKI , G. WALIGÓRA, J . WĘGLARZ. 1998b. Scheduling precedence constrained nonpreemptable jobs on parallel machines with additional, discrete-continuous resources. Proc. Int. Conf. MMAR'98 - to appear.

JÓZEFOWSKA J., AND J. WĘGLARZ. 1998. On a methodology for discrete-continuous scheduling. European Journal of Operational Research, **107**.

KIRKPATRICK S., C.D. GELATT JR., M.P . VECCHI. 1983. Optimization by simulated annealing. Science **220**, 671-680.

VAN LAARHOVEN P.J.M., E.H.L . AARTS. 1987. Simulated Annealing: Theory and Applications. Reidel, Dordrecht.

LAWRENCE C., J.L. ZHOU, A.L . TITS. 1995. Users guide for CFSQP Version 2.3. Available by e-mail: andre@eng.umd.edu.

METROPOLIS N., A. ROSENBLUTH, M. ROSENBLUTH, A. TELLER, E . TELLER. 1953. Equation of state calculations by fast computing machines. Journal of Chemical Physics **21**, 1087-1092.

MICHALEWICZ Z. 1992. Genetic Algorithms + Data Structures = Evolution Programs. Springer Verlag, Berlin - Heidelberg.

WĘGLARZ J. 1982. Modelling and control of dynamic resource allocation project scheduling scheduling systems. In: S. G. TZAFESTAS (ed.), Optimization and Control of Dynamic Operational Research Models. North-Holland, Amsterdam.

14 SCHEDULING OF PROJECTS WITH STOCHASTIC EVOLUTION STRUCTURE

Klaus Neumann

Universität Karlsruhe, Germany

14.1 INTRODUCTION

The so-called classical project networks used by the network techniques CPM, PERT, and MPM, only allow for modelling projects whose evolution in time is uniquely specified in advance (cf. Elmaghraby 1977 and Moder et al. 1983). Here, each project activity is carried out exactly once during a single project execution and it is not possible to return to activities previously performed (that is, no feedback is permitted). Many practical projects, however, do not meet those conditions, for example, R&D projects and projects in production management where quality control is included and thus some feedback may occur.

To deal with more general projects with stochastic evolution structure, *stochastic project networks*, also called *GERT networks*, have been introduced. As compared to CPM and PERT networks, GERT networks use the activity-on-arc representation as well but possess more general arc weights (including the execution probabilities of the project activities), several different node types (containing stochastic elements), and cycles to represent feedback.

If limited resources are required for carrying out the activities of a stochastic project described by a GERT network and the resources are to be assigned to the activities such that a certain objective function (for example, the expected *project duration* or *makespan*) is minimized, we speak of *GERT (network) scheduling*. We always consider the resources to be machines. If there is a single resource of capacity m available, we have a GERT scheduling problem with m identical parallel machines, and if $m = 1$, we speak of a single-machine scheduling problem. When each project activity must be processed on several machines, we have a job-shop or flow-shop scheduling problem.

Whereas the temporal analysis of GERT networks was studied in the seventies and eighties in some detail (see, for example, Neumann and Steinhardt 1979 and Neumann 1984, 1990), GERT network scheduling has been discussed only recently. This chapter is devoted to methods for (approximately) solving *single-machine, parallel-machine, job-shop, and flow-shop scheduling problems with GERT network precedence constraints*. We assume that the reader is familiar with deterministic machine scheduling (cf. Brucker 1995, Pinedo 1995, and Blazewicz et al. 1996).

In Section 14.2, we summarize the most important concepts from the theory of GERT networks needed in what follows and from GERT network scheduling. In particular, we consider so-called *EOR networks* which represent an important class of special GERT networks. Moreover, we introduce the concept of an *action* whose performance corresponds to a certain execution of a project activity, and of an *action list schedule*, i.e. a permutation of the sequence of all actions.

Section 14.3 is concerned with single-machine GERT network scheduling. At first, we list some complexity results. Next, we present polynomial algorithms for scheduling problems with objective functions $E(\sum wC)$ and $\max E(L)$ and EOR network precedence constraints. Finally, we briefly sketch how to solve the general single-machine min-sum GERT scheduling problem by a dynamic programming approach.

In Section 14.4 we deal with GERT network scheduling problems with several identical parallel machines. We propose two types of heuristics for objective functions $E(C_{\max})$, $\max E(C)$ and $E(\sum C)$: generalizations of algorithms for corresponding deterministic scheduling problems and priority-rule methods.

Section 14.5 is devoted to job-shop and flow-shop scheduling with EOR network precedence constraints and objective function $E(C_{\max})$. Since different jobs or activities may have different processing routes, the concept of an action schedule does not make sense in job-shop scheduling. Instead, we introduce the concept of an *aggregate schedule*, which aggregates the sequences of jobs processed on the individual machines and their start times for all project realizations appropriately. The construction of an aggregate disjunctive graph permits us to apply the shifting bottleneck heuristic from deterministic job-shop scheduling. In addition, we briefly sketch a priority-rule heuristic of the Giffler-Thompson type. For flow-shop scheduling, both the concepts of an aggregate schedule and an action schedule are applicable.

14.2 BASIC CONCEPTS FROM GERT NETWORK THEORY AND GERT NETWORK SCHEDULING

14.2.1 GERT networks and EOR networks

We present a brief review of some basic material from the theory of GERT networks in an intuitive way (for more details we refer to Neumann 1990 and for practical applications to Neumann and Steinhardt 1979). A GERT network has exactly one source (which corresponds to the beginning event of the project occurring at time zero) and at least one sink (corresponding to terminal events of the project). A GERT network may contain cycles and thus some activities may be carried out several times during a single project execution.

Each arc $\langle k,l \rangle$ of a GERT network with initial node k and final node l is assigned a *weight vector* (p_{kl}, F_{kl}). $p_{kl} > 0$ is the *conditional execution probability* of the corresponding activity $\langle k,l \rangle$ given that project event k has occurred. F_{kl} is the *conditional distribution function* of the nonnegative *duration* D_{kl} of activity $\langle k,l \rangle$ given that activity $\langle k,l \rangle$ is carried out where $E(D_{kl}) < \infty$. p_{kl} and F_{kl} are assumed to be independent of how many times project event k has occurred or activity $\langle k,l \rangle$ has been carried out before, respectively.

GERT networks possess six different node types resulting from combination of three possible entrance sides and two exit sides of a node. We say that a node has an *AND entrance* if (during a project execution) the corresponding project event occurs (or, equivalently, the node *is activated*) when all incoming activities have been terminated for the first time. A node has an inclusive-or entrance or briefly *IOR entrance* if it is activated when one of the incoming activities has been terminated for the first time. A node has an exclusive-or entrance or briefly *EOR entrance* if it is activated every time when an incoming activity has been terminated. We speak of a *deterministic exit* of a node if all outgoing activities are carried out when the node has been activated. If exactly one outgoing activity is carried out when the node has been activated, we have a *stochastic exit*. A node with at most one predecessor is supposed to have an EOR entrance and a node with at most one successor is supposed to have a stochastic exit (if we speak of a predecessor or successor of a node in what follows, we always mean an immediate predecessor or successor, respectively).

Let Ω be the set of all possible realizations of the underlying GERT network or the corresponding project, respectively. Each *project* or *network realization* $\omega \in \Omega$ begins with the activation of the source at time zero. Subsequently, for each node k with stochastic exit which is activated, project realization ω specifies exactly one activity with initial node k to be carried out. Note that if node k is activated several times during project realization ω, different activations of node k may result in the execution of different outgoing activities. Moreover, ω specifies the realized duration of each activity execution.

To deal with GERT networks, several assumptions have to be satisfied. Since the precise formulation of these assumptions is rather complicated and not needed in what follows, we only loosely state some of them. For more details we refer to Neumann (1990). The most important assumption expresses some *Markov and independence properties* and says in particular that the durations of different activities and of different executions of one and the same activity are independent. Additional assumptions require that each node belonging to *a cycle structure* (i.e. a strong component containing at least two nodes) has EOR entrance and stochastic exit and that each cycle structure "is entered" at most once during any network realization.

A GERT network all of whose nodes have an EOR entrance is called an *EOR network*. An EOR network can be viewed as a superposition of several Markov renewal processes whose states are the nodes of the network (cf. Neumann 1990 and Bücker et al. 1994). A *node* of an EOR network is called *deterministic* or *stochastic* if it has a deterministic or stochastic exit, respectively. For EOR networks we assume that for each deterministic node k and any two distinct successors l_1 and l_2 of k, it holds that $\mathcal{R}(l_1) \cap \mathcal{R}(l_2) = \emptyset$, where $\mathcal{R}(l)$ is the set of nodes reachable from node l. A consequence of that assumption is some *tree-structure property* of an EOR network N, which says that the network N_ω corresponding to a single project realization ω where each activity is taken as often as it is carried out represents an outtree. An example is illustrated in Fig. 14:1, where project realization ω is indicated by bold arrows. Assume that if node 6 is activated for the first time, activity $\langle 6,4 \rangle$ is carried out, and if node 6 is activated for the second time, activity $\langle 6,8 \rangle$ is carried out. Fig. 14:2 shows the corresponding network N_ω.

Let q_k be the probability that node k of an EOR network is activated (during a single project execution). Exploiting the above Markov property, the *activation probabilities* q_k of the nodes k can be computed by solving a system of linear equations (cf. Neumann 1990).

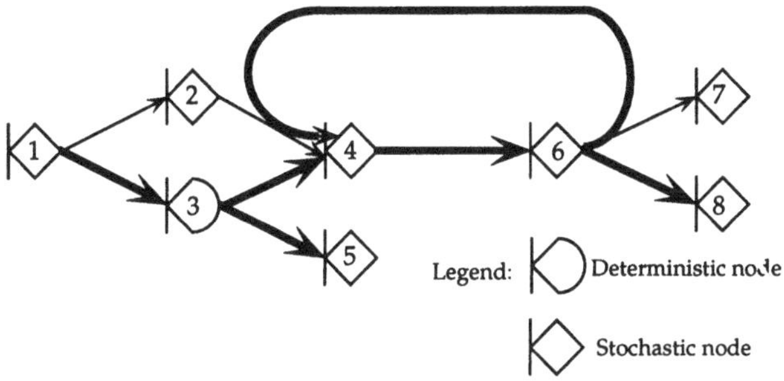

Figure 14:1 EOR network N

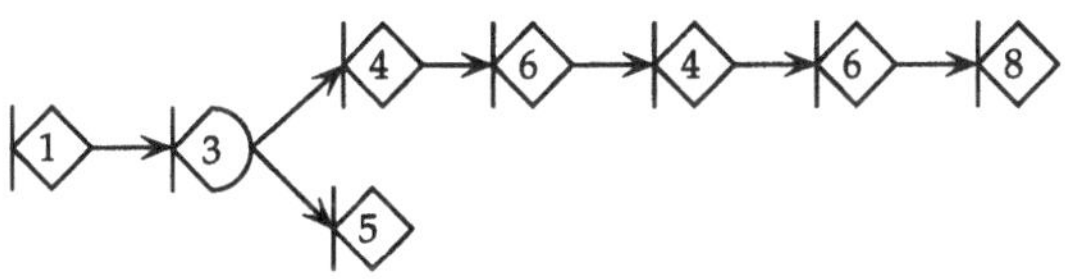

Figure 14:2 Network N_ω

14.2.2 Actions and action list schedules

We introduce the concept of an *action* in GERT network scheduling with one or several parallel machines, which plays the role of a job in deterministic machine scheduling. If more than one activity emanate from a node with stochastic exit, each of these activities is carried out with a probability less than one. To obtain deterministic schedules, we thus combine all activities which emanate from a node with stochastic exit (and whose conditional execution probabilities sum up to unity) in one stochastic action. In Fig. 14:3, the three bold activities emanating from node k with stochastic exit form one *stochastic action* (with *beginning node* k and three *terminal nodes*). Each activity emanating from a node with deterministic exit is carried out surely and represents a *deterministic action.* In Fig. 14:3, the dashed-line and dotted-line actions (both with beginning node l) are deterministic.

An action a is called a *predecessor* of an action a' (and a' is said to be a *successor* of a) if one of the terminal nodes of a coincides with the beginning node of a'. In Fig. 14:3, the bold and solid-line actions are predecessors of the dashed-line and dotted-line actions. An action a' is said to be *reachable* from action a if $a' = a$ or if the beginning node of a' is reachable from a terminal node of a.

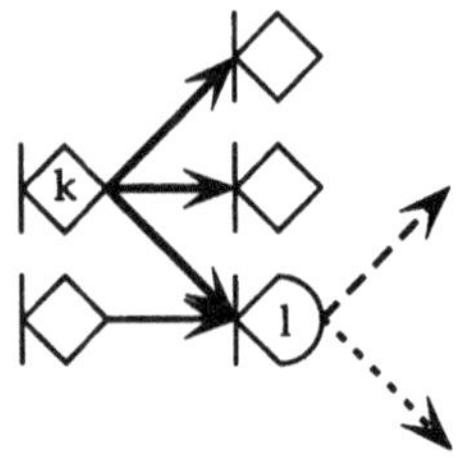

Figure 14:3 Stochastic and deterministic actions

Each *performance* of an action consists of the execution of that of its activities which is carried out during the project realization in question. Given a project realization, an action is said to be *performable* if its beginning node has been activated α times but the action itself has been performed only $\alpha - 1$ times ($\alpha \in \mathbb{N}$), in other words, the action is ready to be performed. The *expected duration* of an action a is $\sum_{<k,l>\in a} p_{kl} E(D_{kl})$.

Let $\mathcal{A} = \{a_1, \dots, a_r\}$ be the *set of all actions* of the underlying GERT network. Then each of the $r!$ permutations of the sequence $(a_1, \dots, a_r)$ represents an *action list schedule* or briefly *list schedule.* The interpretation of a list schedule Q is as follows: Each time a machine is freed, the performable action which has the foremost position in Q among all performable actions is selected for processing. An action list schedule Q is said to be an *action precedence schedule* or briefly *precedence schedule* if, for each project realization, the sequence of actions performed represents a subsequence of Q. Precedence schedules make only sense if the underlying GERT network is acyclic. For each project realization, the performance of any action corresponds to exactly one activity execution, and thus each list schedule specifies a unique schedule of activities to be carried out.

14.2.3 Objective functions for GERT network scheduling

For GERT network scheduling problems, the set of *feasible solutions* is always assumed to be the set of all list schedules if not stated otherwise. Only for the general single-machine min-sum scheduling problem (Section 14.3.3) and for job-shop and flow-shop scheduling (Section 14.5), scheduling policies different from list schedules will be considered.

As to *objective functions* for GERT scheduling problems (which are to be minimized), we distinguish between min-sum and min-max problems. Let $C^{\beta}_{kl}(Q, \omega)$ be the *completion time* of the βth execution of activity $\langle k, l \rangle$ during project realization ω given list schedule Q, where $C_{kl}^{\beta}(Q,\omega) := 0$ if activity $\langle k,l\rangle$ is carried out less than β times ($\beta \in \mathbb{N}$). If the GERT network is acyclic and thus each of its activities $\langle k,l\rangle$ can be carried out at most once, we omit the superscript β. For a *min-sum problem*, the objective function has the form

$$E(\sum f) = E\left[\sum_{<k,l>\in A}\sum_{\beta=1}^{\infty} f_{kl}(C_{kl}^{\beta}(Q))\right]$$

where A is the *set of arcs* of the GERT network and the functions f_{kl} are nondecreasing in C_{kl}^{β}. A special case of that objective function is the *expected weighted flow time* $E(\sum wC)$, where $f_{kl}(C_{kl}^{\beta}) = w_{kl}C_{kl}^{\beta}$ with $w_{kl} \geq 0$ being a weighting factor of activity $\langle k,l\rangle$. As to *min-max problems*, we only consider acyclic GERT networks and special objective functions. Let $\delta_{kl} \geq 0$ be a prescribed *due date* for activity $\langle k,l\rangle$. Then

$$\max E(L) = \max_{a\in\mathcal{A}} E(L_a(Q)) = \max_{a\in\mathcal{A}} \sum_{<k,l>\in a} p_{kl}[C_{kl}(Q) - \delta_{kl}]$$

is the *maximum expected lateness* of actions. Analogously, the *expected maximum lateness* $E(L_{\max})$ is defined. If all due dates are zero, the *lateness* L_a reduces to the *completion time* C_a of action a.

For GERT scheduling problems, we use the *three-field notation* known from deterministic machine scheduling (see Pinedo 1995 or Blazewicz et al. 1996). For example, the symbol $P2|acyclGERT, D \sim G|E(C_{\max})$ stands for a scheduling problem with two identical parallel machines, where the precedence constraints are given by an acyclic GERT network, "$D \sim G$" refers to general distribution of activity durations, and the *expected makespan* or *expected project duration* $E(C_{\max})$ is to be minimized. If we speak of the solution or the time complexity of a GERT scheduling problem in what follows, we always refer to the determination of an *optimal solution* (which minimizes the respective objective function on the set of feasible solutions).

14.3 SINGLE-MACHINE GERT NETWORK SCHEDULING

14.3.1 Some complexity results

We only list a few complexity results. For more details we refer to Bücker (1992) and Zimmermann (1999). First, we consider *min-sum problems*. $1|EOR, D \sim G|E(\sum f)$ and thus $1|GERT, D \sim G|E(\sum f)$ are strongly $\mathcal{NP}$-hard because even the deterministic problem $1||\sum f$ without precedence constraints is strongly $\mathcal{NP}$-hard (cf. Lenstra et al. 1977). $1|EOR, D \sim G|E(\sum wC)$ can be solved in polynomial time as we will see in Section 14.3.2. $1|acyclGERT, D = 1|E(\sum wC)$ and $1|acyclGERT, D \in \{0,1\}|E(\sum C)$, where all activity durations are equal to one or equal to zero or one, respectively, are strongly $\mathcal{NP}$-hard because the same holds for the deterministic problems $1|prec, D = 1|\sum wC$ and $1|prec, D \in \{0,1\}|\sum C$ (see Lenstra and Rinnooy Kan 1978). $1|acyclGERT, D = 1|\sum C$ is trivial because any list schedule is optimal.

Now we turn to ***min-max problems***. $1|acyclEOR, D = 1|\max E(C)$ is equivalent to a special $0-1$ knapsack problem and thus $\mathcal{NP}$-hard (cf. Neumann 1990 and Bücker 1992). $1|acyclGERT, D = 1|\max E(C)$ is strongly $\mathcal{NP}$-hard (see Zimmermann 1999). Problem $1|acyclEOR, D \sim G|\max E(L)$, however, can be solved in polynomial time if we restrict ourselves to precedence schedules as feasible solutions as we will see in Section 14.3.2. $1|GERT, D \sim G|E(C_{\max})$ is trivial because any feasible sequence of activity executions and thus any list schedule is optimal.

14.3.2 Polynomial algorithms for $1|EOR, D \sim G|E(\sum wC)$ *and* $1|acyclEOR, D \sim G|\max E(L)$

We briefly sketch two polynomial algorithms which provide an optimal list schedule for $1|EOR, D \sim G|E(\sum wC)$ and an optimal precedence schedule for $1|acyclEOR, D \sim G|\max E(L)$, respectively. For a detailed description of those procedures we refer to Neumann (1990) and Bücker et al. (1994).

The algorithm for $1|EOR, D \sim G|E(\sum wC)$ exploits the tree-structure property of an EOR network (cf. Section 14.2.1) and represents a generalization of *Horn's method* for the deterministic problem $1|outtree|\sum wC$ (cf. Horn 1972). Let

$$\tau_\lambda := \sum_{\langle k,l\rangle \in a_\lambda} p_{kl} E(D_{kl}) \quad \text{and} \quad v_\lambda := \sum_{\langle k,l\rangle \in a_\lambda} p_{kl} w_{kl}$$

be the expected duration and expected weight, respectively, of action a_λ, and let $\rho_\lambda := \tau_\lambda / v_\lambda$. Find an unscheduled action a_ν with minimum quotient ρ_ν. Then a_ν is inserted at the rear of the current list schedule Q (at the beginning Q is empty) and deleted from the set of unscheduled actions. Moreover, each pair (a_μ, a_ν) where a_μ is any predecessor of a_ν is replaced by a "composite action". The quantities belonging to a composite action, such as expected duration, expected weight, set of terminal nodes etc., can be computed by relatively simple updating formulas. This step is repeated until the set of unscheduled actions is empty. The algorithm can be implemented to run in $O(|A||V|^2)$

time, where V is the *set of nodes* and A is again the set of arcs of the *EOR* network. Notice that the quantities ρ_λ in the algorithm correspond to the so-called dynamic allocation indices or *Gittins' indices* (see Gittins 1989) because the (composite) actions a_λ are scheduled in nondecreasing order of the "priority indices" ρ_λ.

The algorithm for $1|acyclEOR, D \sim G|\max E(L)$ exploits *Lawler's rule* for solving the deterministic problem $1|prec|f_{\max}$, which says : From among all unscheduled jobs without successors, put that job last which will incur the smallest cost in that position.

An action $a \in \mathcal{A}' \subseteq \mathcal{A}$ is said to be a *final action* (with respect to $\mathcal{A}'$) if there is no successor of a in $\mathcal{A}'$. Sets of final actions are indicated by a subscript f. The last element of a precedence schedule is always a final action (this follows immediately from the definition of a precedence schedule). Then the following theorem is a generalization of Lawler's rule.

Theorem 1. Let $a^+ \in \mathcal{A}_f$ be a final action which minimizes the expected lateness, i.e. $E[L_{a^+}(Q_{a^+})] = \min_{a \in \mathcal{A}_f} E[L_a(Q_a)]$, where Q_a is a precedence schedule whose last element is the final action a. Then there is an optimal precedence schedule with last element a^+.

Theorem 1 can be exploited for determining an optimal precedence schedule backwards as follows: Let a^+ be a final unscheduled action with minimum expected lateness. Then a^+ is to occupy the last position among the unscheduled actions (i.e. insert a^+ at the front of the current precedence schedule). Successive application of this device eventually provides an optimal precedence schedule. This algorithm requires $O(|A||V|^2)$ time.

14.3.3 A dynamic programming approach to $1|GERT, D \sim G\,|E(\sum f)$

We sketch a dynamic programming approach to the general single-machine min-sum GERT scheduling problem $1|GERT, D \sim G|E(\sum f)$. For details we refer to Neumann (1990). Instead of list schedules, the feasible solutions now represent general dynamic scheduling policies. Loosely speaking, such a scheduling policy specifies which operation is to be performed next given the project evolution up to the present. To make that concept more precise, we first show how to describe the project evolution up to a specified project state (the so-called *history*) in a succinct form.

The activities of the project have to be carried out one after another because each activity must pass through the single machine available. Let $\langle\kappa\rangle$ denote the κth activity execution in turn ($\kappa = 1, 2, \ldots$), let C_κ be the completion time of activity execution $\langle\kappa\rangle$, and let M be the number of activity executions during a single project execution. Note that C_κ ($\kappa = 1, \ldots, M$) and M are random variables defined on the sample space Ω. We describe the sequence of activity executions $\langle 1\rangle, \dots, \langle\kappa\rangle$ during a project realization $\omega \in \Omega$ (and thus the history up to and including time $C_\kappa(\omega)$) by the following *feasible arc-node set* $X_\kappa(\omega)$: $X_\kappa(\omega)$ contains all activities outside cycles that have been carried out by time $C_\kappa(\omega)$. In addition, from all nodes that belong to any cycle structure and have been activated by time $C_\kappa(\omega), X_\kappa(\omega)$ contains that node in each cycle structure $\mathcal{C}$ which has been activated last provided that $\mathcal{C}$ has not been left.

Let $\mathcal{X} := \{X_\kappa(\omega) | 0 \leq \kappa \leq M(\omega), \omega \in \Omega\}$ be the set of all feasible arc-node sets of the GERT network, where $X_0(\omega) := \emptyset$. Moreover, for given $x \in \mathcal{X}$, let $\mathcal{A}(x)$ be the set of all performable actions. For $x \in \mathcal{X}$, a point in time $s \in \mathbb{R}_{\geq 0}$ is called a *completion time* of x if there exist a project realization ω and a $\kappa \in \{0, 1, \dots, M(\omega)\}$ such that $x = X_\kappa(\omega)$ and $s = C_\kappa(\omega)$. In other words, (x, s) is a realization of a pair of random variables (X_κ, C_κ). If the activity durations are stochastic or the GERT network contains cycles, there are in general several completion times for a given x. A pair (x, s) where $x \in \mathcal{X}$ and s is any completion time of x is then considered a *project state* and $\mathcal{X} \times \mathbb{R}_{\geq 0}$ is the *state space*. Moreover, a mapping $\varphi : \mathcal{X} \times \mathbb{R}_{\geq 0} \to \mathcal{A}$ which assigns a performable action $\varphi(x, s) \in \mathcal{A}(x)$ to each project state (x, s) represents a *scheduling policy*.

This well-known process of enlarging the original state space such that the histories constitute the new (extended) states leads to a Markov problem. Indeed, the selection of a scheduling policy φ induces a Markov renewal process $(X_\kappa, C_\kappa)^{\varphi}_{\kappa \in \mathbb{Z}_{\geq 0}}$, whose transition functions can be expressed in terms of the arc weights of the GERT network (see Neumann 1990). The determination of an optimal scheduling policy (which minimizes the objective function $E(\sum f)$) corresponds to a problem of stochastic dynamic programming, which can be solved by well-known methods such as value-iteration and policy-improvement techniques (cf. Neumann 1990).

14.4 PARALLEL MACHINE EOR NETWORK SCHEDULING

We consider scheduling problems with EOR network precedence constraints for several identical parallel machines. Such scheduling problems occur, for example, in mixed-model assembly lines (see Neumann and Zimmermann 1998).

All EOR scheduling problems for two or more identical parallel machines are $\mathcal{NP}$-hard (cf. Zimmermann 1999). Thus, we only propose heuristic procedures which compute list schedules in polynomial time, where we consider the objective functions $E(C_{\max})$, $\max E(C)$, and $E(\sum C)$. Most of those heuristics can be adapted to general GERT networks (see Zimmermann 1995). Recall that by Jensen's inequality, $\max E(C) \leq E(C_{\max})$. Hence, in principle, each heuristic for objective function $E(C_{\max})$ can be used for the corresponding scheduling problem with objective function $\max E(C)$ as well.

In this section, we sketch two types of heuristics, where we often follow Neumann and Zimmermann (1998) closely: procedures which result from modifying polynomial algorithms for corresponding deterministic scheduling problems and priority-rule methods. We restrict ourselves to acyclic EOR networks. However, all heuristics proposed can be adapted to cyclic EOR networks (see Neumann and Zimmerman 1998).

14.4.1 Modified deterministic procedures

Hu's level algorithm, which solves the deterministic problem $P|outtree, D = 1|\gamma$ with $\gamma = C_{\max}$ or $\sum C$ (see Hu 1961), can be generalized to the EOR scheduling problem $P|acyclEOR, D \sim G|\Gamma$ where $\Gamma = E(C_{\max})$, $\max E(C)$, or $E(\sum C)$ as follows. We define the *level* of an action a to be the maximum number of arcs on a path from a terminal node of a to a sink of the EOR network. Then put the actions in order of nonincreasing levels (highest level first rule or briefly *HL rule*). Actions with the same level are ordered according to nonincreasing expected durations (*LEPT rule*) for objective functions $E(C_{\max})$ or $\max E(C)$ and according to nondecreasing expected durations (*SEPT rule*) for objective function $E(\sum C)$. This algorithm requires $O(|A| \log |A|)$ time.

Fujii's algorithm, which solves the deterministic problem $P2|prec, D = 1|C_{\max}$ (see Fujii et al. 1969, 1971) can be generalized to problem $P2|acyclEOR, D \sim G|\Gamma$ where $\Gamma = E(C_{\max})$ or $\max E(C)$. Fujii's procedure for $P2|prec, D = 1|C_{\max}$ constructs a so-called *compatibility graph* G, whose nodes correspond to the jobs and where there is an edge $[j,l]$ with endnodes j and l exactly if jobs J_j and J_l can be carried out simultaneously. Then a maximum cardinality matching in G provides a maximum number of pairs of jobs that can be processed simultaneously on the two machines (such *job pairs* are also called *compatible*).

If the jobs are subject to deterministic precedence constraints, a job pair (J_j, J_l) is compatible exactly if job node l is not reachable from job node j and j is not reachable from l in the respective precedence graph. The latter statement does not hold any more if the precedence constraints are given by an EOR network. In Fig. 14:4, actions a_j and a_l cannot be performed simultaneously (in other words, the action pair (a_j, a_l) is not compatible) because only one of the two activities $\langle k,j \rangle$ and $\langle k,l \rangle$ emanating from stochastic node k is carried out during any project realization. To find compatible action pairs, we have to direct our attention to deterministic nodes, for example, node k in Fig. 14:5, and the actions $a_\mu, a_\nu, \ldots$ with beginning node k. If for any two actions a_μ and a_ν with deterministic beginning node k, an action a_j is reachable from

a_μ and not reachable from a_ν, and an action a_l is reachable from a_ν and not reachable from a_μ (see Fig. 14:5), then *action pair* (a_j, a_l) is *compatible*, and the compatibility graph G (whose nodes correspond to the actions) contains the edge $[a_j, a_l]$.

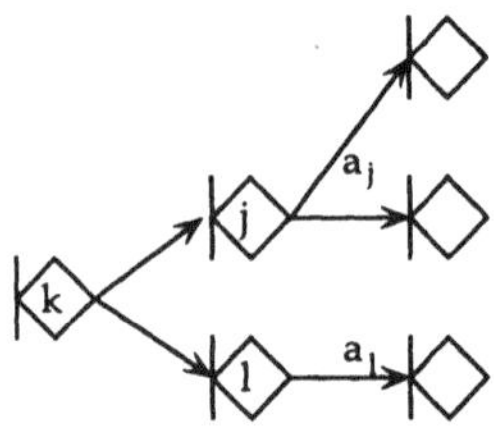

Figure 14:4 Incompatible action pair (a_j, a_l)

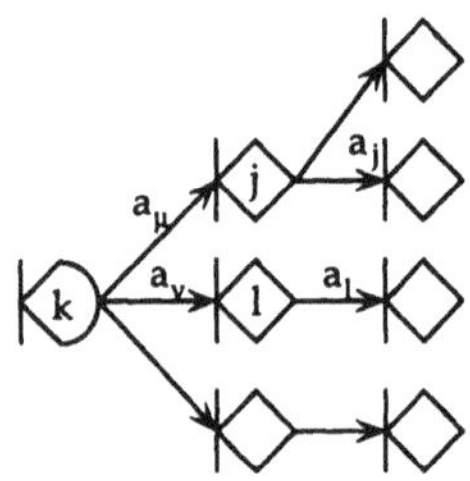

Figure 14:5 Compatible action pair (a_j, a_l)

Moreover, for EOR networks, it is not sufficient to find a maximum cardinality matching in the compatibility graph G because we have to take the performance probabilities pp_{jl} of compatible action pairs (a_j, a_l) into account, which may be less than one. The probability pp_{jl} that the two actions a_j and a_l of a compatible pair are performed simultaneously can be found by means of the activation probabilities q_k of the nodes k of the EOR network. We then weight each edge $[a_j, a_l]$ of graph G by the performance probability pp_{jl} of the compatible pair (a_j, a_l). In Section 14.4.2, we will consider modified edge weights, which additionally take account of the activity durations.

In the weighted graph G, we determine a matching M with largest sum of weights (*weighted matching problem*). Let M_a be the set of compatible action pairs which corresponds to the maximum weight matching M. In the last step of the generalized Fujii algorithm, the compatible action pairs from M_a and the single actions that do not belong to any pair from M_a have to be combined to form a list schedule taking the EOR network precedence constraints into consideration. This process may require some interchanges of actions between compatible pairs from M_a (see Neumann and Zimmermann 1998).

The generalized Fujii algorithm can be implemented to run in $O(|A|^2)$ time. If there are $m > 2$ identical parallel machines instead of two, we have to find (node-disjoint) maximum weight cliques with at most m nodes in the compatibility graph G instead of a maximum weight matching (for details we refer to Zimmermann 1995).

14.4.2 Priority-rule methods

The priority-rule methods are based upon the following general insights into the meaning of the three objective functions $E(C_{\max})$, $\max E(C)$, and $E(\sum C)$. Loosely speaking, minimizing $E(C_{\max})$ corresponds to minimizing idle time. To minimize $\max E(C)$, actions with large expected completion times (caused by the precedence constraints) should be scheduled as early as possible. To minimize $E(\sum C)$, actions with small expected durations should be scheduled first.

First, we consider *objective function* $E(C_{\max})$. To schedule the actions such that idle times of the machines are as small as possible, an action followed by a long sequence of other actions (with large expected durations) should be performed earlier than an action with relatively few successors. Moreover, an action with a large number of deterministic successors should be performed as early as possible (so that several decision alternatives are available as often as possible to avoid idle time).

To do so, we introduce the concepts of the sink distance and outdegree of an action. Let for node k,

$$d_k := \begin{cases} 0, & \text{if } k \text{ is a sink} \\ \sum_{l \in Succ(k)} p_{kl}(E(D_{kl}) + d_l), & \text{if } k \text{ is stochastic and no sink} \\ \max_{l \in Succ(k)} (E(D_{kl}) + d_l), & \text{if } k \text{ is deterministic} \end{cases}$$

where *Succ(k)* is the set of successors of node k. The quantities d_k can be computed backwards beginning with the sinks of the EOR network. We then define the *sink distance* s_a of action a with beginning node k to be equal to d_k. s_a can be regarded as the mean length of a longest path from the beginning node of action a to a sink of the network. Let

$$\delta_a^+ := \begin{cases} \sum_{\langle k,l \rangle \in a} p_{kl}\delta^+(l), & \text{if } a = \{\langle k,l \rangle | l \in Succ(k)\} \text{ is stochastic} \\ \delta^+(l), & \text{if } a = \{\langle k,l \rangle\} \text{ is deterministic} \end{cases}$$

where

$$\delta^+(l) := \begin{cases} 0, & \text{if } l \text{ is a sink} \\ 1, & \text{if } l \text{ is stochastic and no sink} \\ |Succ(l)|, & \text{if } l \text{ is deterministic} \end{cases}$$

δ_a^+ can be viewed as the average outdegree of a terminal node of action a and is called the *outdegree* of action a.

The *priority-rule method for $P|acyclEOR, D \sim G|E(C_{\max})$* then consists of ordering the actions according to nonincreasing sink distances (largest sink distance first rule or briefly *LSD rule*). Actions with the same sink distance are put in order of nonincreasing outdegrees (largest outdegree first rule or briefly *LOD rule*). Alternatively, we may first order the actions according to nonincreasing outdegrees and then put actions with the same outdegree in order of nonincreasing sink distances (called the *alternative priority-rule method*).

In the generalized Fujii algorithm (see Section 14.4.1) it is recommended to use the weights $pp_{jl}(s_{a_j} + s_{a_l})$ instead of pp_{jl} for the edges $[a_j, a_l]$ of the compatibility graph in order to take the expected activity durations into account. This is referred to as the *modified (generalized) Fujii algorithm* in what follows.

For *objective function* $\max E(C)$, actions with large completion times and thus actions with large performance probabilities and preceded by a long sequence of other actions (with large expected durations) should be scheduled as early as possible. For this, we introduce the concept of the source distance of an action in analogy to the concept of the sink distance. Let for node k,

$$\overline{d}_k := \begin{cases} 0, & \text{if } k \text{ is the source} \\ \min_{j \in Pred(k)} (E(D_{jk})) + \overline{d}_j), & \text{otherwise} \end{cases}$$

where *Pred(k)* is the set of predecessors of node k. The quantities $\overline{d}_k$ can be computed forwards beginning with the source of the EOR network. The *source distance* r_a of action a with beginning node k is then defined to be equal to $\overline{d}_k + \sum_{\langle k,l \rangle \in a} p_{kl} E(D_{kl})$. The source distance of action a can be viewed as the mean length of a shortest path from the source of the EOR network to the beginning node of action a plus the expected duration of action a.

Before scheduling an action with large performance probability and large source distance, other actions have to be performed due to the EOR network precedence constraints. The latter actions generally have large sink distances. Thus, we introduce the so-called *source-sink distance* $\sigma_a := r_a + s_a$ of action a. Moreover, we need the *performance probability* of an action a, which equals the activation probability of the beginning node of a.

The *priority-rule method for objective function* $\max E(C)$ then consists of ordering the actions according to nonincreasing performance probabilities (largest performance probability first rule or briefly *LPP rule*). Actions with the same performance probability are put in order of nonincreasing source-sink distances (largest source-sink distance first rule or briefly *LSSD rule*). Again, we may interchange the two priority rules.

For objective function $E(\sum C)$, the SEPT rule should be used. In addition, the outdegrees of the actions should be taken into account to reduce idle time. Thus, the ***priority-rule method for objective function*** $E(\sum C)$ orders the actions according to nondecreasing expected durations (*SEPT rule*). Actions with the same expected duration are put in order of nonincreasing outdegrees (*LOD rule*). Again, we may interchange the two priority rules.

The time complexity of the priority-rule methods is $O(|A| \log |A|)$ for objective functions $E(C_{\max})$ and $E(\sum C)$ and $O(|V|^3)$ for objective function $\max E(C)$.

14.4.3 Experimental performance analysis

An empirical analysis of the heuristics from Sections 14.4.1 and 14.4.2 has been performed by Zimmermann (1995) and summarized in Neumann and Zimmermann (1998). The main results for two-machine problems and acyclic EOR networks with 100 and 200 activities are as follows. For objective function $E(C_{\max})$, the generalized Hu algorithm and the respective alternative priority-rule method have turned out to be best. For objective function $\max E(C)$, the modified Fujii algorithm and the respective priority-rule method have outperformed the remaining heuristics. For objective function $E(\sum C)$, the generalized Hu algorithm and the respective priority-rule method have been superior to the remaining procedures.

Moreover, it has turned out that the accuracy of the approximate solutions for objective function $E(\sum C)$, i.e. the min-sum problem, is worse than for the two min-max problems. One of the reasons for that is that for all three objective functions considered, we have to find a list schedule which, loosely speaking, provides an optimal balance of the load on the two machines and thus minimizes idle times. For objective function $E(\sum C)$, in addition, the idle times have to occur as late as possible because if one action is performed too late, all following actions processed on the same machine are performed too late as well and all errors add up.

14.5 JOB-SHOP AND FLOW-SHOP EOR NETWORK SCHEDULING

In this section, we briefly study the job-shop problem $J|acyclEOR|E(C_{\max})$ with deterministic activity durations, where we often follow Neumann and Schneider (1998) closely. An application of that problem to mixed-model make-to-order production can be found in Neumann and Schneider (1997). For a more detailed discussion of job-shop EOR scheduling, where the case of cyclic EOR precedence constraints is additionally treated, we refer to Neumann and Schneider (1997, 1998). The corresponding flow-shop problem $F|acyclEOR|E(C_{\max})$ is touched upon at the end of this section. Both problems $J|acyclEOR|E(C_{\max})$ and $F|acyclEOR|E(C_{\max})$ are generalizations of the $\mathcal{NP}$-hard deterministic flow-shop problem $F2|outtree|C_{\max}$ (cf. Lenstra et al. 1977) and thus $\mathcal{NP}$-hard as well. Hence, we only present two heuristic procedures: a shifting bottleneck heuristic and a priority-rule method.

14.5.1 Basic concepts

We consider a job-shop problem with job set $\mathcal{J} = \{J_1, \dots, J_n\}$ and machine set $\mathcal{M} = \{M_1, \dots, M_m\}$. Job J_j consists of m operations $o_{i_1 j}, \dots, o_{i_m j}$ which have to be processed in this order on the machines $M_{i_1}, \dots, M_{i_m}$, respectively, where $(i_1, \dots, i_m)$ is some given permutation of $(1, \dots, m)$. The following investigations also hold for the cases where some jobs are not processed on all machines $M_1, \dots, M_m$ and where some jobs are processed on a machine more than once. Let $D_{ij} > 0$ be the (deterministic) ***processing time*** or ***duration*** of operation o_{ij}. Then $D_j := \sum_{i=1}^{m} D_{ij}$ is the processing time or duration of job J_j.

We assume that the jobs are subject to precedence constraints given by an acyclic EOR network, where each job J_j corresponds to exactly one activity or arc $\langle k_j, l_j \rangle$ with ***duration*** D_j and ***conditional execution probability*** p_j. Node k_j is called the ***initial node*** and l_j is the ***final node*** of job J_j. The ***execution probability*** of job J_j is $\pi_j = q_{k_j} p_j$ where q_{k_j} is again the activation probability of k_j. We introduce a strict order $\prec$ in the job set $\mathcal{J}$ by defining $J_j \prec J_{j'}$ exactly if initial node $k_{j'}$ of job $J_{j'}$ is reachable from final node l_j of J_j ($J_j, J_{j'} \in \mathcal{J}$). The strict order $\prec$ in job set $\mathcal{J}$ induces a corresponding strict order in the set $\mathcal{O}$ of all operations (where the operation sequences of the jobs have to be observed in addition), which is also denoted by $\prec$. Jobs which correspond to arcs emanating from one and the same deterministic node can be performed simultaneously and thus compete with each other for the machines (***job competition***). Jobs which correspond to arcs emanating from one and the same stochastic node are performed exclusively, i.e. there is ***no job competition***.

Since in job-shop scheduling, different jobs generally have different processing orders through the machines, the concepts of actions and action (list) schedules cannot be used any more (for a detailed discussion we refer to Schneider 1997). Instead, we introduce a different concept of a deterministic static scheduling policy Φ with the property that for each network realization, we obtain the sequences of jobs processed on the individual machines and their start times in a unique way. Since such a scheduling policy aggregates, in a sense, schedules for all individual network realizations, we speak of an aggregate schedule. More precisely, a mapping $\Phi : \mathcal{O} \to \mathbb{R}_{\geq 0}$ is called an *aggregate schedule* if it is

(a) precedence-consistent, that is, for all $o_{ij}, o_{i'j'} \in \mathcal{O}$ with $ij \neq i'j'$, $\Phi(o_{ij}) + D_{ij} \leq \Phi(o_{i'j'})$ exactly if $o_{ij} \prec o_{i'j'}$

(b) feasible, that is, for all $J_j, J_{j'} \in \mathcal{J}, j \neq j'$, and all $M_i \in \mathcal{M}$, $\Phi(o_{ij'}) \notin [\Phi(o_{ij}), \Phi(o_{ij}) + D_{ij}[$

(c) semiactive (that is, no local left-shift of any operation can be performed).

Let Φ be an aggregate schedule and $\mathcal{J}_\omega \subseteq \mathcal{J}$ and $\mathcal{O}_\omega \subseteq \mathcal{O}$ be the sets of jobs and operations, respectively, carried out in network realization ω. Then a mapping $\Phi_\omega : \mathcal{O}_\omega \to \mathbb{R}_{\geq 0}$ is called a *network realization schedule belonging to* Φ if it is feasible, semiactive, and has the property that for all $J_j, J_{j'} \in \mathcal{J}_\omega, j \neq j'$, and all $M_i \in \mathcal{M}, \Phi_\omega(o_{ij}) \leq \Phi_\omega(o_{ij'})$ exactly if $\Phi(o_{ij}) \leq \Phi(o_{ij'})$. We remark that for given Φ, Φ_ω is uniquely determined for each network realization $\omega \in \Omega$. $\Phi_\omega(o)$ represents the start time of operation $o \in \mathcal{O}_\omega$ in realization ω. Since Φ_ω is assumed to be semiactive, some local left-shifts of operations have to be performed in general to obtain Φ_ω from Φ, after deleting all operations $o \in \mathcal{O} \setminus \mathcal{O}_\omega$.

Table 14:1 Operation sequences and durations

Job (Operation sequence): Durations D_{ij}	Job (Operation sequence): Durations D_{ij}
$J_1\ (o_{21}, o_{11}) : D_{21} = 2, D_{11} = 1$	$J_3\ (o_{13}, o_{23}) : D_{13} = 2, D_{23} = 2$
$J_2\ (o_{12}, o_{22}) : D_{12} = 1, D_{22} = 1$	$J_4\ (o_{24}, o_{14}) : D_{24} = 1, D_{14} = 1$

As an *example*, we consider a job-shop problem with four jobs and two machines where the precedence constraints are given by the EOR network N in Fig. 14:6. The order in which the operations have to be done on the jobs correspond to the order of the operations in Table 14:1, which also contains the operation durations. An aggregate schedule Φ for this problem is given by the Gantt chart in Fig. 14:7, where idle times are indicated by shaded areas. There are two possible realizations of N. In realization ω_1 the jobs J_1, J_2, and J_3 are performed, in realization ω_2 the jobs J_1, J_2, and J_4 are performed. The network realization schedules Φ_{ω_1} and Φ_{ω_2} belonging to Φ are depicted in Figs. 14:8 and 14:9, respectively.

Let $P(\omega)$ be the probability that network realization ω occurs, which is equal to the product of the conditional execution probabilities p_j of all jobs $J_j \in \mathcal{J}_\omega$.

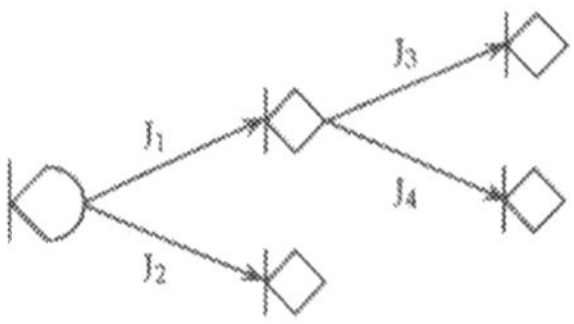

Figure 14:6 EOR network N

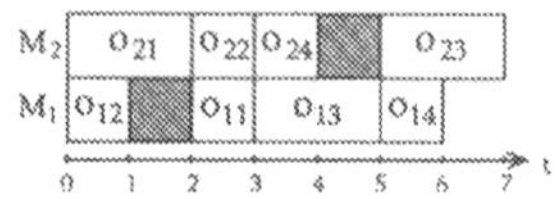

Figure 14:7 Aggregate schedule Φ

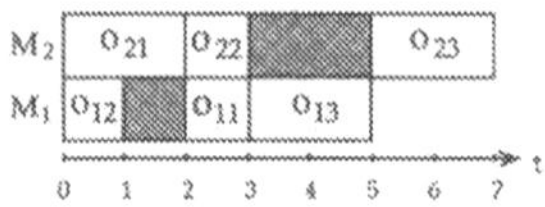

Figure 14:8 Network realization schedule Φ_{ω_1}

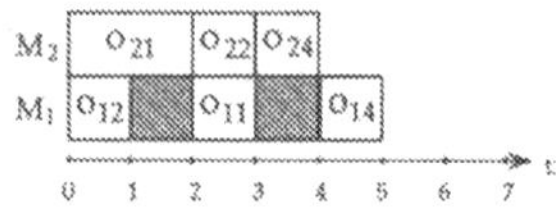

Figure 14:9 Network realization schedule Φ_{ω_2}

Let $C_j(\Phi_\omega)$ be the completion time of job $J_j \in \mathcal{J}_\omega$ in network realization ω and $C_{\max}(\Phi_\omega)$ be the makespan when schedule Φ_ω is applied. Then the *expected makespan* when aggregate schedule Φ is applied is

$$E[C_{\max}(\Phi)] := \sum_{\omega\in\Omega} P(\omega) C_{\max}(\Phi_\omega) = \sum_{\omega\in\Omega} P(\omega) \max_{J_j\in\mathcal{J}_\omega} C_j(\Phi_\omega) \tag{1}$$

Let $\mathcal{S}$ be the set of all aggregate schedules. Then $\Phi^* \in \mathcal{S}$ is called an *optimal aggregate schedule* if its expected makespan is minimum: $E[C_{\max}(\Phi^*)] = \max_{\Phi\in\mathcal{S}} E[C_{\max}(\Phi)]$. Let Ψ_ω^* be an optimal schedule for the deterministic job-shop scheduling problem corresponding to network realization ω. Then

$$E[C_{\max}(\Phi^*)] \geq \sum_{\omega\in\Omega} P(\omega) C_{\max}(\Psi_\omega^*) =: LB \tag{2}$$

where LB stands for lower bound on the minimum objective function value for aggregate schedules. If the EOR network contains only stochastic nodes, there is only one schedule Ψ_ω for each network realization ω and $=$ instead of $\geq$ holds in (2).

14.5.2 Shifting bottleneck procedure

The shifting bottleneck procedure is based upon the disjunctive-graph concept for job-shop scheduling. Let N_ω be that subnetwork of the underlying acyclic EOR network N which belongs to network realization ω. Then the (directed) *disjunctive graph* G_ω for the deterministic job-shop problem $J|N_\omega|C_{\max}$ with deterministic precedence constraints given by network N_ω and objective function $C_{\max}$ is constructed as follows (for details we refer to Neumann and Schneider 1997, 1998). Each operation $o \in \mathcal{O}_\omega$ is assigned a node. In addition,

G_ω contains a source α and a sink β representing the fictitious initial and final operations, respectively, of the problem. So-called *conjunctive arcs* represent the precedence relationships between the operations given by the strict order $\prec$. If two different jobs J_j and $J_{j'}$ are to be processed on the same machine M_i, we introduce a pair of opposite *disjunctive arcs* between the two nodes corresponding to operations o_{ij} and $o_{ij'}$. We consider the *example* from Section 14.5.1 (see Fig. 14:6 and Table 14:1), where network N_{ω_1} corresponding to realization ω_1 is depicted in Fig. 14:10. The associated disjunctive graph G_{ω_1} is shown in Fig. 14:11, where conjunctive and disjunctive arcs are indicated by solid and broken arrows, respectively.

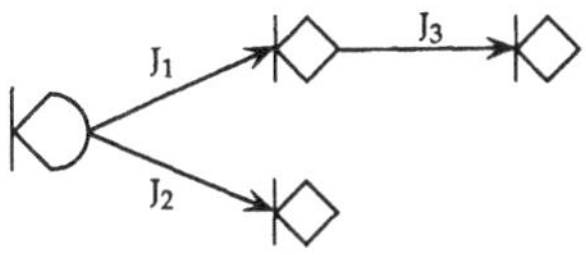

Figure 14:10 Network N_{ω_1}

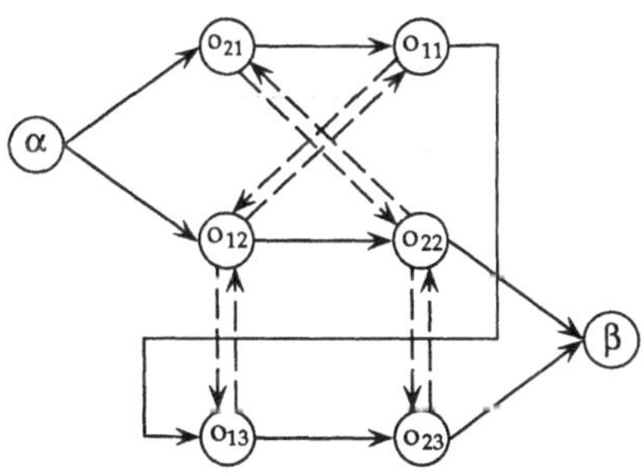

Figure 14:11 Disjunctive graph G_{ω_1}

A subset F of the arc set of the disjunctive graph G_ω is called a *feasible selection* if F contains all conjunctive arcs and one disjunctive arc from each pair of opposite arcs such that the induced directed graph $G_\omega(F)$ is acyclic. The weight of an arc of $G_\omega(F)$ which emanates from operation node o_{ij} equals the duration D_{ij}. Each feasible selection F corresponds to a feasible schedule Ψ_ω for problem $J|N_\omega|C_{max}$, and the length $L[G_\omega(F)]$ of a longest path in $G_\omega(F)$ from source α to sink β is equal to the makespan $C_{\max}(\Psi_\omega)$. Moreover, if $\mathcal{F}$ is the set of all feasible selections and Ψ^*_ω is an optimal schedule for $J|N_\omega|C_{\max}$, then

$$\min_{F\in\mathcal{F}} L[G_\omega(F)] = C_{\max}(\Psi^*_\omega) \tag{3}$$

An optimal schedule Ψ_ω^* can be computed by any branch-and-bound algorithm based upon the disjunctive-graph concept, for example, the algorithms of Carlier and Pinson (1989) or Brucker (1995). To determine an approximate (feasible) schedule Ψ_ω, the well-known shifting bottleneck heuristic can be used (cf. Adams et al. 1988 or Pinedo 1995). For small instances of the stochastic job-shop problem $J|acyclEOR|E(C_{\max})$, the networks N_ω for all realizations $\omega \in \Omega$ can be generated and corresponding optimal schedules Ψ_ω^* can be computed. In addition, the lower bound LB on the minimum objective function value for aggregate schedules defined in (2) can be determined.

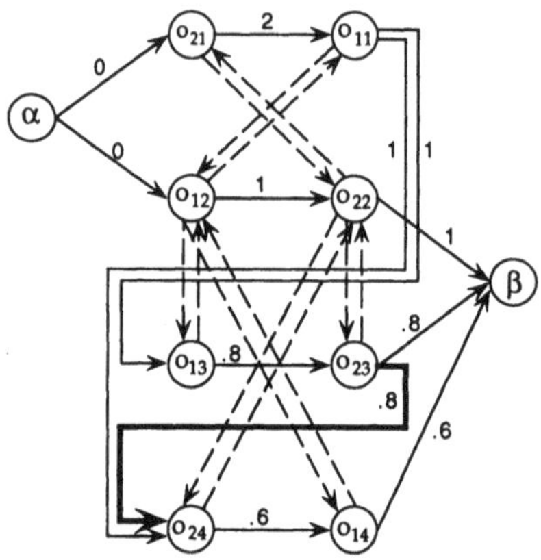

Figure 14:12 Aggregate disjunctive graph G for EOR network N of Fig. 14:6

For the stochastic problem $J|acyclEOR|E(C_{\max})$, we construct an *aggregate disjunctive graph* G which, in a sense, aggregates the disjunctive graphs G_ω for all network realizations $\omega \in \Omega$. G contains all nodes and arcs of the disjunctive graphs G_ω ($\omega \in \Omega$). Moreover, different jobs $J_{j_1}, \ldots, J_{j_r}$ emanating from one and the same stochastic node in the EOR network N, which do not compete for the machines, are linked by introducing so-called *stochastic conjunctive arcs* in G from the last operation of J_{j_ν} to the first operation of $J_{j_{\nu+1}}$ ($\nu = 1, \ldots, r-1$). To take into account that job J_j is executed with probability π_j, the weight of an arc in G emanating from operation node o_{ij} is to be equal to $\pi_j D_{ij}$. We again consider the *example* from Section 14.5.1 (see Fig. 14:6 and Table 14:1), where we additionally assume that $p_3 = \pi_3 = 0.4$ and $p_4 = \pi_4 = 0.6$. The corresponding aggregate disjunctive graph G is depicted in Fig. 14:12, where the weights of the disjunctive arcs are omitted. The pairs of opposite disjunctive arcs result from the competition of job J_2 with jobs J_1, J_3, and J_4 for the machines. Since jobs J_1 and J_3 on the one hand and jobs J_1 and J_4 on the other hand can be carried out consecutively, there are "ordinary" conjunctive arcs $\langle o_{11}, o_{13}\rangle$ and $\langle o_{11}, o_{24}\rangle$. The stochastic node with outgoing jobs J_3 and J_4 in N leads to the stochastic conjunctive arc $\langle o_{23}, o_{24}\rangle$ in G indicated by a bold arrow.

In general, the introduction of stochastic conjunctive arcs to link jobs beyond a stochastic node is neither easy nor unique and may lead to cycles in G all of whose arcs are conjunctive, which does not make sense. Schneider (1997) has proposed an algorithm which constructs an ***admissible aggregate disjunctive graph*** G for each acyclic EOR network N, that is, an aggregate disjunctive graph which is acyclic with respect to conjunctive arcs and where for jobs belonging to a subnetwork of N with only stochastic nodes, there is a sequence of conjunctive arcs in G representing all those jobs.

If the EOR network N contains only stochastic nodes, there are no disjunctive arcs in any corresponding aggregate disjunctive graph G. For each aggregate schedule Φ, the length of the only path in any admissible G from source α to sink β is $L(G) = \sum_{i=1}^{m} \sum_{j=1}^{n} \pi_j D_{ij} = E[C_{\max}(\Phi)]$. If N contains stochastic and deterministic nodes, there is no counterpart to relation (3). Let G be an admissible aggregate disjunctive graph for $N, G(F)$ be the subgraph of G whose arc set is a feasible selection F, and $L[G(F)]$ be the length of a longest path in $G(F)$ from source α to sink β. Then $\min_{F\in\mathcal{F}} L[G(F)]$ can be less or greater than $E[C_{\max}(\Phi^*)]$ where Φ^* is again an optimal aggregate schedule (see Schneider 1997). However, due to the introduction of stochastic conjunctive arcs and the specification of the arc weights in G, the deviation of $L[G(F)]$ with F determined by the shifting bottleneck heuristic from lower bound LB is relatively small in general. F provides a sequence of operations on each machine. Observing conditions (a), (b), and (c) in the definition of an aggregate schedule (cf. Section 14.5.1), such a schedule can then be found.

14.5.3 Priority-rule method and experimental performance analysis

To adapt the well-known priority-rule heuristic by Giffler and Thompson for $J||C_{\max}$ (see Giffler and Thompson 1960 or Blazewicz et al. 1996) to problem $J|acyclEOR|E(C_{\max})$, we replace each node of the EOR network by a node with AND entrance and deterministic exit. The stochastic character of the EOR precedence constraints is taken into account by appropriate priority rules.

In each step of the Giffler-Thompson algorithm , the set $\tilde{\mathcal{O}}$ of all schedulable operations (i.e. operations o such that all $o' \prec o$ have already been scheduled) is determined. Let $C_{\mu\nu}$ be the minimum completion time of all operations from set $\tilde{\mathcal{O}}$. Then an operation $o_{\mu j*}$ from the conflict set $\mathcal{O}^* := \{o_{\mu j} \in \tilde{\mathcal{O}} | S_{\mu j} < C_{\mu\nu}\}$, where $S_{\mu j}$ is the start time of $o_{\mu j}$, is selected for scheduling by some priority rule and $\tilde{\mathcal{O}}$ is updated. Several priority rules have been investigated, for example, *SEPT* (shortest expected processing time), *MEWR* (most expected work remaining), and *FCFS* (first come first served). The MEWR rule, which has turned out to be best in an experimental performance analysis (see Neumann and Schneider 1997, 1998), is as follows: Choose operation $o_{\mu j*} \in \mathcal{O}^*$ such that

$$\pi_{j*} \sum_{M_i \in \mathcal{M}_{\mu j*}} D_{ij*} + \sum_{J_{j'} \in \mathcal{J}_{j*}} \pi_{j'} D_{j'} = \min \left\{ \pi_j \sum_{M_i \in \mathcal{M}_{\mu j}} D_{ij} + \sum_{J_{j'} \in \mathcal{J}_j} \pi_{j'} D_{j'} | o_{\mu j} \in \mathcal{O}^* \right\}$$

where $\mathcal{M}_{ij}$ is the set of machines on which job J_j has to be processed after its processing on M_i including M_i itself and $\mathcal{J}_j$ is the set of jobs $J_{j'}$ such that $J_j \prec J_{j'}$.

The experimental performance analysis in Neumann and Schneider (1997, 1998) has additionally shown that, on the average, the shifting bottleneck procedure provides very good results for small-sized problem instances (i.e. $m \leq 10, n \leq 20$), whereas the Giffler-Thompson heuristic (with MEWR rule) yields more accurate results for large-sized instances. Furthermore, the Giffler-Thompson heuristic requires much less computing time than the shifting bottleneck procedure.

14.5.4 Flow-shop EOR scheduling

In the flow-shop problem $F|acyclEOR|E(C_{\max})$, all jobs have the same processing order through the machines, where we may assume that the machine sequence for each job is $M_1, M_2, \ldots, M_m$. This permits us to adapt the concept of actions known from single- and parallel-machine scheduling (see Section 14.2.2) to flow-shop scheduling as follows. Let k be a stochastic node with outgoing jobs $J_{j_1}, \ldots, J_{j_r}$. Then each set $\{o_{ij_1}, \ldots, o_{ij_r}\}$ ($i = 1, \ldots, m$) is said to be a *stochastic action* with *beginning node* k and (expected) *duration* $\sum_{\nu=1}^{r} p_{j_\nu} D_{ij_\nu}$. If k is a deterministic node with outgoing jobs $J_{j_1}, \ldots, J_{j_r}$, then each set $\{o_{ij_\nu}\}$ ($i = 1, \ldots, m; \nu = 1, \ldots, r$) is said to be a *deterministic action* with *beginning node* k and *duration* D_{ij_ν}. In any network realization, an action is performed (i.e. exactly one of its operations is carried out) when its beginning node has been activated.

Let $\mathcal{A}$ be again the set of all actions of the EOR network. In addition to the sequence of actions on each machine, an action schedule for problem $F|acyclEOR|E(C_{\max})$ should specify the start times of the actions. An *action schedule* $\Upsilon : \mathcal{A} \to \mathbb{R}_{\geq 0}$ is then defined in analogy to the definition of an aggregate schedule in Section 14.5.1, where operations and their durations are replaced by actions and their durations. The concept of a *network realization schedule* Υ_ω *belonging to* Υ (where each action is replaced by that of its operations which is carried out in realization ω), the *expected makespan* $E[C_{\max}(\Upsilon)]$

when action schedule Υ is applied (compare (1)), and the concept of an *optimal action schedule* Υ^* can be introduced as well. Analogously, the shifting bottleneck heuristic and the priority-rule method by Giffler and Thompson can be used for computing an approximate action schedule. The following theorem has been proved by Schneider (1997).

Theorem 2. For $F|acyclEOR|E(C_{\max})$, aggregate schedules dominate action schedules, i.e. $E[C_{\max}(\Phi^*)] \leq E[C_{\max}(\Upsilon^*)]$.

Acknowledgement

This research has been supported in part by the Deutsche Forschungsgemeinschaft (Grant Ne 137/4).

References

ADAMS, J., E. BALAS, AND D. ZAWACK. 1988. The Shifting Bottleneck Procedure for Job Shop Scheduling. *Management Science* **34**, 391-401.

BLAZEWICZ, J., K.H. ECKER, E. PESCH, G. SCHMIDT, AND J. WEGLARZ. 1996. *Scheduling Computer and Manufacturing Processes.* Springer, Berlin.

BRUCKER, P. 1995. *Scheduling Algorithms.* Springer, Berlin.

BÜCKER, M. 1992. Time Complexity of Single Machine Scheduling with Stochastic Precedence Constraints. *ZOR–Mathematical Methods of Operations Research* **36**, 211-225.

BÜCKER, M., K. NEUMANN, AND T. RUBACH. 1994. Algorithms for Single-Machine Scheduling with Stochastic Outtree Precedence Relations to Minimize Expected Weighted Flowtime or Maximum Expected Lateness. *ZOR–Mathematical Methods of Operations Research* **39**, 321-348.

CARLIER, J. AND E. PINSON. 1989. An Algorithm for Solving the Job Shop Problem. *Management Science* **35**, 164-176.

ELMAGHRABY, S.E. 1977. *Activity Networks: Project Planning and Control by Network Models.* John Wiley and Sons, New York.

FUJII, M., T. KASAMI, AND K. NONOMIYA. 1969, 1971. Optimal Sequencing of Two Equivalent Processors. *SIAM Journal of Applied Mathematics* **17**, 784-789, Erratum **20**, 141.

GIFFLER, B. AND G.L. THOMPSON. 1960. Algorithms for Solving Production-Scheduling Problems. *Operations Research* **8**, 487-503.

GITTINS, J.C. 1989. *Multi-Armed Bandit Allocation Indices.* John Wiley and Sons, New York.

HORN, W.A. 1972. Single-Machine Job Sequencing with Treelike Precedence Ordering and Linear Delay Penalties. *SIAM Journal of Applied Mathematics* **23**, 189-202.

HU, T.C. 1961. Parallel Sequencing and Assembly Line Problems. *Operations Research* **9**, 841-848.

LAWLER, E.L. 1973. Optimal Sequencing of a Single Machine Subject to Precedence Constraints. *Management Science* **19**, 544-546.

LENSTRA, J.K. AND A.H.G. RINNOOY KAN. 1978. Complexity of Scheduling under Precedence Constraints. *Operations Research* **26**, 22-35.

LENSTRA, J.K., A.H.G. RINNOOY KAN, AND P. BRUCKER. 1977. Complexity of Machine Scheduling Problems. *Annals of Discrete Mathematics* **1**, 343-362.

MODER, J.J., C.R. PHILLIPS, AND E.W. DAVIS. 1983. *Project Management with CPM, PERT and Project Diagramming.* Van Nostrand Reinhold, New York.

NEUMANN, K. 1984. Recent Developments in Stochastic Activity Networks. *Canadian Journal of Operational Research and Information Processing* **22**, 219-248.

NEUMANN, K. 1990. *Stochastic Project Networks.* Lecture Notes in Economics and Mathematical Systems, Vol. 344. Springer, Berlin.

NEUMANN, K. AND W.G. SCHNEIDER. 1997. Job-Shop and Flow-Shop Scheduling with OR Network Precedence Constraints: Structural Questions and Heuristic Procedures. *Technical Report WIOR-498*, Institut für Wirtschaftstheorie und Operations Research, University of Karlsruhe.

NEUMANN, K. AND W.G. SCHNEIDER. 1998. Heuristic Algorithms for Job-Shop Scheduling Problems with Stochastic Precedence Constraints. Submitted to *Annals of Operations Research.*

NEUMANN, K. AND U. STEINHARDT. 1979. *GERT Networks and the Time-Oriented Evaluation of Projects.* Lecture Notes in Economics and Mathematical Systems, Vol. 172. Springer, Berlin.

NEUMANN, K. AND J. ZIMMERMANN. 1998. Heuristic Procedures for Parallel-Machine Scheduling Problems with Stochastic Precedence Constraints. *Annals of Operations Research*, to appear.

PINEDO, M. 1995. *Scheduling: Theory, Algorithms, and Systems.* Prentice Hall, Englewood Cliffs.

SCHNEIDER, W.G. 1997. *Job Shop Scheduling with Stochastic Precedence Constraints.* Shaker, Aachen.

ZIMMERMANN, J. 1995. Mehrmaschinen-Schedulingprobleme mit GERT-Anordnungsbeziehungen. *Ph.D. Dissertation*, University of Karlsruhe.

ZIMMERMANN, J. 1999. Time Complexity of Single- and Identical Parallel-Machine Scheduling with GERT Network Precedence Constraints. *Mathematical Methods of Operations Research*, to appear.

15 PROJECT SCHEDULING WITH STOCHASTIC ACTIVITY INTERRUPTIONS

Vicente Valls[1]
Manuel Laguna[2]
Pilar Lino[3]
Angeles Pérez[3]
Sacramento Quintanilla[3]

1. Dpto. de Estadística e Investigación Operativa
Universitat de Valencia, Valencia, Spain
2. Graduate School of Business and Administration
University of Colorado, Boulder, USA
3. Dpto. de Economía Financiera y Matemática
Universitat de Valencia, Valencia, Spain

15.1. Introduction

In this chapter we address the problem of scheduling the activities of a resource-constrained project, some of which may be interrupted by an uncertain amount of time. The resources may be, for example, machines in a jobshop, computers with specialized software packages (as those needed for engineering designs), or highly specialized technicians.

We consider that a project consists of a set of activities $A = DA \cup SA$. The DA subset consists of the *deterministic* activities, i.e., those with known duration. The SA subset consists of the *stochastic* activities, i.e., those that are interrupted for an uncertain amount of time. For an activity $i \in SA$, the initial processing time is considered to be known, however, the length of the interruption and the final processing time are uncertain.

An example of such a situation may be a project in which some activities are submitted to an approval process before they can be completed. The time to review and approve the work performed during the initial processing determines the length of the interruption, while the outcome of the approval process may determine the length of the final processing. A purchasing activity, for instance, may require an initial round of negotiations, followed by an approval process. If the initial contract is approved, then the final processing may be shorter than when more negotiating is necessary. In a quality control environment, the interruption may be due to an off-line testing process. If the outcome of the testing is positive, the activity may terminate (reflected by a zero processing time in the final processing).

The duration t_i of an activity $i \in DA$ is simply the processing time of such activity, while the duration of an activity $i \in SA$ is given by:

$$t_i = t_{i1} + w_i + t_{i2},$$

where t_{i1} is the initial processing time, w_i is the waiting time during the interruption, and t_{i2} is the processing time after the interruption. In settings where w_i represents the time needed for an approval process to be completed, the final processing time t_{i2} may depend on the outcome of this secondary process. In this chapter, however, we focus on the case where t_{i2} does not depend on the outcome of the process that originated the interruption. This is the case, for example, when the interruption corresponds to subcontracted work that has an uncertain lead-time. In this situation the final processing time may or may not be certain, but in either case it would not depend on the outcome of the interruption.

In a context where resources are not limited, the interruption can be considered part of the activity's processing time. That is, the activity duration t_i becomes a random variable of the form:

$$t_i = f(w_i, t_{i2}).$$

PERT networks are a special case of this formulation. When resources are limited, however, it may not be convenient to ignore the waiting time caused by the interruption, because resources become available during that time.

In this chapter, we consider the problem of scheduling the set of activities A in order to minimize the expected total weighted tardiness. Since the duration of an activity in SA is uncertain, the scheduling problem is stochastic in nature. In particular, first the scheduler must decide how to schedule the deterministic activities DA and the initial processing of the stochastic activities SA (i.e., the part of the activity that is processed before the interruption occurs). Then, as the uncertainty resolves, the scheduler must adjust the schedule according to the interruptions.

As a way of illustrating the preceding ideas, consider the following example. A project consists of 8 activities (A to H), 3 of which (C, E, and F) must be interrupted for an undetermined amount of time (i.e., C, E, and F $\in SA$). The activities consume resources of two kinds, R_1 and R_2, for which 2 and 3 units are respectively available. The processing times, resource consumption, and precedence relationships are given in Table 15:1.

Table 15:1. Data for illustrative example

Activity	*t*	R_1	R_2	*Immediate predecessor*
A	3	1	0	-
B	4	0	2	A
C		1	2	A
D	2	2	0	B, C
E		1	0	B
F		0	2	D
G	4	2	2	D
H	6	1	1	E, F, G

Suppose that the values for the random variables t_2 and w associated with activities C, E, and F are observed. These values are given in Table 15:2.

Table 15:2. Activity duration for *SA*

Activity	*t*	t_1	t_2	*w*
C	6	1	2	3
E	5	3	0	2
F	8	4	1	3

Using the *t*-values for all activities, the earliest starting times (ES) and latest finishing times (LF) are found (see Figure 15:1).

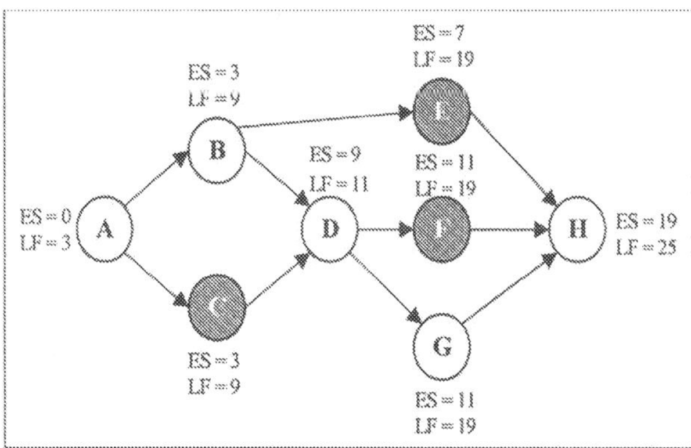

Figure 15:1. Activity network for illustrative example.

Due to the resource constraints, activities B and C cannot be simultaneously performed. (Usage of R_2 would be 4 and there are only three units of this resource available.) If C is considered an activity without interruptions, as we have done so far, then the resource infeasibility must be resolved by delaying either B or C beyond their latest finish time of 9. In fact, the earliest that these activities could finish would be at time 13 (because they would have to be done one after the other). The Gantt charts in Figure 15:2 illustrate this situation.

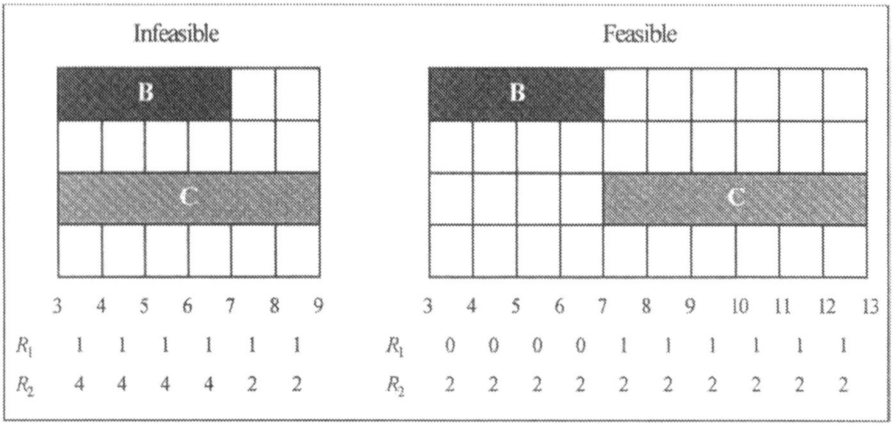

Figure 15:2. Gantt charts ignoring the interruption of activity C.

It becomes apparent from Figure 15:2 that potentially better solutions can be obtained by taking advantage of the interruption associated with activity C. In particular, activities B and C could be finished at time 10, if activity B were scheduled immediately after the interruption of activity C begins. The final processing for activity C then must wait for one time period, in order for the necessary resources to become available (after B is completed), as illustrated in Figure 15:3.

This example illustrates that the policy used to schedule deterministic activities and initial processing of stochastic activities has a significant impact on the quality of the solutions that can be obtained once the uncertainty resolves.

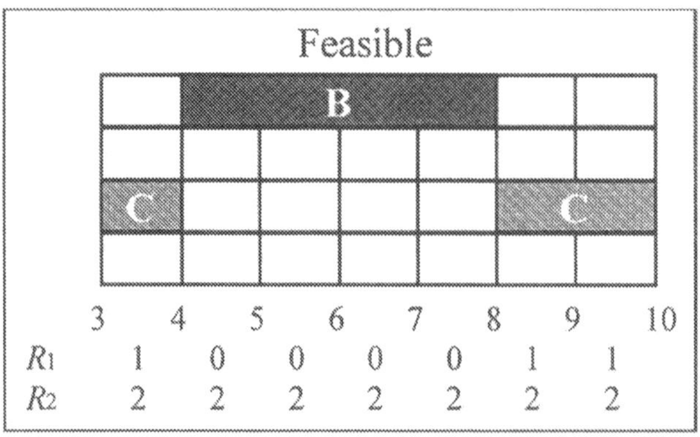

Figure 15:3. Gantt chart when B is initiated during the interruption of C

15.2. Stochastic optimization

The fundamental idea behind stochastic optimization is the concept of recourse. Recourse is the ability to take corrective action after a random event has taken place (i.e., after the uncertainty has been resolved). A simple example of two-stage recourse is the following (NEOS Guide):

- Choose the values for some variables, x, to control what happens today.
- Overnight, a random event happens.
- Tomorrow, take some recourse action, y, to correct what may have gotten messed up by the random event.

In our setting, we deal with a series of two-stage decisions, where in the first stage all the available activities are scheduled, and in the second stage we have the recourse of scheduling additional activities taking into consideration the length of the interruption (i.e., the random event that occurs between the first and the second stage). Consider the example in Table 15:1 (and Figure 15:1). Initially, a scheduler assigns resources and starting times only to activities A, B and the first part of C. (Since activity B is deterministic, the scheduler could also initially schedule activity E, however, this decision may be ultimately affected by the interruption of activity C.) This would be the first stage of the process. The scheduler then waits for the interruption of C to end and takes corrective actions. So in the second stage, the scheduler would schedule D and G and the first parts of E and F. (If E was scheduled during the first stage, then the schedule is revised as needed according to the outcome of the interruption of C.) The recourse, in this case, is the ability to adapt the schedule during the second stage taking into consideration the events that took place during the transition from the first stage to the second.

A common method to characterize uncertainty in stochastic optimization models is to generate a set of scenarios and assign a probability to each of them. In stochastic programming, the scenarios are used to create the so-called deterministic equivalent problem. One important characteristic of these formulations is that the first-stage decision, x, is independent of which second-stage scenario actually occurs. This is called the *nonanticipativity* property. Since the future is considered uncertain, today's decision cannot take advantage of knowledge of the future. In two-stage problems, this property is somewhat simple to enforce. This is customarily done by creating a single set of decisions for the first stage. In other words, there is only one set of x variables regardless of the scenario. Recourse can be introduced by allowing a different set of values for the y variables, depending on the scenario. Therefore a deterministic equivalent formulation of a two-stage stochastic optimization problem consists of a set of x variables and $|S|$ sets of y_s variables (one set for each scenario $s \in S$).

Enforcing the nonanticipativity property in a multistage setting is a more difficult problem. The decision at each stage must be based only on the information known at the time of the decision stage (the state of knowledge at that time). The state of knowledge is dynamic over time and must be determined at each decision stage. The requirement that decisions be based on the state of knowledge is imposed by the presence of nonanticipativity constraints. Solutions that violate these constraints may provide misleading results, since they rely on assumed information (information the project manager may not have known at the time of each decision).

When scenarios are used to model a multistage problem, the decisions made from stage 1 to stage t must be the same for all scenarios that are identical up to stage t. This is an added complexity that can be ignored when dealing with two-stage problems.

Stochastic programming (SP) deals with objective functions based on an expected value calculation. A generalization of this approach has been proposed with the name of *robust optimization* (RO) (Mulvey, et al. 1995). The distinction between RO and SP is important, for each has its own domain of applicability. Applied to personnel planning, for example, a SP solution will design a workforce that can be adjusted (by hiring and layoffs) to meet demand at the least cost. The important consideration of maintaining stability of employment cannot be captured.

The RO model, on the other hand, will design a workforce that will need little adjustment in order to cope with demand for all scenarios. However, the resulting cost will be higher than the cost of the SP solution, since the latter does not balance factors applicable to multiple scenarios. The importance of controlling variability of the solution (as opposed to just optimizing an average response) is, of course, well recognized. The RO framework allows decision-makers to address this issue directly.

Another important distinction between RO and SP is the handling of the constraints in the optimization model. Stochastic programs aim at finding values for the design variables x such that for each scenario a control variable setting (i.e., a value for y_s) is possible that satisfies the constraints. In cases where no feasible solution is possible for every scenario, the SP model is simply declared infeasible. RO, however, explicitly allows this contingency. In engineering applications (e.g., image restoration) the impossibility of global feasibility inevitably arises due to measurement errors. Multiple measurements of the same quantity may be inconsistent with each other. Hence, even if the underlying physical system has a solution (in case an image does exist!) it will not satisfy all the measurements. The RO model, through the use of error terms and the penalty function, will find a solution that violates the constraints by the least amount. The need to make recourse to such a rationale is fairly standard in medical imaging.

The RO framework is very general and Mulvey, et al. (1995) illustrate how this approach includes not only stochastic programming but also other scenario-based techniques such as *scenario optimization* (Dembo, 1991) and *scenario aggregation* (Rockafeller and Wets, 1991). In view of these distinctions, we first propose a model that can be characterized as a stochastic optimization formulation with an objective function makes reference to expected values only. We then perform experiments with a mean-variance model that extends the first one into the robust optimization domain. In our approach, however, a schedule can always be obtained regardless of the scenario, and therefore there is no need to explicitly create a mechanism to handle scenario infeasibility. These points will be further clarified in the Problem Definition section.

15.3. Stochastic project scheduling

The scheduling literature is extensive in regard to the deterministic version of the resource-constrained project scheduling problem (RCPSP). Numerous exact and heuristic procedures have been designed for the RCPSP, dating back to the 1960s. The stochastic version of this problem, however, has been largely ignored. The complexity of modern projects, in particular those related to R&D, makes necessary the development of procedures that can deal with uncertainty in key data (e.g., activity duration).

In a recent publication, Golenko-Ginzburg and Gonik (1997) develop a heuristic procedure for the RCPSP with stochastic activity times, with the objective of minimizing expected project duration. The procedure operates in stages where a decision is made to schedule the next activity based on the precedence constraints and current resource availability. If several activities are fighting for limited resources, a multiple knapsack problem is solved to select the next activities to be scheduled. The objective function of the knapsack formulation represents the maximum total contribution of the selected activities towards the expected project duration. The individual contribution is calculated as the product of the average activity duration and the probability of being in the critical path during the course of the project. The probability values are approximated by frequency values found via simulation. The rational for the design is to give critical activities priority, because of their significant impact on the expected project duration.

Another recent publication dealing with the RCPSP with stochastic activity times is due to Fernandez and Armacost (1996). This article warns users of off-the-shelf scheduling software of the danger of omitting nonanticipativity constraints. Specifically, they mention that Monte Carlo based approaches, such as those in Opera® and @RISK for Project, do not incorporate nonanticipativity constraints and therefore the output may be misleading. The primary output for these software packages generally is the empirical distribution of project duration. Since the procedures in these packages solve each scenario separately without enforcing nonanticipativity constraints, they arrive to non-implementable solutions (i.e., solutions that are based on information that is not available to the decision-maker at the time the decision was made). Fernandez (1995) has extended the nonanticipativity constraints in Escudero, et. al (1993) to the RCPSP with stochastic activity duration. The work by Escudero et. al was done in the context of a multi-period stochastic production planning problem.

Although some publications have dealt with the RCPSP with stochastic activity duration, to the best of our knowledge, the RCPSP with stochastic activity interruptions has never been addressed in the literature. We now proceed to formally define the problem on hand.

15.4. Problem definition

The characterization of the resource-constrained scheduling problem with stochastic activity interruptions that we will address in the remainder of this chapter can be completed as follows:

1. The project can be represented by a directed graph $G = (A, E)$, where A is the set of n activities and E is the set of precedence relationships. These relationships are of the finish-start type with a zero parameter value (i.e., FS = 0). In other words, activity i precedes activity j, iff j cannot start until i has been completed.

2. There are $|R|$ renewable resource types. The availability of each resource type k in each time period is R_k units, for $k = 1, \ldots, |R|$.
3. Each activity i has a due date d_i and tardiness penalty λ_i.
4. Each activity i requires r_{ij} units of resource j for its completion.
5. The two parts of the stochastic activities SA require the same number of units from each resource, although the specific units used for each part do not need to be exactly the same.
6. Both the deterministic activities and each part of the stochastic activities are performed without interruption.
7. The processing time of the second part of a stochastic activity is independent of the length of the interruption.

We assume that the uncertainty in this problem can be captured by a set of scenarios. These scenarios are generated by specifying three time estimates both for the interruption and for the second part of each stochastic activity. The estimates consist of the minimum time, the most likely time and the maximum time. The total number of scenarios is $9^{|SA|}$. (As noted in the Computational Experiments section, we use a sampling procedure to select scenarios to evaluate solutions during the search.)

From the point of view of the scheduler, there are three different situations that lead to a new decision stage:

1. The beginning of the project.
2. After the completion of an activity or the completion of the first part of a stochastic activity.
3. The end of an interruption.

At these stages, the scheduler must consider the available resources, the activities yet to be scheduled and the precedence relationships in order to make scheduling decisions. We assume that the scheduler makes decisions based on a priority system that determines the next activities to be scheduled considering the resources and the precedence relationships. This assumption transforms the problem into a two-stage decision problem, where in the first stage a priority is assigned to each activity and in the second stage, once the uncertainty about the interruptions is resolved, a schedule is constructed using these priorities.

Therefore, a solution to the problem consists of determining the priority for each activity. It is assumed that the scheduler will use the so-called parallel scheduling scheme. Within this scheme, for any priority ordering that does not meet the precedence relationships, there is an equivalent that does meet these relationships. Hence, we only consider topological orderings of G.

A topological ordering (TO) is a function f that maps the vertices in G to the set $\{1, 2, \ldots, n\}$ in such a way that if (i, j) is an arc in G, then $f(i) < f(j)$. A topological ordering can be represented by a vector π of size n, where $\pi(i) = j$ indicates that activity i occupies the jth position and n is the number of activities in the project.

In our solution procedure we seek an ordering π^* such that the expected value of the total weighted tardiness is minimized. In mathematical terms, our objective is:

$$\text{Minimize } E(\pi) = \sum_{s \in S} p_s T_s .$$

Where p_s is the probability associated with scenario s and T_s represents the total weighted tardiness for scenario s, i.e.,

$$T_s(\pi) = \sum_{i \in A} \lambda_i Max\{c_{is} - d_i, 0\},$$

and c_{is} is the completion time of activity i under scenario s. Note that since the completion time of each activity depends on the ordering π, then T_s and the expected tardiness E also depend on π.

Although the weighted tardiness criterion is considered relevant in many practical applications, the literature has been mostly devoted to other objective functions, such as makespan minimization. Mohring (1984) is among the few publications that deal with the measure of tardiness in project management. In this paper, the author considers the problem of minimizing the resource capacity cost in order to meet specified deadlines. Dumond and Mabert (1988) investigate rules of setting project due dates for continuously arriving projects. They evaluate due date assignment policies using simple dispatch heuristics. One criteria used to evaluate due date assignment policies was total (unweighted) tardiness. It should also be noted that the weighted tardiness scheduling problem is known to be NP-complete (Ullman, 1975).

15.5. Scatter search

Our solution procedure is based on the scatter search methodology. In broad terms, the process consists of creating a population of orderings π that „evolves," according to the characteristics of the particular problem instance being solved. This „evolution" continues until a stopping criterion is satisfied yielding the best ordering π_{best}. (Note that in some instances π_{best} may be equal to π^*, but the heuristic nature of the procedure does not guarantee this outcome.)

In this section, we will provide a general description of the scatter search methodology, a more detailed description can be found in Glover and Laguna (1997). Scatter search is a methodology that operates on a population of solution vectors. In our context, we can define this population as the set $POP = \{ \pi_k : k = 1, \ldots, PopSize \}$, where $PopSize$ is the population size. The scatter search process is organized to (1) capture information not contained separately in the original vectors, (2) take advantage of auxiliary heuristic solution methods (to evaluate the combinations produced and to actively generate new vectors), and (3) make dedicated use of strategy instead of randomization to carry out component steps.

Figure 15:4 sketches the scatter search approach in its original form. Extensions are typically created that additionally take advantage of the memory-based designs of tabu search (see Glover and Laguna, 1997). Two particular features of the scatter search proposal deserve mention. The use of clustering strategies is suggested for selecting subsets of points in Step 2, which allows different blends of intensification and diversification by generating new points „within clusters" and „across clusters." Also, solutions selected in Step 3 as starting points for re-applying heuristic processes are not required to be feasible, since heuristics proposed to accompany scatter search include those capable of starting from an infeasible solution. In the current context, structured combinations may be created that are not TOs. These ordering are therefore infeasible but can be readily mapped into feasible TOs.

In sum, scatter search is founded on the following premises.

1. Useful information about the form or location of optimal solutions is typically contained in a (sufficiently diverse) collection of elite solutions. Useful information may also be contained in bad solutions. However, such solutions are usually much more numerous and varied than good ones, and consequently there is less advantage in trying to make use of them. On the other hand, valuable information can be contained in *trajectories* from bad solutions to good solutions (or from good solutions to other good solutions), and such trajectory-based information is one of the elements that tabu search seeks to exploit.

2. When solutions are combined as a strategy for exploiting such information, it is important to provide for combinations that can extrapolate beyond the regions spanned by the solutions considered, and further to incorporate heuristic processes to map combined solutions into new points.

3. Taking account of multiple solutions simultaneously, as a foundation for creating combinations, enhances the opportunity to exploit information contained in the union of elite solutions.

The outline in Figure 15:4 reveals significant differences between classical genetic algorithms and scatter search. While classical genetic algorithms heavily rely on randomization and a somewhat limiting operation to create new solutions (e.g., one-point crossover on binary strings), scatter search employs strategic choices and memory along with structured combinations of solutions to create new solutions. Scatter search explicitly encourage the use of additional heuristics to process selected reference points in search for improved solutions. This is particularly advantageous in settings where heuristics that exploit the problem structure can either be developed or are already available.

1. Apply heuristic processes to generate a starting set of solution vectors (trial points). Designate a subset of the best vectors to be reference points. (Subsequent iterations of this step, transferring from Step 3 below, incorporate advanced starting solutions and best solutions from previous history as candidates for the reference points.)

2. Form structured combinations of subsets of the current reference points to create new points.

3. Extract a collection of the best points generated in Step 2 to be used as starting points for a new application of the heuristic processes of Step 1. Repeat these steps until reaching a specified iteration limit.

Figure 15:4. An outline of the scatter search procedure.

As mentioned in the previous section, our scheduling problem has been conceptualized as a two-stage stochastic optimization problem. The use of scenarios to model the uncertainty compounds the difficulty of the problem, given the combinatorial nature of the deterministic version of the resource-constrained scheduling problem. This was the main motivation for exploring the development of a procedure based on the scatter search technique. We provide details of our implementation in the following section.

15.6. Solution procedure

Our procedure consists of a hybrid algorithm based on scatter search principles, tabu search mechanisms and more traditional operations research techniques. The approach embeds tabu search memory functions to create an interplay between search intensification and diversification. The procedure searches for improved solutions by generating priority orderings and evaluating the quality of the ordering by scheduling the activities under each scenario.

The scenario sub-problems are therefore deterministic and consist of creating the best activity schedule considering the interruption and the final processing times associated with each scenario. In summary, the procedure mimics the two-stage decision making process, where in the second stage, the scheduler has the recourse of adjusting the schedule according to the resolution of the uncertainty. However, the scheduler must obey the priorities set during the first stage, in order to avoid violating the nonanticipativity constraint. Note that although the scenario sub-problems are deterministic, these problems are still computationally hard.

15.6.1. Initial population

The initial population of solutions is generated with the goal of introducing quality and diversity. We measure quality with reference to the objective function value. The scheduling literature is not conclusive in terms of identifying a single priority rule that dominates all others in every situation. In fact, it is customary in scheduling to design procedures that apply a number of rules (both static and dynamic) to the same problem and choose the best result as the outcome. We follow this trend by applying 6 dispatching rules designed for resource constrained multi-project scheduling with tardy costs by Lawrence and Morton (1993). We adapted these rules to the single-project case with activity tardiness penalties. We also use 5 weighted and unweighted versions of well-known sequencing rules. The rules are defined and described in Appendix A. Each rule originates three topological orderings by defining the activity time for each stochastic activity to be:

1. $t_i = t_{i1} + \underset{s \in S}{Min}\{w_{is}\} + \underset{s \in S}{Min}\{t_{i2s}\}$

2. $t_i = t_{i1} + \underset{s \in S}{Max}\{w_{is}\} + \underset{s \in S}{Max}\{t_{i2s}\}$

3. $t_i = t_{i1} + \frac{\sum_{s \in S} w_{is}}{|S|} + \frac{\sum_{s \in S} t_{i2s}}{|S|}$

The activities in the associated RCPSP are scheduled with a parallel scheme following the dispatching rules. The resulting sequencing order becomes a candidate for inclusion in the initial population. The initial population consists of the twenty candidates with the best evaluation.

For ease of implementation, in many cases researchers induce an unknown level of diversification in the initial population by randomly generating solutions. These solutions tend to be of inferior quality and the added diversification cannot be controlled. Our procedure attempts to add a more measurable level of diversification, following the notion of a systematic „Diversification Generator," as proposed in Glover (1997). A diversification generator produces a collection of diverse trial solutions, using an arbitrary seed solution as input. It is worth mentioning that the tabu search methodology does not perceive diversification as being the same as randomization. The goal of diversification is to produce solutions that differ from each other in significant ways, and that yield productive (or „interesting") alternatives in the context of the problem considered. By contrast, the goal of randomization is to produce solutions that may differ from each other in any way (or to any degree) at all, as long as the differences are entirely „unsystematic". From the tabu search viewpoint, a reliance on variation that is strategically generated can offer advantages over a reliance on variation that is distinguished only by its unpredictability (Glover, 1997).

The use of the 11 rules as a basis for creating an initial population achieves the goal of quality and diversity. These rules produce high-quality results in the deterministic setting and because they are guided by different criteria they provide diverse solutions. In preliminary testing we confirmed that an initial population of random sequences represents a poor starting point for our procedure.

15.6.2. Solution evaluation

To evaluate the quality of a solution (i.e., a topological ordering π), the procedure creates a schedule for each scenario as follows:

1. At each time instant τ (initially $\tau = 0$), when an activity has been completed or an interruption has finished, the set $ELG = ELG_1 \cup ELG_2 \cup ELG_3$ of candidate (eligible) activities is constructed, where

   ELG_1 = Deterministic activities whose predecessors have been completed.

   ELG_2 = Second parts of stochastic activities whose interruption has finished.

   ELG_3 = First parts of stochastic activities whose predecessors have been completed.

2. The activities in *ELG* are scheduled according to π considering resource availability. When resources do not permit to schedule the top activity according to π, then the next activity in the ordering is considered.

3. The total tardiness value T_s for each scenario is found. The objective function $E(\pi)$ is then calculated considering the probability of each scenario.

15.6.3. Structured combinations

Scatter search is designed to maintain or identify a set of *reference points* as a basis for creating structured combinations. In our context, a structured combination of *RefSize* topological orderings is defined as follows:

1. Let the set of reference points be $REF = \{ \pi_1, \pi_2, \ldots, \pi_{RefSize} \}$.

2. For each activity i, calculate $\pi'(i) = \pi_1(i) + \pi_2(i) + \ldots + \pi_k(i)$.
3. Then, $\pi' = \{ \pi'(1), \ldots, \pi'(n) \}$ is a structured combination of the reference points. (Note that π' must be transformed into a topological ordering, breaking ties arbitrarily if needed.)

During the intensification phase of the procedure, the new ordering π' replaces the worst currently in the population, only when the quality of π' is superior to the quality of the worst (where quality is measured in terms of the objective function value).

15.6.4. Summary of the procedure

Having described the main components of our procedure in the previous sections, we now provide a pseudo-code summary. Figure 15:5 outlines the procedure at its highest level.

```
1.  Iter1 = Iter2 = 0.
2.  POP = Initial population.
3.  π_globalbest = Best π ∈ POP
4.  While (Iter1 < MaxIter1 and E(π_globalbest) > 0) {
    4.1. While (Iter2 < MaxIter2) {
         4.1.1. π_currentbest = INTENSIFICATION.
         4.1.2. π_currentbest = DIVERSIFICATION(π_currentbest).
         4.1.3. If (E(π_currentbest) < E(π_globalbest)) {
                4.1.3.1. π_globalbest = π_currentbest.
                4.1.3.2. Iter1 = Iter2 = 0.
                    } else {
                4.1.3.3. Iter2 = Iter2 + 1.
                    }
         }
    4.2. π_currentbest = DIVERSIFICATION(π_globalbest).
    4.3. Iter2 = 0.
    4.4. If (E(π_currentbest) < E(π_globalbest)) {
         4.4.1. π_globalbest = π_currentbest.
         4.4.2. Iter1 = 0.
              } else {
         4.4.3. Iter1 = Iter1 + 1.
              }
    }
```

Figure 15:5. Global procedure outline.

The values of *MaxIter*1 and *MaxIter*2 in Figure 15:5 are search parameters that control the termination criterion. The INTENSIFICATION procedure is outlined in Figure 15:6 and the DIVERSIFICATION process is outlined in Figure 15:7. Note that the diversification in step 4.1.2 is applied to the current best solution ($\pi_{currentbest}$) while the diversification in step 4.2 is applied to the global best solution ($\pi_{globalbest}$).

In Figure 15:6, the set of reference points, *REF*, is given a tabu-active classification for a fixed number of iterations. This is usually called the *tabu tenure*, and it is treated as an adjustable search parameter.

If the set of all solutions, *POP*, is updated by the deletion of π_{worst}, the tabu-active classification is dropped from all the *REF* sets that contain π_{worst}. The elements of *REF* during the current iteration *Iter* are probabilistically selected. The probability of belonging to *REF* proportionally increases with the solution quality.

The intensification phase introduces two additional parameters: *RefSize* and *MaxInt*. *RefSize* refers to the size of the set of reference points used to create a structured combination. *MaxInt* is the maximum number of intensification iterations that the procedure is allowed to perform without improving the current best solution ($\pi_{currentbest}$). Note that the current best is not necessarily the best solution found during the search (i.e., it is possible that $\pi_{currentbest} = \pi_{globalbest}$). This is due to the application of the diversification phase, which essentially replaces the current population with a new one. In other words, the solution labeled $\pi_{currentbest}$ is the best since the last application of the diversification phase. This is why we must distinguish between the current best solution ($\pi_{currentbest}$) and the overall best ($\pi_{globalbest}$) in figures 15:5 and 15:6.

```
1. Initialization.
   1.1.  POP = Current population of π's.
   1.2.  πcurrentbest = Best π ∈ POP.
   1.3.  πworst = Worst π ∈ POP.
   1.4.  Iter = 0.
2. While (Iter < MaxInt and E(πcurrentbest) > 0) {
   2.1.  Select REF ⊂ POP of size RefSize, such that REF is not tabu-active.
   2.2.  Make REF tabu active.
   2.3.  Use REF to create a structured combination π'.
   2.4.  If (E(π') < E(πworst) and π' ∉ POP) {
         2.4.1.  Update POP.
         2.4.2.  Update tabu-active status.
         2.4.3.  Update πworst.
         2.4.4.  Iter = 0.
         2.4.5.  If (E(π') < E(πcurrentbest)) πcurrentbest = π'.
         }
   }
```

Figure 15:6. INTENSIFICATION outline.

Figure 15:7 shows the outline of the diversification phase. Note that this phase of the search introduces an additional adjustable parameter, *BlockNum*. Also note that *PopSize* has the same value used during the creation of the initial population. Our diversification procedure relies somewhat on randomization, but it does so within a controlled environment. The goal of this phase is to induce diversification by perturbing solutions in the current population. The process emphasizes systematic diversification over pure randomization to distance itself from the so-called mutation in GA implementations.

The diversification mechanism is better explained by introducing additional notation. Let $\gamma(j)$ be the activity in position j in the ordering π. Therefore, if $\gamma(j) = i$ then $\pi(i) = j$, or $\pi(\gamma(j)) = j$. Hence, for each ordering π there is a corresponding vector γ. The process starts by dividing the γ-vector into *BlockNum* blocks. Starting with the first block, an activity $\gamma(j)$ within the current block is randomly selected. If $\gamma(j-1)$ is not an immediate predecessor of $\gamma(j)$, then we swap their values (i.e., $\gamma(j) \leftrightarrow \gamma(j-1)$). The effect of the swap is to advance activity $\gamma(j)$, so that in the new ordering this activity would have a higher priority. The process continues until the activity cannot be advanced any further. Then a new activity is selected from the following block. This is done until all blocks are considered.

```
1. POP = ∅.
2. Derive γ from π.
3. For (i = 1, ..., PopSize) {
   3.1. For (k = 1, ..., BlockNum) {
        3.1.1. Randomly choose j, such that j is within the k^th block of
               γ.
        3.1.2. Advance γ(j) as much as possible.
        }
   3.2. Create π_i from γ.
   3.3. POP = POP ∪ {π_i }.
   3.4. If (E(π_i) < E(π_currentbest)) π_currentbest = π_i.
   }
```

Figure 15:7. DIVERSIFICATION(π) outline.

The initial value of π in Figure 15:7 depends on when the diversification is called in the global procedure (see Figure 15:5). The output of the diversification phase is a new *POP* and an updated value of $\pi_{currentbest}$. The updated value of $\pi_{currentbest}$ is compared to the overall best (see steps 4.1.3 and 4.4 in Figure 15:5) to store in memory the best solution ever found during the entire search (i.e., $\pi_{globalbest}$).

15.7. Computational experiments

Our computational experiments were designed with the goal of comparing the best solution found by the proposed procedure with the best solution in the initial population. Recall that the initial population is built by executing 33 variants of single-pass heuristics. We employ two project types throughout our experimentation:

T1) Projects for which all activities have due dates.

T2) Project for which only the final activity has a due date. Without loss of generality, we can consider that the penalty factor is one for this project type.

A secondary goal of our experiments is to analyze the robustness of the solutions obtained as the number of scenarios under consideration increases. Before presenting the results of our computational experiments, we consider of importance to discuss how the test problems were generated. The general characteristics of the problem set are:

1. All problems in the set were randomly generated.
2. Both the activity times and the tardiness penalties were drawn from a uniform distribution with parameters (1, 10).
3. The number of immediate successors of each activity was randomly chosen from the interval (1, 5).
4. The number of resources was set to 4.

The due date for each activity i is an integer in the closed interval $[d^i_{\min}, d_{\max}]$. The maximum due date $d_{\max}$ is the same for all the activities and is calculated as follows:

$$RT_j = \sum_{i \in A} r_{ij} \left(t_{i1} + \frac{1}{|S|} \sum_{s \in S} t_{i2s} \right)$$

$$d_{max} = \max_j \left(\frac{RT_j}{R_j} \right) * \left(due\ date\ tightness \right) \text{ for } j \in R,$$

where $t_{i2s} = 0$ if $i \in DA$.

The due date tightness is a factor that allows us to control how restrictive the assigned due dates are. Larger values of the due date tightness result in less restrictive due date values. The lower limit $d^i_{\min}$ is then calculated as follows:

$$d^i_{min} = \max_{j|(j,i) \in E} d_j + t_{i1} + \max_{s \in S} t_{i2s} + \max_{s \in S} w_{is}$$

where $t_{i2s} = w_{is} = 0$ if $i \in DA$.

In addition to the due date tightness, we use two more parameters to vary the characteristics of individual problem instances: the *resource factor* and the *resource strength*. In our experimentation, we consider projects with 50, 100 and 150 activities. We also consider instances with 25%, 50% and 75% of stochastic activities.

The due date tightness was set to 0.65 and 0.7 for instances with 75% of stochastic activities, and to 0.75 and 0.8 for instances with 25% and 50% of stochastic activities. The values for this parameter were empirically chosen so that the best solution generated in the initial population would have an objective function value greater than zero. The rationale behind this selection is that the application of the proposed procedure makes sense only on problems that cannot be optimally solved by simple heuristics. We also avoid extremely tight due dates that would seem unrealistic from a practical point of view. It is easy to verify that smaller values of this parameter result in tighter due dates.

The resource factor (Pascoe, 1966) represents the average percentage of resource types required per each activity. We consider three values for this factor: 0.65, 0.75 and 0.85. The resource strength (Cooper, 1976) measures the relationship between the demand for a resource and its availability. The larger the strength the smaller the demand for resources. We also consider three values for this factor: 2.5, 3 and 3.5. Satisfying the due dates becomes more difficult as the resource factor increases and the resource strength decreases. Hence, a problem generated with a due date tightness of 0.65, a resource factor of 0.85, and resource strength of 2.5 will tend to result in a large optimal expected total tardiness penalty.

Considering all the combinations of factors and sizes (i.e., 4 at 3 levels each and 1 at 2 levels), the set of problems consists of 162 instances of each type (i.e., T1 and T2). Tables 15:3 and 15:4 summarize the results of our first set of experiments. The first column shows the size of the instances in terms of the number of activities (i.e., n). The second column, labeled *%Stoch*, consists of the percentage of stochastic activities in each instance. The third column, *E(Initial)*, consists of the expected weighted tardiness of the best solution in the initial population (i.e., $E(\pi_{globalbest})$ after step 3 in Figure 15:5). In the fourth column, *E(Final)* = $E(\pi_{globalbest})$ after the completion of the search. The *Reduction* value in the fifth column is calculated as follows:

$$Reduction = \frac{E(Initial) - E(Final)}{E(Initial)} * 100$$

We report the average *Reduction* as well as the range (i.e., minimum and maximum) for each set of instances in a row. The *Makespan* corresponding to the best solution found by the procedure is reported in the sixth column. All of these columns report average values over the 18 instances corresponding to each row. Finally, the average CPU seconds is reported in the seventh column labeled *Time*. This time excludes input and output operations and corresponds to a personal computer with a Pentium 166 processor. The following search parameters were used in all of the experiments:

1. MaxIter1 = MaxIter2 = 5
2. MaxInt = 14
3. RefSize = 2
4. |POP| = 20
5. tabu tenure = 7
6. BlockNum = 5 (for n = 50 and n = 100) and 7 (for n = 150).

Also, the results in tables 15:3 and 15:4 were found using a sample size of 30 scenarios.

Table 15:3. Summary of results for T1 instances.

n	*%Stoch*	*E(Initial)*	*E(Final)*	*Reduction*	*Makespan*	*Time*
	25%	809.7	445.0	55.4% 25.4%-100%	199.6	653.2
50	50%	448.6	219.1	74.4% 33.3%-100%	242.3	730.9
	75%	392.3	165.5	82% 33%-100%	263.4	686.8
	25%	1060.8	457.8	78.1% 36.8%-100%	365.1	2188.7
100	50%	421.8	59.6	93.7% 72.9%-100%	456.1	2174.5
	75%	738.5	289.4	84% 44.3%-100%	506.7	2608.0
	25%	1334.3	377.1	83.6% 41.8%-100%	534.5	5054.2
150	50%	412.1	24.6	97.2% 85.6%-100%	685.4	4183.4
	75%	275.1	9.8	98.4% 84.8%-100%	726.5	5004.9

It can be observed in Table 15:3 that the *Reduction* values significantly increase with the problem size. This is particularly true for problems with *%Stoch* = 25. The results in Table 15:3 also indicate that solution time is affected by the problem size (n) but not by *%Stoch*.

Table 15:4. Summary of results for T2 instances.

N	*%Stoch*	*E(Initial)*	*E(Final)*	*Reduction*	*Makespan*	*Time*
	25%	52.8	34.2	49.2% 7.4%-100%	193.2	540.6
50	50%	45.97	24.6	71.7% 24.9%-100%	295.2	1650.9
	75%	25	12.7	69.4% 13%-100%	267.3	636.8
	25%	75.1	40.8	62.4% 15.1%-100%	362.9	2759.4
100	50%	35.8	9.9	82% 27%-100%	452.8	2264.4
	75%	32.3	13.3	86.2% 33%-100%	510.2	2122.3
	25%	91.2	54.1	56% 26.7%-100%	546.8	6745.5
150	50%	43.1	9.53	90% 45.8%-100%	660.48	5725
	75%	30.5	11.4	89.9% 34.9%-100%	762	3864.3

The results in Table 15:4 show that the average *Reduction* obtained for T2 problems is generally less than the one for T1 problems. However, this Reduction is still substantial, warranting the use of the proposed procedure. Other values, such as *Makespan* and *Time,* are similar in both tables.

Table 15:5. Robustness assessment.

		30 Scenarios	50 Scenarios		70 Scenarios		
				30-Scenario Solution		30-Scenario Solution	50-Scenario Solution
n	%Stoch	E(Final)	E(Final)	E(Final)	E(Final)	E(Final)	E(Final)
	25	445	420.8	446.7	442.3	449.7	422.8
50	50	219.1	204.6	216.8	216.4	218.3	194.0
	75	165.5	168.6	171.2	154.2	171.2	168.7

Since the results reported in tables 15:3 and 15:4 were obtained with a sample size of 30 scenarios, we now perform a second experiment to assess the robustness of the method in relation to the number of scenarios used during the search. In this experiment, we employ the problem instances of size 50 from Table 15:3 and increase the sample size to 50 and 70 scenarios. Table 15:5 shows the results of this experiment.

The *30 Scenarios* column simply reports the *E*(*Final*) values from Table 15:3. Under the *50 Scenarios* columns, we report the *E*(*Final*) value obtained by the procedure when the sample size increases by 20 and the solution found in the *30-Scenario* case, as evaluated using 50 scenarios. In all cases, the quality of the solutions found using 30 scenarios decreases when evaluated over 50 scenarios. We then repeat the process for 70 scenarios. In this case, some of the solutions found using 50 scenarios turn out to be of a better quality than those found using 70 scenarios. In particular, problem instances with %*Stoch* = 50 result in an average *E*(*Final*) of 194.0 (when evaluated using 70 scenarios), which compares well against 216.4 found using 70 scenarios.

We perform one final experiment to explore the flexibility in reference to changes in the form of the objective function. In particular, we consider the following mean-variance objective function:

$$\text{Minimize } E(\pi) + Penalty^{*}\, V(\pi),$$
$$\text{where } V(\pi)=\sum_{s\in S} p_s\left(T_s - E(\pi)\right)^2 .$$

The results of this experiment are reported in Table 15:6.

Table 15:6. Mean-variance objective function.

		Penalty = 0		Penalty = 0.01		Penalty = 0.1	
n	%Stoch	E(Final)	V(Final)	E(Final)	V(Final)	E(Final)	V(Final)
	25	445.0	10122.9	490.9	6884.6	644.6	3948.1
50	50	219.1	5472.0	264.3	4812.6	426.9	4735.7
	75	165.5	8330	205.7	5959.1	367.6	5459.2

The procedure performs as expected, the variance consistently decreases as the *Penalty* value increases.

15.8. Conclusions

In this chapter, we have developed a solution procedure for a stochastic project-scheduling problem that to the best of our knowledge has never been addressed in the literature prior to this study. The problem was derived from practical applications and incorporates reality by modeling the uncertainty related to key input data.

The application of scatter search within the framework of stochastic optimization represents an innovation. Our implementation follows the guidelines that separates scatter search from other evolutionary procedures, such as genetic algorithms. In particular, our procedure relies less in randomization and more in systematic diversification to search for improved outcomes. The search operates on a solution representation (topological orderings), that represents a new approach within project scheduling procedures.

The computational experiments indicate that the procedure is effective. In particular, the procedure is able to significantly reduce the average objective function value obtained by the heuristics used to generate the initial population. The procedure is also robust, since good solutions can be obtained using relatively small sets of scenarios. Our experiments also showed that the procedure is capable of handling an objective function that includes a penalty term for the variance.

The encouraging results from this study indicate that the adaptation of meta-heuristic procedures, such as tabu search and its evolutionary cousin scatter search, open the door to the development of procedures for solving complex project scheduling problems found in practice.

Endnotes

1. This research was partially supported by Generalitat Valenciana under contract GV-3310/95.

References

COOPER, D. F. (1976) „Heuristics for Scheduling Resource-Constrained Projects: An Experimental Investigation," *Management Science*, vol. 22, pp. 1186-1194.

DEMBO, R. (1991) „Scenario Optimization," *Annals of Operations Research*, vol. 30, pp. 63-80.

DUMOND, J. AND V. A. MABERT (1988) „Evaluating Project Scheduling and Due Date Assignment Procedures: An Experimental Analysis," *Management Science*, vol. 34, no. 1, pp. 101-118.

ESCUDERO, L., P. KAMESAN, A. KING AND R. WETS (1993) „Production Planning Via Scenario Modeling," *Annals of Operations Research*, vol. 43, pp. 311-335.

FERNANDEZ, A. A. (1995) „An Optimal Solution to the Resource Constrained Project Scheduling Problem with Stochastic Task Durations," Unpublished Dissertation, University of Central Florida.

FERNANDEZ, A. A. AND R. L. ARMACOST (1996) „The Role of the Nonanticipativity Constraints in Commercial Software for the Stochastic Project Scheduling," *Computer and Industrial Engineering*, vol. 31, no. ½, pp. 233-236.

GLOVER, F. (1997) „A Template for Scatter Search and Path Relinking," Working Paper, University of Colorado.

GLOVER, F. AND M. LAGUNA (1997) *Tabu Search*, Kluwer Academic Publishers, Boston.

GOLENKO-GINZBURG, D. AND A. GONIK (1997) „Stochastic Network Project Scheduling with Non-consumable Limited Resources," *International Journal of Production Economics*, vol. 48, pp. 29-37.

LAWRENCE, S. R. AND T.E. MORTON (1993) „Resource-Constrained Multiproject Scheduling with Tardy Costs," *European Journal of Operational Research*, vol. 64, no. 2, pp. 168-187.

LAWRENCE, S. R. AND E. C. SEWELL (1997) „Heuristic, Optimal, Static and Dynamic Schedules when Processing Times are Uncertain," *Journal of Operations Management*, vol. 15, pp. 71-82.

MOHRING, R. H. (1984) „Minimizing Costs of Resource Requirements in Project Networks Subject to a Fixed Completion Time," *Operations Research*, vol. 32, no. 1, pp. 89-120.

MULVEY, J. M., R. J. VANDERBEI, AND S. A. ZENIOS (1995) „Robust Optimization of Large Scale Systems," *Operations Research*, vol. 43, no. 2, pp. 264-281.

NEOS Guide, http://www.mcs.anl.gov/home/otc/Guide/

OPERA (1995), Welcom Software Technology, 15995 North Barkers Landing Rd., Houston, TX 77079.

PASCOE, T. L. (1966) „Allocation of Resources C.P.M.," *Revue Francaise Recherche Operationelle*, vol. 38, pp. 31-38.

@RISK for Project (1995), Palisade Corporation, 31 Decker Rd., Newfield, New York, 14867.

ROCKAFELLER, R. T. AND R. J-B WETS, (1991) „Scenarios and Policy Aggregation in Optimization Under Uncertainty," *Mathematics of Operations Research*, vol. 16, no. 1, pp. 119-147.

ULLMAN, J. D. (1975) „NP-Complete Scheduling Problems," *Journal of Computers and Systems Science*, vol. 10, pp. 384-393.

16 FUZZY MULTI-MODE RESOURCE-CONSTRAINED PROJECT SCHEDULING WITH MULTIPLE OBJECTIVES

Maciej Hapke
Andrzej Jaszkiewicz
Roman Słowiński

Institute of Computing Science, Poznań University of Technology

16.1. Introduction

Classical models of resource-constrained project scheduling (RCPS) problems are very often not adequate to real world problems. For this reason, the classical RCPS models have been extended to deal with multiple-category resources (Węglarz, 1980), multiple performing modes of activities (Patterson et al., 1990) and multiple project performance measures (Słowiński, 1981, 1989). Another realistic aspect of project scheduling is uncertainty of activity time parameters. Many stochastic approaches to solving RCPS problems under uncertainty have been proposed (e.g. Loostma, 1966, 1989; Elmaghraby, 1967; Gaul, 1981). However, the use of new techniques and methodologies in many projects decreases the relevance of past experience. It is obvious that the lack of historical data does not allow for stochastic approach.

A good example of such a project is a software project scheduling. In practice, it appears that software project managers are often able to estimate approximate duration of particular activities, taking into account their experience, good knowledge about new techniques and skills of particular engineers. They can express intervals of possible duration together with the most possible values and decreasing possibilities of other values within the intervals. Such an information about activity duration can be naturally expressed by a fuzzy number.

Scheduling under uncertainty modeled by fuzzy numbers is quite new research field. First papers were concerned with a fuzzy PERT model (Prade, 1979; Chanas and Kamburowski, 1981; Buckley, 1989; Rommelfanger, 1994). Kerr and Walker (1989), McCahon and Lee (1992) and Fortemps (1998) applied fuzzy arithmetics for a job shop scheduling problem. More general project scheduling problem with fuzzy time parameters has been being investigated by Hapke and Słowiński. They have generalized priority heuristics to deal with fuzzy time parameters (Hapke and Słowiński, 1996). Then Hapke (1995) has shown how the fuzzy parallel scheduling procedure can be effectively used in metaheuristic search.

In practical applications it is realistic to consider both fuzzy time parameters and multiple objectives in a common model of the general class of RCPS problems. Hapke et al. (1997) have proposed a solution to a fuzzy multi-objective RCPS problem. It proceeds in two stages. In the first stage a sample of the Pareto-set to the multi-objective fuzzy RCPS problem is generated. Then, in the second stage, an interactive procedure is used to organize the search over the sample for the best compromise schedule. Since the RCPS problems are NP-hard, even in the single-objective case, we use a multi-objective metaheuristic method, called Pareto-Simulated Annealing (PSA) (Czyżak and Jaszkiewicz, 1996) to generate a set of non-dominated schedules. The PSA method is adapted to deal with fuzzy RCPS problems such that the sample is composed of fuzzy non-dominated schedules.

The second stage of the proposed approach consists in application of, so called, "Light Beam Search" (LBS) interactive procedure (Jaszkiewicz and Słowiński, 1997) that allows for interactive analysis of the set of generated solutions and supports the decision maker in selecting the best compromise schedule. The procedure had to be generalized to fuzzy objective measures. The generalizations of scheduling procedure, of the PSA method and of the LBS procedure need specification of appropriate rules for comparisons of fuzzy numbers.

The paper is organized as follows. The second section presents some introductory material: uncertainty modeling using fuzzy sets, some operations on fuzzy numbers, weak and strong comparison rules of fuzzy numbers and characterization of a fuzzy multi-objective combinatorial optimization problem. In the third section the considered model of RCPS with fuzzy time parameters is described and some issues concerned with the fuzzy scheduling procedure are discussed. Next, the adaptation of the PSA method to fuzzy RCPS problem is presented. The fifth section describes a way the best compromise schedule is searched using the "Light Beam Search" procedure. Next section presents an application of the presented approach to a software project scheduling problem. The final section groups conclusions.

16.2. Basic concepts and definitions

16.2.1. Uncertainty modeling

Set A in a base set X can be described by a membership function $\mu_A : X \to \{0,1\}$ with $\mu_A(x)=1$ if $x\in A$ and $\mu_A(x)=0$ if $x\notin A$. If it is uncertain, whether or not element x belongs to set A, the above model can be extended such that the membership function maps into interval [0,1]. A high value of this membership function implies high possibility, while a low value a poor possibility. This leads to the following definition of a fuzzy set (Dubois and Prade, 1980):

$$\tilde{A} = \{(x, \mu_{\tilde{A}}(x)) \mid x\in X\} \tag{1}$$

where $\mu_{\tilde{A}}$ is called the membership function of $\tilde{A}$.

Another useful definition is that of a level set or a level cut. Let $\tilde{A}$ be a fuzzy set in X and $\alpha\in(0,1]$. The α-level set or α-level cut of $\tilde{A}$ is the set

$$\tilde{A}_\alpha = \{ x \in X \mid \mu_{\tilde{A}}(x) \geq \alpha \} \tag{2}$$

Lower and upper bounds of α-cut $\tilde{A}_\alpha$ are equal to $\inf_{x\in X} \tilde{A}_\alpha$ and $\sup_{x\in X} \tilde{A}_\alpha$, respectively.

A flat fuzzy number is a special kind of a fuzzy set. Precisely, a flat fuzzy number is defined as a normalized convex fuzzy subset of real line $\boldsymbol{R}$, i.e. it is defined according to (1), where $X=\boldsymbol{R}$.

The precise form of a fuzzy number can be described by an expert only rarely. A practical way of getting suitable membership functions of fuzzy data has been proposed by Rommelfanger (1990). He proposes that the expert expresses his/her optimistic and pessimistic information about parameter uncertainty on some prominent membership levels α, e.g.:

$\alpha=1$:	$\mu(x) = 1$	means that value x certainly belongs to the set of possible values,
$\alpha=\lambda$:	$\mu(x) > \lambda$	means that the expert estimates that value x with $\mu(x) \geq \lambda$ has a good chance of belonging to the set of possible values,
$\alpha=\varepsilon$:	$\mu(x) < \varepsilon$	means that value x with $\mu(x) < \varepsilon$ has only a very little chance of belonging to the set of possible values, i.e. the expert is willing to neglect the corresponding values of x with $\mu(x) < \varepsilon$.

Thus, a flat fuzzy number in six-point convention is represented by six real numbers. Formally, a flat fuzzy number $\tilde{M}$ is represented by the following list of symbols:

$$\tilde{M} = (\underline{m}^\varepsilon, \underline{m}^\lambda, \underline{m}, \overline{m}, \overline{m}^\lambda, \overline{m}^\varepsilon). \tag{3}$$

We assume that λ = 0.6, ε = 0.1. An exemplary flat fuzzy number in the six-point convention is presented in Figure 16:1.

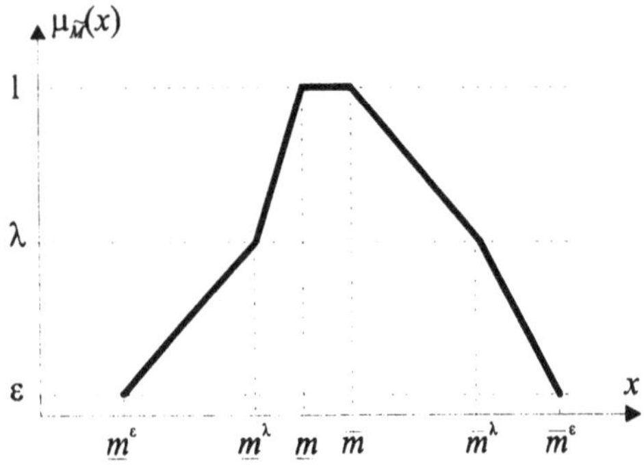

Figure 16:1. Flat fuzzy number $\tilde{M}$ in six-point representation

16.2.2. Fuzzy arithmetics

Let $\tilde{A}$ and $\tilde{B}$ be fuzzy numbers of the universe *X* and *Y*. Let * denote any basic arithmetic operations (+, −, ×, /). Then any operation $\tilde{A} * \tilde{B}$ can be characterized by Zadeh's extension principle (Zadeh, 1978):

$$\mu_{\tilde{A}*\tilde{B}}(z) = \max_{x*y=z} \{\min[(\mu_{\tilde{A}}(x), \mu_{\tilde{B}}(y)]\}. \tag{4}$$

In the case of arithmetic operations on fuzzy numbers in piece-wise representation the equation (4) corresponds to

$$\tilde{C}_\alpha = (\tilde{A} * \tilde{B})_\alpha = \tilde{A}_\alpha * \tilde{B}_\alpha, \text{ for any } \alpha \in (0, 1]. \tag{5}$$

Equation (5) shows that the α-cut on a general arithmetic operation on two fuzzy numbers is equivalent to the arithmetic operation on the respective α-cuts of the two fuzzy numbers. Both $(\tilde{A} * \tilde{B})_\alpha$ and $\tilde{A}_\alpha * \tilde{B}_\alpha$ are interval quantities the operations on which can make use of classical interval analysis (Ross, 1995). Thus, one can define the following arithmetic operations for fuzzy numbers in six-point representation:

$$\tilde{A} \oplus \tilde{B} = \left(\underline{a}^\varepsilon + \underline{b}^\varepsilon,\ \underline{a}^\lambda + \underline{b}^\lambda,\ \underline{a} + \underline{b},\ \bar{a} + \bar{b},\ \bar{a}^\lambda + \bar{b}^\lambda,\ \bar{a}^\varepsilon + \bar{b}^\varepsilon\right) \tag{6}$$

$$\tilde{A} \ominus \tilde{B} = \left(\underline{a}^\varepsilon - \bar{b}^\varepsilon,\ \underline{a}^\lambda - \bar{b}^\lambda,\ \underline{a} - \bar{b},\ \bar{a} - \underline{b},\ \bar{a}^\lambda - \underline{b}^\lambda,\ \bar{a}^\varepsilon - \underline{b}^\varepsilon\right) \tag{7}$$

$$\tilde{A} \times \tilde{B} = \left(\underline{a}^\varepsilon \times \underline{b}^\varepsilon,\ \underline{a}^\lambda \times \underline{b}^\lambda,\ \underline{a} \times \underline{b},\ \bar{a} \times \bar{b},\ \bar{a}^\lambda \times \bar{b}^\lambda,\ \bar{a}^\varepsilon \times \bar{b}^\varepsilon\right) \tag{8}$$

$$\tilde{A} / \tilde{B} = \left(\underline{a}^\varepsilon / \bar{b}^\varepsilon,\ \underline{a}^\lambda / \bar{b}^\lambda,\ \underline{a} / \bar{b},\ \bar{a} / \underline{b},\ \bar{a}^\lambda / \underline{b}^\lambda,\ \bar{a}^\varepsilon / \underline{b}^\varepsilon\right) \tag{9}$$

and moreover:

$$\tilde{\max}(\tilde{A}, \tilde{B}) = (\max(\underline{a}^{\varepsilon}, \underline{b}^{\varepsilon}), \max(\underline{a}^{\lambda}, \underline{b}^{\lambda}), \max(\underline{b}, \underline{a}), \max(\overline{b}, \overline{a}), \max(\overline{b}^{\lambda}, \overline{a}^{\lambda}), \max(\overline{b}^{\varepsilon}, \overline{a}^{\varepsilon})) \quad (10)$$

$$\tilde{\min}(\tilde{A}, \tilde{B}) = (\min(\underline{a}^{\varepsilon}, \underline{b}^{\varepsilon}), \min(\underline{a}^{\lambda}, \underline{b}^{\lambda}), \min(\underline{b}, \underline{a}), \min(\overline{b}, \overline{a}), \min(\overline{b}^{\lambda}, \overline{a}^{\lambda}), \min(\overline{b}^{\varepsilon}, \overline{a}^{\varepsilon})). \quad (11)$$

16.2.3. Comparison of two fuzzy numbers

The definition of dominance, the combinatorial search in a fuzzy objective space as well as decision making require comparing fuzzy scores on objective functions. The way of comparing fuzzy scores is connected with the semantics of fuzzy numbers. Indeed, the fuzzy numbers can be seen as imprecise probability distributions (see Dempster, 1967; Shafer, 1976). In this perspective, the comparison of two fuzzy numbers can be substituted by the comparison of their mean values defined consistently with the well-known definition of expectation in probability theory. The idea exploited by Dubois and Prade (1987) rely on the mathematical fact that, with respect to a fuzzy number, the possibility measure corresponds to an upper probability distribution, while the necessity measure, to a lower probability distribution of the corresponding random variable in the sense of Dempster (1967). Then it is reasonable to define the mean value of a fuzzy number as a closed interval whose bounds are expectations of upper and lower probability distributions. The comparison of two fuzzy numbers boils then down to the comparison of arithmetic means of these bounds, which is computationally equivalent to another intuitive comparison principle based on the area compensation determined by the membership functions of two fuzzy numbers being compared (Kołodziejczyk, 1986; Chanas, 1987; Roubens, 1990; Fortemps and Roubens, 1996).

The main advantage of the area compensation method is that it answers not only which one of two fuzzy numbers is greater but also what is the degree to which one fuzzy number is greater than another one. Working on fuzzy scheduling problems, the authors (Hapke et al., 1997) have used this comparison method under the name *weak comparison rule* (WCR) in order to distinguish it from a more restrictive comparison method used to observe precedence and resource constraints and called *strong comparison rule* (SCR).

Let $\tilde{A}$, $\tilde{B}$ be two fuzzy numbers and $S_L(\tilde{A} \geq \tilde{B})$, $S_R(\tilde{A} \geq \tilde{B})$, $S_L(\tilde{A} \leq \tilde{B})$, $S_R(\tilde{A} \leq \tilde{B})$ the areas determined by their membership functions (Figure 16:2):

$$S_L(\tilde{A} \geq \tilde{B}) = \int_{U(A,B)} (\inf_{x\in\mathbf{R}} A_\alpha - \inf_{x\in\mathbf{R}} B_\alpha) d\alpha \quad (12)$$

$$S_R(\tilde{A} \geq \tilde{B}) = \int_{V(A,B)} (\sup_{x\in\mathbf{R}} A_\alpha - \sup_{x\in\mathbf{R}} B_\alpha) d\alpha \quad (13)$$

where

$$U(\tilde{A}, \tilde{B}) = \{\alpha \mid \inf_{x\in\mathbf{R}} A_\alpha \geq \inf_{x\in\mathbf{R}} B_\alpha, \varepsilon \leq \alpha \leq 1\} \tag{14}$$

$$V(\tilde{A}, \tilde{B}) = \{\alpha \mid \sup_{x\in\mathbf{R}} A_\alpha \geq \sup_{x\in\mathbf{R}} B_\alpha, \varepsilon \leq \alpha \leq 1\} \tag{15}$$

$S_L(\tilde{A} \leq \tilde{B})$ and $S_R(\tilde{A} \leq \tilde{B})$ are defined analogously.

$S_L(\tilde{A} \geq \tilde{B})$ and $S_R(\tilde{A} \geq \tilde{B})$ are the areas that are related to the possibility of $A^* \geq B^*$, where A^* and B^* are potential realizations of $\tilde{A}$ and $\tilde{B}$. The degree $C(\tilde{A} \geq \tilde{B})$, to which $\tilde{A} \geq \tilde{B}$, is calculated as a sum of areas in which $A^* \geq B^*$ minus a sum of areas in which $A^* \leq B^*$.

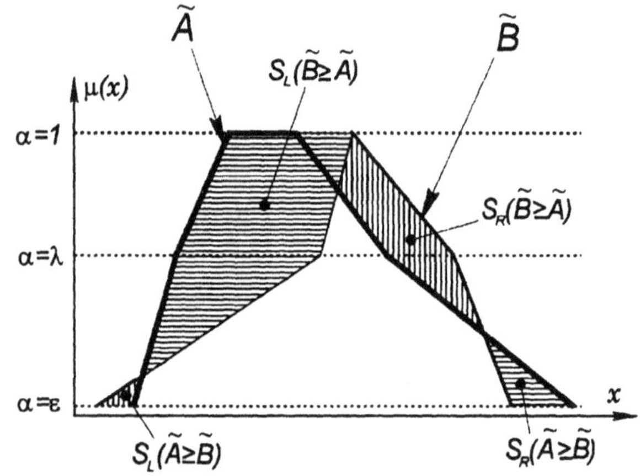

Figure 16:2. The comparison of two fuzzy numbers based on area compensation

Definition 1. According to the Weak Comparison Rule – WCR, the degree C to which fuzzy number $\tilde{A}$ is greater than or equal to $\tilde{B}$ is defined as:

$$C(\tilde{A} \geq \tilde{B}) = \frac{1}{2}\{S_L(\tilde{A} \geq \tilde{B}) + S_R(\tilde{A} \geq \tilde{B}) - S_L(\tilde{B} \geq \tilde{A}) - S_R(\tilde{B} \geq \tilde{A})\} \tag{16}$$

Using WCR, one can define three relations between $\tilde{A}$ and $\tilde{B}$, corresponding to weak inequality, strict inequality and equivalence between fuzzy numbers, respectively:

$$\tilde{A} \geq \tilde{B} \text{ iff } C(\tilde{A} \geq \tilde{B}) \geq 0 \tag{17}$$

$$\tilde{A} > \tilde{B} \text{ iff } C(\tilde{A} \geq \tilde{B}) > 0 \tag{18}$$

$$\tilde{A} \sim \tilde{B} \text{ iff } C(\tilde{A} \geq \tilde{B}) = C(\tilde{B} \geq \tilde{A}) = 0 \tag{19}$$

It is easy to note that $C(\tilde{A} \geq \tilde{B}) = -C(\tilde{B} \geq \tilde{A})$.

The proposed WCR is related to the comparison of "mean values of fuzzy numbers" (Chanas, 1989; Fortemps, 1998), as defined in (Dubois and Prade, 1987) in the framework of the Dempster-Shafer theory. The mean value of a fuzzy number $\tilde{A}$ is the interval [$E_*(\tilde{A})$, $E^*(\tilde{A})$], where $E_*(\tilde{A})$ and $E^*(\tilde{A})$ are the mean values related to the cumulative possibility distribution (left spread), and to the cumulative necessity distribution (right spread), respectively. The natural way of defuzzifying such an interval is to calculate the arithmetic mean of its limits. The defuzzification function $\Im(\tilde{A})$ is defined as follows (Chanas, 1987; Fortemps, 1998):

$$\Im(\tilde{A}) = \frac{E_*(\tilde{A}) + E^*(\tilde{A})}{2} \tag{20}$$

The following properties of the function $\Im(\tilde{A})$ and of the WCR will be useful.

Property 1. (Hapke, 1997) For fuzzy number $\tilde{A}$ in the six-point representation, the function $\Im(\tilde{A})$ can be calculated as:

$$\Im(\tilde{A}) = \frac{1}{4(1-\varepsilon)}\left\{(\lambda-\varepsilon)(\underline{a}^{\varepsilon} + \underline{a}^{\lambda} + \overline{a}^{\lambda} + \overline{a}^{\varepsilon}) + (1-\lambda)(\underline{a}^{\lambda} + \underline{a} + \overline{a} + \overline{a}^{\lambda})\right\} \tag{21}$$

Property 2. (Fortemps, 1997) The degree $C(\tilde{A} \geq \tilde{B})$ to which $\tilde{A}$ is greater than or equal to $\tilde{B}$, calculated according to the WCR, can be calculated using $\Im(\tilde{A})$ and $\Im(\tilde{B})$ as:

$$C(\tilde{A} \geq \tilde{B}) = \Im(\tilde{A}) - \Im(\tilde{B}) = \frac{E_*(\tilde{A}) + E^*(\tilde{A})}{2} - \frac{E_*(\tilde{B}) + E^*(\tilde{B})}{2} \tag{22}$$

From properties 1 and 2 immediately follows the formula for $C(\tilde{A} \geq \tilde{B})$ where $\tilde{A}$ and $\tilde{B}$ are six-point fuzzy numbers.

Application of the WCR is the most suitable when comparing different independent fuzzy scores e.g. fuzzy optimization results. However, its application to scheduling procedure might lead to unfeasible schedules (see p. 16.3). In order to avoid violation of precedence and resource constraints the scheduling procedure will use a stronger comparison rule.

Definition 2. According to the Strong Comparison Rule - SCR,

$$\tilde{A} >>= \tilde{B} \Leftrightarrow max(\tilde{A}, \tilde{B}) = \tilde{A} \tag{23}$$

It is obvious that SCR includes WCR. An application of the SCR will be shown in next section.

16.2.4. Fuzzy multi-objective combinatorial optimization and fuzzy dominance

Definition 3. The general fuzzy multi-objective combinatorial optimization (fuzzy MOCO) problem is formulated as:

$$\max\{\tilde{f}_1(\mathbf{x}),\ldots,\tilde{f}_J(\mathbf{x})\}, \text{ s.t. } \mathbf{x} \in D, \tag{24}$$

where: *solution* $\mathbf{x} = [x_1,\ldots,x_I]$ is a vector of discrete and crisp *decision variables*, D is a finite set of feasible solutions, $\tilde{f}_1,\ldots,\tilde{f}_J$ are fuzzy objective functions (criteria) which, for given **x**, take values called *fuzzy scores*.

Notice that a single solution (schedule) **x** of a RCPS problem is determined by vectors of start times **S** and activity performing modes **M** of activities.
The image of solution **x** in the objective space is a vector $\tilde{\mathbf{f}}^{\mathbf{x}} = [\tilde{f}_1(\mathbf{x}),\ldots,\tilde{f}_J(\mathbf{x})]$ composed of J fuzzy numbers. We will say that $\tilde{\mathbf{f}}^{\mathbf{x}}$ is a *fuzzy point* in the objective space. In order to define a dominance relation in the objective space we will use the WCR for the fuzzy scores. The dominance relation defined in this way will be called *WCR-dominance.*

Definition 4. A fuzzy point $\tilde{\mathbf{f}}^{\mathbf{x}}$ *WCR-dominates* fuzzy point $\tilde{\mathbf{f}}^{\mathbf{y}}$ iff $\tilde{f}_j(\mathbf{x}) \geq \tilde{f}_j(\mathbf{y})\ \forall j$ and $\tilde{f}_j(\mathbf{x}) > \tilde{f}_j(\mathbf{y})$ for at least one j, i.e. iff $C(\tilde{f}_j(\mathbf{x}) \geq \tilde{f}_j(\mathbf{y})) \geq 0\ \forall j$ and $C(\tilde{f}_j(\mathbf{x}) \geq \tilde{f}_j(\mathbf{y})) > 0$ for at least one j.

Definition 5. A fuzzy point $\tilde{\mathbf{f}}^{\mathbf{x}}$ is *WCR-non-dominated* (*WCR-Pareto-optimal*) if there is no other fuzzy point $\tilde{\mathbf{f}}^{\mathbf{y}}$ ($\mathbf{y} \in D$) that WCR-dominates $\tilde{\mathbf{f}}^{\mathbf{x}}$. A solution **x** is *WCR-efficient* if its image in the objective space is WCR-non-dominated. The set of all WCR-efficient solutions will be denoted by N.

Definition 6. The fuzzy *ideal point* is a fuzzy point $\tilde{\mathbf{f}}^{*}$ in the objective space composed of the best attainable fuzzy values of objectives, i.e.:

$$\forall j = 1,\ldots,J\ \forall \mathbf{x} \in D\ \ \tilde{\mathbf{f}}_j^{*} \geq \tilde{\mathbf{f}}_j^{\mathbf{x}}. \tag{25}$$

Definition 7. The fuzzy *nadir point* is a fuzzy point $\tilde{\mathbf{f}}_{*}$ in the objective space composed of the worst attainable fuzzy values of objectives in the set of WCR-non-dominated points, i.e.

$$\forall j = 1,\ldots,J\ \forall \mathbf{x} \in N\ \ \tilde{\mathbf{f}}_{*j} \leq \tilde{\mathbf{f}}_j^{\mathbf{x}}. \tag{26}$$

In definitions 6 and 7, the fuzzy scores are compared with WCR.

Definition 8. *Achievement scalarizing function* in the fuzzy objective space is defined as:

$$s(\tilde{\mathbf{f}}^{\mathbf{o}}, \mathbf{w}, \rho, \tilde{\mathbf{f}}^{\mathbf{x}}) = \max_j \{ w_j C(\tilde{f}_j^{\mathbf{o}} \geq \tilde{f}_j(\mathbf{x})) \} + \rho \sum_{j=1}^{J} C(\tilde{f}_j^{\mathbf{o}} \geq \tilde{f}_j(\mathbf{x})) \tag{27}$$

where $\tilde{\mathbf{f}}^{\mathbf{o}} = [\tilde{f}_1^{\mathbf{o}}, \ldots, \tilde{f}_J^{\mathbf{o}}]$ is a fuzzy reference point composed of fuzzy aspiration levels on particular objectives, $\mathbf{w}=[w_1, \ldots, w_J]$ is a weighting vector, $w_j \geq 0$, $\sum_{j=1}^{J} w_j = 1$ and ρ is a sufficiently small positive number.

16.3. Project scheduling under fuzziness

The multi-objective fuzzy project scheduling problem can be characterized by set of resources $\boldsymbol{R}$ to be allocated among precedence constrained activities from set $\boldsymbol{Z}$, and by set $\boldsymbol{C}$ of project performance measures (objectives, criteria) see (Węglarz 1980, Słowiński 1981).

The set $\boldsymbol{R}$ is composed of resources belonging to three categories:
- *renewable* resources with a limited usage,
- *nonrenewable* resources with a limited consumption,
- *doubly constrained* resources with a limited consumption and usage.

Activities from set $\boldsymbol{Z}$ have discrete resource requirements. They can be performed in several performing modes. Each performing mode of an activity is defined by different multiple category resource requirements and duration. Moreover, for each activity, ready time is known. It is assumed that all the time parameters of activities are, in general, uncertain and modeled by fuzzy numbers (denoted with ~).

The multiple performing modes of activities imply the multiple criteria nature of the problem (time-cost trade-offs). In the project scheduling problem the following performing measures are usually considered: project makespan, resource utilization smoothness, maximum lateness, mean flow time, net present value, project cost. Let us notice that in our case, because of uncertain time parameters of activities, the values of time and time-cost criteria from set $\boldsymbol{C}$ are also uncertain, so, their values are fuzzy numbers.

In solving NP-hard project scheduling problems a parallel scheduling procedure has been found the most efficient heuristic. Its important feature is that allocation of resources to activities is considered in subsequent events, in which the procedure checks if considered resource can be allocated to activities. A generalization of the parallel procedure on fuzzy time parameters has been proposed by Hapke and Słowiński (1996). The generalization consists in solving two questions:

- how to determine the sequence of fuzzy time moments in which new activity (activities) is (are) to be considered,
- how to determine the time moment in which a renewable or doubly constrained resource becomes available for other activities (activity has been completed by current fuzzy time moment)

Our goal is to construct a procedure that would give feasible schedules for *every realization* of fuzzy time parameters. The following example shows that an application of WCR for solving above questions may lead to unfeasible schedules. Assume that the scheduling procedure considers an allocation of resource R to activity Z_2 in its ready time $\tilde{a}_{Z_2}$. Activity Z_2 has an only predecessor Z_1, completed in $\tilde{f}_{Z_1}$. When using WCR for checking if Z_1 has been completed by $\tilde{a}_{Z_2}$ the relation $\tilde{a}_{Z_2} \tilde{\geq} \tilde{f}_{Z_1}$ would be true (see Figure 16:3).

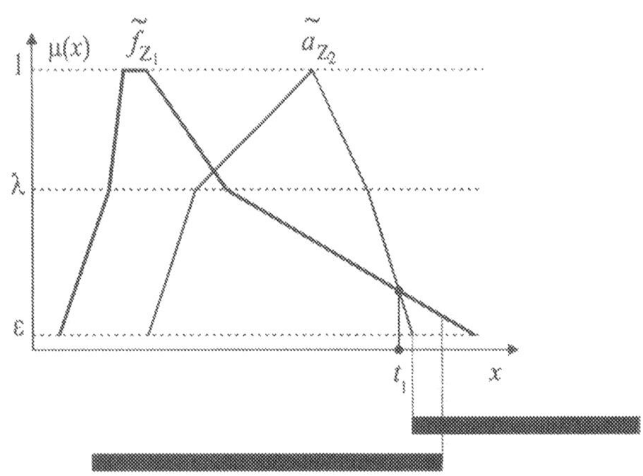

Figure 16:3. Example of an unfeasible fuzzy schedule

Notice, however, that for a very pessimistic realization of the duration of Z_1 (above t_1) the resulting schedule would be unfeasible because activity Z_2 would start before its predecessor Z_1 was completed. The activity Z_2 cannot be therefore scheduled in $\tilde{a}_{Z_2}$. This shows that the use of the WCR for solving the first question may lead to the violation of precedence constraints. Similar example could show that the use of WCR for solving the second question might lead to the violation of resource constraints. In order to avoid the violation of precedence and resource constraints, the WCR has to be substituted by the SCR.

The fuzzy parallel scheduling procedure is as follows:

---

procedure Fuzzy_parallel_scheduling_procedure
begin
    $\tilde{t} := \tilde{t}_0$
    repeat
        Compose a set $Q(\tilde{t})$ of those activities which have not been scheduled yet and whose immediate predecessors have been completed by time $\tilde{t}$;
        for each activity Z_i from $Q(\tilde{t})$, in the order of the priority list **do**
        begin
            if Z_i's resource requirements $\leq$ resource availabilities **then**
            if $\tilde{r}_i <<= \tilde{t}$ **then**
            begin
                Z_i's start time: $\tilde{s}_i := \tilde{t}$
                Z_i's finish time: $\tilde{f}_i := \tilde{s}_i \oplus \tilde{p}_i$
                remove Z_i from $Q(\tilde{t})$
                allocate required resources to Z_i
                insert $\tilde{f}_i$ into set B
            end
            else
                insert $\tilde{r}_i$ into set B
        end
        $\tilde{t} := \max(\tilde{t}, \tilde{l})$
        if $\tilde{l}$ is a finish time of any activity **then**
            update resource availabilities
        remove $\tilde{l}$ from set B
    until all activities from the priority list are completed.
end

---

where:

$\tilde{t}$ - current time moment in which the an allocation of resources to an activity is considered
$\tilde{l}$ - least (in the sense of WCR) value from the set B,
set $B = \{ \tilde{f}_i : \forall Z_i \in S(\tilde{t}) \} \cup \{ \tilde{r}_i : \forall Z_i \in R(\tilde{t}) \}$,
$S(\tilde{t})$ – the set of activities scheduled before $\tilde{t}$, $\tilde{s}_i <<= \tilde{t}$,
$R(\tilde{t})$ – the subset of $Q(\tilde{t})$ of activities not ready in $\tilde{t}$, $\tilde{r}_i \geq \tilde{t}$.

16.4. Fuzzy Pareto simulated annealing - FPSA

In order to solve fuzzy MOCO problems we propose to use adaptation of Pareto-simulated annealing (PSA) procedure proposed by Czyżak and Jaszkiewicz (1996) for crisp MOCO problems. The adaptation of PSA to the fuzzy case will yield a procedure called fuzzy PSA (FPSA).

In the crisp case, PSA generates a set of approximately efficient solutions, considered as an approximate representation of the whole set of efficient solutions. In the fuzzy case, FPSA will generate a set of approximately WCR-efficient solutions *N'*. The goal of FPSA is to generate a set *N'* that would be a good approximation of the set of all WCR-efficient solutions *N*. The term "good approximation" means that the decision maker (DM) should be able to find in *N'* a solution close to the solution that he/she would select if set *N* was known.

Adaptation of PSA to the fuzzy case requires:

- the use of WCR-dominance in updating the set of approximately WCR-efficient solutions,
- calculating the probability of accepting a new solution basing on fuzzy scores,
- updating the average position of a fuzzy point in the objective space.

In each iteration of the procedure, a sample (population) of solutions (fuzzy points in the objective space), called generating sample, is used. The main idea of FPSA is to assure a tendency for approaching the set of WCR-efficient solutions as well as for dispersing the solutions constituting the generating sample over the whole WCR-efficient set *N*. The first tendency is taken into account when calculating the probability of acceptance. The inclination for dispersing the solutions from the generating sample over the whole set *N* is obtained by controlling the weights of particular objectives used in an aggregation formula.

In the case of a single objective SA, a new solution is accepted with probability equal to one if it is not worse than the current solution. Otherwise, it is accepted with probability less than one. In the (fuzzy) multi-objective case, one of the following three situations may appear while comparing a new solution y with the current solution **x**:

- y may (WCR-)dominate **x**,
- y may be (WCR-)dominated by **x**,
- y and **x** may be mutually (WCR-)non-dominated.

In the first situation the new solution may be considered as not worse than the current one and accepted with probability equal to one. In the second situation the new solution may be considered as worse than the current one and accepted with probability less than one. Serafini (1994) and Ulungu (1993) have proposed several multi-objective rules for acceptance probability which in different way treat the third situation. According to some tests (see e.g. Czyżak and Jaszkiewicz, 1998) the following rule gives the best results in PSA:

$$P(\mathbf{x},\mathbf{y},T,\Lambda^{\mathbf{x}}) = \min\left\{1, \exp\left(\sum_{j=1}^{J} \lambda_j^{\mathbf{x}} \left(f_j(\mathbf{y}) - f_j(\mathbf{x})\right) / T\right)\right\}, \tag{28}$$

where **x** is one of the generating solutions, **y** is a solutions obtained from **x** by random selection of a basic move, T is a parameter called temperature, $\Lambda^{\mathbf{x}} = [\lambda_1^{\mathbf{x}},...,\lambda_J^{\mathbf{x}}]$ is the vector of weights associated with generating solution **x**. In other words, the differences on particular objectives are locally aggregated with a simple weighted sum (see Figure 16:4).

In FPSA the probability of acceptance is calculated as follows:

$$P(\mathbf{x},\mathbf{y},T,\Lambda^{\mathbf{x}}) = \min\left\{1, \exp\left(\sum_{j=1}^{J} \lambda_j^{\mathbf{x}} \left(C\left(\tilde{f}_j(\mathbf{y}) \geq \tilde{f}_j(\mathbf{x})\right)\right) / T\right)\right\}. \tag{29}$$

In other words, the distance between two fuzzy scores $\tilde{f}_j(\mathbf{x})$ and $\tilde{f}_j(\mathbf{y})$ is calculated according to the WCR. Note that this rule assures that **y** is accepted with probability equal to 1 if it WCR-dominates **x**.

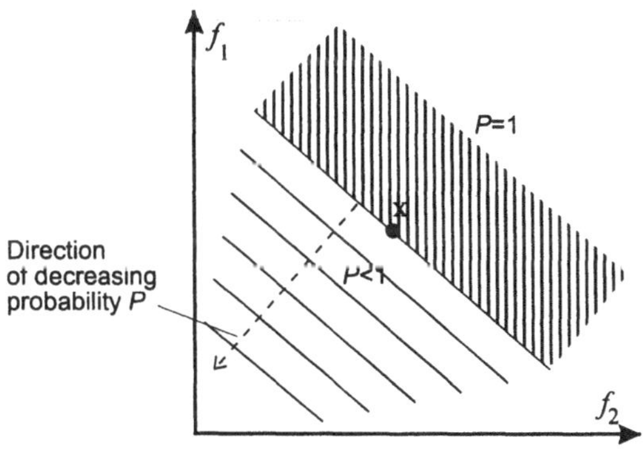

Figure 16:4. Acceptance probability *P* in bi-objective case, maximized objectives

Please note, that the higher the weight associated with a given objective, the lower the probability of accepting moves that decrease the value on this objective and the lower the probability of worsening the value of this objective. Consider for example the situation presented in Figure 16:5a. In this case $\lambda_1^{\mathbf{x}} > \lambda_2^{\mathbf{x}}$. Observe that probability of acceptance decreases rapidly when the value of the first objective decreases. So, by controlling the weights one can control the probability of worsening/improving values of the particular objectives.

The outcome of the procedure is the set of approximately WCR-efficient solutions N'. At the beginning of the procedure, set N' is empty. In order to update the set, the following operations are performed, whenever a new solution **y** dominating **x** is generated:

- remove from set N' all solutions WCR-dominated by **y**,
- add **y** to N' if it is not WCR-dominated by any solution from this set.

The general scheme of the PSA procedure may be written as follows:

---

procedure PSA
begin
  Select a starting sample of generating solutions $S \subset D$
  for each $\mathbf{x} \in S$ **do**
    Update set N' of approximately WCR-efficient solutions with **x**
  $T := T_0$
  repeat
    Update the number of the generating solutions
    Update weights of all generating solutions
      for each $\mathbf{x} \in S$ **do**
      Construct $\mathbf{y} \in V(\mathbf{x})$
      if $\tilde{\mathbf{f}}^{\mathbf{y}}$ is not WCR-dominated by $\tilde{\mathbf{f}}^{\mathbf{x}}$
      then
        Update set N' of approximately WCR-efficient solutions with **y**
      x := **y** (accept **y**) with probability $P(\mathbf{x}, \mathbf{y}, T, \Lambda^{\mathbf{x}})$
    if the conditions of changing the temperature are fulfilled **then**
      decrease T
  until the stop conditions are fulfilled
end

---

where $V(\mathbf{x}) \subset D$ is the set of feasible solutions that can be obtained from **x** by performing a single move.

The goal of updating the weights is to assure a dispersion of generating solutions in the objective space. As, however, position of each generating solution may change significantly in subsequent iterations, it is more important to assure dispersion of average positions of generating solutions. The average position of solution **x** is a point in the objective space, denoted by $\mathbf{v}^{\mathbf{x}}$. In the standard PSA, the average positions are calculated by exponential smoothing and updated whenever the generating solution is changed (a move is accepted). In the FPSA, the fuzzy scores are defuzzified before applying the exponential smoothing:

$$v_j^{\mathbf{x}} := \beta v_j^{\mathbf{x}} + (1-\beta)\Im(\tilde{f}_j(\mathbf{x})),\ \mathrm{j} = 1,\dots,\mathrm{J}, \tag{30}$$

where $\beta < 1$ is a smoothing factor close to one (e.g. β=0.95).

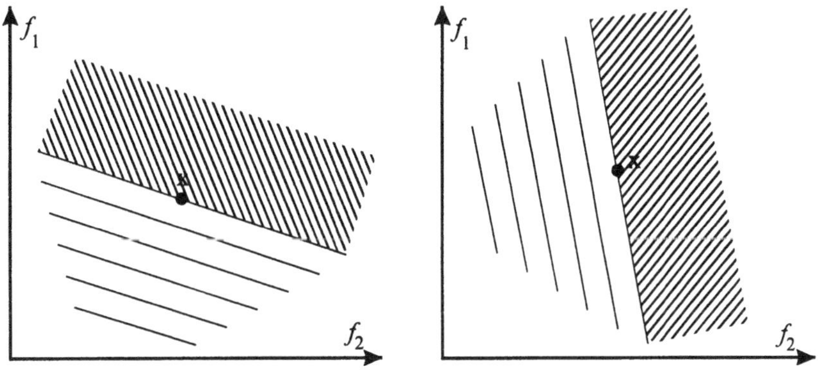

(a) $\lambda_1^x > \lambda_2^x$ (b) $\lambda_1^x < \lambda_2^x$

Figure 16:5. The role of weights in the multi-objective rules for acceptance probability

The algorithm for updating the weights is as follows:

---

procedure **Update_weights**
begin
  for each $\mathbf{x} \in S$ **do**
    Select solution $\mathbf{x}' \in S$ such that $\mathbf{v}^{\mathbf{x}'}$ is closest (e.g. according to Euclidean distance) to $\mathbf{v}^{\mathbf{x}}$ and such that $\mathbf{x}'$ and $\mathbf{x}$ are mutualy WCR-non-dominated
    if there is no such solution **x'** or it is the first iteration with **x** **then**
      Set random weights such that:
      $\forall_j \lambda_j^x \geq 0 \text{ and } \sum_j \lambda_j^x = 1$
    else
      for each objective f_j
        $$\lambda_j^x = \begin{cases} \alpha\lambda_j^x & ,\text{if } v_j^{x'} \geq v_j^x \\ \lambda_j^x / \alpha & ,\text{if } v_j^{x'} < v_j^x \end{cases}$$
      normalize the weights such that
      $\sum_j \lambda_j^x = 1$
end

---

where $\alpha > 1$ is a constant close to one (e.g. $\alpha = 1.01$).

The size of the sample of the generating solutions is not constant but can be updated in each iteration. Some non-interesting solutions may be removed from the sample and some solutions may be duplicated. A solution is removed if in a specified number of iterations it is dominated by some other generating solutions. A solution is duplicated if in specified number of iterations it is neither dominated nor dominates any other generating solution. The starting size of the generating sample is equal to J+1 and typically increases when temperature decreases. Please note that except of weights updating, each of the generating solutions operates independently, so, the algorithm is naturally parallel.

As FPSA is a metaheuristic procedure, it defines only a general scheme of the calculations. This general scheme has to be customized for a given MOCO problem. The customization consists in defining the way a new solution is generated from the neighborhood of the current solution.

Generally, the neighbors of **x** (see 16.2.4) are the solutions that can be obtained from **x** by a basic move. The neighbor of the current schedule is constructed according to the following algorithm:

---

procedure Neighborhood
begin

  draw an activity Z_x at random from $\boldsymbol{Z}$
  copy to a new schedule all the activities Z_i : $\tilde{f}_i <<= \tilde{s}_x$
  create a set W of activities Z_j whose start times $\tilde{s}_i >> \tilde{s}_x$
    and satisfy the precedence constraints in
    moment $\tilde{s}_x$ whose $\tilde{r}_i <<= \tilde{s}_x$
  draw an activity Z_y at random from W, draw its
    performing mode number at random from
    $\{1,2,\ldots, z\}$ and schedule it in moment $\tilde{s}_x$
  schedule the rest of activities in order of their start
    times of the last schedule taking activity Z_x
    instead of Z_y according to the fuzzy scheduling
    procedure

end

---

where: $\tilde{s}_i$ and $\tilde{f}_i$ are the start and finish times of activity Z_i.

PSA and FPSA have been implemented in standard C++ as two libraries of classes. Both the implementations share the same set of most classes. FPSA includes classes that are responsible for comparing, storing, reading and performing mathematical operations on fuzzy numbers. The main aim of the object-oriented implementation was to assure easy adaptation to a given problem.

In order to use FPSA library to solve a given problem one is only required to define two new classes. One of the classes is responsible for reading and storing information about an instance of a given problem. Another class corresponds to a solution of a given problem. It should contain fields to store information about a single solution and operations responsible for finding the first solution, performing basic moves and storing the solutions. More information about the implementation can be found at http://www.cs.put.poznan.pl/~and_j/psa.html.

16.5. Interactive analysis of approximately WCR-non-dominated schedules

After the stage of generating the set of approximately WCR-non-dominated schedules N' an interactive search for the best compromise schedule is organized over N'. This stage requires co-operation with the DM. The DM learns of the possible trade-offs and specifies his/her preferences in order to guide the search towards the best compromise schedule. We propose to use in this stage the interactive procedure called LBS-Discrete ("Light Beam Search") by Jaszkiewicz and Słowiński (1997) adapted to solve decision problems with fuzzy criteria.

Interactive procedures are characterized by phases of decision alternating with phases of computation. In each computation phase, a solution, or a sample of solutions, is selected for examination in the decision phase. As a result of the examination, the DM inputs some preferential information that intends to improve the solution(s) selected in the next computation phase.

The interactive process ceases when among the solutions proposed, the DM finds the one that yields the best compromise. The convergence of this procedure is of a psychological nature rather than mathematical one, since no stable and restrictive model of DM's preferences is assumed throughout the search process. Instead, it is assumed that the DM expresses his/her preferences only locally, with respect to the proposed sample of solutions, and influences the next calculation phase only. This way of organizing the interactive search over a large set of solutions has the following advantages:

- in a single iteration the DM is required to supply relatively simple preference information,
- the DM is involved in the decision process which increases his/her confidence to the final solution,
- the DM is supported in learning of his/her preferences and of the possible trade-offs.

The general scheme of the LBS-Discrete procedure adapted to our case is as follows:

1. Present to the DM the points composed of the best and the worst values of the objective functions in set N', i.e. the approximations of fuzzy ideal and fuzzy nadir points, respectively.
2. Ask the DM to specify a fuzzy reference point or make the fuzzy ideal point the first reference point.
3. Find a starting current solution by projecting the reference point onto set N' using the achievement scalarizing function (27) in the fuzzy objective space.
4. Ask the DM to specify the preferential information defining the size of an outranking vicinity of the current solution in the set N'.
5. Present to the DM a sample of solutions coming from the outranking vicinity. The sample is composed of J solutions optimizing particular objectives within the vicinity.
6. Allow the DM to see other solutions in the outranking vicinity. Terminate the procedure at this step if the best compromise solution has been found, or give the DM the possibility to move the current solution and return to step 4.

The local preference model used to define a vicinity of a current solution is an outranking relation S; **x** S **y** says that solution **x** is at least as good as **y**.

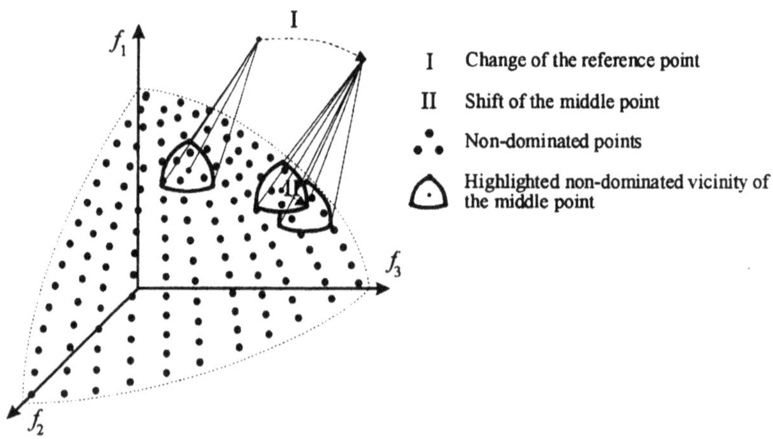

Figure 16:6. Light Beam Search over the set of non-dominated points

The preferential information used to define relation S consists of indifference q_j, preference p_j and veto v_j thresholds on each fuzzy objective $\widetilde{f}_j$. By specifying the value of the indifference threshold q_j, the DM considers two solutions **x** and **y** indifferent with respect to $\widetilde{f}_j$ if:

$$\begin{cases} \tilde{f}_j(\mathbf{x}) \le \tilde{f}_j(\mathbf{y}) + q_j \\ \tilde{f}_j(\mathbf{y}) \le \tilde{f}_j(\mathbf{x}) + q_j \end{cases} \text{and} \tag{31}$$

By specifying the value of the preference threshold p_j, the DM states that he/she strictly prefers **x** to **y** if:

$$\tilde{f}_j(\mathbf{x}) \ge \tilde{f}_j(\mathbf{y}) + p_j \tag{32}$$

The region between q_j and p_j corresponds to the hesitation of the DM between indifference and strict preference. By specifying the value of the veto threshold v_j, the DM states that he/she cannot accept **y** as being not worse than **x** if:

$$\tilde{f}_j(\mathbf{x}) \ge \tilde{f}_j(\mathbf{y}) + v_j \tag{33}$$

(31)-(33) are solved using the WCR.

In LBS-Discrete the preference and veto thresholds are optional. All thresholds have only local meaning and can be updated by the DM in each iteration. Using the above preference information, an outranking relation can be defined in the vicinity of a current solution (see (Jaszkiewicz and Słowiński, 1997) for detailed description of this step). All solutions from *N'* which outrank the current solution create the outranking vicinity presented to the DM in step 5.

In step 6, the method offers to the DM two possibilities to move the vicinity over the set *N'*. The first one consists in specifying a new fuzzy reference point in the objective space which is then projected onto set *N'* giving a new current solution. The second possibility consists in making the new current solution a selected solution in the vicinity of the former current solution. The two possibilities allow the DM free scanning of the whole set *N'* and learning of the possible trade-offs. By presenting samples of solutions at each iteration, the method facilitates and speeds up the learning process.

16.6. Application to software project scheduling

The presented approach may be used in software project management. The hypothetical example is concerned with a software project of building a commercial data base system. The project consists of 51 precedence related activities (see Figure 16:7). The activities require resources belonging to two categories: renewable and non-renewable (see Table 16:1).

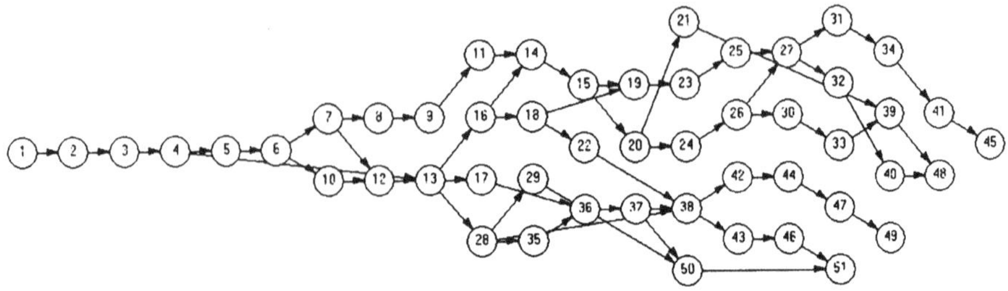

Figure 16:7. Precedence related software project activities

Name	Category	Availability	Cost ($/h)
Business Analyst	renewable	12	30
Technical Editor	renewable	8	25
Product Manager	renewable	8	35
Production Specialist	renewable	6	15
Programmer	renewable	25	28
Senior Programmer	renewable	15	32
Software Engineer	renewable	10	31
Tester	renewable	28	23
Trainer	renewable	8	30
Translation Specialist	renewable	16	25
Writer	renewable	16	24
Money	non-renewable	470.000$	-

Table 16:1. The set of available resources

Table 16:1 includes two groups of resources: programmers and senior programmers. Some activities can be performed both by the former and the latter. Performing an activity by a programmer is cheaper but making the same by a senior programmer takes less time. Thus, the existence of multiple performing modes implies multi-criteria nature of the problem.

Project manager (PM) is interested in minimization of three different criteria:
- the project completion time (**Cmax**) expressed in days,
- the total project cost (**Cost**) expressed in $,
- the resource smoothness rate (**Rs**) calculated as an average deviation from the average resource use expressed in resource units.

First two criteria will take fuzzy values, while the last one real values.

In order to solve the example problem the presented two-stage method has been used. In the first stage the following parameters have been set for the FPSA procedure:

- conditions of changing the temperature (number of iterations at one temperature level) 2000
- the size of the sample of the generating solutions 8

- starting temperature 0.1
- stop condition temperature ≤ 0.001
- the temperature changing factor 0.9

The FPSA procedure has generated a set of 592 WCR-nondominated solutions.
In the second stage the PM analyses the set in the use of LBS procedure. The procedure starts from calculating ideal and the approximation of nadir points giving the PM information about possible ranges at each criterion. Here are the two points (the value of each fuzzy criterion is represented in two ways – by fuzzy number and its defuzzified counterpart according to the function $\Im$):

Point	Cmax	Rs	Cost
Ideal	**546** 525,536,548,548,555,565	**3.5**	**436866** 428391,431835,436656,436656,440788,448365
Nadir	**660** 645,652,659,659,667,682	**7.2**	**453670** 442847,448248,454420,454420,458277,464449

Now the PM can define the reference point that in general can be expressed by means of fuzzy numbers.

Cmax:=590
Rs:=5.5
Cost:=440000

The reference point is then projected onto the set of WCR-nondominated solutions. This results in the starting solution:

Cmax	Rs	Cost
601 589,595,603,603,606,611	**5.7**	**439669** 430831,435220,441071,441071,443266,446923

Then the PM may express his/her preferential information by specifying the indifference and veto thresholds on each criterion.

Criterion	q_j	p_j
Cmax	5	14
Rs	1	2
Cost	300	1000

This means that solutions differing by less than 5 days are treated by the PM as indifferent. A schedule shorter than 14 days is strongly preferred by the PM. In the outranking vicinity, three characteristic points have been found (each point locally minimizes one of three criteria):

	Cmax	Rs	Cost
1.	**601** 589,595,603,603,606,611	5.7	**439669** 430831,435220,441071,441071,443266,446923
2.	**596** 577,587,598,598,602,613	7.0	**439454** 425743,433121,441238,441238,444189,452305
3.	**606** 588,597,608,608,613,624	6.9	**438945** 425868,432386,440353,440353,443974,451941

In comparison with the reference point, the above solutions have low values of the project cost (**Cost**). However the project completion time (**Cmax**) is not satisfactory to the PM. Thus, he/she decides to change the reference point by decreasing **Cmax** at the expense of **Rs** and **Cost** co-ordinates. A new reference point is as follows:

Cmax:=560
Rs:=6
Cost:=450000

As expected, a new current solution is better on **Cmax** criterion.

Cmax	Rs	Cost
562 541,552,564,564,571,580	6.2	**449392** 432621,441417,451013,451013,456611,463808

Then, the PM changes thresholds on **Cost** criterion. He/she finds indifferent two schedules differing by less than 400$ and strongly prefers a schedule being cheaper by 1500$. Moreover, he/she changes the preference threshold on **Cmax** to 10 days.

Criterion	q_j	p_j
Cmax	5	10
Rs	1	2
Cost	400	1500

This results in the three characteristic points:

	Cmax	Rs	Cost
1.	**546** 525,536,548,548,555,565	7.2	**445969** 428751,437735,447535,447535,453251,461418
2.	**567** 547,557,569,569,577,585	6.2	**449136** 433039,440956,450456,450456,456789,463123
3.	**570** 551,562,573,573,578,588	7.1	**441755** 426489,435003,443518,443518,447388,455128

The PM chooses point #1 as the new current point. The corresponding characteristic points are as follows:

	Cmax	Rs	Cost
1.	**546** 525,536,548,548,555,565	7.2	**445969** 428751,437735,447535,447535,453251,461418
2.	**556** 535,546,558,558,564,573	7.1	**443770** 427307,436093,445678,445678,450470,457658
3.	**556** 535,546,558,558,564,573	7.1	**443770** 427307,436093,445678,445678,450470,457658

Then the PM wants to see some schedules being between points #1 and #2 in the current outranking vicinity. Here are the points:

	Cmax	Rs	Cost
1.	**546** 525,536,548,548,555,565	7.2	**445969** 428751,437735,447535,447535,453251,461418
2.	**551** 530,541,553,553,560,571	7.2	**445715** 428576,437471,447175,447175,452835,461730
3.	**552** 531,542,554,554,562,572	7.2	**445691** 428391,437265,446946,446946,453401,461468
4.	**556** 535,546,558,558,564,573	7.1	**443770** 427307,436093,445678,445678,450470,457658

Finally, the PM chooses the point #4 as the best compromise schedule. Taking into account crisp equivalents of each criterion the PM can see that the schedule is about 10 days longer than the shortest possible and about 104 days shorter than the longest possible. It has almost the worst (the highest) resource smoothness rate. The corresponding project cost is about 6904$ higher than the cheapest possible and about 9900$ lower than the most expensive.

Cmax	Rs	Cost
556 535,546,558,558,564,573	7.1	**443770** 427307,436093,445678,445678,450470,457658

The schedule can be presented in the following fuzzy Gantt chart (Figure 16:8):

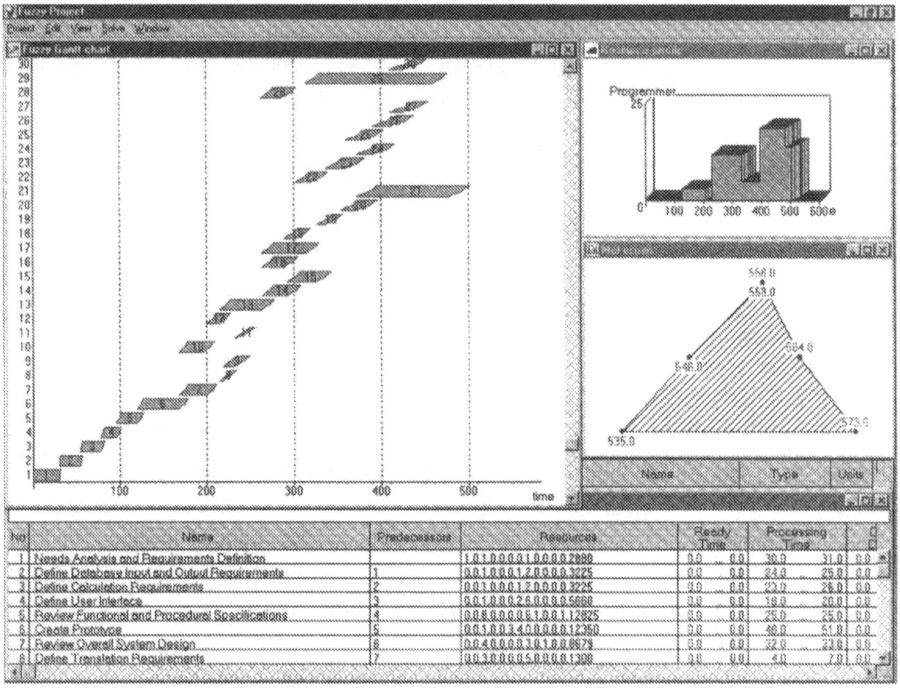

Figure 16:8. Best compromise fuzzy schedule

16.7. Conclusions

The paper presents a two-stage solution procedure for a multiple objective, multi-mode project scheduling problem under multi-category resource constraints and uncertain time parameters of activities modeled by fuzzy numbers. In the first stage, the Pareto-simulated annealing metaheuristic procedure is used to generate a representative sample of the set of approximately WCR-non-dominated schedules. The feasible schedules are constructed using a parallel scheduling heuristic generalized to handle fuzzy time parameters. The generalization consists in the use of special comparison rules for fuzzy numbers and of some arithmetic operations on fuzzy numbers. In particular, the concept of WCR-non-dominated schedules has been defined using one of the proposed comparison rules. In the second stage, an interactive procedure organizes the search for the schedule that yields the best compromise between conflicting fuzzy objectives over the sample of WCR-non-dominated schedules. To do so, an achievement scalarizing function has been extended to the fuzzy objective space. The approach has been applied to a hypothetical software project scheduling problem. The methodology presented is quite general and can be applied for solving any fuzzy multi-objective combinatorial optimization problem.

Acknowledgement

The authors wish to acknowledge financial support from State Committee for Scientific Research, KBN research grant no. 8T11C 013 13 and from CRIT 2 – Esprit Project no. 20288. The calculations were performed in Poznan Supercomputing and Networking Center.

References

BUCKLEY J.J. 1989. Fuzzy PERT, in: *Applications of Fuzzy Set Methodologies in Industrial Engineering*, G.W. Evans, W. Karwowski, M.R. Wilhelm (eds.), Elsevier, Amsterdam.

CHANAS S. 1987. Fuzzy optimization in networks, in: J. Kacprzyk, S.A. Orlovsky (eds.), Optimization models using fuzzy sets and possibility theory, Reidel Publishing Company, Dordrecht, 303-327.

CHANAS S., J. KAMBUROWSKI. 1981. The use of fuzzy variables in PERT, *Fuzzy Sets and Systems* 5, 11-19.

CZYŻAK P., A. JASZKIEWICZ. 1996. Metaheuristic technique for solving multiobjective investment planning problem, *Control and Cybernetics* 25, 177-187.

CZYŻAK P., A. JASZKIEWICZ. 1998. Pareto simulated annealing - a metaheuristic technique for multiple-objective combinatorial optimization, *Journal of Multi-Criteria Decision Analysis* 7, 34-47.

DEMPSTER A.P. 1967. Upper and lower probabilities induced by a multiple-valued mapping, *Annals of Mathematical Statistics* 38, 325-339.

DUBOIS D., H. PRADE. 1987. The mean value of a fuzzy number, *Fuzzy Sets and Systems* 24, 279-300.

DUBOIS D., H. PRADE. 1980. *Fuzzy Sets and Systems: Theory and Applications*, Academic Press, New York.

ELMAGHRABY S.E. 1967. On the expected duration of PERT type networks, *Management Science* 5, 299-306.

FORTEMPS P. 1997. Fuzzy Sets for Modelling and Handling Imprecision and Flexibility, PhD Dissertation, Faculte Politechnique de Mons.

FORTEMPS P. 1998. Jobshop scheduling with imprecise durations: a fuzzy approach, Technical Report, *IEEE Trans. on Fuzzy Systems* (to appear).

FORTEMPS P. 1996. M. Roubens, Ranking and defuzzification methods based on area compensation, *Fuzzy Sets and Systems* 82, 319-330.

GAUL W. 1981. Bounds for the expected duration of the stochastic project planning model, *Journal of Information and Optimization Sciences* 2, 45-63.

HAPKE M. 1995. Two-stage fuzzy optimization for project scheduling, In *Proceedings of the Operational Research of Italy Annual Conference, AIRO'95,* Ancona, 20-22 September 1995, 295-298.

HAPKE M., A. JASZKIEWICZ, R. SŁOWIŃSKI. 1994. Fuzzy project scheduling system for software development, *Fuzzy Sets and Systems* 21, 101-117.

HAPKE M., A. JASZKIEWICZ, R. SŁOWIŃSKI. 1997. Interactive analysis of multiple-criteria project scheduling problem, *Europ. J. Opl. Res.* (to appear).

HAPKE M., A. JASZKIEWICZ, R. SŁOWIŃSKI. 1997. Fuzzy project scheduling with multiple criteria, *Proceedings of Sixth IEEE International Conference on Fuzzy Systems*, FUZZ-IEEE'97, Barcelona, Spain, 1277-1282.

HAPKE M., R. SŁOWIŃSKI. 1996. Fuzzy priority heuristics for project scheduling, *Fuzzy Sets and Systems* 83, 291-299.

JASZKIEWICZ A., R. SŁOWIŃSKI. 1997. The LBS-Discrete interactive procedure for multiple-criteria analysis of decision problems, in: J. Climaco (ed.) *Multicriteria Analysis*, Springer-Verlag, Berlin, 320-330.

KERR R.M., R.N. WALKER. 1989. A job shop scheduling system based on fuzzy arithmetics, in *Proceedings 3rd International Conference on Expert Systems and the Leading Edge in Production and Operation Management*, Hilton Head, S.C., 433-450.

KOŁODZIEJCZYK W. 1986. Orlovsky's concept of decision-making with fuzzy preference relation – further results, *Fuzzy Sets and Systems*, 19, 11-20.

LOOSTMA F.A. 1966. Network planning with stochastic activity durations, an evaluation of PERT, *Statistica Neerlandica* 20, 43-69.

LOOSTMA F.A. 1989. Stochastic and fuzzy PERT, *Europ. J. Opl. Res* 43, 174-183.

MCCAHON C.S., E.S. LEE. 1992. Fuzzy job sequencing for a flow shop, *Europ. J. Opl. Res.* 62, 294-301.

PATTERSON J. H., R. SŁOWIŃSKI, F.B. TALBOT, J. WĘGLARZ. 1990. Computational experience with a backtracking algorithm for solving a general class of precedence and resource constrained project scheduling problems, *Europ. J. Opl. Res,* 49, 68-79.

PRADE H. 1979. Using fuzzy set theory in a scheduling problem, *Fuzzy Sets and Systems* 2, 153-165.

ROMMELFANGER H. 1990. FULPAL: An interactive method for solving (multiobjective) fuzzy linear programming problems, section 5 in: Slowinski R., Teghem J. Eds., *Stochastic Versus Fuzzy Approaches to Multiobjective Mathematical Programming under Uncertainty*, Kluwer Academic Publishers, Dordrecht, 279-299.

ROMMELFANGER H. 1994. Network analysis and information flow in fuzzy environment, *Fuzzy Sets and Systems* 67, 119-128.

ROSS T.J. 1995. Fuzzy Logic with Engineering Applications, McGraw-Hill Inc.

ROUBENS M. 1990. Inequality constraints between fuzzy numbers and their use in mathematical programming, section 7 in: R. Slowinski, J. Teghem Eds., *Stochastic Versus Fuzzy Approaches to Multiobjective Mathematical Programming under Uncertainty*, Kluwer Academic Publishers, Dordrecht, 321-330.

SERAFINI P. 1994. Simulated annealing for multiple objective optimization problems. In: G.H. Tzeng, H.F. Wang, V.P. Wen, P.L. Yu (eds*), Multiple Criteria Decision Making. Expand and Enrich the Domains of Thinking and Application*, Springer Verlag, 283-292.

SŁOWIŃSKI R. 1981. Multiobjective network scheduling with efficient use of renewable and non renewable resources, *Europ. J. Opl. Res.* 7, 265-273.

SŁOWIŃSKI R. 1989. Multiobjective project scheduling under multiple-category resource constraints. section I.7 in R. Słowiński, J. Węglarz J.(eds.): *Advances in Project Scheduling*, Elsevier, Amsterdam, 151-167.

SŁOWIŃSKI R., B. SONIEWICKI, J. WĘGLARZ. 1994. A DSS for multiobjective project scheduling, *Europ. J. Opl. Res.* 79, 220-229.

ULUNGU E.L. 1993. Optimisation Combinatoire Multicrit re: *Détermination de l'ensemble des solutions efficaces et méthodes interactives*, PhD thesis, Université de Mons-Hainaut.

WĘGLARZ J. 1980. On certain models of resource allocation problems, *Kybernetes* 9, 61-66.

ZADEH L. A. 1978. Fuzzy sets as a basis for a theory of possibility, *Fuzzy Sets and Systems* 1, 3-28.

17 KNOWLEDGE-BASED MULTIOBJECTIVE PROJECT SCHEDULING PROBLEMS

Jarosław Nabrzyski[1]
Jan Węglarz[2]

[1]*Poznań Supercomputing and Networking Center*
[2]*Institute of Computing Science, Poznań University of Technology*

17.1. Introduction

Multiobjective Project Scheduling Problems (MPS) have arisen in organizational settings ranging from construction planning to Research and Development. To date several papers have been published on this subject. In Słowiński, Soniewicki and Węglarz (1994) the Decision Support System (DSS) for the MPS was considered. In this system a set of heuristically generated, nondominated schedules was considered by the decision maker (DM) using an interactive procedure finding compromise schedules. Although pretty efficient, it was not satisfactory for one reason - the scheduling process often entails decisions that cannot be integrated into a mathematical model.

In Nabrzyski and Węglarz (1995) we proposed to solve this problem in a knowledge-based way. Such an approach to problem solving emphasizes the role of expert knowledge, which is often experimental and more easily represented symbolically, rather than the role of scientific theory, which is mostly described mathematically and/or algorithmically. Thus, in addition to the approach described in Słowiński, Soniewicki and Węglarz (1994) we proposed to use a knowledge based expert system (ES) which co-operated with a dialog method for finding compromise schedules. The scheduling problem was thought as an interactive search process involving the selection and sequencing of tasks and resources over time to achieve the problem objectives, subject to a set of precedence and resource constraints. The user could make the decisions in all the critical steps. The process of constructing a schedule was divided into two phases: the search phase and the interactive one. In general, during the search phase the system generated plenty of feasible schedules, from among which non-dominated ones were then found. Expert system techniques were used during this step. On the set of the non-dominated schedules the interactive procedure described in Słowiński, Soniewicki and Węglarz (1994) was organized yielding the best compromise solution.

In this chapter we present a new approach in which the process of constructing a schedule is done within one step. The proposed approach relies on the integration of multiobjective decision support, expert systems and metaheuristics in one problem solving environment that is sensitive to the key role played by the DM preferences during the problem solving process. As an example of such an approach we present a modern multicriteria tabu search algorithm (TS) that co-operates with the expert system to solve the MPS problem. Since in the real world there isn't a single expert who has the knowledge of the whole project to be solved, we decided to incorporate several experts to work on one problem. All these experts have a knowledge of different parts of the problem to be solved. The organization of their work is represented by the blackboard architecture, used previously in the HEARSAY system (Erman et al. (1980)). We show how the experts communicate with each other and with the TS algorithm using the blackboard.

In section 2 we formulate the general MPS problem. In next two sections we describe main concepts of the approach. Section 5 contains final remarks.

17.2. Basic notions

17.2.1. Multiobjective project scheduling problem

Generally stated the problem addressed is to find the best compromise schedule or a set of "good" compromise schedules for precedence and discrete resource-constrained activities of a project. Resource constraints can concern: (i) the amount available at every moment, (ii) the consumption over a given time period. If only the constraint (i) is imposed we speak about renewable resources, if only (ii) is defined - about nonrenewable resources, and if both (i) and (ii) - about doubly-constrained resources (Węglarz (1980)). The set of project performance measures is composed of time objectives (e.g. project duration, mean weighted flow time, mean weighted lateness) and cost objectives (e.g. weighted resource consumption, net present value - NPV). The specific form of the scheduling problem examined here permits each activity to be performed in one of several ways called processing modes of activities or simply modes. Each mode represents a different way of combining resources to accomplish a given activity. The duration and resource requirements of each activity mode are known a priori, and thus we can speak about resource-duration interactions. In this model a variety of multiple resource-duration interactions, such as using different technologies or types of labor to accomplish the same activity, can be expressed and evaluated.

The project is characterized by four components: set of resources $\boldsymbol{R}$, set of activities $\boldsymbol{A}$, precedence constraints in set $\boldsymbol{A}$, and set $\boldsymbol{Q}$ of project performance measures. Set $\boldsymbol{R}$ is composed of p,v and u types of renewable, nonrenewable and doubly-constrained resources respectively. Given are *n* activities belonging to ***A*** which have discrete resource requirements. For each activity $A_j \in \boldsymbol{A}$ we have in general *w* performing modes, i.e. feasible assignments of resource amounts to the activity. The performing mode *m* of activity A_j is defined by he vector r_{mj} $(m=1,2,...,w;\ j=1,2,...,n)$, whose elements determine the usage of renewable and doubly constrained resources and consumption per period (constant for every activity) of nonrenewable resources. For each mode *m* the duration D_{mj} of each A_j is also specified.

The precedence constraints are represented by a directed acyclic graph where activities are represented by integer labeled nodes, such that the label of the node is always higher then the labels of all its predecessor nodes. This is the representation called ACTIVITY-ON-NODE (AON). Dummy starting and finishing nodes 0 and n+1 respectively are attached to the graph. The precedence constraints can also be represented in a form of ACTIVITY-ON-ARC (AOA) graph. This method will be used in our computing example.

The set ***Q*** of project performance measures is composed of the following time and cost criteria:

- project completion time: CT,
- smoothness of the resource profile: S_{k_r} (k=1,2,...,p) and S_{dk} (k=1,2,...,u),
- mean weighted lateness: MWL,
- mean weighted flow time: MWFT,
- total resource consumption: TC_{kn} (k=1,2,...,v) and TC_{kd} (k=1,2,...,u),
- weighted resource consumption: WC,
- net present value: NPV.

Scheduling of the project defined in this way is based on the allocation of resources to activities in time in a way that all the activities are completed, all the constraints are satisfied and the best compromise between the criteria is reached.

We assume that the DM working with the problem solving environment does not posses the preference model concerning the solution (final schedule) at the beginning of the decision process.

In the next section we describe the expert system that aims to solve the problem defined above. Attention will be paid to the multicriteria TS algorithm for finding compromise solutions and its cooperation with the expert system.

17.3. The expert system for solving MPS problems

17.3.1. The idea of the system

Our aim was to build a system that would integrate the concepts from ES and DSS to produce a successful decision aiding system that explicitly considers the role of preference in the MPS problem solving process. In our opinion the MPS problems belong to a class of problems that a solution of which very heavily depends on user preference. When we speak of the preferences contained in the knowledge base we mean preferences addressed indirectly by the expert. On the other hand real DM preferences are the most important during the problem solving process. So, by combining both ES and DSS into one system one can describe the problem in a symbolic manner (easy to understand for the DM) and to take into account a real DM preferences introduced to the system within the framework of multicriteria TS procedure having some DSS features. Below we describe the system architecture and the main ideas of the approach.

17.3.2. System architecture

In our system the scheduling problem is thought as an interactive search process involving the selection and sequencing of activities and resources over time to achieve the problem objectives, subject to a set of constraints. In general, in the first step the system generates a set of feasible solutions, that are the start solutions for TS procedure. This procedure, together with the expert system find a solution (or a set of solutions) starting the search in parallel from the initial schedules. The system is composed of two modules (Figure 17:1): (1) expert system which runs as a kernel of the whole system and (2) parallel TS algorithm for multiobjective project scheduling. Exploring the search phase we will show the main concepts and possibilities of employing an expert system to control the behavior of the TS algorithm. This solution gives the DM a very flexible tool for defining his own tabu strategy.

All specific aspects concerning a search strategy for all the criteria are stored in the knowledge base in a form of rules and objects and control the processing of the algorithm. The algorithm is composed of the following modules: (1) definition of a schedule (configuration), (2) definition of a move and (3) generation of the neighbors of the current schedule.

The process of solution finding starts with generating a set of one or several initial schedules. This first phase is based on the project data stored in a database. The initial schedules are further improved using the TS algorithm together with the knowledge of the experts. Both, the TS and expert system generate candidates in the neighborhood of the current solution providing that the new schedules are feasible. In each iteration and admissible move is selected from the candidate list that has no tabu restrictions and has the best compromise between the criteria. Aspiration level criteria allow the tabu restrictions to be overriden to improve the schedule. This new schedule is then returned to the blackboard. The search stops after a specified number of iterations. This is the DM's role to decide if the obtained schedule fits his preferences. If no, the schedule goes back to the system and serves as a new initial schedule for the TS algorithm. On this stage the DM can also modify his preferences. The role of the expert system is crucial. It controls the behavior of the TS algorithm and drives the search into the certain regions of the set of nondominance solutions. When, during a search, some conditions which are stored in the knowledge base are satisfied the inference engine searches the rules to fire. The conclusions which come from the rules are then transferred to the algorithm and the search goes on using some new facts obtained from the expert system.

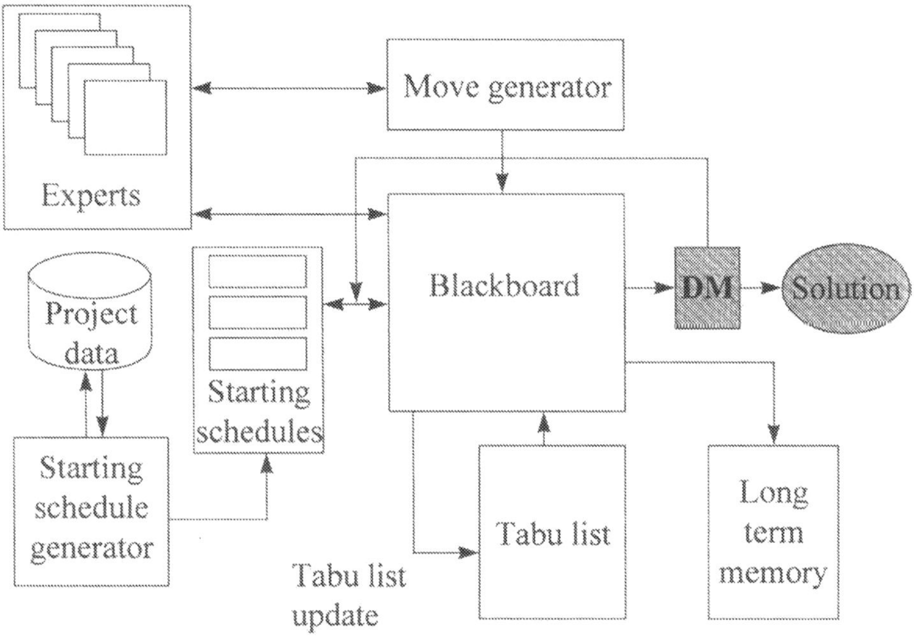

Figure 17:1. System architecture

The system described in this paper is based on the blackboard architecture. The idea of such an architecture is quite simple. The entire system consists of a set of independent modules, called knowledge sources (KS's), that contain the domain-specific knowledge (knowledge base and inference engine) of the system, and a blackboard - a shared data structure to which all the KS's have access (Figure 17:2).

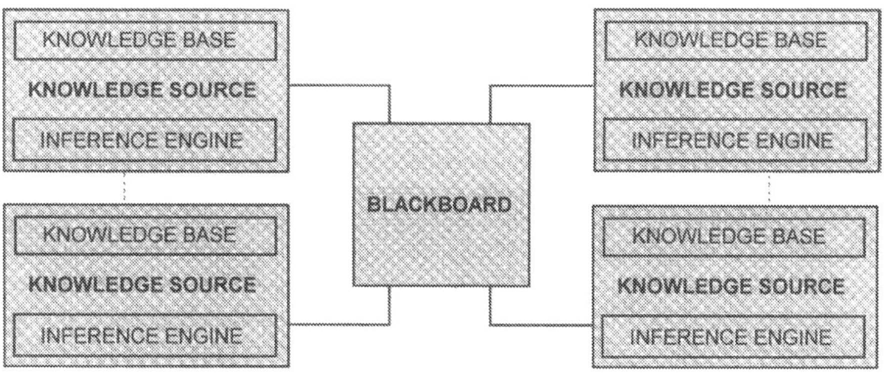

Figure 17:2. Blackboard architecture of the expert system

When a KS is activated it examines the current contents of the blackboard and applies its knowledge either to create a new hypothesis and write it on the blackboard, or to modify an existing one. Associated with each KS is a set of triggers that specify conditions under which the KS should be activated. These conditions are determined by particular experts. As mentioned before there are several types of experts in the system. Entire knowledge which is necessary to solve any problem is stored in KS. Each KS contains:

- activation conditions,
- knowledge base,
- inference engine.

KS can use one of three types of inference methods: forward, backward or mixed reasoning. Because there are different experts co-working on one problem (represented by the KS) every KS has its own independent inference engine. A blackboard, i.e. a data structure that is available to all experts, serves for communication and stores the active knowledge of the state of the problem solving process. The blackboard carries problem observations (symptoms, i.e. objects with their values), local abstractions (data abstractions with their values), hypotheses and intermediate and final solutions. All problem features are known from the beginning and are incorporated into the system as a knowledge bases and the initial contents of the blackboard. When a KS is activated, a particular expert examines the current contents of the blackboard and applies his knowledge either to create a new hypothesis and write it on the blackboard, or to modify an existing one. Observations and data abstractions can be read by all the problem solving experts. We have two typical interactions:

- certain experts write suggested hypotheses on the blackboard which are tested by other experts if there is no contradictory knowledge about the same situations,
- the experts write different intermediate and final results on the blackboard and the decision between them is taken on a global level.

To represent a domain knowledge one needs to choose the type of the knowledge representation and a special formalism to introduce the knowledge into the system. There are some specialized languages for knowledge representations available, just to mention Schema Representation Language (SRL) or Nested Interactive Array Language (NIAL). However these languages have some disadvantages. In the first place they are platform dependent and in the second they are not object oriented. Our system is implemented in the heterogeneous environment of metacomputer at Poznan Supercomputing and Networking Center. This metacomputer consists of SGI PowerChallenge, CRAY J916 and T3E, IBM RS6000/SP2. Thus it was necessary to define a specialized meta-language which would be able to be compiled on any computer architecture in a Unix environment. We also wanted this language to be object-oriented, as it is a natural way of thinking about such complex environment like MPS problems. This is why the special meta-language has been designed for the purpose of the system. It is used for knowledge representation and defining the knowledge sources, the blackboard and other mechanisms of the system. Following two paragraphs describe knowledge representation techniques used in the system and a specialized meta-language.

17.3.3. Knowledge representation

From among the widely used knowledge representation schemes we have chosen frames and production rules. The idea described in this chapter corresponds to Minsky's classical paper on frames (Minsky, (1988)). Frames are used to represent objects related to the description of project scheduling problems. A single frame is a structure with two kinds of slots: attributes and methods. Attributes represent descriptive knowledge, whereas methods are used to represent procedural knowledge. Frames allow to build a multi-layer system for modeling project scheduling problems. The basic concepts are those of states, object and acts (actions). The relations provide time ordering among states and actions. Casual relations define how states enable actions, and actions cause states. A scheduling operation is defined as an action which is linked to other states and actions by time and causality relations. Operations can also be defined on multiple levels of abstractions. Resources are defined as objects. Attributes and physical structure may be defined for an object. Allocation of resources is defined as a possession state of an activity or resource with a specified time relation (e.g. duration). Criteria are also represented with these primitives. The hierarchy of abstractions which is made possible by frame-based representation provides a good model of the problem while decomposing the domain knowledge into reasonable, manageable pieces. There are the following object types defined in the system:

- **Observations and data abstractions**. Observations correspond to the questions of heuristic classification and serve for primary data collection. Data abstractions are derived from the observations with simple rules. This type of object allows the project scheduling problem to be defined as it is,

- **Observation classes.** They group together observations and data abstractions and serve for the simplification of the dialog control or for assigning solutions to groups of data abstractions. They include the question classes of heuristic classification, the materials of the functional models and the set of the parameters of a state in fault models and combine their attributes,

- **Intermediate and final solutions**. They represent the current state of the problem solving process and include the combination of the attributes of the solutions and states of all the KS's. In particular, they can be valued differently by each KS,

- **Rules and constraints**. They express the knowledge of those object-object relationships which cannot be expressed by direct object references. Rules and constraints are connected to the objects concerned by a system of pointers. Thus, inference process can be implemented very efficiently by message passing between the objects. This requires that all the objects appearing in the rules must be known.

An object has only parts or refinements as its successors. A variable number of parts is allowed, but may be used only if the number of parts is of no great importance, so that a case distinction with specializing objects is not required. The number is represented as the parameter "number" of the part. This has the consequence that all parts must be refined in the same way, since they are represented as a single object.

The refinements of objects have the same parameters as their predecessors, but their parts usually have a different parameter structure. The specializing hierarchy therefore corresponds to the usual frame structure with the inheritance, attached procedures and default values. However, the parameters are represented as independent objects rather then attributes, the objects having pointers to their parameters, so that we require only a triple representation (object-attribute-value + parameter-attribute-value) instead of the quadruple representation (frame-attribute-facet-value). A parameter belongs potentially to several objects of a specializing hierarchy. For each refinement of an object the current value range of a parameter may be further restricted.

3.4. Specialized meta-language

The language we talk about is a specialized object-oriented meta-language which is used by experts to represent their knowledge and by knowledge engineers to define the structure of the entire system. It is an original high level programming language, called EXLAN (EXpert system LANguage), that contains some similar instructions to those from Pascal or C and instructions which are characteristic of the project scheduling problem and the system architecture. It uses some ideas of the language defined by Nabrzyski and Węglarz (1994) for the description of expert systems with the blackboard architecture. Below we describe only these elements of EXLAN which are characteristic for MPS problems and expert systems based on a blackboard architecture.

17.3.4.1. Object classes. A class of an object can be defined by giving to it a name (class identifier) and established number of slots. Slots are divided onto two sections: private and public. If a slot belongs to a private section then it cannot be inherited by successor classes. Other slots are available due to all rules of inheritance.

Each attribute is characterized by some features, namely it has its identifier, type and value. There exist three standard kinds of the type in EXLAN: NUMERIC, SYMBOL and BOOLEAN. Methods are characterized by another features. Except identifier they have a parameter list and a body. Identifiers of the methods and attributes belonging to one class are unique.

To define an instance of any class one needs to observe the following definition:

<class_instance>::=<instance_identifier> :
<class_identifier><instance_identifier>::=<identifier>

Access to slots of any object is possible by using a period.

EXLAN contains some special mechanisms to define the blackboard, knowledge sources and other parts of the expert system. The blackboard is divided onto some areas to which particular objects belong. To allocate the object to those areas one needs to declare them as follows:

<area_declaration> ::= STATIC [<object_declaration>]
DYNAMIC [<object_declaration>]
<object_declaration> ::= <object_identifier>{,<object_identifier>}

As we remember the entire knowledge which is necessary to solve any problem is stored in KS's. Each KS contains activation conditions, the knowledge base and the inference engine. Thus each KS is defined by assigning to it an identifier, list of the activation conditions, an inference method and a name of the knowledge base file.

Definition of the KS's is as follows:

<knowledge_source>::=KNOWLEDGE_SOURCE
<knowledge_source_identifier>
PRECONDITION <activation_condition_list>
PATTERN <list_of_connectivity>
METHOD <list_of_inference_methods>
BASE <file_name>
<knowledge_source_identifier> ::= <identifier>
<activation_condition_list> ::= <connection>{;<connection>}
<list_of_inference_methods>::=<identifier>{;<identifier>}
<file_name>::=<identifier>.

<file_name> is the identifier of the file which contains the set of the rules. It is a sequential file of the rules. Semantics of the rule is defined as follows:

<rule> ::= RULE # <rule_identifier>
IF<condition> THEN<action>[{<comment>}].
<rule_identifier> ::= <identifier>
<condition>::=<logical_expr.> {AND/OR <logical_expr.>}
<action> ::= <instruction>{AND<instruction>}

Condition part of a rule is similar to a normal logical expression. One can use AND/OR keywords to combine some logical expressions. Conditions in the system mainly concern such aspects as: availability of resources, activity and system states and schedules. An action part of the rule is a list of instructions separated with the AND keyword. There are some standard actions defined in our system, for example:

- actions that are responsible for calling particular operations to find a solution;
- actions that express object features; as an example we can give changing of a activity performing mode if the actual one is not efficient;
- actions that are responsible for error detection and for coupling with those errors.

Rule identifiers are unique. Each rule can be followed with a comment explaining the meaning of the rule.

17.3.5. Applying an expert knowledge to solve the MPS problems

We propose to use an expert knowledge on some levels of the problem solving process. Such a knowledge could be acquired from experts and from literature and should concern all aspects of the MPS problems.

There are some standard classes of objects defined in the system which are characteristic for project scheduling problems. Figure 17:3 depicts the basic structure of standard classes. As we can see, there are four parent classes defined in the system: TActivity, TResource, TObjective and TSchedule. Other classes are child classes of parent ones. Of course the user can define his own object classes.

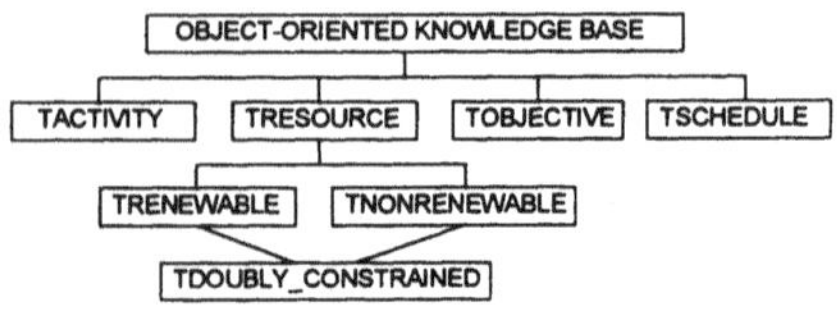

Figure 17:3. Sample classes in the system

The sample class definition for TActivity is presented in the Table 17:1.

Table 17:1. Definition of the TActivity class

TActivity Class
CLASS: *TActivity*;
SLOT: *READY_TIME*: NUMERIC;
SLOT: *DUE_DATE*: NUMERIC;
SLOT: *PREDECESSORS*: *SYMBOL;
SLOT: *PROCESSING_MODE*: *SYMBOL;
SLOT: *STATUS*: SYMBOL;
SLOT: *PRIORITY*:SYMBOL;
METHOD: *INIT(...)*;
METHOD: *SCHEDULE_ACTIVITY(...)*
METHOD: *RESCHEDULE_ACTIVITY(...)*

PREDECESSORS is a list of all the predecessors of the particular activity. The semantics of this slot is as follows:

<*PREDECESSORS*>::=(<ActivityName>,...,<ActivityName>)
<ActivityName>::=<identifier>

PROCESSING_MODE's semantics is as follows:

<*PROCESSING_MODE*>::=<Processing_mode_list>
<Processing_mode_list>::=(([<ResourceRequired>,<Amount>],<Duration>),..., ([<ResourceRequired>,<Amount>],<Duration>)).

For every activity we can define its status, which describes its characteristics during project processing. Possible statuses are presented in the Table 17:2.

Table 17:2. Possible statuses of activities

IF Status(A_i) = 0, then the activity is non-schedulable 1, then the activity is schedulable 2, then the activity is being processed 3, then the activity has been completed 4, then the activity satisfies the first of the two conditions in the definition of activity schedulability 5, then the activity satisfies schedulability condition only from the hard constraint point of view (but doesn't satisfy this condition from the point of view of the soft constraints). Activity A_i is schedulable at time t if all activities preceding A_i have been completed before time t and all resources required by activity A_i are available at time t.

SCHEDULE_ACTIVITY and *RESCHEDULE_ACTIVITY* methods are constructed to assign and reassign any resources to the activity. Reassigning is necessary if system comes into an error point i.e. a system state in which one or more project constraints are not satisfied. *INIT* is a method which is characteristic for every object. One has to use this method to initialize any object he/she wants.

In this way we can define the descriptive knowledge about the problem to be solved. Rules represent another kind of knowledge in the system - procedural knowledge. In fact, rules are the main element of the knowledge base. As it was mentioned before they are thought as objects of special type and are divided into some classes depending on the problem, which particular class handles. We present only some of the rule categories.

17.3.5.1. Rules. We have defined the following groups of rules in the system:

- Rules that transform enhanced constraints to the standard ones without interferencing in scheduling procedure. Using rules from this group one can do the following:

  - create groups of activities, constraints etc. taking into account some common features of these objects (for example resources required, performing modes, expected completion times, ready times and so on)

  Example:
  RULE #1
  IF TACTIVITY.DUE_DATE < 18THEN GROUP(GROUP1)
  {the rule creates a group (GROUP1) of activities whose DUE_DATE<18}

  - assign priorities to the activities

  Example:
  RULE #45
  IF SUC_AMOUNT(Activity1) = MAX(SUC_AMOUNT(TACTIVITY))

THEN Activity1.SCHEDULEACTIVITY(...) { take the activity with the largest number of successors}

- create constraints for groups of activities, resources etc. (for example: activities from A group have to be completed before activities from B group),
- reserve the resources for some activities with the highest priority

Example:
RULE #69
IF Activity4.PRIORITY = HIGH AND
(Resource6 IN(Activity4.PROCESSING_MODES) OR
Resource12 IN(Activity4.PROCESSING_MODES)
THEN Reserved((Resource6, Resource12), Activity4, t1, t2)
{reserve the resources 6 and 12 for activity 4 in the period between t1 and t2}

- determine the period between completion of two activities

Example:
RULE #66 IF Activity1.STATUS = DONE
THEN Activity2.READYTIME :=Schedule1.CURRENT_TIME + 2

- enhance precedence constraints

Example:
RULE #24 IF ActivityNew.PRIORITY = HIGH AND ActivityX.STATUS = 0
THEN ActivityX.PREDECESORS := ActivityX.PREDECESSORS + ActivityNew

- Rules that select an appropriate algorithm to solve the considered problem. There exists the heuristic algorithm base in the system. This base contains some heuristic algorithms such as B&B, simulated annealing, TS to name just a few. All these algorithms are able to interact with the knowledge base of the expert system to solve the problem in a strict cooperation. The rules form this group are also responsible for taking a control over the chosen algorithm.

Example:
RULE #20 IF PROBLEM_TYPE = PS
THEN ALGORITHM = PSA
RULE#21 IF Number_of_activities < 60
AND AMOUNT(TACTIVITY.PROCESSINGMODES) = 1
AND Number_of_objectives = 1
THEN ALGORITHM := MILPF {MILPF = Mixed Integer Linear Programming Formulation}

- Rules from this class are responsible for soft constraints relaxation and dealing with conflicts

Example:
RULE #55
IF Activity12.MODE = (Resource1, 5, 7, 15) AND
Resource1.FREE_USAGE<5
AND Resource.FREE_CONSUMPTION < 7
THEN Activity12.Schedule(...)

- Rules from this class are responsible for estimating and assignment every solution to particular group of solutions. This way we minimize the set of schedules which will be presented to the DM.

We have shown just a few rule classes that are invoked in the knowledge bases. All these rules are used during the search phase of the problem solving process

17.3.6. Multicriteria TS algorithm

The scheduling problems considered here are NP-hard even for single-criteria formulations. In the literature one can find several applications of heuristics that solve single criteria project scheduling problems. The survey of these heuristics can be found in one of the preceding chapters of this book. The benefits of the TS algorithm allow us to incorporate multicriteria decision support into the TS as well as to transfer the TS strategy (finding a neighbor, tabu moves, tabu list) to the knowledge base of the expert system. The final product of this approach is the multiobjective TS procedure together with a specialized expert system and decision support for interaction with DM that generate preference driven solutions.

TS is a metaheuristic procedure which facilitates embedded search procedures to escape from local optima. It was developed by Glover (1990) and some extensions were made by Glover and Laguna (1993 and 1994). The general search algorithm starts with an initial schedule of the problem. This is the first problem that has to be solved - finding one or more initial schedules. The procedure then iteratively selects a best move from the neighborhood of solutions to improve the solution quality. Restrictions are imposed on the moves to differentiate the admissible moves from the tabu moves. The most common tabu moves are those that attempt to reverse or repeat previous moves. By restricting this moves for a fixed time period, called tabu tenure, the search may be driven out of local traps. This restriction can be released at critical moments where this seemingly tabu move may actually lead to better solutions. This is called the aspiration level. By enforcing the tabu restriction, the short term memory structure in TS procedure encourages the search to focus on a more promising region, while the long term memory allows the search to become more diversified. Our decision about applying the TS algorithm to our approach was based on the fact, that it occurred a very efficient procedure for finding good quality solutions for many single objective optimization problems. TS has been successfully applied to hard instances of several well known combinatorial problems like Traveling Salesman Problem (Malek et al. (1989)), quadratic assignment problem (Skorin-Kapov (1990)), and production scheduling such as job shop (Dell'Amico and Trubian (1993)), flow shop (Widmer and Hertz (1989)) etc. Moreover the TS seems to be a very good algorithm to co-operate with expert systems. In the paper we show several levels of such a co-operation, specially when implementing tabu list which is controlled by the expert system.

Multicriteria approach within tabu search has been previously studied by Hansen (1996). He proposes the Multiobjective Tabu Search (MOTS) for multiobjective optimization problems. Within this approach the objective function vectors of the solutions are sought to cover the whole frontier and repeatedly for each solution, an optimization direction is made so that it tends to move away from the other points while moving towards the non-dominated frontier. The algorithm is proposed to be used as a generator of approximations to the non-dominated set to the other programs as for example the light beam search presented in Jaszkiewicz (1996). Thus, in the proposed method the best compromise solutions are found in two-steps.

Some other multiobjective metaheuristics have been presented by Serafini (1992) - simulated annealing, Srinivas and Deb (1995) - genetic algorithms.

The approach presented in this chapter allows to find the best compromise solution within one step, allowing the DM to take part in the decision making process since the very beginning of the system start up. It is achieved by the control and support of the expert system.

The approach described here is based on the concept of partial and complete preorders on the set of alternatives, i.e. the schedules generated in every iteration of TS. The framework for presented TS procedure incorporates efficient candidate move generation, evaluation and selection (based on DM preferences and the knowledge base) to both intensify and diversify the search for a schedule (or a number of schedules) and finally to find a schedule which fits the best the DM preferences. In this approach we merge techniques applied to expert systems, multicriteria decision support and metaheuristics.

In this paragraph we discuss some critical elements of the multiobjective TS procedure that impact solution quality and its application to MSP. These problems are discussed in the following.

17.3.6.1. Initial schedule. The choice of a good initial schedule is critical to the successful performance of TS. An initial schedule is generated by the application of some simple heuristic rules. We consider the following rules: Shortest Processing Time (SPT), Longest Processing Time (LPT) etc.

The TS approach implemented in the system uses some different solutions as starting points from the schedules generated by the expert system using the priority rules and evaluates them in parallel on different processors or parallel computers.

All the solutions are represented in the form of an activity list:

$$\lambda = [A_i, A_j, ..., A_n]$$

in which each activity must have a higher index than each of its predecessors. Additionally to the activity list, the corresponding activity mode list:

$$ML = [m_i, m_j, ..., m_n]$$

and activity completion time list:

$$CTL = [ct_i, ct_j, ..., ct_n]$$

are given.

17.3.6.2. Neighborhood generation strategy. At each iteration, a candidate list of moves must be identified. Previous studies have shown that a choice of a neighborhood has a significant impact on the schedule quality and generation time (Glover (1990)). While a large number of candidate moves generated from multiple rules may increase the likelihood of obtaining good solution, the computational effort for selecting and updating these move attributes will increase as well. The tradeoff between efficiency of search and effectiveness of solution quality calls for a simple neighborhood structure and a good move selection strategy. There are many ways to generate a neighborhood of a move. The objective is to generate a small enough candidate list of alternatives that are likely to be superior to the current move (solution). Neighborhood generation strategy used in our approach is based on the restricting the number of possible moves by choosing such moves that cause only some changes between the activities with a better trade-offs between all the criteria taking into account the DM preferences. Such a strategy allows to explore a short term memory in the sense of TS memory strategies. The second possible strategy can limit the number of possible moves by applying the considered moves only when the new schedules will give the same or better outranking positive and negative flows for the best alternative in comparison to the current schedule. The idea here is based also on using some specialized rules (expert knowledge) to help the TS algorithm to restrict a neighborhood search to the smallest possible level. The possibility to transfer the TS strategy outside the algorithm and program itself and placing it in a knowledge base of the ES in a special text form makes the system more flexible. To change the behavior of TS algorithm one needs just to change the knowledge state of the system. So all the changes to the strategy can be made without compiling the program after every change. Below we present a sample rules that restrict the moves enlarging unfeasibility of a schedule. Let λ be a current solution, $v(\lambda)$ be a measure of unfeasibility of λ, λ' be the neighbor solution for which one of the criteria functions, let's say $c(\lambda')$, is minimized (hence, typically $v(\lambda')>0$) and λ'' be the neighbor solution for which $v(\lambda'')-v(\lambda)$ is minimized. Then the best neighbor can be defined using the rules from the Table 17:3:

Table 17:3. Possible best neighbor policy

RULE#12

IF $v(\lambda)>0$ THEN *BestNeighbor*$(\lambda):=\lambda''$

RULE#13

IF $v(\lambda)=0$

AND $v(\lambda'')>0$ THEN *BestNeighbor*$(\lambda):=\lambda'$

RULE#14

IF $v(\lambda)=0$ AND $v(\lambda'')=0$ AND $c_j(\lambda'')\leq t_j$

THEN *BestNeighbor*$(\lambda):=\lambda''$

RULE#15

IF $v(\lambda)=0$ AND $v(\lambda'')=0$ AND $c_j(\lambda'')>t_j$ THEN *BestNeighbor*$(\lambda):=\lambda'$

As we can see in the example, when a current solution is infeasible, the best neighbor is the one that reduces unfeasibility the most. When the search is inside the feasible region but no neighbor exists within this region, the search moves to the infeasible neighbor with the hardest positive and negative flows within outranking approach. Nonimproving moves within the feasible region are accepted as long as the particular objective function values do not exceed the threshold t_j. In opposite, when the objective function values of the neighboring feasible solution exceed t_j, the search is forced into unfeasibility. Every time the search enters the feasibility, the state is kept in memory to get the best solution found so far.

17.3.6.3. Neighborhood structure. In the general MPS problem, a move can be generated by the modification of a current schedule. This modification can be based on a pairwise interchange (swapping) of the start time of two activities, by inserting an activity in front of another activity or by choosing another processing mode for a particular activity. We incorporate all these types of moves in generating the neighborhoods by the definition of some operators which transform the current schedule to the new one. These operators, namely SWAP, INSERT and CHANGE_MODE are called in every iteration of the TS procedure to generate different solutions from the current one. The operators are applicable only if the resulting schedule is feasible. One needs to observe that making one SWAP, INSERT or CHANGE_MODE operation it is possible that some others will be required.

In order to limit the size of the candidate list, we do not allow violation of precedence and resource constraints. However, we allow to define some soft and hard constraints while defining the problem to be solved. Soft constraints can be violated using special group of rules described in paragraph 17.3.5.

17.3.6.4. Evaluation functions. After the neighborhood is defined, a move needs to be selected from the set of admissible moves or candidate moves. The move is evaluated using tabu restrictions, and the PROMETHEE method embed into the procedure. In every iteration of TS we deal with the following multiobjective optimization problem:

$$\text{Min (Max)}\ \{f(\lambda) \mid \lambda \in \mathbf{N}\},$$

where $f(\lambda)$ stands for criteria functions to be optimized and **N** is a finite set of feasible schedules belonging to the neighborhood of the current schedule.

On every step of TS algorithm we deal with selecting a single move or as restricted as possible subset of moves containing the most satisfactory actions. So it is almost intuitive to choose the approach based on the outranking approach (Brans and Mareschal (1992)). The approach relies on building an outranking relation to represent the DM preferences. The second step aims to exploit of this relation in order to help the DM in his choice. Because the decision should be made on every iteration of TS algorithm, we distinguish between a real DM (the user) and the semi-automatic, artificial DM incorporated into the TS algorithm and making decisions basing on real DM preferences. In every iteration of the TS procedure we incorporate the PROMETHEE method. We denote it as a function:

$$\text{CHOOSE\_THE\_BEST\_MOVE}(N, \lambda_{S'}),$$

where $\lambda_{S'}$ is a chosen move.

17.3.6.5. Choosing the next move. The PROMETHEE method is particularly appropriate to treat multicriteria problems of the type given above. It allows to consider the amplitudes of the deviations between the criteria values. The method eliminates the scaling effects and allows the DM to express his incomparability relations among criteria. The method is understandable by the DM because it includes no technical parameters without economical significance. PROMETHEE allows also the conflictual analysis of the criteria. Thus, the DM has the opportunity to appreciate the criteria expressing the same, independent or opposite preferences. The method is based on three main steps: (1) enrichment of the preference structure, (2) enrichment of the dominance relation, (3) exploitation for decision aid. The DM can use PROMETHEE I, which provides a partial relation on A, including possible incomparabilities or PROMETHEE II, which provides a complete ranking, is more efficient but the information used is more disputable.

Taking into account that we incorporate PROMETHEE into TS procedure, some modifications are required. Here again some features of PROMETHEE are transferred to the knowledge base of the expert system in a form of rules. First, the preference function $P(\lambda_a, \lambda_b)$, giving the intensity of preference of λ_a over λ_b as a function of deviation d is represented in the form of rules. Having $d= f(\lambda_a)-f(\lambda_b)$ these rules are as follows:

Indifference rule:

RULE #1

IF $d \leq 0$ THEN $P(\lambda_a, \lambda_b)=0$

Weak preference rule:

RULE #2

IF $d > 0$ THEN $P(\lambda_a, \lambda_b) \approx 0$

Strong preference rule:

RULE #3

IF $d >> 0$ THEN $P(\lambda_a, \lambda_b) \approx 1$

Strict preference rule:

RULE #4

IF $d >> > 0$ THEN $P(\lambda_a, \lambda_b) = 1$.

A generalized criterion is also transferred in the form of rules to the knowledge base. Graphical form of generalized criterion presented in Figure 17:4, is transferred to rules, as shown in the example below for usual criterion:

Usual criterion rule:

RULE #1 IF d=0 THEN H(d) = 0 ELSE H(d)=1.

Usual criterion $H(d) = 0, d=0$ $H(d) = 1, \lvert d \rvert=0$	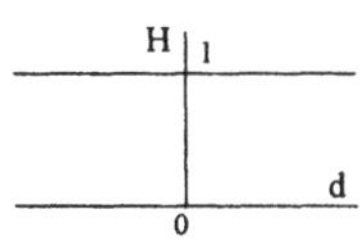
V-shape criterion $H(d) = \lvert d \rvert/p, \ \lvert d \rvert \leq p.$ $H(d) = 1 \ , \ \lvert d \rvert > p.$	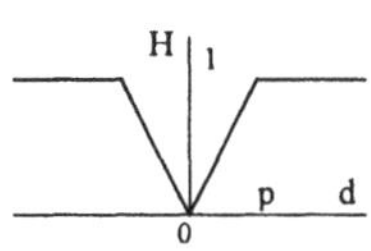
Gaussian criterion $H(d)=1-\exp(-d^2/2s^2)$	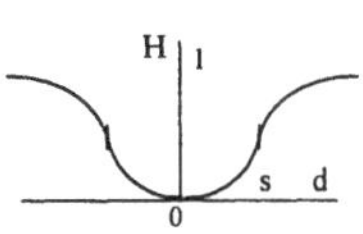

U-shape criterion H(d) = 0, $\|d\| \leq q$ H(d) = 1, $\|d\| > q$	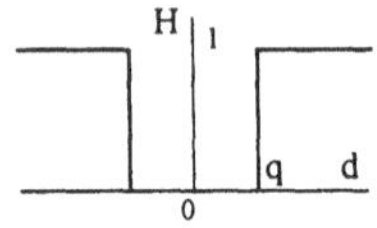

Figure 17:4. Generalized criteria

The user simply adjusts the p, q and s parameters in a graphical based way by dragging the appropriate line and then the drawing is transformed into a rule readable by the system. During the next step, after for each criterion a generalized criterion has been defined, the outranking graph is being built. Within the PROMETHEE a preference index $\pi(\lambda_a, \lambda_b)$ of alternative λ_a over λ_b over all the criteria is calculated. For each pair of possible moves and for each criterion j we dispose on:

$$\forall \lambda_a, \lambda_b \in \mathbf{N}, \forall j = 1,2,...,k, \{ f_j(\lambda_a), f_j(\lambda_b), d_j = f_j(\lambda_a) - f_j(\lambda_b), P_j(\lambda_a, \lambda_b)\} \text{ and}$$

$$\pi(\lambda_a, \lambda_b) = \Sigma\, w_j\, P_j(\lambda_a, \lambda_b),\ j = 1,2,...,k\ \ \Sigma\, w_j = 1,$$

where w_j , j =1,2,...,k are weights associated to each criterion. In general these weights are given by the DM at the beginning of the TS procedure.

If the DM doesn't posses any model of preference, we consider equal weights. The weights can be changed by the DM during the problem solving process. In the case of the same weights we have:

$$\pi(\lambda_a, \lambda_b) = (\Sigma\, P_j(\lambda_a, \lambda_b))/k,\ j=1,2,...,k.$$

For each pair λ_a, λ_b of possible moves from **N** the values $\pi(\lambda_a, \lambda_b)$ and $\pi(\lambda_a, \lambda_b)$ are calculated. This implies a complete outranking graph on the set of possible moves. For each possible move the positive (a) and negative (b) outranking flows are constructed:

$$\text{(a) } \Phi^+(\lambda_a) = \Sigma\, \pi(\lambda_a, \lambda_i) / (n-1),\ i=1,2,...,|\mathbf{N}|$$

$$\text{(b) } \Phi^-(\lambda_a) = \Sigma\, \pi(\lambda_i, \lambda_a) / (n-1),\ i=1,2,...,|\mathbf{N}|$$

These outranking flows express how each possible move is outranking all the others and how it is outranked by all the others respectively. Basing on the outranking flows the partial preorder is then constructed:

$$\lambda_a S^+ \lambda_b \text{ if } \Phi^+(\lambda_a) > \Phi^+(\lambda_b) \text{ and } \lambda_a I^+ \lambda_b \text{ if } \Phi^+(\lambda_a) = \Phi^+(\lambda_b)$$

and

$$\lambda_a S^- \lambda_b \text{ if } \Phi^-(\lambda_a) < \Phi^-(\lambda_b) \text{ and } \lambda_a I^- \lambda_b \text{ if } \Phi^-(\lambda_a) = \Phi^-(\lambda_b)$$

The PROMETHEE I partial relation is the intersection of these two preorders:

$$\begin{cases} \lambda_a P^I \lambda_b \text{ if } \begin{cases} \lambda_a S^+ \lambda_b \text{ and } \lambda_a S^- \lambda_b \\ \lambda_a S^+ \lambda_b \text{ and } \lambda_a I^- \lambda_b \\ \lambda_a I^+ \lambda_b \text{ and } \lambda_a S^- \lambda_b \end{cases} \\ \lambda_a I^I \lambda_b \text{ if } \quad \lambda_a I^+ \lambda_b \text{ and } \lambda_a I^- \lambda_b \\ \lambda_a R \lambda_b \text{ otherwise} \end{cases}$$

where P, I and R state respectively for preference, indifference and incomparability. On the basis of such an information, the artificial DM can make the following conclusions: (1) λ_a outranks λ_b, (2) λ_a is indifferent to λ_b or (3) λ_a is incomparable to λ_b. In the first case the information is sure. In the second one both positive and negative flows of both possible moves are respectively equal. In the latter case the artificial DM is not able to decide which alternative is better. But in such a case the complete preorder is automatically generated. It is not possible to ask a real DM in every iteration of the TS procedure what to do in such a situation. Thus, we decided to generate the complete preorder. Net outranking flow is first calculated as:

$$\Phi(\lambda_a) = \Phi^+(\lambda_a) - \Phi^-(\lambda_a)$$

The higher the net flow the better the alternative. The complete ranking is then defined by:

$$(P^{II}, I^{II}): \lambda_a P^{II} \lambda_b \text{ if } \Phi(\lambda_a) > \Phi(\lambda_b) \text{ and } \lambda_a I^{II} \lambda_b \text{ if } \Phi(\lambda_a) = \Phi(\lambda_b)$$

In this way incomparability is removed.

The next move chosen by PROMETHEE method is then projected on the tabu list of the TS algorithm and the decision is made on the basis of its consistence. If the move is not marked as a tabu or any other restrictions do not apply the move is evaluated as the current best schedule. Otherwise the succeeding move from the partial relation or the complete ranking is taken on the same basis. The TS procedure runs a certain number of iterations (see *Termination conditions*) and it collects all the schedules that fit into the DM preferences.

We use a GAIA visual modeling technique (Brans and Mareschal (1992)) for visualizing the schedules that are presented to the DM. GAIA provides a clear view of the alternatives being particularly good (or not) on the different criteria. All the alternatives and criteria are divided into the clusters of similar alternatives and the DM can choose the one that fits his preferences. A decision axis along which the DM is invited to select the best alternatives, according to the weights allocated to each criteria is also provided. When the criteria are strongly conflicting, the decision axis is rather short and good compromise schedules are located close to the origin. In the example below, the schedules λ_3 and λ_4 should be the best compromises.

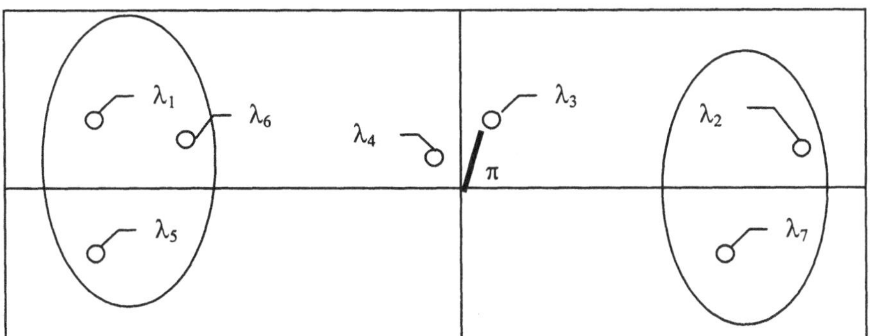

Figure 17:5. Visualizing solutions with the GAIA plane

17.3.6.6. Move attributes. An attribute of a move from λ to λ_N can encompass any aspect that changes as a result of the move. A single move evidently can cause multiple attributes to be assigned to it.
In the Table 17:4 we present some TS rules that represent move attributes for a move λ_1 to λ_N, where λ_N is a tentative solution.
Move attributes are used to impose tabu restrictions, i.e. constraints that prevent moves from being chosen if the choice would reverse the changes represented by these attributes.

17.3.6.7. Tabu restrictions. A major feature of the TS procedure is the set of restrictions imposed on candidate moves to prevent cycling. Many successful applications recommend using a simple neighborhood strategy and appropriate evaluation functions with more complex candidate selection and evaluation criteria. A move is evaluated based on the tabu status in addition to the evaluation functions described above. There are three kinds of tabu restrictions corresponding to the three types of neighborhoods defined earlier (Swap, Insert, ChangeMode). The basic idea is to try to avoid a reversal of moves. For example, suppose an activity A_i is just swapped with activity A_j, then a move which attempts to swap A_i and A_j back in the very next iteration is abandoned. Similarly, if an activity A_j is inserted in front of activity A_i, in one iteration, the attempt to move it back immediately thereafter is a forbidden move and another mode for an activity is chosen, the attempt to choose back the previous mode is also forbidden for a certain period of time (number of iterations).

There are special rules designed in the knowledge base of the expert system that control the strategy of tabu restrictions. These rules are transformed to data on the blackboard which is readable also by TS algorithm, as explained in the system architecture picture. In Table 17:5 we present some sample rules that represent simple tabu restrictions.

Table 17:4. Move attributes

RULE #1
IF *Changed(VARIABLE1, 0, 1)* THEN *MoveAttribute:=A1;*
{*Changed(VARIABLE1, 0, 1)* means change of selected *VARIABLE1* from 0 to 1}
RULE #2
IF *Changed(VARIABLE2,1, 0)* THEN *MoveAttribute:=A2;*
RULE #3
IF *Changed(VARIABLE1, 0, 1)* AND *Changed(VARIABLE2,1, 0)* THEN *MoveAttribute:=A3;*
RULE #4
IF *Changed(level_value1, CRITERIA_FUNCTION(λ_1), CRITERIA_FUNCTION(λ_N))* THEN *MoveAttribute:=A4;*
RULE #5
IF *Changed(level_value2, CRITERIA_FUNCTION(λ_1), CRITERIA_FUNCTION(λ_N))* THEN *MoveAttribute:=A5;*

17.3.6.8. Aspiration level. Tabu restrictions limit move selection to avoid cycling behavior. However under certain circumstances, it is possible to find the move which is classified as tabu but actually it is an excellent move that may lead to a better solution than is currently available. An aspiration criterion allows the tabu status of a move to be overridden if a solution is better than the current best solution. If the aspiration level is achieved, the tabu restriction of the candidate schedule is removed and the move is executed.

Proper use of such criteria is very important for the performance of the TS algorithm. One can differentiate two kinds of aspirations: *move aspirations* and *attribute aspirations*. When a move aspiration is satisfied it revokes the move's tabu classification. When an attribute aspiration is satisfied it revokes the attribute-active status. Let us remind that an attribute is tabu-active when its associated reverse attribute has occurred within a stipulated interval of regency or frequency in the past moves. There is a special group of rules in the knowledge base that define these aspirations. The user can define such aspiration criteria like aspiration by default, aspiration by objective, aspiration by influence etc. As an example we show the aspiration by objective in a global form which is the most commonly used. It is also implemented in the form of knowledge base rules.

Table 17:5. Sample tabu restrictions

RULE #6
IF *Changed(VARIABLE1, 0, 1)*
AND *PreviousChanged(VARIABLE1, 1, 0)* {x_j previously changed from 1 to 0}
THEN *Move*(λ, λ_N)*:=Tabu;*
RULE #7
IF *Changed(VARIABLE2,1,0)* AND *PreviousChanged(VARIABLE2,0, 1)*
THEN *Move*(λ, λ_N)*:=Tabu;*
RULE #8
IF *(Changed(VARIABLE1, 0, 1)* AND *PreviousChanged(VARIABLE1, 1, 0))*
OR *(Changed(VARIABLE2, 1, 0)* AND *PreviousChanged(VARIABLE2, 0, 1))*
THEN *Move*(λ, λ_N)*:=Tabu;*
RULE #9
IF *(Changed(VARIABLE1,0,1)* AND *PreviousChanged(VARIABLE1,1, 0))*
AND *(Changed(VARIABLE2,1,0)* AND *PreviousChanged(VARIABLE2, 0, 1))*
THEN *Move*(λ, λ_N)*:=Tabu;*

17.3.6.9. Candidate list size and tabu tenure. The number of moves recorded and evaluated at each iteration is not limited in our study. The size of the tabu list restricts the number of iterations that a particular move is prohibited. Tabu list structure is controlled in our system by special rules that check the behavior of the DM and his preferences as well as the influence of the tabu list size on the process of solution finding. Basing on this knowledge the tabu list size changes dynamically between the suggested values.

17.3.6.10. Termination condition. Given that TS is an iterative procedure, many stopping criteria can be used to terminate the search process. Three stopping criteria are possible to be defined in our system: the first specifies the number of iterations; the second specifies the number of iterations there is no improvement for a certain group of criteria due to the DM preferences; and the third applies a time limit for each iteration.

17.3.6.11. Memory structure. TS allows the use of both short term and long term memory structures through the use of tabu lists and multiple initial schedules, respectively. As discussed earlier, we use both structures in our design and experimentation.

17.3.6.12. The algorithm. General concept of the multicriteria TS algorithm is presented in the Table 17:6.

Table 17:6. Multicriteria TS algorithm

---

begin

counter:=0

Initialise(KnowledgeBase)

choose an initial schedule $\lambda_S \in \mathbf{S}$; {**S** is a set of feasible solutions}

TL:={}

$\pi(\lambda_a, \lambda_b) = (\Sigma P_j(\lambda_a, \lambda_b)/k, \; j=1,2,...,k$

Get the generalized criteria from the DM for all the criteria

while DM is not satisfied (or any other stop condition is not fulfilled)

inc(counter);

Generate_Neighborhood(TL, λ_S) = **N**;

CHOOSE_THE_BEST_SOLUTION(N, $\lambda_{S'}$);

$\lambda_S := \lambda_{S'}$;

update $\pi(\lambda_a, \lambda_b)$; {possible only in a case of user interaction}

update TL; {Add new tabu move, remove the oldest tabu move}

end while

end

The algorithm is nothing else but the iterative process of going from one solution to the next. We start with one or some initial schedules, which are feasible. Using the properties of the MPS problem contained in the knowledge base(s) in a form of rules as well as the operators defined before we generate the neighborhood of the current schedule. The search is driven by the DM and its preferences concerning particular criteria. The search goes on until stopping condition is fulfilled. When the stopping condition is fulfilled the DM is asked either to accept the solution(s) found so far or to modify his preferences. If no final decision is made the search starts again.

The TS algorithm is still controlled by the rules from the knowledge base. The outranking method finds a partial or global preorder in the set of the alternatives - possible moves. This preorder is mixed with the TS rules to decide (on the basis of tabu lists, memory status etc.) which move should be performed in the next step.

The algorithm is a standard TS procedure which is used by replacing the solution with the best (in the sense of the dominance relation) neighbor solution. Within this framework it is possible to present the DM one or several solution at each step. On one hand there should be so many solutions, that the non-dominated set / frontier is filled on a reasonable level. On the other hand the DM has his preferences and may want to see only one schedule, which covers his preference model the best.

At the beginning of the procedure we reset the iteration counter. Further lines initialize the knowledge base (getting the facts from the knowledge bases of different experts), choose the initial schedule from the set of generated ones, reset the tabu list and set the weights of criteria equal to each other, and get the generalized criteria for all the criteria. In every **while** loop the counter is incremented. Next, for the current solution its neighborhood is generated by the move generator module (see Figure 17:1). In the next line of the TS algorithm the PROMETHEE method is called. As a result we get the new current solution λ_S'. When stopping conditions are met the solutions found so far are presented to the DM. During this phase the DM has opportunity to change the weights of the criteria as well as the generalized criteria.

17.4. Computational experiments - application to an agriculture problem

The system was applied to the agriculture problem (Słowiński, Soniewicki and Węglarz, (1994)) composed of 40 tasks, listed in the Table 17:7. Precedence constraints are given in a form of the AOA graph depicted on the Figure 17:6. Two resources are available: man power (r_1) and agrimotors (r_2). The three following criteria are taken into consideration:

- (min) project completion time - Cmax
- (min) total consumption of resource r_1 - TC1
- (min) total consumption of resource r_2 - TC2

Table 17:7. Project details

Activity number	completion time	ready time	amount of r_1 required	amount of r_1 required
1	8	0	4	4
2	8	0	13	13
3	9	0	20	11
4	11	0	25	10
5	10	1	6	0
6	16	0	33	12
7	8	0	10	10
8	11	0	10	10
9	20	0	23	23
10	10.5	0	16	16
11	17	0	26	8
12	12	10	12	12
13	20	0	56	50
14	12	45	14	10
15	40	42	65	9
16	6	8	20	12
17	20	8	12	12
18	6	15	23	23
19	4.4	16	10	10
20	5.5	26	4	4
21	20	2	12	7
22	29	23	46	42
23	40	22	36	20
24	24	22	106	6
25	12	58	19	19
26	12	57	14	10
27	4	84	6	5
28	19	70	15	15
29	13	78	14	10
30	8.5	85	20	16
31	23	68	30	17
32	24	68	44	44
33	24	93	48	38
34	24	93	13	13
35	9.5	90	6	6
36	6	63	19	15
37	12.5	62	56	6
38	13	69	6	6
39	20	60	58	48
40	19	60	20	20

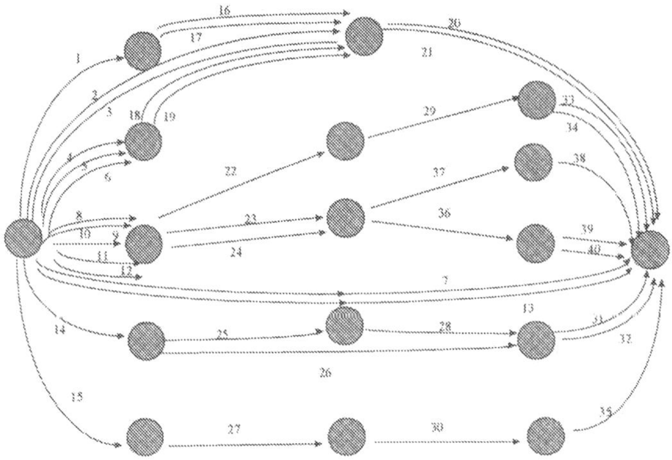

Figure 17:6. Precedence constraints of the project

We have decided to run the computation with only one initial schedule. Since the influence of the initial schedule on the solution finding process is not a matter of our investigations in this work we have taken the random generated feasible schedule as the initial schedule for further evaluation by the TS procedure. The schedule was characterized by the following values of the criteria:

Table 17:8. Initial schedule

Criteria	Value
Cmax (days)	280
TC1	199
TC2	127

The allocation of weights to the different criteria is a crucial problem in all the multicriteria techniques. At the beginning we have assigned the same weights to all the criteria (equal to 1/3). The parameters of the TS algorithm were the following:

- **Tabu List Length (TLL)** beginning value = 9 - The number of iterations that the particular move is prohibited. The TLL parameter can vary from time to time during the problem solving process and is controlled by the expert (knowledge base). The bracket values defined here were <7, 11>.

- **Termination Condition (TC)** - The number of iterations of the TS algorithm was given as a rule:

  RULE # 123
  ; Number of iterations
  IF Counter = 1000
  ; Time of the TS operations
  OR ElapsedTime = 300
  ;The number of solutions which will be presented to DM

OR Number_Of_Solutions_Found_So_Far = 10
; Exit TS and present the solutions to the DM
THEN Exit.

- **Memory structure** - We use both memory structures for the example: short term (in a form of tabu list) and long term memory, where previously visited moves are stored. The long term memory is also used to collect the knowledge about problem solving process.

- **Types of generalized criteria given by the DM** - The DM was asked to give the generalized criteria for every criterion. The proposal is given in the Table 17:9.

Table 17:9. The generalized criteria given by the DM

Criteria	Generalized criterion	Parameters
Cmax	V-shape	p=20
TC1	V-shape	p=8
TC2	V-shape	p=3

V-shape criterion means that as soon as non zero deviation is observed, an intensity of preference, increasing linearly with d, is considered. Above a deviation p the preference is strict. The DM fixes only p parameter. If the DM sees that the search doesn't give the expected results, or if his/her model of preferences changes the parameters and the generalized criteria for every criterion can be changed in whatever phase of the interaction with the system.

- **Knowledge base** - During solving the project being described in this paragraph it occurred that the exhaustive knowledge about the problem was the crucial aspect for solving the project. However it is very hard to acquire it. The most important knowledge, the dynamic one, comes from previous iterations of the TS algorithm. The time of tuning the application and equipping it with the proper knowledge was very long, and we can say that not long enough. The knowledge base was built out of 157 rules. Most of them determined the behavior of the TS algorithm in certain situations, like what has to be done, if there is no feasible neighborhood of the current solution.

17.4.1. Computing

The system was working about one hour on the Silicon Graphics Power Challenge (12 R8000 processors) and found several solutions which have been presented to DM after every termination of the TS procedure. The DM always has the following options:

1.Take the solution (just save it and come to it later), then:
   1.1.Exit
   1.2.Continue the search from the point of last termination,
2.Discard the solution - then again two options are available:
   2.1.Completely discard

2.2.Put the solution for the starting point of the TS procedure for further investigation

3.Continue the search from the point of last termination.

In Table 17:10 we present the absolute iteration numbers (counted from the beginning of computations) and the solutions found in these iterations.

We generated two sessions of the program. In the first one the DM did not change his preferences during the search. The solutions were projected on the GAIA plan and the DM had a possibility to choose the solution that fitted his preferences. In the second session after finding the solution number 7 the decision maker changed his preferences and gave the values 0.4 for Cmax and 0.3 for CT1 and CT2, which means that completion time was more important for the DM than other criteria. Moreover the DM puts the solution number 7 as the initial schedule for further investigations.

Table 17:10. The solutions found during first 4000 iterations

Iteration	Solution number	Cmax	TC1	TC2	DM action
137	1	192	123	82	1.2
697	2	190	140	74	1.2
931	3	198	139	77	1.2
1008	4	188	136	69	1.2
1476	5	191	141	72	1.2
2322	6	201	150	93	1.2
2657	7	187	144	75	1.2
3269	8	193	125	80	1.2
3826	9	199	137	76	1.1

The generalized criteria stay the same. In such a situation we have:

Table 17:11. Second initial schedule.

Criteria	Value
Cmax (days)	187
TC1	144
TC2	75

The following are the succeeding schedules found during next iterations:

Table 17:12. The final solutions and the one chosen by the DM.

Iteration	Solution number	Cmax	TC1	TC2	DM action
4270	10	169	145	84	1.2
5134	**11**	**155**	**170**	**88**	1.2
5679	12	176	145	77	1.1

Now the DM decides to view the schedules obtained after the change of criteria weights and projects them onto the GAIA plan. Then the DM decides to take the schedule number 11 as the best compromise one obtained so far.

17.5. Final remarks

In this chapter knowledge based approach to the multiobjective project scheduling problems was presented. We have focused the investigation on a novel multicriteria Tabu Search algorithm (supported by the expert system) for the MPS problems which uses the PROMETHEE method in every iteration to choose the solution for the next move.

The DM does not have the view of all non-dominated solutions. The search is driven only into the region of DM interest. It depends on the DM if the system has to generate one, two or more solutions. The time of computation is of course crucial.

Experimental tests have shown that the procedure is very sensitive to the DM preferences, but also to the knowledge about the problem and its solving environment. Such a knowledge is necessary to avoid the traps. However, the system requires quite a long time to tune it for the considered problem. It is the disadvantage of this system. The methods for knowledge acquisition for such problems will be a matter of future works.

Future work will also be directed towards the implementation of the system in the heterogeneous supercomputing environment with full support of several distributed experts. Such aspects as effects of different tabu sizes, different neighborhood strategies, termination criterion, initial schedule quality and memory structure will be investigated.

17.6. References

SŁOWIŃSKI, R., SONIEWICKI, B. AND WĘGLARZ, J. 1994. DSS for Multiobjective Project Scheduling , in: European Journal of Operational Research, Volume 79, Number 2, pp.220-229.

NABRZYSKI, J., WĘGLARZ, J. 1994. A knowledge-based multiobjective project scheduling system , International Journal of Decision Systems, vol 3 n° 3, pp.185-200.

NABRZYSKI, J., WĘGLARZ, J. 1995. On an Expert System with Tabu Search for Multiobjective Project Scheduling, Proceedings of INRIA/IEEE Symposium on Emerging Technologies and Factory Automation, III, pp. 87-94.

GLOVER, F., LAGUNA, M. 1993. Tabu Search, Modern Heuristic Techniques for Combinatorial Problems, Colin R. Reeves (Ed.), 70-150, Blackwell Scientific Publications, Oxford.

GLOVER, F. 1994. Tabu Search fundamentals and uses, Technical Report, Graduate School of Business and Administration, University of Colorado at Boulder.

MALEK, M., GURUSWAMY, M., PANDYA, M. AND OWENS, H. 1989. Serial and parallel simulated annealing and tabu search algorithms for the traveling salesman problem. Annals of Operations Research, 21(1), pp. 59-84.

SKORIN-KAPOV, J. 1990. Tabu Search applied to the quadratic assignment problem, ORSA Journal on Computing, 2(3), pp. 33-45.

DELL'AMICO, M., TRUBIAN, M. 1993. Applying tabu search to the job-shop scheduling problem, Annals of Operations Research, 41(1), pp. 231-252.

LAGUNA, M., WESLEY-BARNES, J., GLOVER, F. 1989. Tabu search methods for a single machine scheduling problem. Technical Report OR89-06, Graduate Programs in Operations Research, The University of Texas at Austin.

WIDMER, M., HERTZ, A. 1989. A new method for the flow sequencing problem, European Journal of Operational Research, 41(2), pp. 186-193.

GLOVER, F. 1990. *Tabu Search* , ORSA Journal on Computing.

ULUNGU, E. L. AND TEGHEM, J. Multi-objective combinatorial optimization problems: a survey, Journal of multiple-criteria decision analysis, Vol. 3, pp. 83-104.

ROY, B. The outranking approach and the foundations of Electre methods in Multiple Criteria Decision Aid, Carlos A. Bana e Costa and Philippe Vincke (eds.),

BRANS, J.B., MARESCHAL, B.: The PROMETHEE methods for MCDM; The PROMCALC, GAIA and Bankadviser software, in Multiple Criteria Decision Aid, Carlos A. Bana e Costa and Philippe Vincke (eds.).

MORTON, T.E. AND PENTICO, D.W. 1993. Heuristic Scheduling Systems: With Applications to Production Systems and Project Management , John Wiley and Sons.

MATARAZZO, B.: A pairwise criterion comparison approach: the MAPPAC and PRAGMA methods , in Multiple Criteria Decision Aid, Carlos A. Bana e Costa and Philippe Vincke (eds.).

ERMAN, L.D., HAYES-ROTH, F., LESSER, V.R., REDDY, D.R. 1980. The Hearsay-II Speech-Understanding System: Integrating Knowledge to Resolve Uncertainty , Computing Surveys, Vol. 12, No. 2.

WĘGLARZ J. 1980.On certain models of resource allocation problems, Kybernetes9, pp.61-66.

SŁOWIŃSKI R., WĘGLARZ J. (eds.). 1989. *Advances in Project Scheduling*, Elsevier.

18 NEW MODELLING CONCEPTS AND THEIR IMPACT ON RESOURCE–CONSTRAINED PROJECT SCHEDULING

Andreas Drexl
Jan Juretzka
Frank Salewski
Andreas Schirmer

Universität Kiel

18.1. Introduction

A recurring problem in project management involves the allocation of scarce resources to the individual activities comprising the project. In many situations such as audit–staff scheduling or timetabling, the resources correspond to individuals (skilled labour). This naturally leads to an assignment–type project scheduling problem, i.e. a project has to be performed by assigning one or more of several individuals (resources) to each activity.

In this paper we consider nonpreemptive project scheduling models with mode identity and partially renewable resources. Mode identity refers to a generalization of the multi-mode case where the set of all activities is partitioned into disjoint subsets while all activities forming one subset have to be processed in the same mode. Time and cost incurred by processing a subset of activities depend on the resources assigned to it. Partially renewable resources are capacitated over arbitrary subsets of periods. Both concepts represent substantial and non-trivial extensions of common scheduling concepts, the former of the multi-mode case of project scheduling, the latter of all types of scarce resources, containing renewable and nonrenewable resources as special cases.

Traditional resource-constrained project scheduling approaches (cp., e.g., Demeulemeester and Herroelen 1992, Patterson 1984) have been restricted to the case where each activity may be performed in only one predefined way. In this context, the resource–constrained project scheduling problem (RCPSP) is one of the most important problems (cp., e.g., Baar et al. 1997, Brucker et al. 1998, Demeulemeester and Herroelen 1997, Klein and Scholl 1997, Kolisch 1996a, Kolisch 1996b, Kolisch and Drexl 1996, Mingozzi et al. 1997, Sprecher 1996). More recently efforts have been made to formulate and solve the more general preemptive project scheduling problem where activity durations are functions of consumed resources (cp., e.g., Błażewicz et al. 1986, 1996). Meanwhile efforts have been documented (cp., e.g., Drexl 1991, Elmaghraby 1977, Kolisch 1995, Kolisch et al. 1995, Patterson et al. 1990, Sprecher 1994, Sprecher and Drexl 1998, Sprecher et al. 1997, Talbot 1982) which address the formulation and solution of a variety of nonpreemptive project scheduling problems where activity durations are discrete functions of activity performance modes (Multi–Mode Resource–Constrained Project Scheduling Problem, MRCPSP). However, the case of mode identity where the set of all activities is partitioned into a number of disjoint subsets where all activities belonging to the same subset have to be processed in the same mode is considered only in Salewski et al. (1997). We refer to this problem as the Mode Identity Resource–Constrained Project Scheduling Problem (MIR-CPSP). Similarily, project scheduling under partially renewable resource constraints, where capacity constraints for subsets of periods have to be observed, have been investigated only in Böttcher et al. (1996), Drexl et al. (1993), and Schirmer and Drexl (1997). To this problem we refer to as project scheduling problem under partially renewable resource constraints (RCPSP/π).

The paper is organized as follows: Section 18.2. covers resource–constrained project scheduling with mode identity. Section 18.3. is devoted to project scheduling under partially renewable resources. In Section 18.4. we stress the relevance of both new modelling concepts by presenting applications to audit–staff scheduling, assembly management, and school timetabling, respectively. Finally, Section 18.5. gives a summary and outlines future work.

18.2. Mode Identity

As indicated by the name, the MIRCPSP belongs to the field of resource–constrained project scheduling, to which it adds a new mode concept being more general than the currently known one. The MIRCPSP has been recently introduced into the project scheduling literature in Salewski et al. 1997. Some of the contributions of Salewski et al. (1997) will be recovered in what follows.

18.2.1. Background and Motivation

In the multi–mode case of project scheduling all mode–activity–assignments are mutually independent in the sense that assigning a mode to one activity j of a project consisting of a set n of nonpreemptable activities does not necessarily force any other activity to be processed in a specific mode. Consider the following example: Let denote $\mathcal{M}_j$ $(1 \leq j \leq n)$ the set of modes in which activity j may be performed, and m_j the specific mode in which it is actually processed. (Note that we introduce m_j for the sake of simplicity; we could also employ the binary variable x_{jmt} defined below. Also, the example disregards any resource or temporal constraints.) In addition, let be given mode–activity–assignments as shown in Table 18:1 where each activity j has to be processed in a specific mode m_j from a set of modes $\mathcal{M}_j$, regardless of all other assignments.

Table 18:1. Exemplary Mode–Activity–Assignments

j	$\mathcal{M}_j$	m_j
1	{1,2,3,4}	3
2	{1,2,3,4}	2
3	{1,2,3}	2
4	{1,2,3,4}	2
5	{1,2,3,4}	4
6	{1,2,3}	3
7	{1}	1

Imagine, however, a situation in which certain activities belong together in the sense that they must be executed in the same way. Consider, e.g., the activities 1, 2, and 5 on the one hand as well as the activities 3 and 6 on the other hand, which were originally assigned to different modes (cp. Table 18:1). Now, all activities belonging to the same subset must be processed in the same mode. In order to formulate this situation in a scheduling model we partition the set of all activities into U disjoint subsets of activities $H_u (1 \le u \le U)$ as shown in Table 18:2. Let $\mathcal{M}_u^I$ denote the set of modes in which the activities of H_u may be performed; w.l.o.g. all activities $j \in H_u$ are assumed to have the same number of modes. Finally, let m_u^I denote the specific mode in which all activities $j \in H_u$ are actually processed. Then the so–called mode identity — explicitly formulated in constraints (3) below — guarantees that all activities belonging to the same subset H_u are processed in the same mode m_u^I from a set of modes $\mathcal{M}_u^I$.

18.2.2. Model Formulation

Now we give a formal description of the MIRCPSP. The problem parameters are summarized as follows:

c_{um}	:	cost of processing subset u in mode m
$\bar{d}_j$	:	deadline of activity j
$\underline{d}_j$	:	release date of activity j
H_u	:	specific nonempty subset of activities
M_u	:	number of modes of subset u, indexed by m
$\mathcal{N}$	:	set of nonrenewable resources, indexed by k
n	:	number of activities, indexed by j
p_{jm}	:	time required to process activity j in mode m
$\mathcal{P}_j$	:	set of all immediate predecessors of activity j
$q_{j'm'j}$	:	mode–dependent time lags (finish–to–start) for $j' \in \mathcal{P}_j$, if j' is processed in mode m'
$\mathcal{R}$	:	set of renewable resources, indexed by k, $\mathcal{R} \cap \mathcal{N} = \emptyset$
R_k	:	total availability of nonrenewable resource $k \in \mathcal{N}$
$R_k(t)$	:	availability of renewable resource $k \in \mathcal{R}$ in period t
r_{jmk}	:	total consumption of nonrenewable resource $k \in \mathcal{N}$ required to execute activity j in mode m
r_{jmk}	:	per period usage of renewable resource $k \in \mathcal{R}$ required to execute activity j in mode m
T	:	number of periods (planning horizon), indexed by t
U	:	number of disjoint subsets of activites, indexed by u

Note that — in addition to the "standard" parameters of project scheduling models — the parameters $H_u\,(1 \le u \le U)$ and $M_u\,(1 \le u \le U)$ are employed.

Table 18:2. Partitioning the Set of all Activities

u	H_u	$\mathcal{M}_u^I$	m_u^I	m_j
1	{1,2,5}	{1,2,3,4}	3	$m_1 = m_2 = m_5 = 3$
2	{3,6}	{1,2,3}	1	$m_3 = m_6 = 1$
3	{4}	{1,2,3,4}	2	$m_4 = 2$
4	{7}	{1}	1	$m_7 = 1$

W.l.o.g. the parameters $\mathcal{N}$ and $\mathcal{R}$ as well as all $c_{um}, p_{jm}, \underline{d}_j, \bar{d}_j, r_{jmk}$, $R_k(t)$, and R_k are assumed to be nonnegative integers, while n, M_u, T, and U are assumed to be positive integers. All parameters $q_{j'm'j}$ may take arbitrary, i.e. positive, zero, or negative, integer values. In addition, we assume for all activities j and j' $(1 \leq j \leq n; 1 \leq j' \leq n)$ that $j' \in \mathcal{P}_j$ implies $j' \neq j$. This means, loosely speaking, that no activity can be a direct (immediate) predecessor of itself. Note, however, that this assumption does not exclude the case where $j' \in \mathcal{P}_j$ and $j \in \mathcal{P}_{j'}$ simultaneously hold; thus, an activity may be its own indirect predecessor.

In order to reduce the number of binary variables in the following formulation of the MIRCPSP, we compute from the time–lags $q_{j'm'j}$, the release dates $\underline{d}_j$, and the deadlines $\bar{d}_j$ the earliest start time ES_j and the latest finish time LF_j for each activity j. The activity j must be completed within the interval $[ES_j + p_{jm}, LF_j]$ or at least one of the temporal constraints will be violated. Further, let f_u denote the activity with the smallest index of subset H_u, i.e. $f_u = \min\{j \mid j \in H_u\}$.

Defining variables $x_{jmt} = 1$, if activity j is processed in mode m and finished in period t (0, otherwise), allows to formulate a binary program using the general framework given in Pritsker et al. (1969) as follows:

$$\min \sum_{u=1}^{U} \sum_{m=1}^{M_u} c_{um} \sum_{t=ES_{f_u}+p_{f_u m}}^{LF_{f_u}} x_{f_u mt} \tag{1}$$

$$\text{s.t.} \sum_{m=1}^{M_u} \sum_{t=ES_{f_u}+p_{f_u m}}^{LF_{f_u}} x_{f_u mt} = 1$$

$$u = 1, \ldots, U \tag{2}$$

$$\sum_{t=ES_{f_u}+p_{f_u m}}^{LF_{f_u}} x_{f_u mt} - \sum_{t=ES_j+p_{f_u m}}^{LF_j} x_{jmt} = 0$$

$$u = 1, \ldots, U, \forall j \in H_u \setminus \{f_u\}, m = 1, \ldots, M_u \tag{3}$$

$$\sum_{m'=1}^{M_{u'}} \sum_{t=ES_{j'}+p_{j'm'}}^{LF_{j'}} (t + q_{j'm'j})\, x_{j'm't} -$$

$$\sum_{m=1}^{M_u} \sum_{t=ES_j+p_{jm}}^{LF_j} (t - p_{jm})\, x_{jmt} \leq 0$$

$$u' = 1, \ldots, U, u = 1, \ldots, U, \forall j \in H_u, \forall j' \in \mathcal{P}_j \cap H_{u'} \tag{4}$$

$$\sum_{u=1}^{U} \sum_{m=1}^{M_u} \sum_{j \in H_u} r_{jmk} \sum_{\substack{q=t \\ q \in \{ES_j+p_{jm}, \ldots, LF_j\}}}^{t+p_{jm}-1} x_{jmq} \leq R_k(t)$$

$$\forall k \in \mathcal{R}, t = 1, \ldots, T \tag{5}$$

$$\sum_{u=1}^{U} \sum_{m=1}^{M_u} \sum_{j \in H_u} r_{jmk} \sum_{t=ES_j+p_{jm}}^{LF_j} x_{jmt} \leq R_k$$

$$\forall k \in \mathcal{N} \tag{6}$$

$$x_{jmt} \in \{0, 1\}$$

$$u = 1, \ldots, U, \forall j \in H_u, m = 1, \ldots, M_u, t = ES_j + p_{jm}, \ldots, LF_j \tag{7}$$

In (1), x represents the vector of all binary variables; $Z(x)$ is the objective function value for any feasible vector x with respect to (2)–(7). One can verify that the objective function (1) covers makespan minimization as a special case.

By virtue of the mode identity constraints (3), only the activity with the smallest index of each subset of activities needs to be regarded for maintaining the corresponding mode. The activity completion constraints (2) stipulate that the activity with the smallest index of each subset of activities is completed exactly once in one of its modes. The mode identity constraints (3) guarantee for each subset of activities that if the activity with the smallest index is completed then all the other activities will be completed as well, and in the same mode as that activity. Thus, (2) and (3) combine to ensure that all the activities of each subset of activities will be processed in the same mode, and that all of them will be completed.

The temporal constraints (4) enforce respection of the time–lags — and thus the precedence relations — between the activities. The temporal constraints will be dealt with in more detail in the following section.

The capacity constraints (5) and (6) represent per–period and total availability restrictions of renewable and nonrenewable resources, respectively.

18.2.3. Classification of Precedence and Temporal Relations

Note from Section 18.2.2. that the formulation of the temporal constraints (4) is general enough to cover not only the standard type of time–lags, namely minimum time–lags, but also the less common maximum time–lags, since these can easily be converted into minimum time–lags using the transformation introduced in Bartusch et al. (1988). It is quite obvious that the presence of maximum lags tends to render scheduling problems even more difficult. In fact, to our knowledge there are only very few results so far on the performance of solution procedures for this kind of problems (cf., e.g., Brinkmann and Neumann 1994, Neumann and Zimmermann 1995).

In order to establish a terminology as a starting point for the analysis of such problems, we propose three criteria which allow to distinguish between different cases of precedence and temporal relations. These criteria are connectivity, cyclicity, and domain. In particular, the first two criteria pertain to both precedence and temporal relations, while the last applies to temporal relations only.

For notational brevity in what follows, let $\mathcal{J} = \{1, \ldots, n\}$. Also, let denote j and $j'\,(1 \leq j \leq n, 1 \leq j' \leq n)$ activities belonging to subsets H_u and $H_{u'}\,(1 \leq u \leq U, 1 \leq u' \leq U)$ while m and $m'\,(1 \leq m \leq M_u, 1 \leq m' \leq M_{u'})$ denote arbitrary modes in which j and j' may be processed. Finally, all maximum time–lags are assumed to have been transformed into minimum ones. Note that in the definitions to come the more restricted case is always given first.

Definition 1 (Connectivity) A precedence relation is called *isolating* iff only activities belonging to the same subset of activities H_u are related, i.e. for all activities $j' \in H_{u'}, j \in H_u, j' \in \mathcal{P}_j$ implies $u' = u$. It is called *connecting* otherwise, i.e. iff at least two activities from different subsets of activities are related, i.e. there exist activities $j' \in H_{u'}, j \in H_u, j' \in \mathcal{P}_j$ such that $u' \neq u$.

Definition 2 (Cyclicity) A precedence relation is called *acyclic* iff it forms no cycle, i.e. there exists no subset of activities $\{j_1, \ldots, j_n\} \subseteq \mathcal{J}$, where $j_1 \in \mathcal{P}_{j_2}, j_2 \in \mathcal{P}_{j_3}, \ldots, j_{n-1} \in \mathcal{P}_{j_n}$, and $j_1 = j_n$. It is called *cyclic* otherwise, i.e. iff it forms at least one cycle, i.e. there exists a subset of activities $\{j_1, \ldots, j_n\} \subseteq \mathcal{J}$, where $j_1 \in \mathcal{P}_{j_2}, j_2 \in \mathcal{P}_{j_3}, \ldots, j_{n-1} \in \mathcal{P}_{j_n}$, and $j_1 = j_n$.

Definition 3 (Domain) A temporal relation is called *non–negative* iff each time–lag has a non–negative value, i.e. for all activities $j'\,(1 \leq j' \leq n)$ and $j\,(1 \leq j \leq n)$ with $j' \in \mathcal{P}_j$ the inequality $q_{j'm'j} \geq 0$ holds. It is called *arbitrarily–valued* otherwise, i.e. iff at least one time–lag has a negative value, i.e. there exist activities $j'\,(1 \leq j' \leq n)$ and $j\,(1 \leq j \leq n)$ with

$j' \in \mathcal{P}_j$ where $q_{j'm'j} < 0$.

These definitions are illustrated in terms of an example in Salewski et al. (1997). There we also prove that the MRCPSP is a special case of the more general MIRCPSP. Moreover, it is shown that the (feasibility variant of the) MIRCPSP is strongly ($\mathcal{NP}$–complete) $\mathcal{NP}$–hard.

18.3. Partially Renewable Resources

In what follows we consider another generalization of the RCPSP. We introduce so–called partially renewable resources by assuming for each resource a capacity to be defined on arbitrary subsets of periods. The concept of partially renewable resources is a fundamental tool in order to make e.g. timetabling and shift scheduling aspects amenable to project scheduling. Additionally, partially renewable resources serve to model complicated labor regulations. Furthermore, they cover traditional renewable and non-renewable resource constraints as special cases. For short we refer to the problems as RCPSP/π. Some of the contributions of Böttcher et al. (1996) will be recovered in what follows.

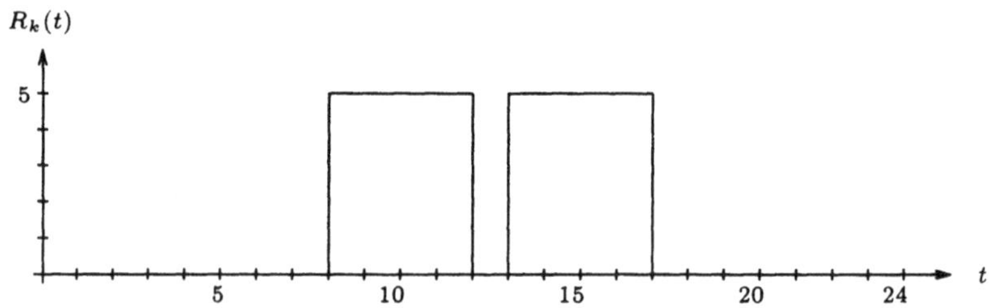

Figure 18:1. Availability of Staff — Variant 1

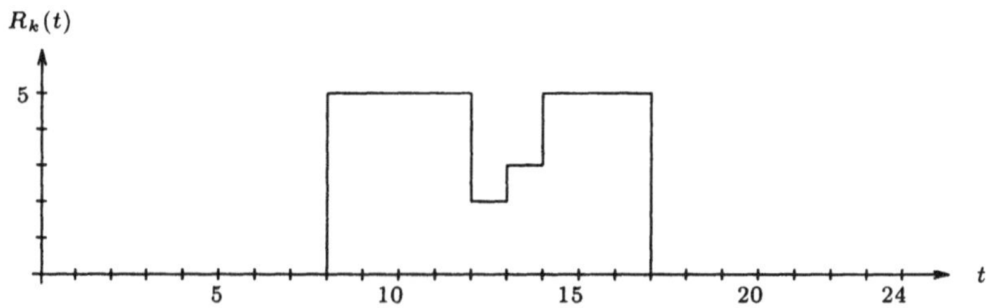

Figure 18:2. Availability of Staff — Variant 2

18.3.1. Background and Motivation

In order to motivate the RCPSP/π, we consider the case where the resource type "staff" has to be scheduled. Figure 18:1 depicts that no worker is available before period 9 and after period 17, respectively. Moreover, 5 workers are available during periods 9 to 12 and 14 to 17, respectively. In period 13 all the workers have their lunch break. Then, e.g., no non-preemptable activity with a duration of at least 3 periods is allowed to start in period 11.

More flexibility is offered if each of the workers could have his break either in period 13 or in period 14. In Figure 18:2 e.g. 3 (2) workers have their lunch break in period 13 (14). This would now allow to start an activity with the duration of 3 periods in period 11 if it requests less than 3 workers per period. Unfortunately, there exist several possibilities to specify in advance how many workers should have their lunch break in period 13 and how many in period 14. Clearly, this is impractical and does not provide the degree of flexibility which is necessary when we want to schedule a one hour lunch break for each worker either in period 13 or in period 14, without fixing the period in advance.

The "skyline" (cp. Elmaghraby 1977) of resource profiles in both figures illustrates that the modelling capabilities of renewable resources are not sufficient to tackle more general and complicated situations. The RCPSP/π allows to accommodate such requirements: Let the periods 13 and 14 be element of the subset $\pi = 9$ of periods and define a resource constraint with an associated capacity of five workers.

18.3.2. Model Formulation

Now we give a formal description of the RCPSP/π. The problem parameters are summarized as follows:

n	:	number of nondummy activities, ordered topologically, indexed by j
0	:	dummy source activity
$n+1$	:	dummy sink activity
p_j	:	time required to process activity j
$\mathcal{P}_j$	:	set of all immediate predecessors of activity j
$\mathcal{R}$	:	set of partially renewable resources, indexed by k
$P_k(\pi)$	:	subset of periods in which resource k is available with resource capacity $R_k(\pi)$
$\Pi(k)$	:	set of subsets of periods $P_k(\pi)$, $\pi = 1, \ldots, \lvert\Pi(k)\rvert$
$R_k(\pi)$	:	resource capacity of resource k
r_{jk}	:	number of units of resource $k \in \mathcal{R}$ required by activity j every period of its processing time p_j
$\mathcal{T}$	:	set of periods during which the activities must be processed, indexed by t

The objective of the RCPSP/π is the minimization of the makespan such that precedence and resource constraints are met.

We derive earliest and latest finish times EF_j and LF_j, respectively, by critical path analysis. Let denote $E_j = \{EF_j, \dots, LF_j\}$ and $Q_{jt} = \{t, \dots, t+p_j-1\}$. Now, based on the decision variables $x_{jt} = 1$, if activity j is finished in period t (0, otherwise), the RCPSP/π can be modelled as follows:

$$\min \sum_{t \in E_{n+1}} t \cdot x_{n+1,t} \tag{8}$$

$$\text{s.t.} \sum_{t \in E_j} x_{jt} = 1 \qquad j = 0, \dots, n+1 \tag{9}$$

$$\sum_{t \in E'_j} t \cdot x_{j't} - \sum_{t \in E_j} (t - p_j)\, x_{jt} \leq 0$$

$$j = 1, \dots, n+1, \forall j' \in \mathcal{P}_j \tag{10}$$

$$\sum_{j=1}^{n} r_{jk} \sum_{t \in P_k(\pi)} \sum_{q \in Q_{jt} \cap E_j} x_{jq} \leq R_k(\pi)$$

$$\forall k \in \mathcal{R}, \pi = 1, \dots, |\Pi(k)| \tag{11}$$

$$x_{jt} \in \{0,1\} \qquad j = 0, \dots, n+1, \forall t \in E_j \tag{12}$$

The objective function (8) minimizes the completion time of the unique sink and hence the makespan of the project. Equations (9) are activity completion constraints. Constraints (10) take into consideration the precedence relations between each pair of activities (j', j), where j' immediately preceeds j. Finally, constraints (11) limit the total usage of the partially renewable resources to the available amount. Notice, that outside of the period subsets, resources are not capacitated.

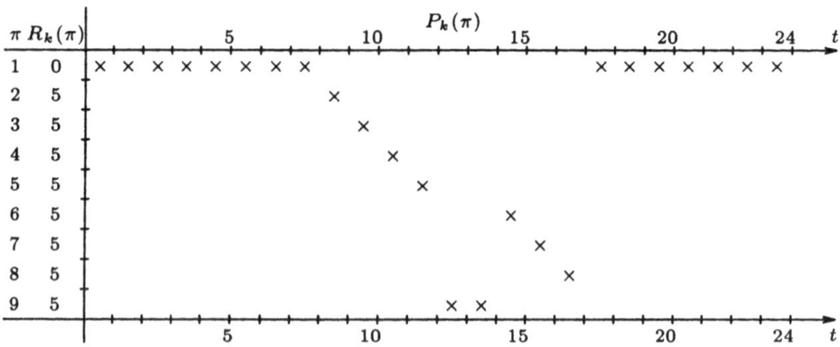

Figure 18:3. Illustration of Partially Renewable Resources

The example introduced in the previous section serves to illustrate the RCPSP/π. The following parameter instantiations appropriately define the problem under consideration in terms of the RCPSP/π and allow to schedule a one hour break for each of the five workers without fixing in advance whether the break is in period 13 or in period 14.

$$\begin{array}{rl} P_1(1) = \{1,\ldots,8,18,\ldots,24\} & R_1(1) = 0 \\ P_1(2) = \{9\} & R_1(2) = 5 \\ P_1(3) = \{10\} & R_1(3) = 5 \\ P_1(4) = \{11\} & R_1(4) = 5 \\ P_1(5) = \{12\} & R_1(5) = 5 \\ P_1(6) = \{15\} & R_1(6) = 5 \\ P_1(7) = \{16\} & R_1(7) = 5 \\ P_1(8) = \{17\} & R_1(8) = 5 \\ P_1(9) = \{13,14\} & R_1(9) = 5 \end{array}$$

Figure 18:3 serves as an illustratation. For each of the nine period subsets π we provide the resource capacity $R_k(\pi)$ and the covered periods $P_k(\pi)$ also. The period subset $P_k(9)$ contains the periods 13 and 14 (indicated by $\times$) and has a capacity of $R_k(9) = 5$ — and this is just what we want to achieve.

Apparently, in Figure 18:3 only special cases of what partially renewable resource in general comprise to model are covered: The period subsets may consist of unit time periods which are not labeled consecutively. Moreover, the period subsets must not be pairwise disjoint.

18.3.3. Assessment

In order to position partially renewable resources in the context of traditional resource types, recall $\mathcal{R}$ ($\mathcal{N}$) to denote the set of renewable (nonrenewable) resources:

- A nonrenewable resource $k \in \mathcal{N}$ with capacity R_k can be expressed in terms of a partially renewable resource with period subset $P_k(1) = \mathcal{T}$ and $R_k(1) = R_k$. In other words, each period subset covers the whole planning horizon.

- A renewable resource $k \in \mathcal{R}$ with capacity $R_k(t)$ can be expressed in terms of a partially renewable resource with period subsets $P_k(t) = \{t\}$ and $R_k(t)$. In other words, each period subset covers exactly one single period.

- Obviously, the RCPSP/π is a generalization of the RCPSP, where we have $|\Pi(k)| = |\mathcal{T}|, P_k(1) = \{1\}, P_k(2) = \{2\}, \ldots, P_k(|\Pi(k)|) =$

$\{|\mathcal{T}|\}$ for all $k \in \mathcal{R}$ as well as $R_k(\pi) = R_k(t)$ for all $k \in \mathcal{R}$ and $\pi \in \{1, \ldots, |\Pi(k)|\}$, respectively.

As a consequence, partially renewable resources cover the case of doubly–constrained resources as well. Hence, when modelling resource–constrained (project) scheduling problems it suffices to use partially renewable resources only since all other types can be expressed in terms of them. That is, this resource type may be seen as unifying since it combines the expressive power of the traditional resource types.

Finally, we refer to some important characteristics of the RCPSP/π:

- A new modelling approach for the RCPSP has recently been proposed in Icmeli and Rom (1996). In this paper, the authors define resources in a way to which they refer as partially renewable resources. In Schirmer and Drexl (1997) it is shown, however, that these resources are in effect traditional renewable resources, modified appropriately to accommodate the non–standard dsicretization of the planning horizon the authors use.

- In Schirmer and Drexl (1997) it is shown that partially renewable resources can be used to express several kinds of logical relations between the scheduling of activities. In addition, a number of practical requirements on activities' scheduling can be formulated such as maximum or minimum quotas, as well as issues of calendarization. This underscores the expressive power of partially renewable resources.

- In order to avoid the tedious addressing of period subsets via a resource *and* a period subset index the distinction between the different resources with which period subsets are associated can be dropped. This can be done by replacing each combination of a resource and a period subset by one separate, 'new' resource which is capacitated over that subset. This process is called 'normalization' in Böttcher et al. (1996). Afterwards, each normalized resource is defined over exactly one period subset such that the number of subsets equals the number of resources.

- Unfortunately, the RCPSP/π belongs to the class of strongly $\mathcal{NP}$–equivalent problems. In fact, in Schirmer and Drexl (1997) it is proven that the (feasibility variant of the) RCPSP/π is strongly $\mathcal{NP}$–equivalent (strongly $\mathcal{NP}$–complete).

18.4. Applications

18.4.1. Audit–Staff Scheduling

When scheduling its audit–staff, the management of an auditing firm encompasses a number of decisions. These may be grouped into several categories which differ markedly in terms of organizational echelon involved, length of the planning horizon and the planning periods, degree of aggregation of the audit tasks, degree of detail of the required information, and decision objective. However, traditional audit–staff scheduling models (cp., e.g., Dodin and Elimam 1997, Dodin et al. 1997) are single–level models which try to construct a direct assignment of auditors to tasks and periods. To facilitate algorithmic treatment, all these models are more or less gross simplifications of practical planning situations. These observations led us to conduct a survey among the 200 biggest certified public accountant (CPA) firms in Germany (cp. Salewski 1995). Based upon its results we formulated an hierarchical model in Salewski (1995) comprising three levels: The medium–term planning assigns teams of auditors to the engagements; it constructs a schedule by determining the workload per auditor and week over a planning horizon of between three and twelve months. The medium–to–short–term planning disaggregates the results of the first level for one week and all auditors; the outcome is a schedule for each auditor that covers — on the basis of periods of four hours — all engagements in which he is involved in the considered week. The short–term planning is based upon the results of the second level for one week and one engagement; it assigns the auditors involved in the auditing of that engagement to the corresponding audit tasks and schedules these tasks to periods of one hour. Here, we will focus on the first, that is, the medium–term level.

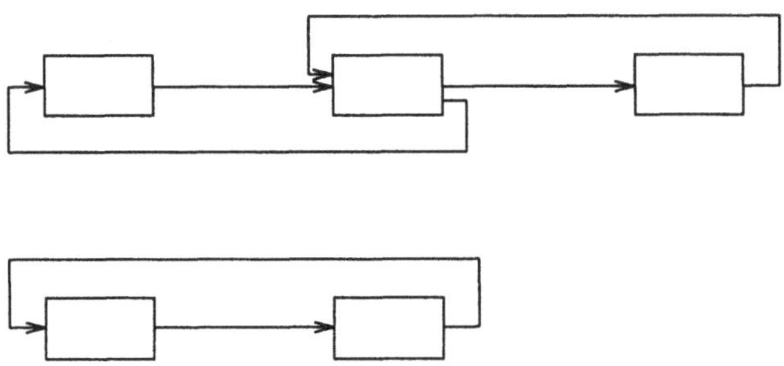

Figure 18:4. Characterisation of Precedence Relations

The assumptions of the Medium–Term Audit–Staff Scheduling Problem (MASSP) which relate to mode identity may be characterized as follows (cp. Salewski et al. 1997 also):

- Consider a set of engagements each of which consists of several jobs.
- The engagements have to be assigned to alternative teams of auditors.
- The jobs of one engagement have to be processed by the auditors of one team.

Note, it is easy to show that the MIRCPSP can be restricted to the MASSP; for details cf. Salewski et al. (1997). Moreover, in terms of definitions 1 to 3 the precedence and temporal relations of the MASSP can be classified as isolating, acyclic, and arbitrarily–valued, respectively (cp. the illustration given in Figure 18:4 for the case of two engagements).

The presence of minimum and maximum time–lags between subsequent jobs, along with their special sequence, which arises from the above mentioned decomposition process, implies for each engagement a chain structure of the time lags as illustrated in Figure 18:4 for the case of two engagements. Clearly, this facilitates the algorithmic treatment substantially. Greedy randomized algorithms for the solution of the MASSP are presented in Salewski et al. (1997).

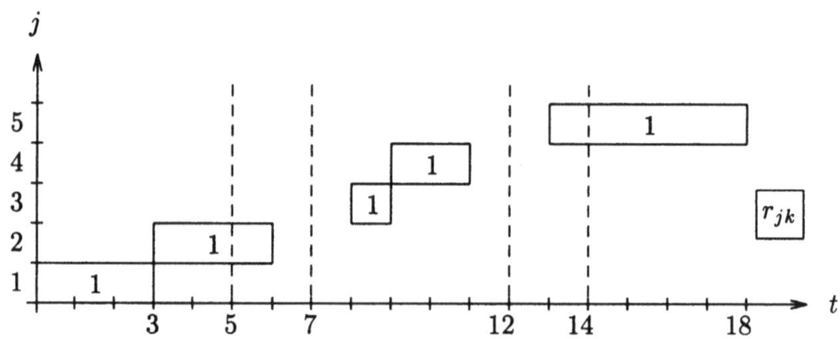

Figure 18:5. Shift Scheduling in Assembly Management

18.4.2. *Assembly Management — Shift Scheduling*

Consider the following example which frequently arises in the context of assembly management. Assume that staff is available on at most eight day hours within two consecutive days (periods $1, \ldots, 5, 8, \ldots, 12, 15, \ldots, 19$). Moreover, assume that staff is not allowed to work on more than two of the corresponding four night hours (periods $6, 7, 13, 14$). Without reproducing all the details of an instance then, clearly, the parameter instantiations

$$
\begin{array}{lllll}
P_k(1) & = & \{1\} & R_k(1) & = 1 \\
P_k(2) & = & \{2\} & R_k(2) & = 1 \\
\vdots & & & \vdots & \\
P_k(14) & = & \{14\} & R_k(14) & = 1 \\
P_k(15) & = & \{6,7,13,14\} & R_k(15) & = 2 \\
P_k(16) & = & \{1,\ldots,5,8,\ldots,12\} & R_k(16) & = 8 \\
\vdots & & & \vdots &
\end{array}
$$

allow to compute the Gantt–chart provided in Figure 18:5 which satisfies the labor requirements of a single worker. Note, that activity $j = 5$ must not start before period 14 because the (partially renewable) resource 'night hour' is not available in more than two units in the respective period subset $P_k(15) = \{6, 7, 13, 14\}$.

18.4.3. School Timetabling

The aim of this section is to show that the modelling concepts described previously are very useful in order to come up with an advanced school timetabling model.

The school timetabling problem requires to schedule a set of meetings between teachers and pupils over a set of time periods, where some resources must be available and several additional constraints have to be met. Among the constraints which have to be taken into account are: No class, subject, room, and teacher overlaps are allowed; one has to take care of class, subject, room, and teacher availabilities; compactness constraints have to be regarded; preassignment of lectures must be possible; some lectures have to be given simultaneously; lunch breaks have to be observed, etc. Note that compactness constraints typically require no free time (other than lunch breaks or time to move from one building to another) between lessons for the pupils. Most of the constraints mentioned can be modelled using the unifying framework of partially renewable resources. Moreover, it is necessary to distribute lessons of different lengths. These requirements can be easily accommodated by the use of the multiple mode concept with mode identity constraints.

Beside other parameters, there are Basic Teaching Units (BTUs). Each BTU is characterized by the attributes length, required teacher, pupils to be taught, kind of subject (main/subsidiary), and requested type of room (see below), respectively. BTUs in general are large teaching units (e.g. six hours math per week) where we have to decide how to split them into smaller ones (called blocks) and how to schedule them. This splitting and scheduling will be called "distributing". For the distribution of BTUs we introduce modes, blocks, and activities as in the example of Table 18:3.

Table 18:3. Distribution of BTUs

BTU	length	block			activity			mode	costs
1	2	□□	∅		1	2		1	0
		□	□					2	5
2	3	□□□	∅	∅	3	4	5	1	5
		□□	□	∅				2	0
		□	□	□				3	1

For BTU 1 with a length of 2 hours we have two possible modes of distribution: First, two consecutive hours (□ $\simeq$ 1 hour, ∅ $\simeq$ 0 hours), i.e. one block of a two-hour lecture and one dummy block (i.e. no distribution), and second two blocks with a duration of one hour each. Analogously, for BTU 2 with length of 3 hours three modes of distribution are provided. Moreover, notice that blocks will be denoted as activities and that costs are associated with each mode of processing the activities of a specific BTU. Finally, it is noteworthy to mention that the definition of the dummy blocks secures an identical number of blocks for each BTU irrespective of the modes introduced.

Using the mode identity concept introduced in Section 18.2.2. it is straightforward to formulate the distribution requirements described above. More precisely, we have to minimize the sum of costs of processing the first block of each BTU in the chosen mode. To take the first block w.l.o.g. is valid in conjunction with mode identity-constraints similiar to what has been done in equations (1)–(7). For the sake of shortness we restrain from going into details and refer the reader to Drexl and Salewski (1997) where models and methods are presented.

In our framework, teachers, pupils, subjects, and rooms are treated as resources. Relabeling the resources then allows to formulate capacity constraints for subsets of periods.

Table 18:4. Days and Hours

Monday	1	2	3	4	5	6
Tuesday	7	8	9	10	11	12

Consider the simple instance given in Table 18:4 where we have two days with four a.m.–hours and two p.m.–hours as periods. If we define the period subsets $P_k(1) = \{1\}, \ldots, P_k(12) = \{12\}, P_k(13) = \{1, \ldots, 12\}, P_k(14) = \{1,2,3,4\}, P_k(15) = \{5,6\}, P_k(16) = \{7,8,9,10\}$, and $P_k(17) = \{11,12\}$, respectively, it is straightforward to formulate different kinds of partially renewable resource constraints which are relevant in school timetabling.

18.5. Summary and Future Work

In this paper we consider nonpreemptive project scheduling models with mode identity and partially renewable resources. Mode identity refers to a generalization of the multi–mode case where the set of all activities is partitioned into disjoint subsets while all activities forming one subset have to be processed in the same mode. Time and cost incurred by processing a subset of activities depend on the resources assigned to it. Partially renewable resources are capacitated over arbitrary subsets of periods. Both concepts represent substantial and non–trivial extensions of common scheduling concepts, the former of the multi–mode case of project scheduling, the latter of all types of scarce resources, containing renewable and nonrenewable resources as special cases.

We analyze the impact of both new modelling concepts on resource–constrained project scheduling, audit–staff scheduling, assembly management, and school timetabling, respectively.

Enterprises are and have been facing mounting pressure to exercise reductions in costs arising from producing their goods or services and to make better use of existing staff or equipment. It is well–known from practical experience that this pressure can be met, at least in part, by more efficient and intelligent planning. Successful application of these methods, however, depends to a large degree on the ability to unambiguously and efficiently model the relevant specifics of the problems tackled. This ability, in turn, calls for expressive modelling concepts, which allow to capture a wide range of requirements appearing in real world problems. What we presented in this paper so far clearly demonstrates that both new modelling concepts, viz. mode identity and partially renewable resources, can play an important role in this process.

This process could be enhanced by encorporating additional modelling concepts — just to mention a few: Multi–mode resource–constrained project scheduling with minimal and maximal time lags (cp., e.g., De Reyck and Herroelen 1997), resource–constrained project scheduling with net present value maximization (cp., e.g., Icmeli and Erengüc 1994, 1996), resource investment and resource levelling (cp., e.g., Demeulemeester 1995, Zimmermann 1997), respectively. Finally, we point to Franck et al. (1997) where it is shown that a tool box equipped with various different project scheduling models and methods is a prerequisite for successful applications.

Acknowledgement

This research has been partially supported by the Deutsche Forschungsgemeinschaft.

References

BAAR, T., P. BRUCKER, AND S. KNUST. 1997. Tabu–search algorithms for the resource–constrained project scheduling problem. Working Paper, Universität Osnabrück.

BARTUSCH, M., R.H. MÖHRING, AND F.F. RADERMACHER. 1988. Scheduling project networks with resource constraints and time windows. *Annals Oper. Res.* **16**, 201–240.

BŁAŻEWICZ, J., W. CELLARY, R. SŁOWIŃSKI, AND J. WĘGLARZ. 1986. *Scheduling under Resource Constraints — Deterministic Models, Baltzer, Basel* (Annals Oper. Res. **7**).

BŁAŻEWICZ, J., K.H. ECKER, E. PESCH, G. SCHMIDT, AND J. WĘGLARZ. 1996. *Scheduling Computer and Manufacturing Processes*, Springer, Berlin.

BÖTTCHER, J., A. DREXL, R. KOLISCH, AND F. SALEWSKI. 1996. Project scheduling under partially renewable resource constraints. Manuskripte aus den Instituten für Betriebswirtschaftslehre der Universität Kiel, No. 398.

BRINKMANN, K., AND K. NEUMANN. 1994. Heuristic procedures for resource–constrained project scheduling with minimal and maximal time lags: The minimum project–duration and resource–levelling problems. Report WIOR–443, Universität Karlsruhe.

BRUCKER, P., S. KNUST, A. SCHOO, AND O. THIELE. 1998. A branch & bound algorithm for the resource–constrained project scheduling problem, *European J. Oper. Res.* (to appear).

DEMEULEMEESTER, E. 1995. Minimizing resource availability costs in time–limited project networks. *Mgmt. Sci.* **41**, 1590–1598.

DEMEULEMEESTER, E., AND W. HERROELEN. 1992. A branch–and–bound procedure for the multiple resource–constrained project scheduling problem. *Mgmt. Sci.* **38**, 1803–1818.

DEMEULEMEESTER, E., AND W. HERROELEN. 1997. New benchmark results for the resource–constrained project scheduling problem. *Mgmt. Sci.* **43**, 1485–1492.

DE REYCK, B., AND W. HERROELEN. 1997. The multi–mode resource–constrained project scheduling problem with generalized precedence constraints. Working Paper, Katholieke Universiteit Leuven.

DODIN, B., AND A.A. ELIMAM. 1997. Audit scheduling with overlapping activities and sequence dependent setup costs. *European J. Oper. Res.* **97**, 22–33.

DODIN, B., A.A. ELIMAM, AND F. ROLLAND. 1996. Tabu search in audit scheduling. Working Paper, University of California/Riverside.

DREXL, A. 1991. Scheduling of project networks by job assignment. *Mgmt. Sci.* **37**, 1590–1602.

DREXL, A., J. JURETZKA, AND F. SALEWSKI. 1993. Academic course scheduling under workload and changeover constraints. Manuskripte aus den Instituten für Betriebswirtschaftslehre der Universität Kiel, No. 337.

DREXL, A., AND F. SALEWSKI. 1997. Distribution requirements and compactness constraints in school timetabling. *European J. Oper. Res.* **102**, 193–214.

ELMAGHRABY, S.E. 1977. *Activity Networks — Project Planning and Control by Network Models*, Wiley, New York.

FRANCK, B., K. NEUMANN, AND C. SCHWINDT. 1997. A capacity-oriented hierarchical approach to single-item and small-batch production planning using project-scheduling methods. *OR Spektrum* **19**, 77–85.

ICMELI, O., AND S.S. ERENGÜC. 1994. A tabu search procedure for the resource constrained project scheduling problem with discounted cash flows. *Computers & Oper. Res.* **21**, 841–853.

ICMELI, O., AND S.S. ERENGÜC. 1996. A branch and bound procedure for the resource constrained project scheduling problem with discounted cash flows. *Mgmt. Sci.* **42**, 1395–1408.

ICMELI, O., AND W. ROM. 1996. Solving the resource constrained project scheduling problem with optimization subroutine library. Working Paper, Cleveland State University.

KLEIN, R., AND A. SCHOLL. 1997. Computing lower bounds by destructive improvement — An application to resource-constrained project scheduling. Working Paper, Technische Hochschule Darmstadt, *European J. Oper. Res.* (to appear).

KOLISCH, R. 1995. *Project Scheduling under Resource Constraints — Efficient Heuristics for Several Problem Classes*, Physica, Heidelberg.

KOLISCH, R. 1996a. Efficient priority rules for the resource constrained project scheduling problem. *J. Operations Management* **14**, 179–192.

KOLISCH, R. 1996b. Serial and parallel resource-constrained project scheduling methods revisited — Theory and computation. *European J. Oper. Res.* **90**, 320–333.

KOLISCH, R., AND A. DREXL. 1996. Adaptive search for solving hard project scheduling problems. *Naval Res. Logist.* **43**, 23–40.

KOLISCH, R., A. SPRECHER, AND A. DREXL. 1995. Characterization and generation of a general class of resource-constrained project scheduling problems. *Mgmt. Sci.* **41**, 1693–1703.

MINGOZZI, A., V. MANIEZZO, S. RICCIARDELLI, AND L. BIANCO . 1997. An exact algorithm for the resource constrained project scheduling problem based on a new mathematical formulation, *Mgmt. Sci.* (to appear).

NEUMANN, K., AND J. ZHAN. 1995. Heuristics for the minimum project-duration problem with minimal and maximal time lags under fixed resource constraints. *J. Intelligent Manufacturing* **6**, 145–154.

PATTERSON, J.H. 1984. A comparison of exact approaches for solving the multiple constrained resource, project scheduling problem. *Mgmt. Sci.* **30**, 854–867.

PATTERSON, J.H., F.B. TALBOT, R. SŁOWIŃSKI, AND J. WĘGLARZ. 1990. Computational experience with a backtracking algorithm for solving a general class of resource constrained scheduling problems. *European J. Operational Research* **90**, 68–79.

PRITSKER, A.A.A., W.D. WATTERS, AND P.M. WOLFE. 1969. Multiproject scheduling with limited resources: A zero–one programming approach. *Mgmt. Sci.* **16**, 93–10.

SALEWSKI, F. 1995. *Hierarchische Personaleinsatzplanung in Wirtschaftsprüfungsgesellschaften*, Physica, Heidelberg.

SALEWSKI, F., A. SCHIRMER, AND A. DREXL. 1997. Project scheduling under resource and mode identity constraints: Model, complexity, methods, and application. *European J. Oper. Res.* **102**, 88–110.

SCHIRMER, A., AND A. DREXL. 1997. Allocation of partially renewable resources — Concept, models and application. Manuskripte aus den Instituten für Betriebswirtschaftslehre der Universität Kiel, No. 455.

SPRECHER, A. 1994. *Resource-Constrained Project Scheduling — Exact Methods for the Multi-Mode Case*, Springer, Berlin.

SPRECHER, A. 1996. Solving the RCPSP efficiently at modest memory requirements. Manuskripte aus den Instituten für Betriebswirtschaftslehre der Universität Kiel, No. 426.

SPRECHER, A., AND A. DREXL. 1998. Solving multi–mode resource–constrained project scheduling problems by a simple, general and powerful sequencing algorithm. *European J. Oper. Res.* (to appear).

SPRECHER, A., S. HARTMANN, AND A. DREXL. 1997. An exact algorithm for project scheduling with multiple modes. *OR Spektrum* **19**, 195–203.

TALBOT, F.B. 1982. Resource–constrained project scheduling with time–resource tradeoffs: The nonpreemptive case. *Mgmt. Sci.* **28**, 1197–1210.

ZIMMERMANN, J. 1997. Heuristics for resource–levelling problems in project scheduling with minimum and maximum time lags. Report WIOR–491, Universität Karlsruhe.

19 INTEGRATING QUALITY AS A MEASURE OF PERFORMANCE IN RESOURCE-CONSTRAINED PROJECT SCHEDULING PROBLEMS

S. Selcuk Erenguc[1]
Oya Icmeli Tukel[2]

1. University of Florida
2. Cleveland State University

19.1. Introduction

A survey conducted in the United States in 1994 (Icmeli-Tukel and Rom 1998), revealed that the issues Project Managers (PMs) face today are more complicated than those studied in the project scheduling literature. The study indicates that for over 90% of PMs the objective is to maximize the quality of projects and their outcomes. They usually measure quality by the degree to which a project's outcome conforms to the customer's requirements and the degree to which the project completes within budget and on schedule. In order to achieve these goals, PMs are trying to eliminate the rework associated with non-conforming activities. A job is considered to be non-conforming when it fails to meet customer/design requirements and therefore needs some rework (Ireland 1991). Non-conformities usually result in increased project implementation time and cost. In the traditional Resource Constrained Project Scheduling Problem (RCPSP) formulations, however, the general assumption is that work will be done as scheduled and that rework will not be required. This assumption clearly makes the RCPSP models more amenable to solution yet less realistic. PMs report (Icmeli-Tukel and Rom 1998) that most of the projects they implement have at least one non-conforming activity which needs to be reworked before the project moves to the next stage. Because of these reworks, they have a difficult time anticipating delays and cost overruns when using traditional techniques. If an activity is (partially) reworked to satisfy customer requirements, then the start time of all immediately succeeding activities are delayed, which, in turn might delay the project completion time. This in turn contributes to the consistently poor record of project performance (Cooper 1994).

In this paper we present a mathematical formulation and an exact solution algorithm for the resource constrained project scheduling problem in which activities may require rework due to non-conformities. This part of the paper closely follows Icmeli-Tukel and Rom (1997). Next, we introduce a dynamic procedure to monitor a project's progress and if necessary, reschedule activities in order to reduce the project makespan. The dynamic procedure uses the optimal schedule obtained by using the Icmeli-Tukel and Rom(1997) algorithm as a baseline.

The plan of this paper is as follows. In section 2, we discuss quality issues in project management and review the relevant literature. In section 3, we discuss and highlight the important features of the Resource Constrained Project Scheduling Problem with Reworks and present an illustrative example. A mixed integer programming (MIP) formulation of the planning model is presented and computational experience with MIP is discussed in the last part of section 3. Section 4 is devoted to the introduction of a dynamic procedure for monitoring the progress of the project and rescheduling activities in order to improve project makespan. Also included in section 4 is our computational experience with the dynamic makespan improvement procedure. Some concluding remarks are given in section 5.

19.2. Literature review

Recently, there has been an interesting parallel development in the fields of project management and quality management. On the one hand, project management tools were adopted to implement Total Quality Management(TQM) programs, and on the other hand, more and more PMs have started using the TQM principles to manage their projects. Among the works which promote the use of project management tools to implement TQM programs we see Leavitt and Nunn(1994) and Stamatis(1994). The importance of using the TQM principles in project management is demonstrated by the articles by Fletcher(1996), Samaha(1996) and Walker(1996), all of which appeared in a special issue of **PMNetwork** devoted to TQM in project management. In these articles the authors discuss the ways to implement the TQM philosophy in project management and highlight the gaps between theory and practice.

A natural starting point in implementing the TQM principles in project management is to pose the question: "How should quality of a project be defined and more importantly how should it be measured?" Pinnell and Busch (1993) suggested that the quality of a project can be measured by time (how long it takes), cost (is it under budget) and how well the project conforms to specifications. They further emphasize the PMs role and responsibility in improving quality, although no quantitative measures of quality are suggested. Many studies (Pinto and Slevin 1987, 1988, 1989, Baker, Murphy and Fisher 1983) consider these performance indicators (time, cost and customer satisfaction) to be important criteria for the successful completion of a project. In a later study, Short (1993) advocated the use of schedule variance as a measure of project performance, and also a possible measure of quality. Schedule variance is defined as the difference between the actual cost of work performed and the budgeted cost of the work scheduled. Continuous monitoring of this variance allows PMs to detect any discrepancies early in the project life. Other important contributions to the definition/measurement of project quality are due to Cooper(1993) and Ireland (1991), both of whom adopt the well-known principle of "do it right the first time." Based on his consulting experience, Cooper judges the number of reworks as the main determinant of project quality. Reworks require additional time to correct the errors due to non-conforming activities. Accordingly, he defines quality as the proportion of work done which does not require rework and empirically derives the value of quality based on the real development projects he was involved with. There are other possible causes for rework, in addition to non-conforming activities. Sometimes customer specifications were never made clear and thus require corrections as the project is being completed. In other cases design changes occur after a project is initiated and thus require rework of completed activities. He identifies the effects of rework (or discovery) time on project progress and concludes that testing for quality at earlier stages of project life increases testing cost, but reduces project makespan considerably and increases the probability of meeting the customer's specifications. Ireland(1991) suggested that a good measure of project quality is the cost of non-conformance which includes cost of rework, scrap, warranties and liability judgements. In all the studies we report here, measures of project quality are discussed conceptually. They are not quantified, generalized or operationalized. It must be clear from the ongoing discussion that an operational and practical measure for quality should include cost of reworks, project completion time and customer satisfaction. In an effort to present a quantitative and operational measure of quality, we incorporate total rework cost and time explicitly and customer satisfaction implicitly in our model which will be discussed in the sequel.

19.3. The project planning model

During the implementation of a project, activities are performed at varying but usually less than perfect quality (Cooper 1993). If a PM's objective is 100% customer satisfaction, many of these activities need to be totally or partially reworked. If the requirement for rework is discovered right after an activity is completed, the effect on the general project performance will often be marginal. Reworks are considered to be a natural part of an iterative process of accomplishment and will be in the form of an extension of the normal activity durations. Reworks, however, will require additional resources, usually in the form of overtime. Overtime is more expensive than regular time and furthermore, since productivity often goes down as the amount of overtime increases, corresponding cost increases at a faster rate when the rework time gets longer (Cooper 1994). This cost is considered to be the rework cost. In addition, rework time and rework cost are affected by how much the start time of an activity is delayed beyond its earliest start time and by several other factors such as the number of activities delayed by reworking a particular activity. All these factors should be taken into consideration during the planning stage.

In addition, there are more complicated situations where not all reworks can be discovered immediately. The longer the discovery time the larger the gap between the planned schedule and the actual progress. In order to close this gap, additional resources are required. But, new hiring or transferring people unfamiliar with the project lowers the average skill level of the staff, which in turn causes delays in project completion time, due to lower productivity. As a result, overall costs can increase considerably. In addition, the effects of reworks can be more severe on the completion time of a project, since most of the slack time might be consumed by rework.

Clearly, reworks have considerable negative impacts on planned project schedules. The challenge for PMs is to develop project schedules that are minimally affected by reworks. In this study we address this concern. The problem we define here is an extension of the traditional RCPSP, which we call the Resource Constrained Project Scheduling Problem with Reworks (RCPSPR). In addition to the resource and precedence constraints of RCPSP, RCPSPR has two additional constraints to limit the total amount of rework time and the total amount of rework cost. The objective for the problem is maximizing quality where the quality of a project is defined as the degree to which the project completes within budget and on schedule. Following are the important characteristics of RCPSPR.

1. As in RCPSP, each activity in RCPSPR has a known normal duration, consumes known, deterministic and constant level of resources of each type during processing, and cannot be preempted.

2. In RCPSPR, each activity will require rework, consuming nonnegative amounts of its rework time and cost. For certain activities rework time, and rework cost may be zero. During the planning stage, PMs do not know exactly which activities will require rework, thus they develop a schedule based on no-rework. This schedule should be flexible so that when reworks occur during project implementation, the schedule will provide minimum additional rework time and additional rework costs.

3. Rework time for an activity increases as its scheduled time is delayed from its early start time. The rework cost is an increasing function of the rework time.

4. Rework required by an activity (if there is any) is discovered immediately after the activity is completed. Thus, there is no undiscovered rework for activities that have previously been completed.

5. Rework times are bounded by activity durations. That is, in the worst case the activity will be reworked completely. Therefore, the maximum rework time is equal to the activity duration. Similarly, rework costs are bounded by the money available for the project and in the worst case a PM would be willing to allocate an additional 100% of the budget for rework costs.

6. The objective of RCPSPR is to maximize the quality of a project by minimizing a function of the total rework time and total rework cost, rather than minimizing makespan, or maximizing NPV, as is typical in the RCPSP.

Before we present the mathematical models for RCPSPR, we give an example to illustrate why it is beneficial to consider RCPSPR rather than RCPSP.

19.3.1. An example

Figure 19:1 represents the activity-on-node network of a small project with six activities. The network is adapted from (Kolisch 1994). The numbers of the nodes are the activity numbers. The two numbers at the top of each node represent the duration (days) and the resource usage per day, respectively. There is one resource type and the availability of that resource is 4 units per time period. Consider the following scenario: A project manager wants to determine a minimum makespan schedule which satisfies the resource constraints and precedence relationships. He employs an exact solution algorithm for RCPSP (for example the one developed by Demeulemeester and Herroelen (1992)) and finds the optimal solution of seven days (see Figure 19:2). During the implementation of the project, however, he realizes that activity 3 does not conform to customer's specifications and needs to be reworked. It takes an additional three days to complete this activity. This delay, based on the optimal RCPSP schedule, delays the project completion time by 3 days and the project completes on the 10th day instead of the 7th day. Now, consider the feasible schedule given in Figure 19:3. Clearly, it is not an optimal schedule to RCPSP. However, it would provide a better schedule for the project manager since with the additional rework time the project could have been completed at the 8th day, exceeding optimal project completion by only one day. This example demonstrates that an optimal schedule to RCPSP might not necessarily be optimal when non-conformities occur. Note that resource usage per day is the same during the rework time as it was during the original execution.

During the planning stage, PMs need to prepare schedules for their projects. However, they cannot predict which activities will require reworking and how long the corresponding rework time will be. The question then is: How to incorporate any additional rework time/cost into planning? They could use a pessimistic approach and assume that each activity will require rework, and append expected rework times (independent of when activities are scheduled to occur) to activity durations and prepare the schedules with these longer than normal durations. But during the project implementation, if several activities were to conform 100% to specifications, which is not uncommon, especially for standard activities, then there will be unnecessary waiting periods between the completion time of an activity and the start times of its successors. This could lead to substantial delays in project completion time. Another approach would be to completely ignore rework times and costs at the planning stage and deal with reworks as they occur during the implementation stage. This approach may, of course result in significant variances between the planned and actual project costs and completion times. A compromise and arguably a better planning approach which we suggest here is to employ a procedure that would schedule the activities with their normal times while making aggregate allowances for rework times and costs. Once a project plan is obtained using this approach, the project would be closely monitored and schedule adjustments would be made during the implementation stage. In section 3.2, we present the planning model as a Mixed Integer Program (MIP).

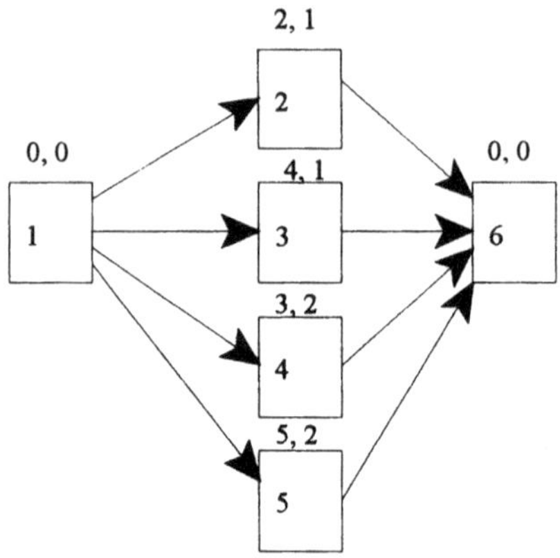

Figure 19:1. Example Project

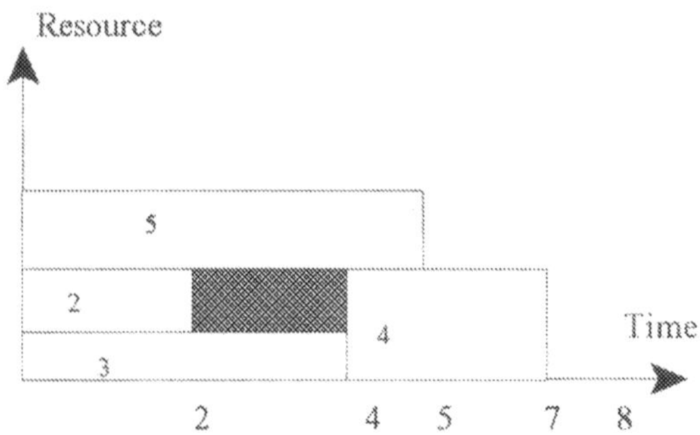

Figure 19:2. Optimal solution to example problem

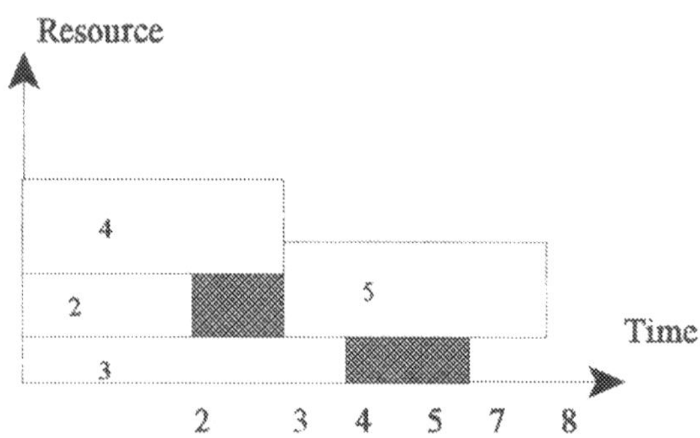

Figure 19:3. A feasible solution to example problem

19.3.2. A mixed integer programming formulation of RCPSPR

In this section we present the mathematical formulation for the RCPSPR which first appeared in Icmeli-Tukel and Rom (1997) . In an earlier study Icmeli and Rom (1996) presented a new formulation of RCPSP, which made the larger scale project scheduling problems more amenable to solution with a general-purpose optimization package like the Optimization Subroutine Library (OSL). In contrast to the traditional integer programming formulations in the literature the assumption on the integrality of the activity durations and project time line is relaxed. The project timeline is divided into equally spaced time intervals using time milestones, and the resources are allocated to these time intervals. Computational results revealed that this new model is not only more practical (since in practice activity durations vary in length. e.g. 0.5 hours, 2 weeks, 4 days) but is also computationally superior to the integer formulations that appeared in the literature. The following model is an extension of that model with the quality variables included in the objective function and the first two constraints.

$$(MIP) \qquad Min \; Q_1 + Q_2$$

subject to

$$\sum_{i=1}^{I} \sum_{j=1}^{N} P_{ij} \, \alpha_i^j \leq D \, Qsub1 \tag{1}$$

$$\sum_{i=1}^{I} \sum_{j=1}^{N} (RWC)_{ij} \, \alpha_i supj \leq B \, Q_2 \tag{2}$$

$$\alpha_1^j \leq y_1^j \quad \forall' j \tag{3}$$

$$x_j = \sum_{i=1}^{I} \alpha_i^j \, T_i, \; \sum_{i=1} tol \, \alpha_i^j = 1, \; \alpha_i^j \geq 0 \quad \forall j \tag{4}$$

$$\alpha_i^j \leq y_{i-1}^j + y_i^j \;\; i = 2,..,T-1 \quad \forall j \tag{5}$$

$$\alpha_T^j \leq y_{T-1}^j \quad \forall j \tag{6}$$

$$\sum_{i=1}^{I} y_i^j = 1 \quad \forall j \tag{7}$$

$$\sum_{j=1}^{N} a_{kj} \, y_i^j \leq R_{ik} \quad \forall k, i \tag{8}$$

$$x_j \geq x_i + d_j \qquad \forall (i,j) \in H \tag{9}$$

$$y_i^j \in \{0,1\} \;\; \forall i,j \, , \;\; x_j \geq 0 \;\; \forall j \tag{10}$$

$$0 \leq Q_1, Q_2 \leq 1 \tag{11}$$

where:

decision variables:

x_j: completion time of task j, j=1,.., N;

α_{ij}: weight assigned to activity j at milestone T_i;

y_{ij}: 1, if activity j completes in the time interval between T_i and T_{i+1}, and 0, otherwise;

Q_1: ratio of total rework time used to total rework time allowed;

Q_2: ratio of total rework cost expended to total rework cost allowed;

parameters:

P_{ij}: rework time of activity j if it is completed in the time interval between T_i and T_{i+1};

$(RWC)_{ij}$: rework cost of activity j if it is completed in the time interval between T_i and T_{i+1};

I: total number of milestones;

N: last activity in the project;

d_j: normal duration of task j;

T_i: time milestones, i=1,.., I;

R_{ik}: the amount of resource k available for the interval between T_i and T_{i+1}, k=1,..,K;

a_{jk}: amount of resource k needed by task j;

D: sum of the normal activity times of all the activities in the project;

B: total budget allocated to the normal functioning of the project;

H: set of precedence relationships.

The completion time of an activity is represented as a convex combination of time milestones, where the α_i^j 's are the weights assigned to each milestone (constraint set 4). With constraint sets 3, 5 and 6 we ensure that at most two consecutive the α_i^j 's are positive. This forces the completion time of an activity to occur within an interval. The seventh set of constraints guarantees that each activity completes exactly once. In this model, resources are allocated to time intervals and the total resource consumption of all the activities completing in that interval can not exceed the resource availability (constraint set 8). The first two constraints impose rework time and rework cost limitations and the objective function implicitly minimizes the sum of the left hand sides constraints by minimizing the sum of Q_1 and Q_2. The objective function of MIP is a function of the total rework time and the total rework cost. More specifically, it is the sum of two ratios: the ratio of the total rework time used to the total available(allowed) rework time ; and the ratio of the total rework cost expended to the total available budget. The objective is to minimize this sum. The first constraint ensures that the total rework time does not exceed sum of the activity durations. The second constraint guarantees that the total rework cost does not exceed the allocated budget. The P_{ij}'s and RWC_{ij}'s are the rework times and the rework costs when they occur at time milestones. If an activity is completed between T_i and T_{i+1}, the amount of rework time and rework cost acquired from each milestone is proportional to closeness of the activity completion to that milestone.

Note that it is possible to change the assumptions concerning the way quality is measured. For example, we assumed that Q_1 and Q_2 are between 0 and 1. This interval can be reduced to make the first two constraints more constraining or can be enlarged to make them less constraining or even redundant. Furthermore, the assumptions regarding the way we construct P_{ij} and RWC_{ij} can also be modified. Possible modifications will be discussed and tested in the computational results section.

19.3.3. Computational results with MIP

There are some aspects of the MIP which are especially suited to the use of OSL. The constraints which involve the sum of a number of 0-1 variables are handled separately by OSL, as special ordered sets. Rather than branching on individual variables, the entire set of variables for an activity are branched on, by setting all the variables before the branch point equal to zero or one. In addition, OSL has a preprocessor which analyzes the consequences of setting various variables to 0 or 1, and then adds them as additional constraints. This often contributes to an early solution of the problem. We found that in many cases the preprocessor either solved the problem or required few branching operations before an optimal solution was found.

OSL has other features which allow for specialized approaches in solving problems. One feature allows the user to specify a priority for each SOS set to determine which set to branch at a node in the branch and bound tree. In the computational results, we based the priority on where in the network a node occurred. We started by branching on the early activities, then periodically, every 150 branching operations, we reversed the priorities and branched on the later activities, and then back again to the early activities. We also tried basing priorities on the number of succeeding activities, but this did not lead to a significant improvement. OSL also allows the user to specify the node in the branch and bound tree to be evaluated next. The default is a depth first search. We modified this to follow a depth first search for 100 nodes and then restart at the node with the best continuous solution. Once a feasible solution was found, any node which did not have a value at least 5% better than the best current solution was pruned.

The first two constraints of the model require the parameters P_{ij} and RWC_{ij} to be determined. As will be discussed later, the problem generator we employed here does not contain routines to generate these parameter values. For a project manager, the best way to estimate activity rework time and cost would be to determine the factors that effect the rework time and cost, and develop a corresponding function. Here, we suggest five methods to generate these values. In all these methods, normal activity cost, activity duration, and activity schedule time are assumed to be the dominant factors affecting the rework time and costs. We suggest several linear and quadratic functions as possible ways of anticipating rework times and costs. Generally these functions have the following properties: rework times and costs are larger as they occur later in the project duration; rework costs are an increasing function of rework times; and rework time increases as the number of predecessors of an activity increases. Note that these are only general guidelines to be adapted to specific situations. The following five alternatives were used in the computational tests.

1. $(RWC_{ij})_1$: The following formulas are used for generating rework times and rework costs.

$$(RWC_{ij})_1 = (i*cost_j /I) * DFj \qquad \text{(equation 1)}$$

$$P_{ij}= (i/I)* d_j * DF_j \qquad \text{(equation 2)}$$

where I is the total number of time intervals, i is the time interval activity j completes in, and DF_j is the difficulty factor for activity j. DF_js are randomly generated from a uniform distribution over [0.5,1.5]. $Cost_j$ is the projected cost of activity j and is generated from a uniform distribution over [1000,10000]. As the formulas indicate, the activity rework time and cost increase as the activity is completed later in the project duration. DF_j values are introduced into the formulas so that activities which are completed in the same time period do not necessarily have the same rework time. Although in this study DF_js are randomly generated, in practice they should be determined by the judgement of the PMs.

Taguchi introduced a commonly used quality improvement tool in which conformance(quality) is measured by the deviation of the value of a quality characteristic from its target value. Specifically, Taguchi introduced the quadratic quality loss function in which quality loss is proportional to the square of the deviation from the target value (Gitlow, Oppenheim and Oppenheim 1989). In cases where the ideal value of the quality characteristic y is zero, then the loss function L(y) has the form $L(y)= ky^2$, where k is the proportionality constant. In the next method for computing rework cost, we incorporate this loss function where the quality characteristic y is the rework time P_{ij}.

2. $(RWC_{ij})_2$: The following formula for the loss function is used. P_{ij} is the same as in $(RWC_{ij})_1$.

$$(RWC_{ij})_2 = k*(P_{ij}**2) \qquad \text{(equation 3)}$$

The constant k is chosen as shown in equation 4.

$$(RWC_{ij})_2 = [(B/N)/((D/N)**2)]*(P_{ij}**2) \qquad \text{(equation 4)}$$

Here B is the sum of the normal costs and D is the sum of normal durations. The term inside the brackets corresponds to the constant k. This was found to give roughly the same importance to the rework time and rework cost constraints.

3. $(RWC_{ij})_3$: The number of predecessors, in addition to the rework time, is used to determine the rework cost. P_{ij}s are calculated using the same formula as in $(RWC_{ij})_1$.

$$(RWC_{ij})_3 = ([(B/N)/((D/N)**2)]*P_{ij}**2)*NOPRED(j) \qquad \text{(equation 5)}$$

where NOPRED(j) = the number of predecessors for task j.

4. $(P_{ij})_2$: The rework time depends on the number of predecessors in addition to a random factor, RWC is the same as in $(RWC_{ij})_1$:

P_{ij} = $(i*d_j/I)$*Uniform(0.5,1.0) If the number of predecessors is ≤ 3 (equation 6)

P_{ij} = $(i*d_j/I)$*Uniform(0.75,1.5) If the number of predecessors is > 3 (equation 7)

5. RWC&P: For the last case the rework costs are generated using $(RWC_{ij})_3$ and rework times are generated using $(P_{ij})_2$.

To investigate the practical applicability of the planning model represented by MIP, we experimented with 110 test problems, generated using PROGEN, a problem generator developed by Kolisch et al.(1995). These problems have the following characteristics.

. *Number of Activities:* Each project network has 32 activities.

. *Density of a Project Network:* The average problem density in the data set is 1.80. Problem density is defined as the ratio of the number of precedence relationships to the number of activities.
. *Number of Resource Types*: Each project utilizes four resource types.
. *Due Dates*: due dates are not given in the original Kolisch data set. Thus, we generated a feasible due date for each problem as follows:

due date= 1.25(feasible project completion time found by a heuristic procedure)

Note that in our computations a parallel scheduling scheme (Kolisch 1994) is employed for generating a resource feasible schedule.
. *Resource Usage Factor (UF):* UF is used as a measure of resource constrainedness. It is the ratio between the resource usage of the activities and the availabilities. It is calculated as follows:

$$UF_r = \frac{K_r(\text{due date})}{\sum_j (k_{jr} d_j)} \qquad \text{(equation 8)}$$

where K_r is the resource availability of resource type r, k_{jr} is the resource usage per time period by activity j, and d_j is the duration of activity j. The average UF values for four resource types in the test problems are 3.88, 3.85, 3.88 and 3.80, respectively.

The computational results are given in Tables 19:1-4. For each problem we set an LP iteration limit of 30000. Table 19:1 gives the computational results for the 110 test problems using the five alternative methods specified above. The second column gives the number of problems successfully solved within the iteration limit. For all different methods of generating rework costs and rework times, at most 7.3% of the problems (8 out of 110) terminated prematurely, by reaching the iteration limit of 30000. In the third column the average additional time required for reworks is reported. This is given as the percentage of total time spent performing activities at their normal times. The results indicate that rework times are approximately 39% of the total activity durations, except for the last parameter setting, where this percentage is down to 28.5%. This gives valuable information to managers in terms of tracking the expected overtime and overtime cost. Managers should add tolerance factors to their time and budget estimates to compensate for possible rework. Similarly, in column 4 we report the average additional costs incurred due to rework. It is given as a percentage of the total budget. When the rework costs depend not only on the rework time but also the density of the project network (the third method) almost 50% additional budget is required. Although we report, in column three, the additional time needed for reworks, this does not give us the information about how this additional rework time will affect the project completion time. This is reported in column five, where we give the average percent makespan extension. For each problem the percent extension is calculated as

$$\frac{(PCT_2 - PCT_1)}{PCT_1} \times 100 \qquad \text{(equation 9)}$$

where PCT_1 is the project completion time determined by solving MIP with normal durations. When the expected rework times determined from (the original) MIP are explicitly added to activity durations and MIP is resolved, the resulting project completion time thus obtained is PCT_2 . This gives the manager an idea of how much of a delay to expect if reworks occur for each task. As seen in columns 3-5 the possibility of reworks can change the plans quite severely during implementation. For example, if rework costs occur in real life according to the third method, the project will require 38.30% additional rework time, 48.80 % additional funds, and 53.84% due date extension due to reworks. Once again this confirms the importance of integrating expected reworks into the project planning stage. The last column gives the average and median time (in seconds) required to solve the problems on a PC with a 100 Mhz Pentium processor. The first entry in each cell corresponds to the average CPU time and the second entry is the median. The reason for reporting median values is to point out that while a few problems require excessive computational time, most of the problems can be solved within 10 seconds.

In general, results in Tables 19:1-3 indicate that the use of quadratic loss functions produced schedules with reduced rework times and costs, and shorter makespan extensions. This result was independent of the objective function used. Specifically, the average minimal additional rework time and the average minimal makespan extensions are obtained by the RWC&P method, and the average minimal cost of rework is obtained by RWC_2 method.

Next we wanted to determine how these results would be affected if we were to use the objective functions commonly used for RCPSP. We solved the same set of problems with the same five choices for the rework costs and times, but with the objective of minimizing the project makespan and of maximizing the Net Present Value (NPV) of the project's cash flows. Tables 19:2 and 19:3 give the summary results. Since the data set we worked with does not contain data on cash flows for the activities, additional data were generated. The cash flows were generated randomly and uniformly between [-500,1000] and the nominal discount rate used was 0.01. As can be seen in column 2 of Tables 19:2 and 19:3, the number of problems successfully solved decreased; 11.8% of problems when minimizing makespan and 14.5% of problems when maximizing NPV terminated prematurely by reaching the iteration limit. Furthermore, the time required to solve each problem increased moderately when the objective was minimizing makespan, and considerably when the objective was maximizing NPV. A similar pattern appeared when we compared the results in columns 3-5 of Tables 19:2 and 19:3 with the same columns in Table 19:1. Additional rework time and costs increased moderately for the minimizing makespan objective and considerably with the maximizing NPV objective. However, it seems the average due date extension is slightly less for both of these objectives compared to maximizing quality. We also investigated the quality of the solutions obtained with the three objectives. Specifically, we wanted to determine if scheduling the project with the objective of quality maximization yielded a "good schedule" in terms of the makespan and the net present value objectives. To determine this, we compared the makespans and net present values of the schedules obtained with the quality maximization objective with the optimal makespan and optimal NPV obtained when the problems were solved with the makespan minimization and the NPV maximization objectives, respectively. This analysis is summarized in Table 19:4. The first row gives the makespan optimality gap for each method. For example, with the first method, the makespans obtained with the quality maximization objective are on the average within 0.9% of the optimal makespans. When the rework costs and time depend on the number of predecessors this gap is the largest (14.74%). Similarly, the second row gives the optimality gap in NPVs. In the worst case, the NPVs obtained with the quality maximization objective are on the average within 4.67% of optimality. Note that the five parameter settings we generated here are based on the discussions in the literature. For example, studies suggest that activity rework time and rework costs increase with the activity delay. One can argue that due to this assumption early start schedules should provide better project quality in addition to shorter makespans. Thus, makespan optimality gaps are as small as 0.9%. Thus, one should expect changes in the size of the optimality gaps if the project rework times and costs do not follow these practical assumptions.

Table 19:4 also provides valuable information to practitioners. When quality is chosen as an objective, the schedules are not only providing a minimum of additional rework time and cost but also they are within 5% of the optimal makespans (in most cases), and within 5% of the optimal net present values. The results confirm the practicality of the quality objective, as in the survey results (Icmeli-Tukel and Rom 1998). Managers report that they want their projects to be on time, within budget, and with high performance. It appears that they can accomplish these multiple objectives by minimizing rework time and rework costs. The planning model discussed in this paper will allow them to integrate this objective into their traditional scheduling methods.

19.4. A dynamic makespan improvement procedure

The optimal solutions obtained by solving the MIP model can be used as baseline schedules during the monitoring of the project progress. That is, project progress can be tracked by comparing the baseline solutions, which contain activity schedules, resource estimates and cost estimates, against actual progress. In this section we offer a monitoring procedure that will improve the makespan of the baseline solutions. Specifically, it is a dynamic procedure a project manager can employ during the course of a project to reduce the effects of the reworks on baseline schedules. The procedure involves monitoring the project progress at certain time intervals, called time milestones. At each milestone the activities that had to be reworked are determined and the corresponding rework times are added to the expected activity durations which constitutes actual work completed. Then, the non completed activities are rescheduled based on their finish time variance. For each activity, its finish time variance is calculated as the difference between the expected finish time as a result of the changes occurred at the current milestone and the baseline finish time provided by the MIP model. The tasks with longer delays from their baseline finish times will have higher priority to be scheduled earlier.

<u>The Dynamic Procedure:</u>

The following is a formal statement of the procedure:

For i =1,.., M do

{Update the project progress}
for each $j \in S_i$ calculate RWT_j and
let $d_j \leftarrow d_j + RWT_j$
for each activity $j \in C_i$ calculate FV_{ij}
update the scheduling priority of each $j \in C_i$ based on FV_{ij}
reschedule C_i
$i = i+1$
end do

where:
M: number of milestones;
RWT_j: rework time of activity j;
d_j: duration of activity j;
S_i : set of activities completed between milestone i-1 and i; $S_1 = \varnothing$;
C_i: set of activities in progress or yet to start at milestone i;
FV_{ij}: finish time variance of activity j at milestone i.

Activities with larger FV_{ij} have higher priorities to be scheduled earlier.

19.4.1. Computational results with the dynamic procedure

In order to determine whether we can reduce the effects of rework times on baseline completion times, we applied the dynamic procedure to the 110 test problems. Microsoft Project 98 was used to track the project progress and reschedule the activities in accordance with the priorities established using the finish time variances. We monitored each project at 5 time milestones, which correspond to five time intervals. At each milestone, for the activities that are completed earlier than their baseline completion times, their rework times are adjusted. This adjustment is necessary since rework time for an activity is proportional to the interval in which the activity is completed. Based on finish time variances, scheduling priorities are set as the highest, higher, high, medium,..etc. At the last milestone, which signals the completion of the project, we determine project makespan. The actual project makespan for each project is compared to the baseline project makespan. The results are reported in Table 19:5. Recall that in Tables 19:1-3, we report the average percent makespan extensions which are calculated using equation 9. Note that in general these extensions are maximal since PCT_2 is the project completion time when rework times are explicitly added to the model. Thus, one would expect the average percent makespan extension to be smaller when rework times are reduced by rescheduling at the milestones. In Table 19:5, we report the smallest gaps which are produced when **RWC&P** is used for calculating rework time and cost. As can be observed from the table, when the dynamic procedure (with Microsoft Project 98) was applied, in 21% of the test problems makespan did not improve. However, for 79% of problems monitored we observed an average 14.73% makespan extension compared to 30.82 % reported as the best gap so far.

19.5. Conclusions

This study provides valuable insights to Project Managers about the criticality of rework in project scheduling. As stated by Cooper (1993, p.21) "Reworks must be recognized, monitored and minimized..". With this study we initiated an effort to achieve this goal. We offer a project planning model which recognizes the possibility of reworking activities and generates schedules which minimize a function of the total rework time and total rework cost. The model is formulated as a mixed integer program and is easy to effectively implement for large networks. Computational tests were performed using a data set generated by the PROGEN problem generator. As part of the computational testing, an analysis of the quality of the schedules obtained using different functions for rework times and costs was also undertaken.

Subsequently, a dynamic makespan improvement procedure was introduced to adjust the baseline schedule obtained by the planning model. This procedure uses the well-known Microsoft Project 98 software to monitor and reschedule project activities as reworks for activities become necessary. We also reported computational results with the dynamic makespan improvement procedure.

The recent emphasis given to quality management, continuous improvement, and customer satisfaction in project management mandate the development of new tools and techniques that integrate these concepts into scheduling decisions. This study constitutes an effort to address some of these important issues. We expect to see a good portion of future research developing in this area.

References

BAKER, B.N., MURPHY, D.C. AND FISHER, D. 1983. Factors Affecting Project Success. *Project Management Handbook.* Van Nostrand Reinhold Company, New York.

COOPER, K.G. 1993. The Rework Cycle: Benchmarks for the Project Manager. *Project Management Journal.* 24, 1, 17-21.

COOPER, K.G. 1994. The $2,000 Hour: How Managers Influence Project Performance Through the Rework Cycle. *Project Management Journal.* 25, 1, 11-24.

DEMEULEMEESTER, E. AND HERROELEN W. 1992. A Branch and Bound Procedure for the Multiple Resource Constrained Project Scheduling Problem. *Management Science.* 38, 1803-1818.

FLETCHER, C. 1996. Total Quality Management: A Practical Guide. *PMNetwork.* 10, 2, 13-17.

GITLOW, H., OPPENHEIM A. AND OPPENHEIM R. 1989. *Quality Management:Tools and Methods for Improvement.* Burr Ridge, Illinois.

ICMELI-TUKEL O. AND ROM, W. O. 1996. Solving The Resource Constrained Project Scheduling Problem with Optimization Subroutine Library. *Computers and Operations Research.* 23, 8, 801-817.

ICMELI-TUKEL O. AND ROM, W. O. 1997. Ensuring Quality in Resource Constrained Project Scheduling. *European Journal of Operational Research.* 103, 3, 483-496.

ICMELI-TUKEL O. AND ROM, W. O. 1998. Analysis of the Characteristics of Projects in Diverse Industries. *Journal of Operations Management.* (forthcoming)

IRELAND L.R. 1991. *Quality Management for Projects and Programs.* Upper Darby, PA: Project Management Institute.

KOLISCH, R. 1994. Efficient Priority Rules for the Resource Constrained Project Scheduling Problem. Working Paper, Christian Albrechts Universitat zu Kiel, Kiel, Germany.

KOLISCH R., SPRECHER A. AND DREXL A. 1995. Characterization and Generation of a General Class of Resource Constrained Project Scheduling Problems:Easy and Hard Instances. *Management Science*. 41, 1693-1703.

LEAVITT, J.S. AND NUNN P.C. 1994. *Total Quality Through Project Management*, New York, McGraw-Hill.

PINNELL, S.S. AND BUSCH, J.S. 1993. How Do You Measure the Quality of Your Project Management? *PMNetwork*. December, 35-36.

PINTO, J.K. AND SLEVIN, D.P. 1987. Critical Factors in Successful Project Implementation. *IEEE Transactions on Engineering Management*. EM-34, 22-27.

PINTO, J.K. AND SLEVIN, D.P. 1988. Project Success: Definitions and Measurement Techniques. *Project Management Journal*. XIX, 1, 67-72.

PINTO, J.K. AND SLEVIN, D.P. 1989. Critical Success Factors in R&D Projects. *Research Technology Management*. January- February, 31-35.

SAMAHA, H.E. 1996. Total Quality Management: A Barrier to Innovation. *PMNetwork*. 10, 2, 23-24.

SHORT, J.W. 1993. Using Schedule Variance as the Only Measure of Schedule Performance. *Cost Engineering*. 35, 10, 35-40.

STAMATIS, P.H. 1994. Total Quality Management and Project Management. *Project Management Journal*. 25, 3, 48-54.

WALKER R.A., 1996. Traditional PM and TQM: Inconsistent in Theory and Practice? *PMNetwork*,10, 2, 19-21.

Optimization Subroutine Library-Guide and Reference. 1991. IBM, Kingston, New York.

Table 19:1. Computational results when the objective is Maximizing Quality

	No. of Problems successfully solved	*Additional rework time (percentage of total activ. durations)*	*Additional cost of rework (percentage of total budget)*	*Average % makespan extension*	*Average and Median CPU time (sec.)*
RWC_1	108	38.50	37.60	39.29	16.04 4.72
RWC_2	104	38.20	22.80	54.48	8.44 3.15
RWC_3	104	38.30	48.80	53.84	14.57 3.21
P_2	106	38.12	37.78	39.76	21.07 7.42
RWC&P	102	28.51	28.00	34.60	15.26 7.69

Table 19:2. Computational results when the objective is Minimizing Makespan

	No. of Problems successfully solved	*Additional rework time (percentage of total activ. durations)*	*Additional cost of rework (percentage of total budget)*	*Average % makespan extension*	*Average and Median CPU time (sec.)*
RWC_1	97	39.10	38.10	41.32	19.25 5.33
RWC_2	100	39.00	23.60	40.66	28.64 6.15
RWC_3	98	39.20	50.90	41.94	33.05 5.60
P_2	99	39.17	38.28	41.66	32.02 5.99
RWC&P	98	29.52	29.85	31.30	28.32 5.65

Table 19:3. Computational results when the objective is Maximizing the Net Present Value

	No. of Problems successfully solved	*Additional rework time (percentage of total activ. durations)*	*Additional cost of rework (percentage of total budget)*	*Average % makespan extension*	*Average and Median CPU time (sec.)*
RWC_1	98	44.90	44.60	42.28	61.02 11.07
RWC_2	94	45.10	30.00	42.24	73.99 13.82
RWC_3	94	45.03	60.84	42.28	100.27 17.35
P_2	94	44.87	44.64	42.04	97.31 19.53
RWC&P	97	33.73	35.50	30.82	64.36 16.23

Table 19:4. Comparison of the performances of objective functions

	RWC_1 (%)	RWC_2 (%)	RWC_3 (%)	P_2 (%)	$RWC\&P$ (%)
makespan	0.90	5.10	2.29	1.92	14.74
NPV	2.00	2.57	2.28	2.29	4.67

Table 19: 5. The comparison of makespan extensions when RWC&P rule is used

% solutions improved	*Ave. % makespan extension Quality obj.*	*Ave. % makespan extension Makespan obj.*	*Ave. % makespan extension NPV obj.*	*Ave. % makespan extension Dynamic procedure*
79	34.60	31.30	30.82	14.73

20 COGNITIVE SCIENCE AND PROJECT SCHEDULING: MORE REALISTIC REPRESENTATION

Pierre Baptiste
Olivier Grunder

Laboratoire d'Automatique de Besançon, UMR CNRS 6596, ENSMM, UFC, Institut de Productique, 25 rue Alain Savary, 25000 Besançon, France.

20.1. Notion of project

Project scheduling represents the biggest part of project modeling research. More and more hypotheses have to be done on the project to apply scheduling algorithms. From an other point of view, detailed scheduling is not the only problem in project modeling.

The notion of project itself has to be defined clearly. So it seems important to propose, even briefly, a taxonomy of projects. This will allow to define for which class of projects the existing tools are proposed. We can easily distinguish two main classes of project:

20.1.1. "Realization" projects

When the area of the project is well defined, the project is usually nothing else than a new instance of an already known project. Consequently, most possibilities for the realization of all sub-projects are known a priori. This kind of problem will happen mainly in civil engineering, machine tool, and so on. The different technologies are well known, but the project is a new assembly of existing tiles.

Then, the pre-project is usually nothing more than a succession of choices among several well-known possibilities. Those choices allow to build a perfectly defined project.

When the project begins, all tasks are perfectly defined, in term of duration, resource consumption, and so on. The modeling is detailed, and its aim is mainly to schedule and to evaluate the cost and the delay.

The multi-resource scheduling is a very important tool at this stage. Models are mainly deterministic, at least for the structure of the project (not always for the durations).

During the realization, the same model is used for the project monitoring. The aim is to control and anticipate any deviation from the forecasted delay, or any deviation from the initial cost.

Globally, the evolution of the knowledge concerning this class of project can be represented as in figure 20:1. In the initial pre-project, all the knowledge on possible realizations are already present, and as soon as the modeling begins, all choices are done. The uncertainty on what will happen is very small. Of course, durations are hardly forecasted.

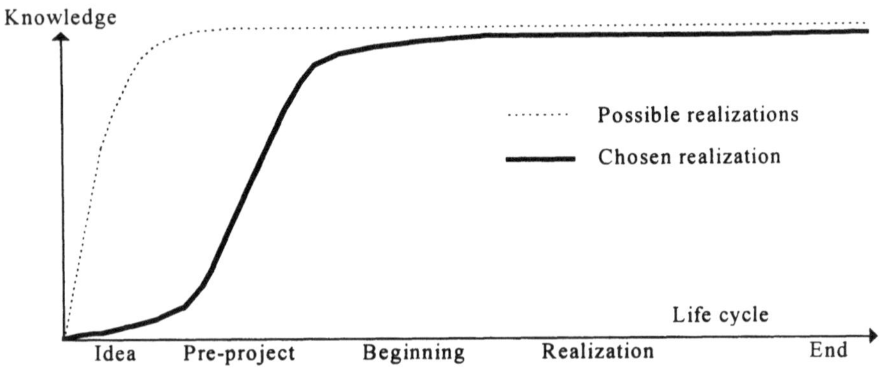

Figure 20:1. Knowledge evolution in realization project

20.1.2. "Innovation" projects

A project is an innovation project as soon as research is needed to reach the target. It means that several tasks have to be performed, but with no insurance of success. Then, the project is a complicated assembly of risky tasks, failure anticipation, and alternative solutions. This is the case in new product development, certain one-of-a-kind production, Research and Development projects, etc.

In the area of innovation, the creation of the knowledge is completely different. A great amount of possible futures can be defined, and it is very difficult to choose among those possibilities.

Moreover, the choice can not be done for most of them before the realization itself, depending on the result of previous sub projects.

The pre-project consists in the definition of all possible futures, in term of description, duration and cost.

At the beginning of the project (starting date), most of the real realizations are unknown. The choice between possible futures (already defined and described as possible instances of the project) is still not done. The preoccupations of the project manager at this point are:

- evaluation of the risk,
- evaluation of the cost,
- evaluation of the delay.

The actions he can use are:

- to anticipate a failure by adding secure alternatives,
- to put several sub-projects in competition,
- to define the limit duration of several tasks.

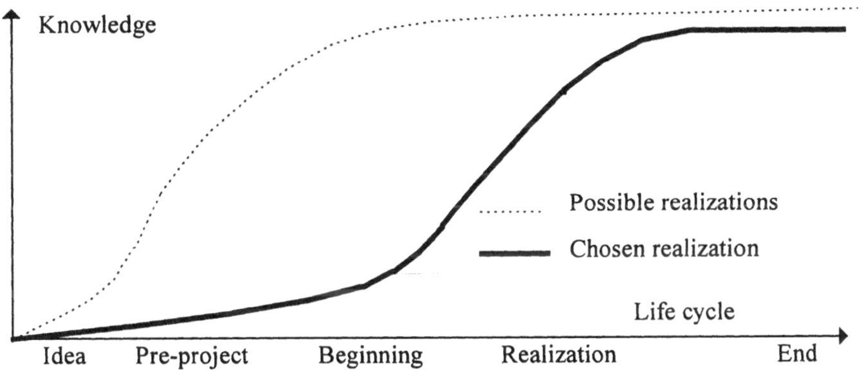

Figure 20:2. Knowledge evolution in innovation project

But while the risk is not completely absent, cost and delay have not the same signification. In fact, the failure of most of the risky tasks can be either forecasted, (and then an other task can be planned simultaneously), or just considered (alternative task at the moment the failure is decided), but not at the same cost.

The aim of the project modeling and project scheduling in this type of problems is an help for the definition of the structure of the graph itself. The parameters are more the decision variables, and the objective functions are the risk and the cost. The resource scheduling is usually a non-sense: so many parameters are unknown that the probability of even the most probable future is very low.

During the realization, the knowledge concerning already done and current sub-projects permits to choose among the different possibilities, by using the organization rules defined in the pre-project. The decisions are very important and concern the organization of the project itself.

The evolution of the knowledge on the project can be model schematically as in figure 20:2. Most of the final organization decisions have to be taken during the project itself, but according to the rules defined at the beginning of the project. The main difficulty in this kind of project is that the modeling tools are defined for modeling a single future, but not for modeling the knowledge about the choice among different futures.

20.1.3. Nature of the knowledge

Particularly in "innovation" projects, the knowledge is not limited to a simple graph. The knowledge can be viewed as structural knowledge and strategic knowledge.

20.1.3.1. Structural Knowledge. This knowledge concerns the description of the different possible realizations of the tasks: It answers the question "How this task can be performed ?". In realization, most of the tasks have only one possible description (except for example, if the realization depends upon the weather, or the nature of the grown in civil engineering), however in most innovation projects, the task may have several descriptions. For example, a car factory may study the adaptation of an old motor on a new car as well as the development of a new one. In most cases, it can be modeled as alternative sub-projects, usually exclusive, but seldom unconnected.

20.1.3.2. Strategic Knowledge. This kind of knowledge concerns the choice of the different possible futures. It defines first on which information the choices will be based on, and it is a set of rules that allow the choice of a particular realization for each task, depending of the context.

The context can be connected to the result of earlier tasks (sub-projects), or connected to external variables that describe the environment of the project. The knowledge as this level should explicit what to do if any of the risky tasks fail, and also define the real definition of failure. Most of modeling tools do not take into account the second kind of knowledge, except through a set of probabilities.

20.2. Existing models for "innovation" project modeling

The basic CPM graph is not sufficient as soon as several futures can be considered. Early, authors have been interested in more efficient models. The first generation deals with graph based model, but extended with logical operators (AND-OR) and probabilistic computing. The second generation is connected to the more recent research in cognitive modeling.

20.2.1. Probabilistic models

Obviously, deterministic models as CPM and PERT are not useful for this kind of projects. In PERT, the durations are stochastic, but the structure of the graph is completely deterministic, and neither alternative nor failure can be modeled.

Soon after the first deterministic model, it appears that the single future was not adequate. The first extension is proposed in Eisner (1962) with the introduction of the decision boxes.

Next, the well known GAN model has been proposed in Elmaghraby (1964, 1966, 1977). The GAN activity network is based on a set of reception/transmission nodes. This network allows the representation of alternatives, AND-OR relations (Figure 20:3) , etc., but in term of probability.

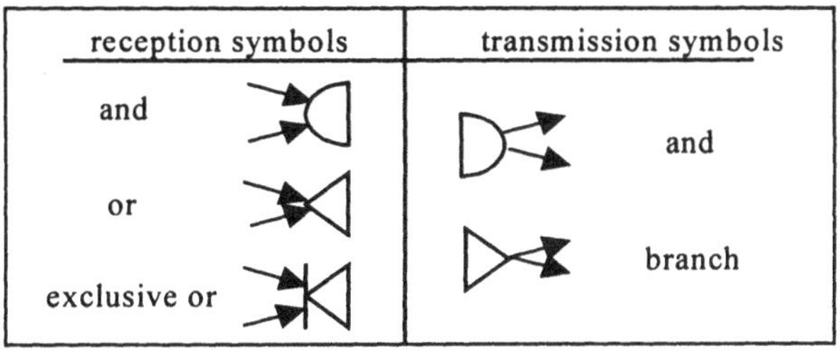

Figure 20:3: connection for the GAN activity network

The method provides several procedures to reduce a general network in order to analyze the properties of the network.

The GERT method (*Graphical Evaluation and Review Technique*) in Pritsker and Happ (1966a), Pritsker and Whitehouse (1966b) and Whitehouse (1969) and after the Q-GERT model in Pritsker (1979) uses the same kind of paradigm, for the simulation of stochastic activity networks.

The VERT model has been proposed later, mostly for military applications in Moeller (1972) and Moeller and Digman (1981), as an extension of GERT and Q-GERT.

The GAN model is the base of the most recent developments in the area proposed in Ramat, Lente and Tacquard (1995), Ramat et al. (1996) and Ramat (1997). Their approach consists in a hierarchical modeling of the project in terms of elementary activities and exploring activities.

A probability of failure, and two durations (known as probabilistic functions); the duration in case of success and the one in case of failure describe each elementary activity. So an elementary activity can be risky or not (see figure 20:4).

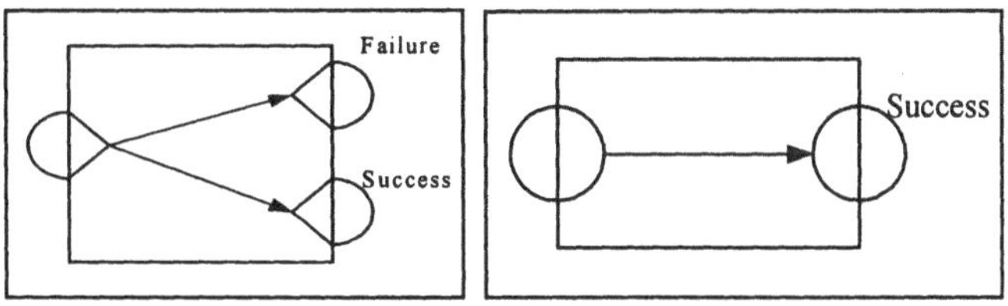

Figure 20:4. Elementary activity of the RAIH model, with and without failure

Exploring activities are predefined graphs. Basically, the authors define 4 classes of activity: parallel, cyclic, linear and conditional. (see figures 20:5a, 20:5b, 20:5c, and 20:5d).

For each of them, they show how to compute recursively the probability of failure, and an approximation of the two probabilistic distributions (duration in case of failure and in case of success).

The limit of such a model is that only the tactical knowledge is considered. If the different futures are correctly described, nothing is said on the choice of one among the others.

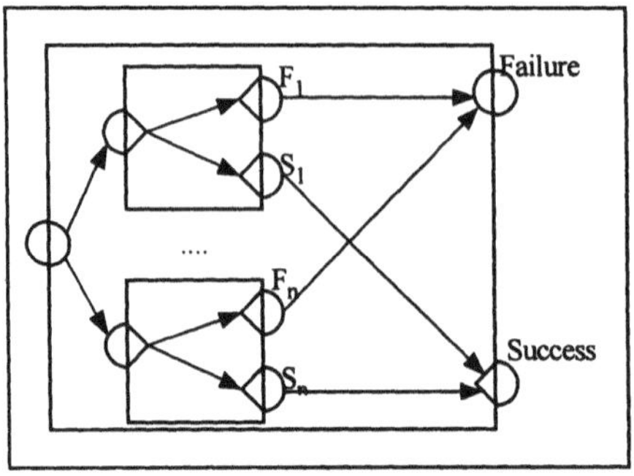

Figure 20:5a. Parallel activity

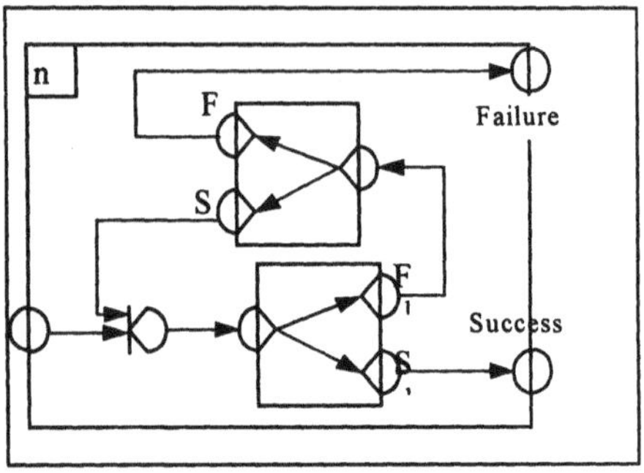

Figure 20:5b. Cyclic activity

In the 12 elements pointed out by Pinedo to explain the lake of real applications of scheduling in industry one of the more important aspect is that models assume the independence of variables. This is the most important defect of the probabilistic model. While strategic knowledge is presented with probability, it is implicit that choices and failures are independent, witch is obviously seldom true. So the risk of failure is not correctly estimated, neither the duration. That is why it is necessary to express more correctly the strategic knowledge.

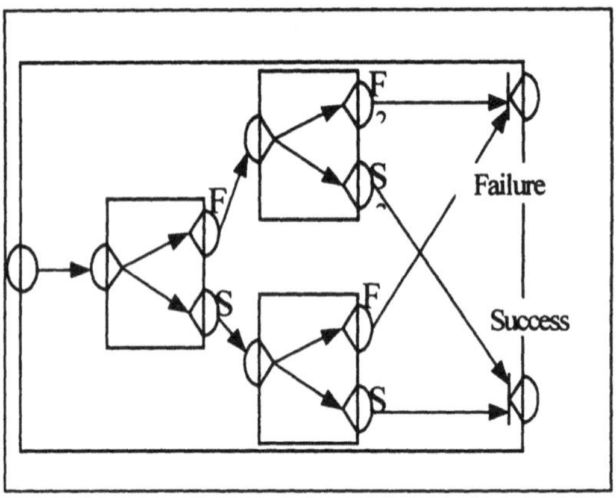

Figure 20:5c. Sequential activity

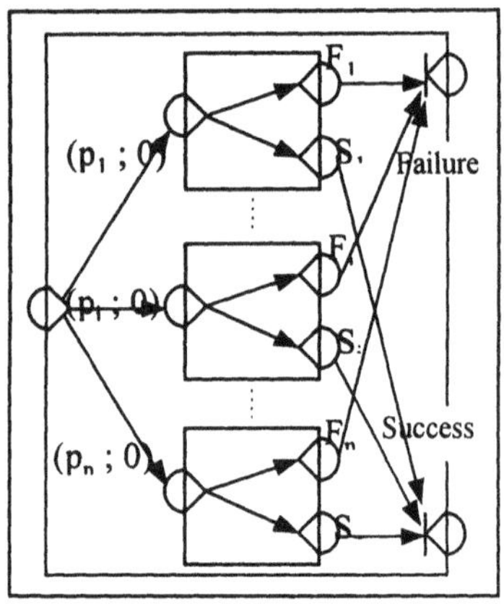

Figure 20:5d. Conditional activity

20.2.2. Cognitive model in problem solving

If the probability is not adequate to represent the strategic knowledge, then the rules have to be explicit themselves. This is the topic of cognitive science modeling, and particularly of problem solving modeling. This large area of research has been explored in many directions.

20.2.2.1. The three levels of knowledge. Most researches in knowledge modeling and particularly in problem solving, insist on the three levels of knowledge: the operational level , the tactical level and the strategic level. Those three levels are necessary in problem solving to reach a goal. Schematically, those levels can be described as bellow :

Operational level. This concerns the physical realization of elementary tasks. It can be assimilated as elementary algorithms, that perform a given task and that can not be shared. The operational knowledge can be considered as the tiles that can be used for solving a problem.

Tactical level. Solving a problem need the cooperation of several elementary algorithms, so the activation of several operational knowledge. The description of one possible way to solve a problem concerns the tactical level. So tactical knowledge is more a structural knowledge. It gives more or less the recipe to solve a problem (reach a goal) and is usually represented by a precedence graph.

Strategic level. For the same problem, several methods can be defined at the tactical level. They are different ways to see and to solve the same problem. The difficulty is to choose the more accurate one in a given context. This is the aim of strategic knowledge. This level considers the context of the resolution and can indicate which tactical knowledge to use.

Of course, the problem solving modeling can be recursive, and include several levels of problem and sub-problems. Then, an operational task for a given goal can be viewed as a goal itself with the three levels of knowledge.

The relationship between those three levels of knowledge and the knowledge in project modeling is obvious. Operational tasks can be connected to the elementary tasks that can be met in the project. If several futures are possible, then the tactical knowledge corresponds to the tactical level in problem solving modeling: each future is assimilated to a possible way to realize this task. Then, the strategic knowledge corresponds to the strategic level.

This comparison is very interesting. It is clear that in project modeling, strategic level have often been neglected. While deterministic models consider a single future (no choice, no alternative), probabilistic models consider that several futures can be defined, but limit the strategic knowledge to a probability.

The main differences between project modeling and problem solving are the following:

- The aim is different. In problem solving, it is to solve, then to reach the target in any situation. In project modeling, it is to evaluate the risk, the duration and the cost of the "resolution".
- Elementary task are internal (under control, part of the process) in problem solving, but external in project modeling. So the uncertainty is of a different nature.

20.2.2.2. The resolution cycle. The second important notion in problem solving that can be extended to project management is the cycle of resolution. Most models and particularly the one proposed in Tacquard (1993) represents the strategic knowledge in problem solving as a cycle of three different steps: planning, reaction, and synthesis.

Those three steps are necessary to control the activation of the tactical knowledge during the resolution itself. Briefly, each of those steps can be described as bellow:

Planning. The planning of a task in problem solving consists in choosing the best tactic to reach the goal, among the set of known tactics. This is done by using the context of the resolution, and the already solved sub-problems. The quality of this step insures the efficiency of the resolution.

Reaction. The reaction step is in fact the monitoring of the resolution itself. While the tactic is chosen, it is necessary to apply it, and simultaneously to check if it works. In case of a failure, the reaction should propose an other way to reach the goal. In problem solving, the best the strategic step is, the less useful the reaction is. The reaction insures the success of the resolution

Synthesis. Finally, in problem solving, the synthesis allows the to produce a synoptic of the resolution. This synoptic is the set of all produced information that can be used by other resolutions (resolution of high-level sub problems)

This resolution cycle is closed to the project management. When alternatives and risk occurs, the manager have to plan (choose among several possible futures, at least when a task have to begun), he needs to react in case of a failure, and he needs to select the accurate information for the rest of the project. In most cases, the knowledge associated to each step is not clearly defined in project management, but only implicitly known.

20.3. The GrAC model: General Framework

The GrAC model is a project model that comes from the cognitive modeling in problem solving, and that try to make a compromise between the semantic representation of the strategic knowledge, and the possibility of formal calculus on the model.

The bases of the model are:

- A management of information produced by the tasks
- A hierarchical representation of the project in term of tasks, where:
    - The leaves of the project are the elementary tasks,
    - The intermediate nodes are the cognitive tasks.
- A description of each task with:
    - A description of all possible realizations for each task,
    - Three cognitive systems to model the choice of a given planning, the reaction in case of failure, and the production of new information.

The base of the system is to consider an alternative in a project as a decision, and not a probability. So the choice of one particular decision have to be explicitly described.

20.3.1. Management of information

A semantic approach of the project model supposes to take into account two kind of information.

20.3.1.1. Internal information. During the project, some information are (or can be) produced by a task, and those information are Key Information for the different choices that can occur in the project. The result of a prototype, the cost of a prototype, the nature of a rock, the performance of a system, etc. A good modeling of the strategic knowledge should be based on that information.

From a task point of view, an information can be produced systematically if the task is performed, or not. For a given task t,

$Iexs_t$, is the set of systematically produced information,

$Iexo_t$, is the set of non systematically produced information,

$Iex_t = Iexs_t + Iexo_t$ is the set of all information that can be produced by t.

20.3.1.2. External information. Some choices can depend upon variables that are external to the project. For example, the weather can decide of a particular way to realize a task in civil engineering, the cost of row material can modify a strategy, etc. This kind of information has to be defined a priori, and can be used in all cognitive systems that describe the choices.

20.3.2. Elementary tasks

The elementary task is in the project a task that is not (or not yet) decomposed itself. It is characterized by:

- $Dmin_t$ the lower bound of the duration of t,
- $Dmax_t$ the upper bound of the duration of t,.

Then an elementary task can be certain, or can be uncertain. If it is uncertain, the following data are defined:

- $DEmin_t$ lower bound of duration before the failure of t,
- $DEmax_t$ upper bound of duration before the failure of t,.

Those data indicate that the decision of failure is taken between $[T+DEmin_t$, $T+DEmax_t]$, if T is the starting date of the task. This means that the failure is a decision, and that it is not linked to the duration of the task. This is in fact the same notion that the one used in the RAIH model in Ramat (1997). The set of certain tasks is denoted C, the set of uncertain task is denoted UC.

20.3.3. Cognitive tasks

The cognitive tasks are the intermediate nodes of the hierarchical representation. At each cognitive task are associated:

- a set of decompositions
- a set of plannings
- cognitive system for planning, reaction and synthesis
- a set of information produced

Globally, the cognitive task can be represented as in figure 20:6. In this example, the task can be perform with three different decompositions (d1, d2, d3) but the manager have decided that if decomposition d1 is not chosen (planning p1) then decomposition d2 and d3 will be performed simultaneously (planning p2). The rules that will induce the choice between p1 and p2 have to be clearly defined in the cognitive system for planning.

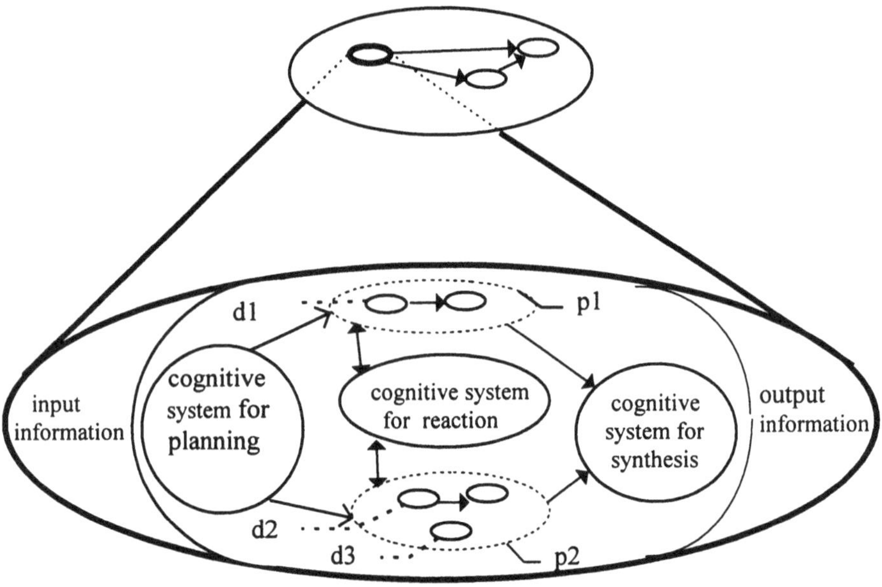

Figure 20:6. A cognitive task

In case of a failure of one of those planning, the rule that explained what to do will be defined in the cognitive system for reaction.

20.3.3.1. Set of decompositions. A decomposition is a graph of tasks (from the lower level) that can be used to perform the current task. It corresponds to the tactical knowledge. Several different decompositions can be assigned to the task.

<u>Notation</u>: $Decomp_A$ = the set of all decompositions of task A

20.3.3.2. Set of planning. A single decomposition is sufficient to perform the task. Nevertheless, in a given project, the manager can choose to operate several decompositions simultaneously to minimize the failure risk, or the expected duration. A set of decompositions that can be selected by the manager is a planning.

As for the decompositions, several plannings can exist. One only will be chosen when the task will physically begin.

Notation: $Planif_A$ =the set of all decomposition of task *A*.

20.3.3.3. Cognitive system for planning. When a task is to be performed, one planning has to be chosen among the set of plannings. This is the rule of the planning cognitive system. It will answer the question "how shall we perform this task if the current project is in this state, and if the previous results are those one?".

The variables used in this system are produced by the predecessor tasks (that are already performed when the observed task is to be performed).

20.3.3.4. Cognitive system for reaction. When a planning is chosen to perform a task, one or several decompositions are performed simultaneously (sub-projects). Those tasks have to be controlled, to detect a failure if it happens. In this case, an other decomposition can be performed. This is the rule of the reaction cognitive system. It will answer the question "what should we do if the current project fails?»

When the current decomposition p_{cour_r} is performed, all associated decompositions are simultaneously performed. Then the task can be in five different stages:

Partial failure: one decomposition of p_{cour} at least is not finish, but none of the finish decompositions have been successful.

Partial success: at least one decomposition of p_{cour} is successively finished, but at least one is not finished,

Success: All decompositions are finished, and at least one succeeds.

Rescheduled: All decompositions of p_{cour} are finished and failed, but an other planning can be chosen.

Failure: All decompositions of p_{cour} are finished and failed, no other planning can be applied.

The state graph is presented in figure 20:7. As soon as a planning is chosen by the cognitive system for planning, the task become in the partial failure state (no decomposition have succeed, but at least one is being proceed). The transition cans be detailed as follows:

From partial failure:

- If a decomposition ends on a success, the state becomes partial success,
- If the last decomposition ends on a success, the state becomes success,

- If one of the remaining decomposition fails, no change,
- If the last decomposition fails, the state becomes rescheduled,

From reschedule:

- If no other planning can be planned, the state becomes not failure,
- Otherwise, a new planning is chosen, the state becomes partial failure,

From partial success:

- The last decomposition ends, the state becomes success,
- An other decomposition ends, the state is unchanged,

From success or failure:

- The synthesis system can produce the corresponding new information,

The management of those transitions is modeled with the cognitive system for reaction.

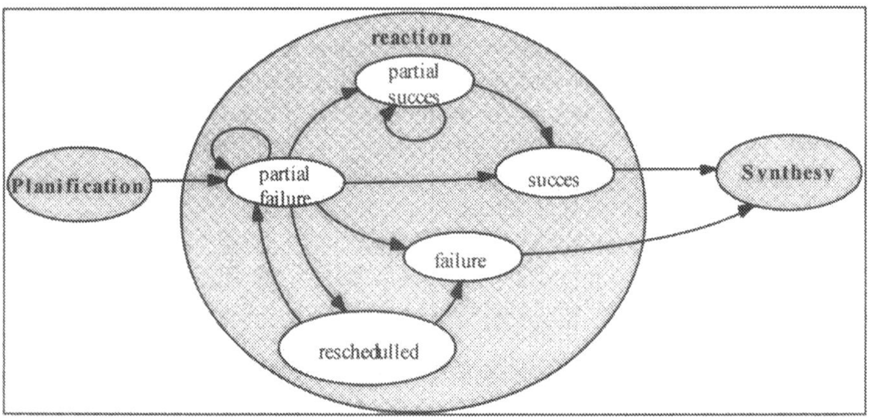

Figure 20:7. State graph concerning a cognitive task

20.3.3.5. Cognitive system for synthesis. The cognitive system for synthesis is only a kind of filter that produces a sub set of information available for the rest of the graph (for the cognitive system of planning of following tasks). This system can also provide new information from the ones that have been produced during the execution of the task.

20.3.3.6. Properties of Cognitive tasks. A cognitive task *t* can be characterized by the same variables than an elementary one:

- $[Dmin_t, Dmax_t]$ the duration bounds of *t*.
- $[DEmin_t, DEmax_t]$ the duration before failure bounds of *t*.

Those data will be computed recursively in the graph with a bottom up approach, as shown in the next section.

20.4. The GrAC model: an operational model

The previous part presents a general framework for the definition of cognitive tasks. With this kind of model, it is possible to express a very precise semantic, but few calculus can be insured. Computing several pieces of information can only be insured by the restriction of the model. Then, two limitations are proposed: first boolean variables are introduced for the information and second, boolean functions are used for the cognitive systems.

Moreover, a typology of cognitive tasks has been defined. Five different types of cognitive task exists in this typology (figure 20:8), each of them with specific cognitive systems:

- Deterministic task (no choice),
- Alternative task (exclusive choice),
- Concurrent task (competition between different realizations),
- Protected alternative task (exclusive choice, with guarantee of success),
- Protected concurrent task (competition with guarantee of success).

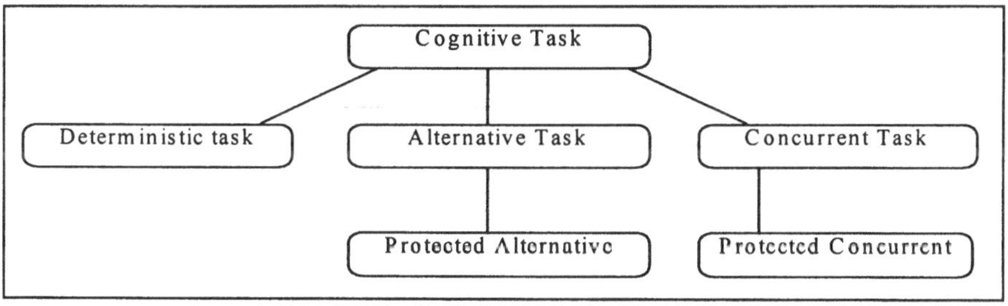

Figure 20:8. A typology of cognitive tasks

The two last ones are specially defined because the failure is a decision, and so in certain circumstances, being sure of the success modifies the decision to stop an active task.

20.4.1. Boolean expressions

To allow some evaluations on the GrAC model, some limitations have been introduced concerning the knowledge systems. For a Knowledge Based System (KBS) one cannot say in general in which case a solution may be found. That is why the GrAC cognitive systems have a simplified structure. Each cognitive systems (planning, reaction and synthesis) is organized as a set of basic logical rules which result is either the start of a planning or the end of a planning.

20.4.1.1. Planning system. This cognitive system has a set of logical rules such that each produces a planning to start. If no rule has been verified, a default planning is chosen. This particular structure for the planning system insures that a planning is always chosen given any context of the project.

20.4.1.2. Reaction system. The reaction system has the same structure than the planning system except that it has no default planning. The reaction system is activated on the failure or on the success of any decomposition. The rules are then checked to see if another planning can be chosen or if the considered planning has to be stopped.

20.4.1.3. Synthesis system. The structure is slightly different because this time, the result of a rule is not the start or stop of a planning but the production of an information. Consequently the action of a rule is to create an information with a given value. At this moment, only boolean pieces of information are considered in the model.

20.4.2. Deterministic task

Definition 1. A deterministic task is a task that contains only one decomposition, so only one planning:

- Card ($Decomp_A$) = 1
- $Planif_A = \{\{d\}\} \;/\; Decomp_A = \{d\}$

20.4.2.1. Cognitive system for planning. The planning is absent. When the task is to be proceeded, only one possibility exists, so it is taken. This planning is equivalent to the general PERT planning.

20.4.2.2. Cognitive system for reaction. The planning is unique, so the reaction is also absent. The state of deterministic task can be either success, total fail or partial fail, but the reaction system cannot change the situation. It only depends upon the sub-tasks of the decomposition.

20.4.3. Alternative task

An alternative task is a mutual exclusion of several decompositions. The task can be performed with several different decompositions (alternative) but a single one have to be chosen (each planning is a single decomposition).

Definition 2. An alternative task A is a task such that:

- Card($Decomp_A$) ≥ 2
- Card($Planif_A$) = Card($Decomp_A$)
- $\forall\, p \in Planif_A,\ \exists!\ d \in Decomp_A \;/\; p = \{\{d\}\}$
- $\forall\, d \in Decomp_A,\ \exists!\ p \in Planif_A \;/\; p = \{\{d\}\}$

The alternative can express that several realizations can be considered, but only one is chosen. In case of failure of the chosen decomposition, the cognitive system of reaction will describe the knowledge that explicit what to do.

20.4.3.1. Cognitive system for planning. This cognitive system will just describe the necessary conditions for the choice of all the possible decompositions.

20.4.3.2. Cognitive system for reaction. The cognitive system for reaction is very important in the alternative task, because it models the fact that the failure of a task has been correctly considered by the manager.

For each alternative task, a reaction graph defines the cognitive system for reaction. It gives for each planning the new direction to try according to the value of several variables.

Definition 3. Given an alternative task A, the reaction graph $Gr_A = [X, U, C]$ of A is defined as follows :

$$\begin{cases} X = Planif_A \text{ set of the nodes of the graph,} \\ U = \text{set of the edges of the graph,} \\ C = \text{set of theconditions on the information such that :} \\ C = \left\{c_{i,j} \,/\, \exists (p_i, p_j) \in X^2, (p_i, p_j) \in U\right\} \\ \text{if } p_i \text{ miscarries and if } c_{i,j} \text{ is true then start } p_j \end{cases} \quad (1)$$

The interpretation is to swap from planning i to planning j if an arc exists between both, and if the condition is verified.

20.4.4. Protected alternative task

An alternative task may fail, depending on the nature of the reaction graph. But from a management point of view, this characteristic will be propagate in all the tasks of higher level. So it can be important to insure, by forecasting a particular way to react, that this task is safe.

Definition 4. An alternative task A is a protected alternative task if A is an alternative one and A can not miscarry.

20.4.5. Concurrent task

A concurrent task has several different decompositions. At least one planning has several different decompositions, and when a planning is chosen, all decompositions are simultaneously performed.

Definition 5. A concurrent task A is such that:

- $\text{Card}(Decomp_A) \geq 2$
- $\exists\, p \in Planif_A \,/\, \text{Card}(Planif_A) \geq 2$

Usually, in project management, and particularly in new product development, this situation occurs when several solution are possible, and have to be tested. In a preliminary phase, a concurrent task should test several possibility, and it will be followed by an alternative task with a single technology chosen.

20.4.5.1. Cognitive system for planning. The cognitive system for planning is the same as the one of the alternative task.

20.4.5.2. Cognitive system for reaction. For a concurrent task, the problem is to define when the task is a success or when it is a failure. The rescheduling is a non-sense, because all decompositions have be done simultaneously.

20.4.6. Protected concurrent task

The protected concurrent task has been created to avoid the failure of a concurrent task. To build a protected concurrent task from a concurrent one, the first step is to avoid the failure of each planning. Any planning has to be certain in order to obtain a protected concurrent task. To transform each planning in a certain one, at least one certain decomposition has to be included in the planning. Consequently, to obtain a protected concurrent task, at least one certain decomposition has to be inside each planning (a given certain decomposition may be the same for all the plannings).

Definition 6. A protected concurrent task if defined by:

- $\text{Card}(Decomp_A) \geq 2$
- $\exists\, p \in Planif_A \,/\, \text{Card}(Planif_A) \geq 2$
- $\forall\, p \in Planif_A$, $\exists\, d \in \text{p} \,/\, d$ is certain

20.5. Model analysis

The model is analyzed mainly from a structural point of view, that is to say that the characteristics of the cognitive activities and of the project are computed according to their structures and the elementary tasks that are part of them. This section will describe this part of the model analysis.

However, once this step is achieved, a more accurate analyze can occur in collaboration with the project manager and which consists in analyzing the influence of a given set of elementary information on the global project behavior. This analyze is called the dynamistic analyze. It is a kind of simulation of the model apart that the variables are not randomly computed, but defined by the project manager. From a theoretical point of view, this analyze is less interesting than the structural analysis, but from a practical point of view, this approach is more interesting for the project manager.

20.5.1. Structural analysis

The structural analysis of the GrAC model - Grunder, Baptiste and Barakat (1996a, 1996b) - has to determine the different expressions of the characteristics of the cognitive activities and of the project. This step is achieved by considering the type of the task, its structure (decompositions, plannings) and the characteristics of the activities and tasks that are part of it.

The structural analysis is a bottom-up approach that starts from the attributes of the elementary tasks to end to the attributes of the cognitive activities and the project. The following attributes are concerned : the minimal and maximal duration, the possibility of miscarriage and the minimal and maximal durations before a miscarriage occurs. The expressions presented in this section are freed from the knowledge systems in that way that only the minimal and maximal values are considered.

First, a notion is defined which is the "plannability" of a cognitive task. The plannability of a cognitive task is its ability to select always one planning given any context of the project at the moment of the start of the task. This important notion concerns the validity of the other expressions presented because this notion characterizes the cognitive activities for which at least one planning may be chosen.

The expressions concerning a decomposition and a planning are presented to facilitate the expressions of the different types of cognitive activities.

20.5.1.1. "Plannability". When a cognitive task starts, a planning is chosen and all decompositions of this planning are started in competition. The basic assumption that is made to express, for example, the minimal duration is that at least one planning has to be chosen.

Definition 7. A cognitive task is strongly planned if one and only one planning can be chosen by the planning system for any context of the project.

The structure of the cognitive system of planning that has been introduced implies that all activities in the GrAC model are strongly planned.

20.5.1.2. Analysis of a decomposition. In the PERT/CPM model, a task is mainly defined by the following durations: the earliest and latest start times and the earliest and latest finish times. As the GrAC model considers a minimal and a maximal duration, each task *t* has :

- a minimal and maximal earliest start time ($ESmin_t$ and $ESmax_t$),
- a minimal and maximal earliest finish time ($EFmin_t$ and $EFmax_t$),
- a minimal and maximal latest start time ($LSmin_t$ and $LSmax_t$),
- a minimal and maximal latest finish time ($LFmin_t$ and $LFmax_t$).

Lemma 1. Given a task *t* of a decomposition *d*, the earliest starting and finish dates are given by :

- if *t* has at least one predecessor:

$$\begin{cases} ESmin_t = \underset{x \text{ predecessor of } t}{\text{Max}} \left(ESmin_x + Dmin_x \right) \\ ESmax_t = \underset{x \text{ predecessor of } t}{\text{Max}} \left(ESmax_x + Dmax_x \right) \end{cases}$$

- if t has no predecessor:

$$\begin{cases} ESmin_t = ESmin_d \\ ESmax_t = ESmax_d \end{cases}$$

- and finally:

$$\begin{cases} EFmin_t = ESmin_t + Dmin_t \\ EFmax_t = ESmax_t + Dmax_t \end{cases}$$

Proof: These expressions are straightforward.

Lemma 2. The minimal and maximal durations of a decomposition d are defined by the following expressions:

- $$\begin{cases} Dmin_d = \underset{t \in d}{\text{Max}}\left(ESmin_t + Dmin_t\right) - ESmin_d \\ Dmax_d = \underset{t \in d}{\text{Max}}\left(ESmax_t + Dmax_t\right) - ESmax_d \end{cases}$$

Proof: These expressions are straightforward because a decomposition is completed when all its tasks and activities have been completed.

Lemma 3. Given a task t of a decomposition d, the latest start and finish times of t are given by :

- if t has at least one successor:

$$\begin{cases} LFmin_t = \underset{x \text{ successor of } t}{\text{Min}}\left(LSmin_x\right) \\ LFmax_t = \underset{x \text{ successor of } t}{\text{Min}}\left(LSmax_x\right) \end{cases}$$

- if t has no successor:

$$\begin{cases} LFmin_t = Dmin_d \\ LFmax_t = Dmax_d \end{cases}$$

- and finally:

$$\begin{cases} LSmin_t = LFmin_t - Dmax_t \\ LSmax_t = LFmax_t - Dmin_t \end{cases}$$

Proof: The expressions of the latest finish times are straightforward. To obtain the latest start time of a task, the duration of the time has to be subtracted from the latest finish time. Consequently the minimal latest start time is obtained by subtracting the maximal duration of the task from the minimal latest finish time. The same deduction applies for the maximal latest start time.

Lemma 4. The possibility of miscarriage of a decomposition d is given by the following formula :

$$Ech_d = \underset{t \in d}{\text{Max}}\, Ech_t$$

where: Ech_d is the possibility of miscarriage of the decomposition d and Ech_t is the possibility of miscarriage of the task t.

Proof: This expression is straightforward because a decomposition miscarries when any task of this decomposition miscarries.

Lemma 5. Given an uncertain decomposition d, the minimal and maximal duration before miscarriage are given by the following expressions:

$$\begin{cases} DEmin_d = \underset{t \text{ task of } d \text{ that may miscarry}}{\text{Min}} \left(ESmin_t + DEmin_t \right) \\ DEmax_d = \underset{t \text{ task of } d \text{ that may miscarry}}{\text{Max}} \left(LSmax_t + DEmax_t \right) \end{cases}$$

Proof: A decomposition miscarries as soon as a task miscarries. Consequently, the above expressions are straightforward.

20.5.1.3. Analysis of a planning. A planning is mainly a set of decompositions and consequently, its characteristics are computed with the characteristics of its decompositions.

Lemma 6. Given a planning p, the minimal and maximal durations of p are:

$$\begin{cases} Dmin_p = \underset{d \in p}{\text{Min}} \left(Dmin_d \right) \\ Dmax_p = \underset{d \in p}{\text{Max}} \left(Dmax_d \right) \end{cases} \quad (2)$$

Proof: A planning can stop after either the success or the miscarriage of any of its decompositions. Consequently, the above expressions are straightforward.

Lemma 7. A planning is uncertain if and only if all its decompositions are uncertain:

- $Ech_p = \underset{d \in p}{\text{Min}} \, Ech_d$

Proof: When a planning is started, all the decompositions of the planning are started simultaneously. Only one decomposition has to be realized and the planning is realized. Counterwise, all the decompositions have to miscarry for the planning miscarries.

Lemma 8. If the planning p may miscarry, the minimal and maximal durations before miscarriage are defined as follows :

- $$\begin{cases} DEmin_p = \underset{d \text{ uncertain decomposition of } p}{\text{Max}} \left(DEmin_d \right) \\ DEmax_p = \underset{d \text{ uncertain decomposition of } p}{\text{Max}} \left(DEmax_d \right) \end{cases}$$

Proof: The lemma 7 says that a planning can only miscarry if all its decompositions miscarry. To decide of the miscarriage of p, one has to decide of the miscarriage of all the decompositions of p. That is why the minimal duration before miscarriage is the maximum of the minimal duration before miscarriage of all the decompositions of p. The expression of the maximal duration before miscarriage is straightforward.

20.5.1.4. Analysis of a deterministic task. The specific structure of a deterministic task implies that all its parameters are identical to the corresponding parameters of its single planning.

20.5.1.5. Analysis of an alternative task. First, the expressions of the durations of an alternative task are expressed. The possibility of miscarriage and the durations before miscarriage are then established.

Duration. To express the durations of an alternative task, all the paths of the reaction graph have to be considered.

Definition 8. Given a graph $G=[X,U]$, the set of all the paths of G is denoted Ch_G.

Lemma 9. Given an alternative task A, the minimal and maximal durations of A is expressed by :

- $$Dmin_A = \operatorname*{Min}_{p \in Planif_A} (Dmin_p) \quad (3)$$
- $$Dmax_A = \operatorname*{Max}_{l=(p_1,p_2,\ldots,p_n) \in Ch_{Gr_A}} \left[\left(\sum_{i=1}^{n-1} DEmax_{p_i} \right) + Dmax_{p_n} \right] \quad (4)$$

Proof: The expression of the minimal duration of an alternative task is straightforward because if the first selected planning succeeds then the alternative task is a success and is completed.
For the maximal duration of an alternative task, any path of the reaction graph can be taken. Given a path of the reaction graph, if all the plannings of this path are realized one after the other, then it means that the first plannings of the path fail. Only the last planning of the path may be a success. But nothing allows to know if the duration of the first planning (in case of a success) of the path is greater or less to the duration of the whole path (that is to say the sum of the duration before miscarriage of the first plannings and the duration of the last planning of the path). Consequently, the maximal duration of all the paths of the reaction graph have to be evaluated in order to establish the maximal duration of an alternative task.

Possibility of miscarriage. An alternative task fails when the current planning fails and there is no more planning to start.

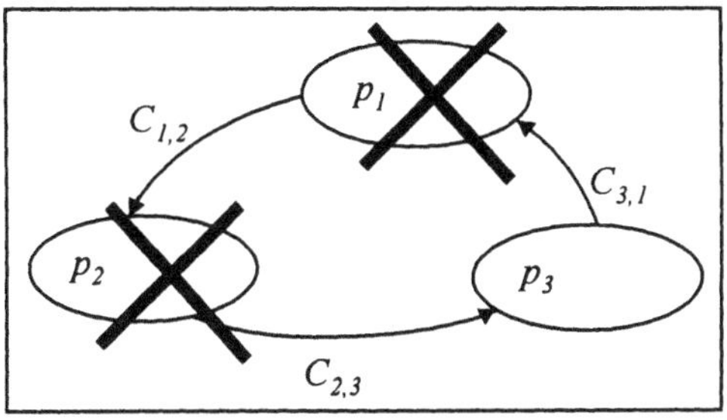

Figure 20:9. An example of a loop in a reaction graph (The execution order of the plannings is p_1, p_2, then p_3)

If the conditions on the edges are always true, an alternative task will fail if the current planning fails and there is no more planning available. This is possible in two cases:

- There is no edge in the reaction graph that starts from the planning that fails. Consequently this planning is a terminal planning of the reaction graph.
- All the edges that start from the failing planning lead to plannings that have been already tested and thus that have failed. This case can occur only if there is a loop inside the reaction graph (example of p_3 in the figure 20:9).

To characterize the alternative tasks that may fail, the notion of terminal planning is now defined. The notion of reaction graph that is robust to information is then introduced.

Definition 9. Given an alternative task *A*, a planning *p* of *A* is said to be terminal if the reaction graph has no edge that starts from *p*. The set of the terminal plannings is denoted $PTerm_A$.

- $p \in PTerm_A \Leftrightarrow \forall\, q \in Planif_A, (p,q) \notin Gr_A$

Corollary 1. A sufficient condition for a planning for being terminal is that the planning is certain.

Proof: By definition, the reaction system of an alternative task *A* is used when a miscarriage occurs. If a planning is certain, no more planning of *A* will be planned after. Consequently, *A* is a terminal task.

For a reaction graph that is robust to information, given any uncertain planning for which at least one alternative is forecaster in case of miscarriage, then all cases have to be forecasted and a planning has to be associated to each case. That is to say that if one decides to react to the miscarriage of a planning then each case of miscarriage has to be considered and a planning has to be associated with.

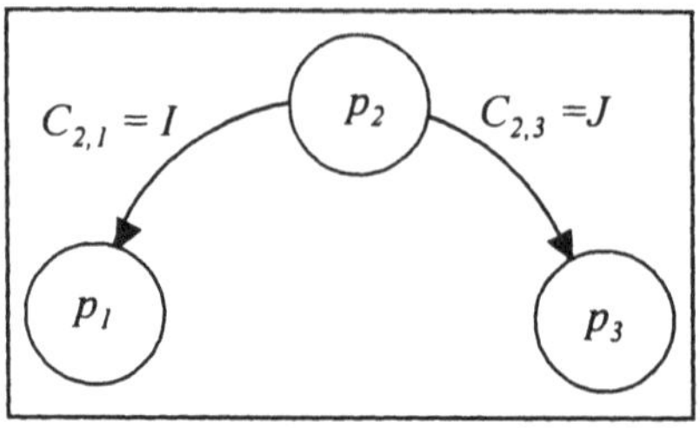

Figure 20:10a. Example of a reaction graph that is not robust to information

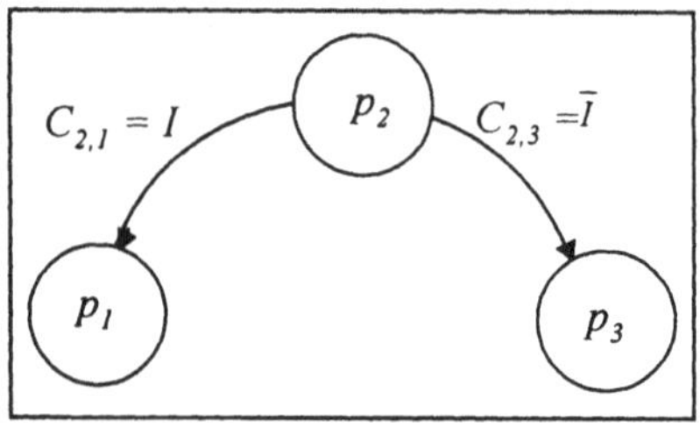

Figure 20:10b. Example of a reaction graph that is robust to information

Example:

Consider the reaction graph of the figure 20:10a. If the planning p_2 miscarries and information I is true, then the planning p_1 has to be chosen, and if information J is true, then p_3 has to be chosen. However if I and J are both not true, then nothing is forecasted. This reaction graph is not robust to information.

Consider now the reaction graph of figure 20:10b. The two cases "I is true" and "$\bar{I}$ is not true" have been considered. All cases are thus considered. This graph is a reaction graph robust to information.

Definition 10. Given an alternative task A, its reaction graph Gr_A=(X, U, C), and $PEch_A$ the set of the uncertain plannings of A, Gr_A is "robust to information" if:

$$\forall p \in X \Rightarrow \left(\sum_{q_i \in X / (p,q_i) \in U} c_{pq_i} \text{ is true} \right) \text{ or } \left(p \in PTerm_A \right)$$

According to this definition, each planning p of Gr_A is either a terminal one or there is always a true condition among the conditions tied to the edges that start from p.

Theorem 1. An alternative task A is certain if the following four points are satisfied :

- $\forall\, p \in PEch_A, p \in Gr_A$,
- the reaction graph Gr_A is robust to information,
- there is no loop inside the reaction graph Gr_A.
- $\forall\, p \in PTerm_A, p \notin PEch_A$.

Proof: For any planning that fails, we will show that a new planning can be selected until the selected planning is a certain one.

Consider a planning p that fails:

(a) $\Rightarrow p \in X \,/\, Gr_A = (X, U, C)$

If $p \in PTerm_A$, then:

(d) $\Rightarrow p \notin PEch_A$, this is impossible.

Consequently:

$p \in X, p \notin PTerm_A \Rightarrow \exists\, q \in X / (p,q) \in U$

$$(b) \Rightarrow \left(\sum_{q_i \in X/(p,q_i) \in U} c_{pq_i} \right) \text{ is true}$$

$\Rightarrow \exists\, r \in X / c_{pr}$ is true

If the planning r has already been tested, then the reaction graph has a loop inside of it, so this is impossible (c). Consequently the planning r has not yet been tried.

Consequently, given any planning that fails, there is always a remaining planning that can be started.

If all the plannings that are selected fail, one terminal planning will be finally selected because the number of planning of an alternative task is a finite one. (d) implies thus that the alternative task will be a success.

The conditions that are listed in the theorem are not necessary ones. In fact, a necessary and sufficient condition for the possibility of miscarriage of an alternative task is difficult to establish. This is a consequence to the difficulty to take into account the information that are generated by the different plannings. However, in real applications, alternative tasks will not have a lot of plannings and the conditions of the theorem will consequently give complete satisfaction.

Accurate expressions of the minimal and maximal durations before are hard to obtain. A general expression may be easily established by saying that the minimal duration before miscarriage of the alternative task is the minimum of the minimal duration before miscarriage of the different decompositions of the task. In fact if the first planning selected fails, a new one will be usually chosen in order to avoid the miscarriage of the task. Consequently, the minimal duration before miscarriage of the task will be greater than the minimum of the minimal duration before miscarriage of the different decompositions of the task.

For a protected alternative task, the minimal and maximal durations are the same than those for an alternative one. This class of tasks has been introduced to facilitate the modeling of a project with anticipated and monitored risks.

20.5.1.6. Analysis of a concurrent task. The durations of a concurrent task are first expressed and second the characteristic of uncertainty is given.

Lemma 10. The minimal and maximal durations of a concurrent task A are given by:

- $Dmin_A = \underset{p \in Planif_A}{\text{Min}} \left(Dmin_p \right)$ (5)
- $Dmax_A = \underset{p \in Planif_A}{\text{Max}} \left(Dmax_p \right)$ (6)

Proof: These expressions are straightforward because the realization of a concurrent task needs the realization of the selected planning that may be any planning among the possible plannings of A.

Lemma 11. A concurrent task A is uncertain if at least one of its planning is uncertain.

- $Ech_A = \sum_{p \in Planif_A} Ech_p$ (7)

Proof: If at least one planning of A is uncertain, then A is uncertain.

Lemma 12. The minimal duration before miscarriage of a concurrent task A is :

- $DEmin_A = \underset{p \in Planif_A}{\text{Min}} \left(DEmin_p \right)$ (8)

- $DEmax_A = \underset{p \in Planif_A}{\text{Max}} \left(DEmax_p \right)$ (9)

Proof: The duration before miscarriage of A corresponds to the duration before miscarriage of the planning that will be selected. Consequently the minimal (respectively maximal) duration before miscarriage of A corresponds to the minimum (respectively maximum) of the minimal (respectively maximal) duration before miscarriage of the different plannings of A.

The protected concurrent task has the same minimal and maximal durations than the concurrent task. Being certain, the protected concurrent task has no minimal or maximal duration before miscarriage.

The above expressions are concerned with the durations of the tasks. Similar results can be obtained with the costs of the tasks or even any quantitative characteristic of the tasks.

20.6. Conclusion

In project modeling, the strategic knowledge has usually been neglected, and most researches focus on the detailed schedule of a perfectly defined project. Uncertainty and alternative have been introduced via probability, but mainly in order to allow scheduling in an uncertain paradigm.

In innovation projects, the definition of the project itself is difficult. Even if all possible futures are well defined, the management of the project can define parallelism, competition, failure anticipation, etc. Management rules have consequently to be precisely modeled, and the effects of those rules have to be to evaluated in any possible future.

Some models begin to appear, such as the GrAC model. This model allows the representation of the different possible executions of a task, like any other stochastic graph modeling, but also the modeling of the management knowledge that permits to choose among the possible realizations, depending on the context.

This kind of tools have to be included in Decision Support Systems, to define the rules to apply in the choice of the different possibilities. Those rules are as important as the definition of the possible futures. They will define the risk, the bounds of cost and the bounds of duration. The aim is no longer to have precise information on one future, but to evaluate craftily the effects of several organizations or management decisions. This direction is very important for the future: if the risk cannot be avoided, it is nowadays essential to control it, and decisions of management have thus to be forecasted and evaluated.

20.7. References

EISNER, H. 1962. A generalized network approach to the planning and scheduling of a research project. Opns. Res. 10, 115-125.

ELMAGHRABY, S. E. 1964. An algebra for the analysis of Generalized Activity Networks. Management Science. 10, 494-514.

ELMAGHRABY, S. E. 1966. On Generalized Activity Networks. The Journal of Industrial Engineering. 17, 621-631.

ELMAGHRABY, S. E. 1977. *Activity networks : project planning and control by network models*. John Wiley, New York.

GRUNDER, O. ; P. BAPTISTE ; AND O. BARAKAT. 1996a. An Approach to Model and Evaluate Innovation and Concurrent Engineering Projects. Proceedings of the Euro Working Group on Project Management and Scheduling. Poznan (Poland). 94-97.

GRUNDER, O. ; P. BAPTISTE ; AND O. BARAKAT. 1996b. An introduction to the Cognitive Activity Network Model. Proceedings of the Computer Integrated Manufacturing and Automation Technology. Grenoble (France). 256-261.

MOELLER, G. L. 1972. VERT - A Tool to Assess Risk. Proceedings of the 23rd Conf. Amer. Inst. Indust. Engineering. 1, 211-222.

MOELLER, G.L. ; AND L. A. DIGMAN. 1981. Operations planning with VERT. Opns. Res. 29, 676-697.

PRITSKER, A. A. B. ; AND W. W. HAPP. 1966a. GERT : Graphical Evaluation and Review Technique - Part I : Fundamentals. The Journal of Industrial Engineering. 17, 267-274.

PRITSKER, A. A. B. ; AND G. E. WHITEHOUSE. 1966b. GERT : Graphical Evaluation and Review Technique - Part II : Probabilistic and Industrial Engineering Applications. The Journal of Industrial Engineering. 17, 293-301.

PRITSKER, A. A. B. 1979. *Modeling and Analysis using Q-GERT Networks*. John Wiley, New-York.

RAMAT, E. ; AND C. LENTE ; AND C. TACQUARD. 1995. Modélisation et planification des projets d'innovation : le modèle RAIH. Proceedings of the 5ème congrès international de Génie Industriel. Grenoble (France). 193-198.

RAMAT, E. ; ET AL. 1996. The RAIH Model: application on Product Realization Process example. Proceedings of the Workshop on Production planning and control. Mons (Belgium). 178-183.

RAMAT, E. 1997. Modélisation et planification de projets complexes à contraintes de ressources : le modèle RAIH. Doctoral Thesis, University of Tours (France).

TACQUARD, C. 1993. Proposition d'un modèle cognitif axé sur l'auto-organisation de la connaissance pour la résolution de problèmes. Doctoral Thesis, University of Franche-Comté (France).

WHITEHOUSE, G.E. ; AND A. A. B. PRITSKER. 1969. GERT : Part III - Further Statistical Results ; Counters Renewal Times, and Correlations. AIIE Transactions. 1, 45-50.

21 ON PAYMENT SCHEDULES IN CONTRACTOR CLIENT NEGOTIATIONS IN PROJECTS: AN OVERVIEW OF THE PROBLEM AND RESEARCH ISSUES

N. Dayanand[1]
and R. Padman[2]

1.Dialogos Inc., Boston, MA 02210
2.Carnegie Mellon University, Pittsburgh, PA 15213

21.1 INTRODUCTION

Over the last several decades, a significant body of knowledge has been developed to assist in planning, organizing, monitoring and controlling project activities. The best documented uses of these techniques are in the construction industry; it can even be argued that modern methods of project management developed on the basis of experiences with these techniques in the construction industry (Proceedings of the Annual Seminar/Symposium of the Project Management Institute (PMI) 1984-1994, Hendrickson and Au 1989, Clough and Sears 1991). While construction is a stereotypical application of project management techniques, there is growing evidence of fruitful application in other functional areas and across industries, such as telecommunications and software development (Whang 1992, PMI Proceedings 1984-1994). Market and organizational pressures to control costs and improve productivity have led corporate managers in many organizations to turn software development into a more controlled production process by the use of project management theory and supporting software tools (Shattow 1990, Chuman 1988).

Since projects are collections of well-defined activities often requiring specialized skills and/or resources, contracting entire projects, or parts of it, is an accepted practice in many industries (Martin and Webster 1987). The relationship between a contractor and his client is crucial to the success of a project. In an attempt to reduce conflict and disputes, many organizations have adopted the concept of partnering in the management of projects. Partnering brings together all parties involved in the management of a project very early in the project and creates synergies that benefit everyone involved. Partnering can be facilitated by improved contracting procedures between a contractor and his client (Eckert 1994, Hartman 1994).

Project finances, payment methods and schedules are a significant component of project contracts. The objective of this chapter is to summarize recent research on developing decision tools to support the contractor and client in the negotiation of payment and activity schedules. These tools fit into the overall scheme of partnering in the management of projects. The payment scheduling problem described here addresses several issues associated with analyzing trade-offs in contract prices, profit margins, project deadlines and other payment parameters by developing models and solution methods for simultaneously determining the amount, timing, and location of payments in projects. Furthermore, these approaches are developed from multiple perspectives, that of the contractor *and* the client. A natural outcome of this development is a joint view that can potentially address the requirements of both parties. In this chapter, we present several research issues that are significant in resolving the differences in the objectives and desired payment and activity schedules of the two parties to the project contract.

The next section provides an overview of the literature emphasizing the financial aspects of the project management problem, and their implications for determining activity schedules. Section 21.3 discusses the major issues in modeling the payment scheduling problem, such as contract and schedule parameters that impact payment schedules, and problem representation, mathematical formulation and solution interpretation. Sections 21.4 and 21.5 summarize current research on the payment scheduling problem from the individual perspectives of the contractor and the client, respectively. Section 21.6 presents several research issues in developing the joint perspective, and Section 21.7 concludes with suggestions for future research on the payment scheduling problem.

21.2 BACKGROUND

The significance of financial obligations in projects has been emphasized by several researchers and practitioners (Bey, Doersch and Patterson 1981, Shattow 1990). Progress payments, a major component, serve many important functions and are a critical issue in the management of project finances (Elmaghraby 1990). In many cases, contractors rely heavily on these payments to support project activities, and it is not uncommon to observe contractors facing the prospect of bankruptcy on account of poor planning for cash flow requirements over the duration of the project (Sanders and Cooper 1991). Progress payments also contribute directly to the contractor's revenues and hence profits. From a client's view point, progress payments to the contractor are expenses. They also allow the client to offer incentives to the contractor for achievement of milestones. In data processing and software development contracts for example, payment is often tied to specific completion points or the achievement of some agreed level of performance by the contractor (Chuman 1988, Whang 1992). For both the contractor and the client, the timing of these payments has a significant impact on the interest cost for the project. Properly structured progress payments can serve to negate much of the financial expenses associated with a project (Hamburger 1989).

Prior research in project scheduling has recognized the impact of cash flows on the schedule of project activities. Since A.H. Russell's pioneering work which established the link between cash flows and activity schedules with the use of the net present value (NPV) criterion as a scheduling objective (Russell 1970), considerable effort has been devoted to the development of schedules that take financial expenses into account (Bey, Doersch and Patterson 1981, Grinold 1972, Elmaghraby and Herroelen 1990). Furthermore, many survey papers have documented the use of these models in solving problems with resource constraints, bonus/penalty functions and other extensions (Herroelen, Dommelan and Demeulemeester 1997, Icmeli, Erenguc and Zappe 1993).

Nearly all of this effort is based on models of projects with cash flows in which project activities are associated with *fixed* cash flows. However, when bids are submitted for projects, the amount and timing of payments are important variables that can be negotiated to improve financial performance. As rightly pointed out by Elmaghraby (1990), the use of network models as aids in the preparation of project bids has received little research attention, even though cost estimation and bidding have been popular topics with practitioners for a fairly long time. In his paper, Elmaghraby suggests a method for estimating project cost based on the expenses associated with each activity in the project and the activity schedule. Each milestone event in the project is allocated some of the total cost of all activities that precede the event, and the activity schedule is used to adjust for the time value of money at this event.

From a practical view point, while expenses associated with project activities are estimated even when contractors bid on the project, payment terms, such as the amount and timing of progress payments, can be negotiated to improve the financial performance of the project. Furthermore, the contractor's role as decision maker has received far more emphasis than that of the client, even though experience has shown that successful projects demand at least an equally active role for the client in the management of the project. All these factors are considered in developing the payment scheduling problem discussed in this chapter.

A contractor's bid on any project includes his estimate of the costs of undertaking the project, and a margin that represents his assessment of the risk involved in the project. In competitive bidding situations where projects are awarded to the lowest bidder, contractors try to keep profit margins low in order to increase their chances of receiving the contract. Once the contract is awarded, the contractor is committed to the bid price, and therefore relies on proper cash flow control to retain profitability (Hendrickson and Au 1989).

The contractor incurs expenses for all activities in a project and customarily receives progress payments as parts of the project are completed. Progress payments serve many purposes. While they often represent payment for completed work, they are sometimes the primary source of financing for future project activities. A contractor's ability to continue work may depend on the amount received in progress payments over the duration of the project. Since these payments are typically tied to the completion of activities, they often serve as an incentive for the contractor to complete the corresponding activity as soon as possible.

In an ideal situation, the contractor's best option is to receive a single payment at the start of the project. From the client's point of view, this is a very risky proposition since it may induce neglect and loss of interest on the contractor's part. Thus, the client's best option is make a single payment at the end of the project. However, this shifts the entire financial burden on the contractor, and may not be acceptable in many project environments. In addition, a policy of late payment may adversely affect the client's ability to attract bids from responsible contractors in the future. As a compromise, the two parties attempt to schedule payments over the duration of the project.

Both the contractor and client wish to achieve their respective financial objectives subject to some limitations imposed by the other party on the amount and/or timing of each payment. For example, the contractor's objective is to maximize project NPV. The client's objective, on the other hand, is to maximize benefits that begin to accrue on successfully completing the project. Since projects often consist of hundreds of activities, span long durations, and involve large monetary outlays, an optimization model that determines payments and their timing can lead to improved financial performance for both the contractor and the client. It serves as a decision tool in the preparation and evaluation of bids for projects.

The next section discusses the major issues in modeling the payment scheduling problem from multiple perspectives. This includes an understanding of the contracting process and identification of the various parameters to be negotiated by the two parties before project activities and progress payments can be scheduled.

21.3 MODELING THE PAYMENT SCHEDULING PROBLEM

Modeling the payment scheduling problem requires an understanding of the objectives, constraints and process characteristics that enable the various parties to arrive at a satisfactory contract for accomplishing the goals of the project. In this section, we begin with a review of the contracting process that identifies the players, the major steps and issues in the negotiations, and the specific parameters that impact the determination of the payment and activity schedules.

21.3.1 The Contracting Process

Contract management begins when a client perceives the need to obtain outside services, and can be viewed as consisting of three major phases - contract planning, formation, and administration (See Figure 21:1 adapted from Gilbreath (1992)). In the planning stage of the process, much of the decision making involves strategic issues such as project organization (the organizational form that best suits the client's control objectives), contractual strategies (the number of contracts to be awarded for any particular project), contract pricing alternatives (the type of contract to be awarded with specific attention to risk sharing and cost, schedule and technical incentives), and contract packaging and scheduling (the subdivison of work into contract or subcontract packages and the sequencing of their formation and award activities over time) (Fisk 1992, Gilbreath 1992).

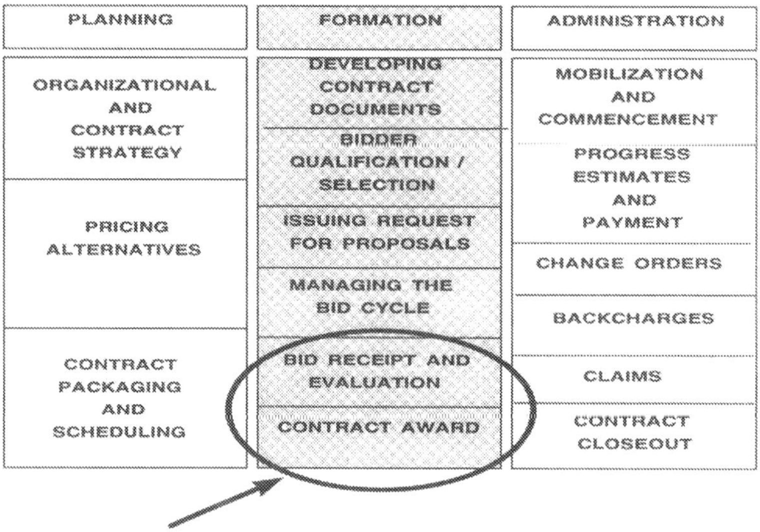

Figure 21:1 **Contract Management**

The final product of the contract formation stage is a written agreement signed by the contractor and client. This stage involves one or more of the following activities - bidder qualification and selection, issuing requests for proposals, management of the bid cycle, bid receipt/evaluation, and contract award. Considerable effort is devoted to the development of clear, enforceable contract documents in order to reduce the possibility of misunderstanding, undue compensation, major changes in project specifications, and claims and litigation (Hughes 1986). Activities in the contract formation stage are therefore geared towards writing a contract that reflects good legal practices, and effective management controls. Whereas public contracts are typically awarded in full and open competition, clients in the private sector generally have complete liberty in choosing prospective contractors from whom they wish to invite bids.

The object of the bidder qualification and selection stage is to shortlist contractors who can best meet the performance requirements of the project. Each qualified bidder is issued a request for a proposal based on detailed technical specifications prepared by the client. The bid cycle begins at this point, and terminates on receipt of completed bids. The cycle typically involves frequent communication with contractors to clarify technical specifications, and other relevant bidding information such as the closing date. In construction projects, bidders are sometimes invited to inspect the construction site in order to enable construction risk assessment. After all the bids have been received, they are evaluated under three criteria - commercial terms, technical considerations, and cost. A fair price estimate (internally prepared bid estimate) is a valuable control and checking technique used during bid evaluations. Preparation of this estimate often reveals bidder errors or misunderstandings, highlights issues for preaward negotiation, identifies unbalanced bids, and provides control against bidder collusion or rigging. Selection of a clear bid winner is generally not possible because of the multitude of issues involved in determining the best bid. Thus, the client often holds several preaward meetings to resolve specific issues before the contract is awarded. When no unresolved issues remain, contract documents are executed signifying the end of formation, and commencement of the administration phase.

The documentation on construction contracts suggests that issues related to payment are resolved during the numerous preaward conferences between the contractor and client, i.e, somewhere between the stages marked "Bid receipt and evaluation" and "Contract award" in Figure 21:1. Issac (1990) also indicates that scheduling decisions are made late in the contracting process, generally after the two parties have agreed on the contract price.

21.3.2 Negotiation Parameters

Many important payment-related parameters are negotiated before the contract is signed. These include the *type of contract*, *contract price*, *number or frequency of payments*, *method of distribution of payments*, *retention factor*, and *procedures for allocation of expenses and payments to various time periods and/or activities in the project*. All these factors can affect the number of payments as well as their amounts and timing. Although the contractor incurs expenses for every activity, it is neither realistic nor desirable to request progress payments at the end of every activity. The client may choose to pay the contractor equal amounts on achievement of some designated milestones. The amount of each payment may be determined by expenses incurred up to the time of payment, estimated expenses for future work, or value of work that has already been completed. The client may also make payments that are proportional to time elapsed on the project. We expand on these critical parameters in the following subsections.

21.3.2.1 The Type of Contract. The term "type of contract" refers to the method of compensating the contractor for undertaking the project. The selection of the type of contract is generally a matter for negotiation since it determines the risk sharing arrangement between the contractor and the client (Hearn 1990, Weidman 1992). There are two basic types of contracts: Fixed-Price and Cost-Reimbursement contracts. At one extreme, Fixed-price contracts transfer all project risks to the contractor while the reverse is true for Cost-reimbursement contracts. Between these two extremes, several incentive type contracts have been designed to provide other ways of allocating project risks to the two parties.

In firm fixed-price contracts, the contractor and the client agree on a price before a definitive contract is awarded. This arrangement is generally preferred when the scope of the project and associated tasks can be clearly specified. These contracts may be either lump-sum contracts or unit-price contracts. A lump-sum contract requires the client to pay a fixed total amount to the contractor for the entire project. The price of the contract includes estimated cost and some markup that represents the contractor's profit (Hendrickson and Au 1989). Unit-price contracts require the contractor to bid on unit prices for all items required for the completion of project activities. The actual price of the contract is obtained by multiplying estimated quantity required of each item with the corresponding unit price. Unit-price contracts are appropriate when there is uncertainty about quantities of various items that are required for contract performance (Hendrickson and Au 1989).

Cost-reimbursement contracts allow the contractor to recover all costs incurred during completion of project activities. In some cases, the contractor receives a fee for his services. The fee may be a fixed amount or a percentage of the total cost depending on whether the type of contract negotiated is a cost-plus-fixed-fee or a cost-plus-percentage-fee contract. This arrangement is typical when the scope of the project is not clearly defined at the time of negotiating the contract, and substantial changes are anticipated as the project makes progress.

Both fixed-price and cost-reimbursement contracts are extreme options in a risk sharing arrangement. While the contractor assumes complete responsibility for all cost over-runs in fixed-price contracts, a cost-reimbursement contract provides no incentive for him to control costs, and therefore forces the client to undertake all risks. To overcome these limitations, various types of incentive contracts have been designed. Fixed-price incentive contracts include a provision for adjustment of profit and establishment of the final contract price based on the relationship between the actual cost and negotiated target cost. The contractor and the client share any cost savings by a predetermined or negotiated formula. In the absence of cost savings, the contract is treated as a fixed-price contract. On the other hand, with cost-plus-incentive-fee contracts, both parties agree to share cost savings or cost over-runs in some negotiated ratio. Thus, both types of incentive contracts provide incentive for the contractor to control project costs (Martin and Webster 1987).

All models described in this chapter assume that the contractor receives a fixed total amount for his services. Although this suggests that the models are only applicable for fixed-price contracts of the lump-sum variety, they can be adapted to suit a wide variety of contracting arrangements (Dayanand and Padman 1995). Furthermore, the primary purpose of these tools is in the planning stage. Planning a schedule of progress payments requires the contractor and client to begin with some estimate of the total amount to be paid for the project.

It is worthwhile mentioning that fixed-price contracts are very popular for several reasons and are the mainstay in many contracting environments (even if the prime contract is not a fixed-price contract, the subcontracts awarded as part of the prime contract are often fixed-price contracts). For one thing, they force the client to lay out project specifications as clearly as possible (Isaac 1990). Poor specifications lead to poor estimates which in turn result in a large number of change orders, and claims later on. Clients also believe that the bidding mechanism by which contractors are selected for such contracts results in a contract price that is determined by market forces, and is a fair reflection of the true worth of the project (Martin and Webster 1988). From a contractor's viewpoint, fixed price contracts are preferred because he stands to gain from any cost savings or performance efficiencies. In the words of Martin and Webster (1986), "*The Fixed-Price Contract is considered the ultimate incentive contract, if the seller can find ways to cut costs while maintaining requisite schedule and technical performance*".

21.3.2.2 The Method of Distribution. The method of distribution specifies the criterion by which the amount of each payment is determined. As the term "progress payment" suggests, these payments are based on some measure of progress on the project (Gilbreath 1992). Commonly used criteria are (1) Time, (2) Cost, and (3) Actual performance or progress (Clough and Sears 1991, Whang 1992). Time-based payments assume that progress on the project is proportional to time elapsed since commencement. Paying the total amount due to the contractor in equal instalments, at regular time intervals over the duration of the project, is an example of a time-based payment scheme. Such schemes offer no incentive to the contractor for achievement of intermediate milestones. Cost-based payment schemes reimburse the contractor for expenses incurred until the time of payment under the assumption that the level of expenses indicate the extent of progress. Cost-based payments are more closely related to progress than time-based payment schemes. However, they sometimes encourage the contractor to incur expenses much earlier than actually needed to complete the project on schedule (Gilbreath 1992, Dayanand and Padman 1993). In contrast to both these criteria, payments based on actual progress are contingent on satisfactory completion of prespecified project activities. Actual performance is measured by the value of work completed at the time of payment. Payments based on actual performance are most preferred from the client's point of view since they allow the client to ensure schedule and quality compliance from the contractor.

Both cost-based and performance-based models of the payment scheduling problem have been reported in the literature to date. Models with the contractor as decision maker have assumed cost-based payments (Dayanand and Padman 1993, Dayanand 1996), while client models have been developed for performance-based payments (Dayanand and Padman 1995, 1998). Performance-based payments are determined by incorporating a mutually agreeable "schedule of values" into the terms of the contract (Clough and Sears 1991, Gilbreath 1992). In practice, the term "schedule of values" is specific to lump-sum contracts. We use the term to indicate the value of work associated with tasks that comprise the project in all types of contracts. Development of the schedule of values varies by type of contract. The discussion that follows examines some issues in determining the schedule of values for different types of contracts. A project activity is assumed to be the basic unit of work, and the client is assumed to pay a total of Z dollars to the contractor. Activities are identified by the starting and ending events.

In lump-sum contracts, the value, W_{ij} of each activity, (i, j), is determined as a proportion of the total value of the project. If r_{ij} denotes the proportion of work completed with activity (i, j),

$$W_{ij} = r_{ij} Z \quad \forall\, (i, j) \tag{1}$$

With unit-price contracts, the schedule of values is obtained from estimated quantities of each item required for completion of each activity. Let Q be the total number of items in the bid, and q_{ij}^k, the quantity of item k required for completion of activity (i, j). The bid price for each item, r_k, typically includes all overheads and profits, so that

$$W_{ij} = \sum_{k=1}^{Q} r_{ik} q_{ij}^k \quad \forall\, (i, j) \tag{2}$$

In cost-based contracts, the contractor is reimbursed all costs associated with the project. Thus, distribution issues arise only with the contract fee. In cost-plus-fixed-fee contracts, the contractor's fee is distributed over project activities in the same proportion as the value of the activity in the entire project. The cost of completing an activity typically represents its value. Thus, if a_{ij} is the cost of each activity, and F the contractor's fee for the entire project,

$$W_{ij} = \frac{a_{ij}}{\sum_{i=1}^{M} \sum_{j=1}^{M} a_{ij}} F \quad \forall\, (i, j) \tag{3}$$

When the contractor's fee is a percentage of the total cost, the total fee, F, is an estimate based on estimated project cost, $\sum_{i=1}^{M} \sum_{j=1}^{M} a_{ij}$. W_{ij} can then be obtained as shown above for fixed-price incentive contracts.

Models of the client's payment scheduling problem presented in this chapter assume cost-based payments. However, Dayanand and Padman (1995, 1998) have also addressed the requirement that payments be determined by "earned value" at the time of payment (Fleming and Koppelman 1994a, 1994b, 1995). The term "earned value" represents the value of work completed at the time of measurement. It is the normal practice to allocate some of the value of each activity, to the beginning of the activity and the rest to the end, i.e., some proportion of W_{ij} is earned when activity (i, j) commences, and the rest is earned when it is completed. There is considerable variation in accounting practices across project environments with respect to the allocation of W_{ij}. We may assume that f_{ij}, $(0 \leq f_{ij} \leq 1)$ represents the proportion of W_{ij} earned on commencement of activity (i, j), so that $(1 - f_{ij})$ is the proportion of W_{ij} earned on completion of the activity. When f_{ij} is 1, the entire value of an activity is earned on commencement. In contrast, when f_{ij} is zero, the contractor may claim payment for the activity only when it is completed (Dayanand and Padman 1995, Dayanand 1996).

21.3.2.3 Retention. In order to ensure that the contractor completes all requirements of the contractual agreement, most contracts also include a provision for retention of some portion of the payment due to the contractor (Hendrickson and Au 1989). The amount of each payment may be based on the contractor's expenses. We refer to this criterion as recovery of expenses. With this method, each payment is a fixed proportion of the expenses incurred until the time of payment. Both parties negotiate how much of the expenses can be recovered with each payment. The contractor ensures that the negotiated percentage, denoted by β, is sufficient to make a profit on the project. The client, however, has a distinct advantage in keeping progress payments as small as possible because she can keep the contractor's interest in the project alive by making a large payment on completion of the project. Thus, the client prefers that the contractor recover small proportions over the duration of the project so that the final amount is large enough to give the contractor incentive to complete the project as scheduled.

Allowing the contractor and the client to negotiate the percentage recovery implies that the client has sufficient information about the contractor's expenses. From the expenses and the total amount paid to the contractor she can determine the profit margin and ensure that the negotiated proportion is appropriate. In many project environments, progress payments are based on the value of work completed at the time of payment and cost information may not be available to the client. In such cases, the client may estimate the expenses and hence the profit margin. For the most part, this is a routine exercise when bids have to be verified to guard against front loading.

When payments ae determined by the earned value criterion, each payment is the total value earned at the time of payment, less some retention, denoted by η, which is greater than 0. The contractor and client negotiate the maximum percentage retention, which may remain constant over the duration of the project. Without retention, every payment covers earned value for all completed activities. If the contractor loses interest in the project, activities in the tail end of the project may never be completed. The most significant consequence of retention is that the final payment is always scheduled only on completion of the project (Clough and Sears 1991).

Observation: If the contractor's profit margin is π, then $\beta = (1 + \pi)(1 - \eta)$.

21.3.2.4 Additional Assumptions. Models of the payment scheduling problem discussed in this chapter make several additional assumptions that are typical for problems of this type. However, none of these assumptions are as restrictive as they appear to be. The models can be applied in a wide variety of project environments. Suitable modifications to the models are explained in appropriate sections of this chapter. These assumptions are as follows:

- Project activities, their durations, and their precedence relationships are known.
- The project must be completed before a specified deadline.
- Expenses to be incurred or value of work to be completed over the duration of the project are known.
- The contractor incurs expenses more often than he receives payments.
- Expenses are incurred on commencement of each activity.
- Payments are distributed using the criterion of recovery of expenses.
- The client and the contractor negotiate the maximum number of payments, the method of payment, and the retention factor, over the duration of the project.

21.3.3 Problem Representation and Early Solutions

21.3.3.1 Representation of Project Data. Representation of project data is an important aspect of cash flow modelling in projects due to the multiple representation formats and their associated interpretations. Dayanand (1996) and Dayanand and Padman (1997) discuss the different representations and their advantages and disadvantages in detail. Projects can be represented as activity-on-arc (AOA) or activity-on-node (AON) networks. AOA representations are event-based while AON representations are activity-based. Representation of project data varies for activity and event based models of the same project and has significant implications for modeling the payment scheduling problem. For example, AOA representations can represent expenses associated with *groups* of activities whereas AON networks can represent expenses for *every* activity. Using AOA networks in which expenses for an activity are associated with the commencement of the activity typically means that (1) the contractor recovers some of the expenses before completing some of the activities that commence at any event, and (2) it is not possible to model the case in which payments are received only for completed work. A third representation, called the modified AOA model described in (Dayanand and Padman 1997) is very flexible in terms of data representation since it combines the features of both AOA and AON networks.

21.3.3.2 Mathematical Models and Solutions. Solutions to the payment scheduling problem include the amount of payment, the event(s) with which the payment is associated, and the occurrence time for all events in the project network. Occurrence times for an event indicate the timing of the associated payment. They also specify the completion time for all activities ending at the event and the starting time for activities commencing at that event. Project scheduling problems with NPV objective have been formulated as linear programs (Grinold 1972), binary integer programs (Doersch and Patterson 1977, Talbot 1982) and nonlinear programs with linear constraints (Russell 1970). These mathematical programs can be extended to model the payment scheduling problem by treating the positive cash flows, and the event with which they are associated, as variables. All these models assume that each event has either a positive or a negative cash flow associated with it. The structure of the mathematical models can be greatly simplified by considering the cash flow at every event as the net cash flow resulting from expenses associated with that event and the level of progress payment received on occurrence of that event. When the progress payment is higher than the expenses incurred at the time of payment, a net positive cash flow occurs and vice versa. Models with indicator variables that restrict the number of payments have nonlinear distribution constraints which can be linearized with appropriate substitutions. Associating payments and expenses with time periods instead of events further simplifies model structure since it permits formulation of linear distribution constraints. The resulting models are either mixed integer linear programs or nonlinear programs, which make them computationally intractable for projects of reasonable size. These extensions and their implications are discussed in Dayanand and Padman (1997), where it is shown that characteristics of the project environment and tractability issues are significant factors in the choice of a model.

Table 21:1 An Example Project

Activity	Predecessor	*Dur.*	*Expenses*	*Activity*	*Predecessor*	*Dur.*	*Expenses*
1	-	5	600	10	6,9	8	1500
2	-	3	400	11	6,9	7	1000
3	-	2	1000	12	4,8,10	5	800
4	1	6	1800	13	4,8,10	4	400
5	2	2	250	14	7,11	3	800
6	2	10	150	15	7,11	12	200
7	2	8	100	16	12,14	6	1000
8	3,5	4	1400	17	13,15	5	300
9	3,5	8	800				

21.3.4 Illustrative Example

Table 1 presents the relevant data on activities, their precedences, durations and expenses for an example project that we use to illustrate the models and methods discussed in this chapter. The total expenses for the project are $12,500. The contractor receives a total of $14,000 in progress payments, so that his profit margin is 12%. The annual cost of capital is assumed to be 20%. The project has an earliest completion time of 37 weeks, and a deadline of 42 weeks. Throughout this chapter, we use the event based representation in formulating the models. All expenses are allocated to the commencement of project activities.

21.4 THE CONTRACTOR'S VIEW

In this section, we review the contractor's model and features of its optimal and heuristic payment schedules (Dayanand and Padman 1993, Dayanand 1996). Consider a project with M events and N activities. Each event i has an expense N_i associated with it. The contractor receives a total of Z dollars in K progress payments for the project and must complete the project before the deadline D. π represents the contractor's profit margin on the project such that

$$Z = (1 + \pi) \sum_{i=1}^{M} N_i \tag{4}$$

Expenses associated with each project event are derived from expenses associated with activities commencing at that event. In general, expenses associated with project events can be determined from the boolean adjacency matrix of the project network ($\mathcal{G} = [g_{ij}]$), expenses (a_{ij}) associated with each activity (i, j), and the proportion of expenses (f_{ij}) allocated to the commencement and completion of each activity as shown below

$$N_i = \sum_{j=1}^{M} g_{ij} f_{ij} a_{ij} + \sum_{k=1}^{M} (1 - f_{ki}) g_{ki} a_{ki} \quad i = 1, \ldots, M \tag{5}$$

In the contractor's model, payments are distributed using recovery of expenses criterion. All expenses are assumed to incur on commencement of an activity so that $f_{ij} = 1$ for all activities. β represents the proportion of expenses recovered by the contractor with each payment. The value of β places an upper bound on the amount of progress payment. When β is greater than 1, the contractor makes a profit with each payment. Values of β less than 1 imply that the client withholds some of the funds expended by the contractor until the project is completed. β is bounded above by $(1+\pi)$, since values greater than $(1+\pi)$ imply that the final payment for the project will be made before the contractor completes the project and is atypical in most project environments. Furthermore, in order to have some control over project completion, the client must ensure that there is a sizeable payment at the end of the project. The notation shown below is used to formulate the contractor's payment scheduling problem.

Notation :

(v, w) = Activity starting at event v and ending at event w
d_{vw} = Duration of activity (v, w)
T_i = Occurrence time for event i
E_i = Earliest occurrence time for event i
L_i = Latest occurrence time for event i
α = Discount rate per time period
c_t = Discount factor for time period t, $c_t = \frac{1}{(1+\alpha)^t}, t = 0, \ldots, D$
N_{it} = NPV of cash flow N_i in time period t, $N_{it} = N_i c_{T_i}, T_i = E_i, \ldots, L_i$
A = Set of activities in the project

The model has three sets of decision variables. P_t is a continuous variable specifying the amount of progress payment in time period t. P_t is greater than zero when there is a payment in time period t and 0 otherwise. x_{it} are binary variables indicating the occurrence of event i in time period t and δ_t are binary variables indicating time periods in which progress payments are made. The model is as follows:

Maximize

$$\sum_{t=0}^{D} c_t P_t - \sum_{i=1}^{M} \sum_{t=E_i}^{L_i} N_{it} x_{it} \tag{6}$$

Subject to

$$\sum_{t=E_i}^{L_i} x_{it} = 1 \quad i = 1, \ldots, M \tag{7}$$

$$\sum_{t=E_w}^{L_w} tx_{wt} - \sum_{t=E_v}^{L_v} tx_{vt} \geq d_{vw} \quad \forall\ (v,w) \in A \tag{8}$$

$$\sum_{t=0}^{D} P_t = Z \tag{9}$$

$$\sum_{t=0}^{D} \delta_t \ leq\ K \tag{10}$$

$$\sum_{t=0}^{T} P_t \leq \beta \sum_{i=1}^{M} \sum_{t=0}^{T} N_i x_{it} \quad T = 0, \ldots, D \tag{11}$$

$$\sum_{i=1}^{M} x_{it} \geq \delta_t \quad t = 0, \ldots, D \tag{12}$$

$$P_t - Z\delta_t \leq 0 \quad t = 0, \ldots, D \tag{13}$$

The objective function maximizes the net present value of profits from the project. This is the present value of progress payments less the present value of the expenses associated with each event. Constraints numbered (7) assign an occurrence time to every event; (8) represent precedence constraints for project activities; (9) constrains the total payment to Z; (10) controls the number of payments; (11) are distribution constraints; (12) ensures that at least one event occurs at the time of payment; and (13) ensures that no payment occurs when the indicator variable for payments takes a value of 0.

Optimal payment schedules for the example project presented in the previous section are shown in Figures 21:2 and 21:3 for values of β greater than 1 and less than 1, respectively. Figure 21:2 shows the optimal schedule with 4 progress payments for β = 110%. This schedule suggests that the contractor should receive the first payment of \$7,150 at the end of the fifth week, a second payment of \$2,750 at the end of the thirteenth week, a third payment of \$2,420 at the end of the twenty-first week and a final payment of \$1,680 at the end of the thirty-eighth week when the project is completed. The first payment may be associated either with event 2 or 4. The second payment is associated with event 5. The third payment may be associated with event 6 or 7, and the final payment, with event 10. Activities (1,2), (1,3), (1,4) and (3,4) should be completed by the time the first payment is received. He can receive the second payment on completion of activities (3,5) and (4,5). The third payment requires completion of activities (2,6), (4,6), (5,6), (3,7) and (5,7), and the remaining activities are completed before he receives the final payment. The project has an NPV of \$1,333.54.

Figure 21:3 shows the contractor's optimal payment schedule with 4 payments for β = 70%. Observe that the contractor still receives payments at the end of the fifth, thirteenth and twenty-first weeks. However, the amounts are \$3,290, \$3,010, and \$1,540. This leaves a final payment of \$6,160 to be received when the project is completed in the thirty-seventh week. Observe also that in order to receive the first payment, he only has to complete activities (1,3), (1,4) and (3,4). The second payment can be received when he completes activities (1,2), (3,5) and (4,5). The third payment

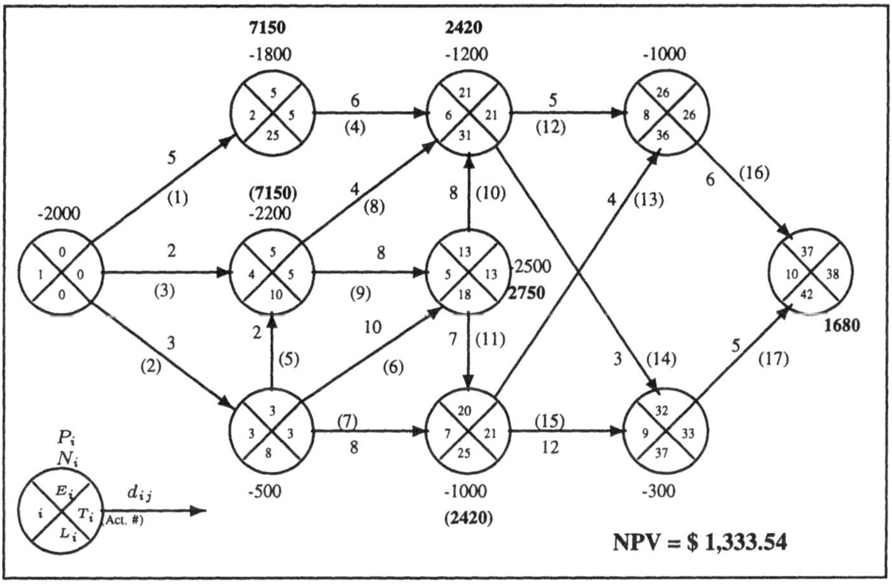

Figure 21:2 The Contractor's Optimal Payment Schedule (β = 110%)

requires completion of activities (2,6), (4,6), (5,6), (3,7) and (5,7). Event 2 is now scheduled at week 13 and event 7 at week 20. The project is completed in the shortest possible time and well before the deadline.

Several small example projects were solved to optimality in (Dayanand and Padman 1993) to obtain insights into the features of optimal solutions for the contractor's payment scheduling problem. These are summarized in the subsection below.

21.4.1 Features of Optimal Payment Schedules for the Contractor

Optimal payment schedules for values of β greater than 1, and values of β less than 1 show similarities and differences. Some of the similarities are as follows:

- The project is completed well before the deadline, and payments over the duration of the project are scheduled at the same time (the final payment is an exception).
- The amount of payment is always the maximum amount that the contractor can receive at the time of the payment and equals β times the cumulative expenses incurred until the time of payment, less any previous payments.
- Project NPV can be increased by receiving larger amounts early on in the project by scheduling as many events as possible at the time of payment.

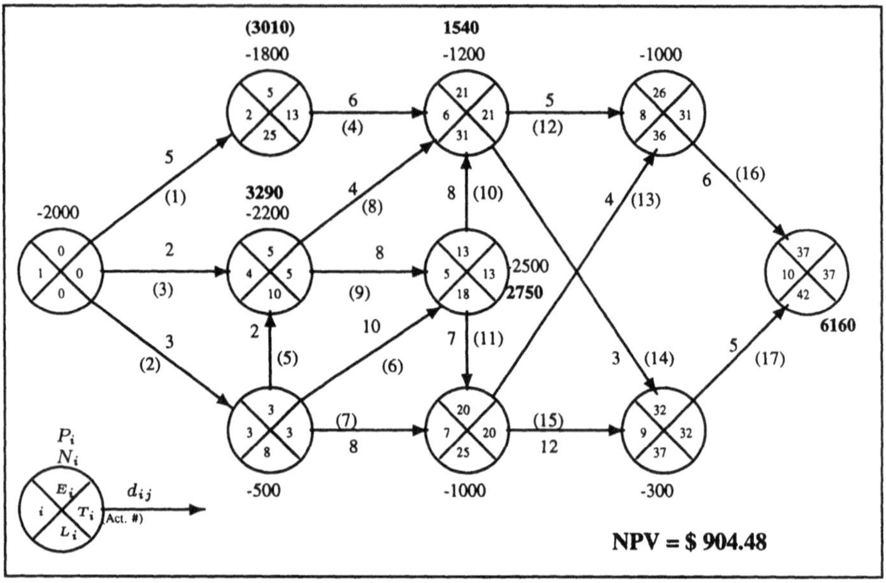

Figure 21:3 The Contractor's Optimal Payment Schedule (β = 70%)

- Events with expenses are delayed until the time of payment, thus decreasing interest costs.

Differences can be observed in the scheduling of slack events and project completion time, and can be summarized as follows:

- Since values of β greater than 1 always result in a net positive cash flow at the time of payment, slack events are scheduled as early as possible.

- Values of β less than 1 can result in net negative cash flows. Thus, optimal schedules for β less than 1 show later scheduling of slack events.

- When β is greater than 1, the contractor makes a profit with each payment. Depending on the number of progress payments, the contractor recovers much of his expenses while the project is in progress. This reduces the amount of final payment and could lead to delays in project completion.

- When β is less than 1, progress payments are relatively smaller and the final payment is large. To receive this large final payment as soon as possible, the contractor must schedule the project for early completion.

With *activity-based* models in which payments are made only for *completed work*, the contractor can improve NPV by scheduling multiple activities for completion at the time of payment. The NPV that can be realized from the project in such environments is smaller than the NPV for the same project when *advance payments* are allowed. This is because the contractor necessarily incurs expenses, for each activity j, at least d_j time periods before he receives the corresponding payment. Thus, at the very minimum, the contractor finances expenses for project activities for the duration of the activity and has to bear the associated interest cost leading to reduced profitability.

21.4.2 Heuristic Solutions

The combinatorial nature of the contractor's payment scheduling problem precludes obtaining optimal solutions for projects of reasonable size. Drawing on observations from optimal solutions to small projects, several heuristics have been reported for the contractor's problem (Dayanand 1996, Dayanand and Padman 1993, 1994). These heuristics use parameters such as the contract price, number of payments, and project duration to determine a time range for each payment, and/or a threshold level of cumulative expenses at which to schedule a payment. An optimal solution to this problem requires two sets of decision variables to be determined simultaneously - the amount of each payment and the event times. They cannot be determined independently because the amount of each payment depends on expenses incurred upto the time of payment. Since expenses are associated with events, in order to assign a payment at an event, we must determine all events that precede the event in question. This implies that the schedule of events must be known. Thus, the heuristic procedures start with an activity schedule, assign payments to events, and then revise the activity schedule based on the payments.

Six heuristics have been reported, three for β greater and three for β less than 1, by first fixing the schedule for some of the events. The schedule for the remaining events was adjusted by considering the effect of the schedule on project NPV. The heuristics were tested on two sets of randomly generated projects available in the literature (Padman and Smith-Daniels 1993, Padman, Smith-Daniels and Smith-Daniels 1997, Russell 1986). The results indicate that heuristic performance depends largely on the expense profile generated by the time schedule for the project (Dayanand and Padman 1993). Futhermore, heuristics that schedule payments whenever expenses exceed a threshold level, determined by contract price and number of payments, produced the best performance. Using the insights that payment patterns that closely match the expense pattern produce the best results, a multi-stage simulated annealing procedure to schedule payments and activities has also been proposed (Dayanand and Padman 1994). Experimental evaluation indicates that this meta-heuristic is better able to adapt to the changing characteristics of the payment scheduling problem than the heuristics based on parameters of the problem. Another notable advantage with the meta-heuristic is that it can be easily adapted for the client's problem also.

21.5 THE CLIENT'S VIEW

Many researchers and practitioners observe that successful projects typically involve a major role for the client as well (Collings 1987, Cleland and King 1983). Progress payments are expenses for the client and may involve substantial cash outflows that may be financed by borrowed funds. Planning a schedule of progress payments enables the client to manage and plan her cash flow requirements. Furthermore, the schedule of progress payments determines the client's interest costs for the project. Thus, project costs can be contained by reducing financial expenses associated with the project. Finally, the client uses progress payments as incentives to encourage the contractor to complete the project on schedule.

Despite common concerns about cash flows, several aspects of the problem suggest that the contractor and client view the problem differently. For the contractor, the project represents regular work and revenues from the project contribute directly to his profit. The client, however, typically undertakes a project with the expectation of revenues and profits from the project *after* it is completed (Cukierman and Shiffer 1976). Compared to its duration, the project has a longer life. Thus, she incurs expenses in the short term and obtains benefits only in the long term. In contrast to the contractor's ideal schedule of progress payments, the client prefers to pay small amounts and delay payments as long as possible. Her best option is to make a single payment at the end of the project to cover the entire cost of the project. Since this is not always acceptable, progress payments should be distributed over the duration of the project by some criterion.

21.5.1 Models and Solutions

The client's payment scheduling problem can be formulated in a manner similar to the contractor's problem. The client's objective depends on whether or not benefits from the project are taken into account. If the client ignores benefits from the project, she can plan a schedule of progress payments that minimizes her expenses for the project. Since projects invariably have deadlines that extend beyond the early completion time, this would suggest that the client should delay project completion until the deadline. However, completing the project early is a very important factor for the client because of potential benefits from the project. Hence the client's objective is to maximize benefits from the project, which may be assumed to be measured in monetary terms. Since the client will undertake the project only if the benefits from the project far exceed its cost, benefits from the project are represented by a large cash flow at the end of the project. This cash flow is the present value of all future costs and revenues discounted to the end of the project. It is important to note that regardless of how benefits are measured, estimating project benefits is an essential part of the project planning process that typically precedes the determination of progress payments and schedules (Schnieders and Jameyson 1987).

Three models have been proposed for the client's payment scheduling problem (Dayanand and Padman 1995). The Basic Client model (BC) allows the optimal timing and level of payments to be determined based only on the contract price, retention and number of payments to be made. The Equal Time Intervals model (EQ) forces these payments to occur at regular time intervals as is common in practice. The Periodic Payment model (PP) allows the client and contractor to negotiate the frequency of payment instead of the number of payments so as to accommodate the client's payment cycle. Models reported in (Dayanand and Padman 1995) use the earned value criterion for distributing payments. For consistency, and to enable comparison, models presented in this chapter are based on recovery of expenses. Using the notation defined for the contractor's problem, models for the client's payment scheduling problem are formulated as shown in the subsections that follow.

21.5.1.1 The Basic Model. As in the case of the contractor's model, the client's model described here is an event-based model in which payments are distributed by recovery of expenses. Whether or not advance payments are allowed can be captured by appropriate representation of project data in the network. In either case, this model assumes that the client has necessary information about the expenses incurred for each activity or groups of activities. Both parties negotiate the number of progress payments and the proportion of expenses to be recovered with each payment. Using the notation defined for the contractor's model, the basic model for the client's payment scheduling problem can be formulated as follows:

Maximize

$$B \sum_{t=E_M}^{L_M} c_t x_{Mt} - \sum_{t=0}^{D} c_t P_t \tag{14}$$

Subject to

$$\sum_{t=E_i}^{L_i} x_{it} = 1 \quad i = 1, \ldots, M \tag{15}$$

$$\sum_{t=E_w}^{L_w} t x_{wt} - \sum_{t=E_v}^{L_v} t x_{vt} \geq d_{vw} \quad \forall\ (v, w) \in A \tag{16}$$

$$\sum_{t=0}^{D} P_t = Z \tag{17}$$

$$\sum_{t=0}^{D} \delta_t = K \tag{18}$$

$$\sum_{t=0}^{T} P_t \geq \beta \sum_{i=1}^{M} \sum_{t=0}^{T} N_i z_{it} \quad T = 0, \ldots, D \tag{19}$$

$$\sum_{i=1}^{M} x_{it} \geq \delta_t \quad t = 0, \ldots, D \tag{20}$$

$$P_t - Z\delta_t \leq 0 \quad t = 0, \ldots, D \tag{21}$$

In this model, z_{it} is a binary variable which takes a value 1 if a payment is scheduled in time period t and an event also occurs in time period t, i.e., z_{it} is 1 if $\sum_{i=1}^{M} x_{it} \geq 1$ and δ_t is 1. B is the total benefit from the project and is a cash flow associated with the final event in the project. x_{Mt} is the binary variable indicating occurrence time for the terminal event. The objective function , numbered (14), represents the client's net benefit from undertaking the project, i.e., the present value of all future benefits less

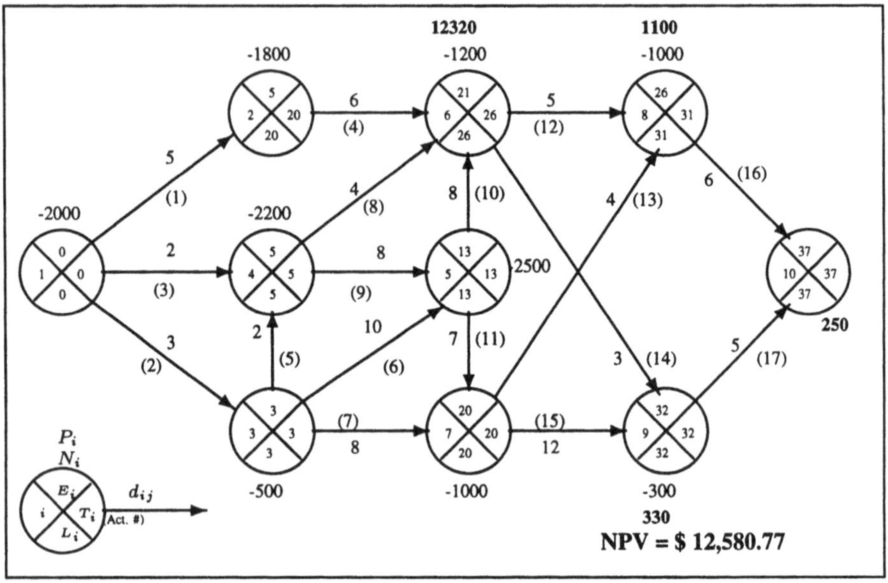

Figure 21:4 The Client's Optimal Payment Schedule (β = 110%)

the present value of all costs discounted to the beginning of the project. Constraint (15) ensures that every event is assigned an occurrence time; (16) are precedence constraints; (17) restrict the total amount of progress payments to Z dollars; (18) specifies the number of progress payments; (19) represents the distribution constraints, where at any time T, the total amount of progress payments must be at least β times the expenses incurred until the time of payment; (20) ensure that at least one event occurs at the time of payment; and (21) ensures that δ_t takes a value 1 when there is a nonzero payment in time period t.

Figure 21:4 shows the client's optimal payment schedule for our example project. Payments are based on expenses and advance payments are allowed. Benefits associated with project completion ensure that it is scheduled for early completion. The client can minimize her expenses by making payments of $12,320 in the twenty-sixth week, $1,100 in the thirty-first week, $330 in the thirty-second week and a final payment of $250 in the thirty-seventh week. In other words, payments are made at the last 4 events in the project. These payments and their timing result in a net present value of $12,580.77 which is considerably less than the expenses of $13,292.98 incurred if the client makes payments according to the contractor's best schedule. On the other hand, the contractor has to wait until the twenty-sixth week for the first payment and receives all the payments over a time span of eleven weeks.

21.5.1.2 Extensions: The Equal Time Intervals Model. Long delays with the first payment impose considerable financial burden on the contractor. Thus, in many project environments, paying at regular time intervals in an extremely popular practice (Chuman 1988, Abernathy 1990). The basic model for the client's payment scheduling problem can be adapted to obtain a payment schedule with payments at equal time intervals. Since the client prefers to have the project completed as soon as possible, the minimum duration for the project and the number of payments can be used to estimate the time interval between payments. If g represents this interval, $g = \lfloor E_M/K \rfloor$. With 4 payments and a minimum duration of 37 weeks for the example project, the client may make a payment every 9 weeks. The basic model can be modified to schedule K payments at equal time intervals by setting δ_t equal to 1 whenever $\text{mod}(t, g) = 0$. Recall that δ_t is the indicator variable for non-zero payments, and therefore, forcing appropriate values for δ_t guarantees payments in those time periods. The client's optimal schedule with this model is shown in Table 21:2.

21.5.1.3 Extensions: The Periodic Payment Model. One of the key assumptions for both client models discussed earlier is that the contractor and the client negotiate the number of payments. In many project environments, the two parties negotiate the frequency of payment instead of the number of payments, and the client makes payments at regular time intervals until the project is completed. The significant difference between this model, called the Period Payment Model, and the Equal Time Intervals model is that the time interval between payments is negotiated by both parties instead of estimating it from the minimum duration for the project. This is also a popular practice in many project environments. Abernathy (1990) recommends that contractors submit bills for payment to coincide with the client's payment cycle to avoid delays in realization of funds.

Again, the basic model can be modified to obtain the periodic payments model. As before, let g denote the time gap between payments and define two sets $S1$ and $S2$ as follows. Let $t \in S1$ if $\text{mod}(t, g) = 0$ and $S2 = D \setminus S1$. The following constraints must be included with the basic model to ensure that δ_t takes value 1 only when the project is in progress.

$$t - T_M + .5 \geq -D\delta_t \qquad t \in S1 \tag{22}$$

$$t - T_M + .5 \geq -D(1 - \delta_t) \qquad t \in S2 \tag{23}$$

The correction factor of .5 in these constraints is necessary to avoid satisfying the constraints as an equality since this leads to ambiguities in the value of δ_t. These constraints force δ_t to take a zero value for all time periods in $S1$ that extend beyond scheduled completion time. K, the number of payments is now given by $\sum_{t=0}^{D} \delta_t$. The optimal payment schedule is shown in Table 21:2.

21.5.1.4 Extensions: Periodic Benefits. All the three models can be easily extended to consider the case in which the client obtains some benefit on completion of intermediate stages of the project by associating every event i in the project with positive cash flow B_i. B_i can be regarded as the net present value of future cash flows associated with the completion of that event. The objective function can then be expressed as

$$\mathcal{Z} = \sum_{i=1}^{I} B_i \sum_{t=0}^{D} c_t x_{it} - \sum_{t=0}^{D} c_t P_t \tag{24}$$

This model bears a strong resemblance to models of the contractor's payment scheduling problem. The client's net benefit at each event is the difference between the benefit associated with that event, and the payment associated with that event. When this net benefit is positive, the client prefers to schedule the event as soon as possible. On the other hand, if her expenses on occurrence of the event are greater than the benefit associated with the event, she may prefer to schedule the event as late as possible. The extent of delay is determined by two important factors. The first is the net benefit associated with events that follow the event in question including the terminal event. The second factor is scheduling slack associated with the event (Elmaghraby and Herroelen 1990, Dayanand and Padman 1993).

21.5.1.5 Extensions: Earned Value Criterion. When the client has to determine the amount of payment on the basis of the value of work completed, V_i, the amount of payment can be limited by defining a maximum percentage retention, $\eta, 0 < \eta \leq 1$. V_i, the value associated with each *event*, is a critical parameter in the earned value model. Since an event denotes the commencement of some activities, and the completion of others, V_i is determined by the schedule of values, precedence relationships between project activities and accounting procedures with respect to earned value (Dayanand and Padman 1995). For balanced bids, it can be calculated by incorporating profit margin in equation (5) as shown below.

$$V_i = (1+\pi)\left(\sum_{j=1}^{M} g_{ij} f_{ij} a_{ij} + \sum_{k=1}^{M} (1 - f_{ki}) g_{ki} a_{ki}\right) \quad i = 1, \ldots, M \tag{25}$$

The distribution constraints may now be modified as follows

$$\sum_{t=0}^{T} P_t \geq (1-\eta) \sum_{i=1}^{M} \sum_{t=0}^{T} V_i z_{it} \qquad T = 0, \ldots, D \tag{26}$$

Illustrations with earned value in this chapter use a value of f_{ij} equal to zero for all activities so that the value of an activity is earned only on completion.

21.5.2 Features of Optimal Payment Schedules for the Client

Project characteristics have a significant impact on optimal payment schedules for the client as well. The three models result in significantly different payment and activity schedules for a given project. Some of the features of optimal payment schedules follow from properties of the models. Others result from project characteristics such as network structure, contract price, the distribution of earned value over project events, the number of payments and maximum percentage retention. Observations based on a study of each of these models with payments based on value of work completed until the time of payment, on 10 sample problems with 12 activities each, provide several interesting insights (Dayanand and Padman 1995). A sample of observations from model properties include:

- Increases in contract price always increases the client's expenses.
- The client can reduce the amount of payment by scheduling only one event to occur at the time of any payment.
- Expenses can also be reduced by increasing retention, or offering fewer payments over the duration of the project.
- For a fixed time gap between payments, the number of payments in a periodic payment schedule is nondecreasing in project duration.
- Expenses associated with a BC, EQ or PP schedule decrease with increasing retention over the duration of the project.

The study of example projects included analyzing the pattern of payments by considering time until the next payment, as a percentage of the project duration, for payment schedules with 2, 4 and 6 payments. It also examined the amount of each payment as a percentage of the total value of the project. A sample of observations include:

- There is considerable variation in the pattern of payment for the BC model. With the EQ and PP models, the time between payments is fixed. The analysis therefore shows that that there is little variation in the mean time to each payment but significant variation in the amount of each payment.
- Depending on the level of expenses and activity durations, the client's optimal schedules can result in payments that are clustered at the beginning or the end of the project.
- Since payments are associated with events, and the timing of these payments is fixed in the EQ model, the project will be delayed beyond early completion, unless (1) a slack event can be scheduled at the time of every payment, or (2) the early occurrence time of a critical path event coincides with the time of payment.
- Feasible BC schedules can always be guaranteed, unlike EQ and PP schedules, though they require the client to modify payment practices to meet the recommendations of the model.

21.5.3 Heuristics for the Client's Problem

Models of the client's payment scheduling problem are also combinatorial in nature. For small networks, the client's problem is relatively easier to solve than the contractor's problem. This is primarily because the project is generally scheduled for completion in the shortest possible time. Thus, the timing of all critical path events including the terminal event is fixed, thereby eliminating many feasible schedules from consideration. Furthermore, slack events are scheduled late except in cases where earlier scheduling of a slack event allows the client to make a smaller payment. Taking these factors into consideration, three simple rules have been proposed for solving the client's problem on large networks (Dayanand and Padman 1998). The heuristics were evaluated on two sets of randomly generated projects, and compared against a bound that identifies the client's preferred solution over all project environments.

Computational results indicate that the client benefits significantly by scheduling progress payments at the beginning of the project when these payments result in lower expenses. In addition, the simple scheme of assigning payments to the final events of the project and identifying the resultant set of activity schedules is dominated by the heuristic that optimally determines the payment schedule based on a fixed activity schedule. The results also show that the heuristics perform within 2% of the lower bound for small projects, and within 10% for larger projects where expenses are distributed over the duration of the project and many more payments are scheduled. Detailed results are reported in (Dayanand and Padman 1998).

21.5.4 Effect of Client's Payment Schedules on the Contractor's Schedule of Activities

The client's payment schedules are designed to maximize her benefit. Thus, activity schedules are planned to minimize expenses. Whether activities are actually completed according to the client's schedule depends on the effect of the payment schedule on the contractor's schedule of activities. In other words, when the client offers the contractor a payment schedule, the contractor determines his schedule of activities to maximize NPV of the project for the specified payment schedule.

Consider the optimal payment schedule obtained with the basic client model for the example problem, shown in Figure 21:4. The first payment of $12,320 is scheduled in the twenty-sixth week. The amount of payment is very large and results in a net positive cash flow at event 6 for the contractor. To maximize his NPV, the contractor will schedule event 6 to occur in the twenty-first week. Note that even with this change in schedule, the contractor completes all activities specified by the client's payment schedule. However, the client incurs higher interest cost on account of having to pay five weeks earlier than planned. Depending on the amount of payment, the increase in interest cost can be substantial.

The effect of the optimal schedules for each model on the contractor's schedule of activities was examined for the set of 10 sample problems (Dayanand and Padman 1995). The important point with the contractor's schedule is that activities completed by the time of each payment must correspond with the requirements of the client's schedule. This will ensure that the amount of each payment is for expenses already incurred so that the distribution constraints are not violated. In some cases, additional precedences must be forced in the project network to maintain feasibility of the distribution constraints.

Table 21:2 presents the contractor's optimal schedule of activities with the client's Basic, EQ, and PP models using the *recovery of expenses* model of payment distribution. Table 21:3 details the same using the *earned value criterion* as the method of distribution. There are significant differences, as expected, in the payment and activity schedules generated by the two approaches, and the two tables clearly indicate

Table 21:2 The Contractor's & Client's Activity Schedule (Recovery of Expenses)

Event	*Basic Model*			*Equal Time Intervals*			*Periodic Payments*		
		Schedule			*Schedule*			*Schedule*	
	Pymnt	*Contr*	*Client*	*Pymnt*	*Contr*	*Client*	*Pymnt*	*Contr*	*Client*
1		0	0		0	0		0	0
2		15	20	4730	13	18		15	17
3		3	3		3	6		3	5
4		5	5	5170	5	9	5170	5	8
5		13	13		13	17	2750	13	16
6	12320	21	26	2420	21	27	4400	21	24
7		21	20		21	24		21	23
8	1100	26	31		32	34	1100	26	32
9	330	33	32		33	36		33	35
10	250	38	37	1680	38	41	580	38	40
NPV		916.83	12580.77		1327.99	12986.75		1307.49	13046.02

Table 21:3 The Contractor's & Client's Activity Schedule (Earned Value)

Event	*Basic Model*			*Equal Time Intervals*			*Periodic Payments*		
		Schedule			*Schedule*			*Schedule*	
	Pymnt	*Contr*	*Client*	*Pymnt*	*Contr*	*Client*	*Pymnt*	Contr	Client
1		0	0		0	0		0	0
2	604.80	13	12	1562.40	15	18		13	16
3	403.20	3	3		3	3		3	6
4	1260.00	5	5	1663.20	5	9	1663.20	5	8
5		13	13		13	17	957.60	13	16
6		26	26	5846.40	21	27	6541.20	21	24
7		20	20		20	24		20	23
8		31	31		31	35	1612.80	26	32
9		32	32		32	36		32	35
10	11732.00	37	37	4928.00	37	41	3315.20	37	40
NPV		441.13	12327.94		834.36	12488.74		941.95	12613.79

that a methodical approach needs to be pursued to resolve these differences in the contractor's and client's preferred schedules.

The most significant observation is that even if the contractor and the client agree on β and K, every model indicates differences in the contractor's and client's schedule of activities. While the EQ and PP models show relative improvement in the contractor's NPV compared to the basic model, schedule disagreements suggest that the two parties must negotiate a schedule of progress payments that is mutually agreeable. This suggests the development of joint models for the payment scheduling problem. In the next section, we summarize some important issues to be addressed before modeling the joint problem.

21.6 THE JOINT VIEW

A joint view of the payment scheduling problem raises several issues that do not arise when we consider the contractor's and client's individual perspectives. Many models have been developed to support negotiation in various contexts. The approach used for these models depends greatly on the characteristics of the problem under consideration. In general, analysis of buyer-seller type transactions are based on the utility function (which captures risk attitudes) for the parties involved in the negotiation, and each party's reservation price. Such analyses typically focus on the pareto-optimal frontier of negotiated settlements (Raiffa 1982) and information asymmetries have a significant impact on the outcome.

Game theory is another popular approach to these types of situations (Gibbons 1982). In game theoretic models, the buyer and seller make simultaneous choices and receive payoffs that depend on the combination of actions chosen by the two parties. Kohli and Park (1989, 1994), for example, examine the case in which a monopolist seller offers incremental quantity discounts and the buyer adjusts the lot size in response. The two parties negotiate over order quantity and average unit price so that the final outcome maximizes joint gain. Contracting relationships have also been studied using the framework of agency theory (Ross 1973). In an agency relationship between two parties, one of the parties (designated the principal) seeks to motivate the other (the agent) to choose actions that would be advantageous to the principal. The agents action, together with a random state of nature, determine the final outcome. Both parties negotiate a fee to be paid to the agent for his services. The fee generally depends on the outcome, the agent's action, and his/her attitude towards risk. Since agency theory considers fee schedules associated with uncertain outcomes, the principal agent model is particularly appropriate for investigating risk sharing in the design of incentive contracts (Cummins 1977, Lambert 1983).

In formulating a joint model, several basic issues have to be considered. The nature of the problem and the number of parameters involved introduce sufficient complexity to require a detailed analysis of the characteristics of the joint payment scheduling problem before a model can be formulated. Some of these issues are discussed here in the context of the joint payment scheduling problem.

- Single versus multiple evaluation criteria:

  For both the contractor's and client's models, the payment schedule which comprises both the amount and timing of payments has been represented by the net present value (NPV) of the schedule. The desirability of a particular schedule is determined by this value. The contractor, for example, always prefers schedules with higher net present value than those with lower net present value. Likewise, the client is always assumed to prefer schedules with higher net benefit or lower expenses. However, as discussed earlier, payment schedules are determined by a number of factors. In many project environments, several features of the schedule may be considered in the negotiation, in addition to NPV. For example, project completion time may be a critical issue in addition to financial performance.

- Symmetry of information:

  Many of the results in contracting and negotiation are driven by informational assymetries. This is a significant issue for the joint payment scheduling problem as well. Consider, for example, the method of distribution. Both cost-based and performance-based distribution methods are popular in many project environments. In formulating the individual models, it has been assumed that appropriate information on expenses and profit margins is available to the "other" party. Even if it is not available, it can be estimated with reasonable accuracy so that each party can examine the impact of payments on activity schedules. However, it can be argued that the contractor will not reveal information on project expenses to the client. In such cases, the client relies on the value of work as revealed by the contractor. If the contractor's bid is unbalanced, the impact of the payment schedule on the activity schedule as seen by the two parties is signficantly different. Thus, if the two parties negotiate on inaccurate information, the negotiated and actual outcomes may be quite different.

- Factors to be negotiated: For both the contractor and client models, the two parties negotiate several factors prior to determining the schedule. Negotiated factors include the contract price, number and/or frequency of payment, and the amount to be retained with each payment. Preferred payment schedules for each party were determined from these parameters and project characteristics. In a joint situation, additional trade-offs may have to be considered. Both parties may be limiting their options by negotiating on the basis of fixed parameters. For example, the contractor may improve financial performance of the project by negotiating a lower contract price, if the client can be persuaded to make payments more often, or a larger number of payments over the duration of the project. Thus, joint views of the payment scheduling problem must consider what factors the two parties negotiate and the sequence in which these factors are negotiated.

- The joint objective: Formal models for joint decision making have been proposed for many buyer-seller type situations (Kohli and Park 1989, 1994, Arcelus and Srinivasan 1993). Thus, the joint payment scheduling problem can be considered one of the class of buyer-seller problems. In many of these models, the objective is to maximize the joint gain of the two parties. This naturally raises two questions in the context of the problem discussed in this chapter - are they joint gains to be achieved in negotiating payment schedules, and, if there are, how can the joint gain be expressed with the project and schedule parameters? For simplicity, let us assume that both parties evaluate the quality of the payment and activity schedule by its net present value. We may consider the joint gain as the sum of the contractor's and client's NPV. Since the monetary value of the benefits from a project are typically much larger than the cost of the project (the client generally wants to undertake only projects in which this is the case), the expression for joint gain is dominated by the value of the benefits. Payment schedules from this objective are biased in favor of the client.

- Differences in cost of capital: Although modeling the problem is somewhat simplified by examining only the financial impact of payment and activity schedules, differences in cost of capital for the two parties may introduce some complexity. As suggested earlier, the objective function must be formulated to avoid biased solutions which can occur when the cost of capital is different for the contractor and client.

21.7 CONCLUSIONS AND FUTURE WORK

With outsourcing as a key business strategy for the 21st century there is an increasing need to address issues in contracting and negotiation. This chapter summarizes a significant step in developing formal decision tools to support negotiation of financial terms in projects. Negotiating payment schedules can be a very complex task due to a number of factors, significant among them being (1) payment terms and activity schedules are tightly interlinked; payment terms are determined by progress on the project, but both parties make operational decisions on scheduling activities based on payment structures (2) the parties involved often have conflicting objectives so that a final settlement must consider the requirements of both parties to the contract, and (3) payment terms are determined by a number of parameters such as the type of contract, the method of distribution, the number/frequency of progress payments, retention, the contractor's profit margin, and the client's benefits from the project, among others. It is critical for the decision makers involved to systematically assess trade-offs in these parameters in order to successfully negotiate payment terms. This research can be viewed as bringing together macro and micro perspectives on project contracting. For the most part, the macro aspects of contract negotiation dealing with contract types, price and risk sharing arrangements have been examined in the economics literature. The operations management literature has examined micro aspects extensively by focusing on scheduling decisions. This chapter combines the two perspectives by examining the impact of some of the macro aspects on operational decision making in projects.

The literature survey clearly indicates a need to better integrate theory and practice in project management. While there is considerable documentation on project management practices in the construction industry, little attention has been directed to other functional areas (such as research and development), or industries (software development and telecommunications, for example). We believe that the quality and relevance of academic research in project management can be greatly enhanced by a survey of project management practices in various functional areas and across industries. Specifically, the survey will serve two purposes. Firstly, it can validate models developed in academic research such as those described in this chapter. These models are based on the existing literature on project scheduling which place a great emphasis on construction projects. Secondly, results from the survey can be used to develop better negotiation support tools that meet the needs of project managers in various industries.

Extensions of the current models can also be developed to assist project managers deal with several issues that have not been included in this research. The practical utility of these tools can be greatly enhanced by considering the impact of resource constraints on payment schedules. Tremendous research effort has been directed at development of both exact and heuristic methods for resource constrained projects with fixed cash flows. Some of these approaches can also be explored for the payment scheduling problem. Models discussed in this chapter also ignore the impact of bonus/penalty structures on payment schedules. Uncertainties associated with project costs, resource availability, and activity durations make real world project management a challenging and difficult task. Uncertainty in activity durations is an issue that has been addressed by some researchers. However, its impact on payment schedules and project profitability have yet to be examined.

The issues associated with the joint view outlined in this chapter focused on negotiating payment schedules using the financial impact of the schedule as an evaluation criterion. In practice, contractors and clients may evaluate the desirability of a schedule using other criteria as well. Project completion time and net cash flow position, for example, are critical factors. Thus, a multiattribute approach to the negotiation of payment schedules is both practical and relevant.

References

ABERNATHY, E. S. 1990. Managing a contractor's billings. *Journal of Accountancy*, 107–110.

ARCELUS, F. AND SRINIVISAN, G. 1993. Delay of payments for extraordinary purchases. *Journal of the Operational Research Society*, **44**, 8, 785–795.

BEY, R. B., DOERSCH, R. H., AND PATTERSON, J. H. 1981. The net present value criterion: Its impact on project scheduling. *Project Management Quarterly*, **12**, 2, 35–45.

CHUMAN, D. 1988. The use of project management when purchasing and implementing a hardware/software system. *Proceedings of the Annual Seminar/Symposium of the Project Management Institute*.

CLELAND, D. I. AND KING, W. R. (Editors) 1983. *Project Management Handbook*. Van Nostrand Reinhold company, NY.

CLOUGH, R. A. AND SEARS, G. A. 1991. *Construction Project Management*. John Wiley and Sons, NY.

COLLINGS, J. L. 1987. An owners approach to project management. *Proceedings of the Annual Seminar/Symposium of the Project Management Institute*.

CUKIERMAN, A. AND SHIFFER, Z. F. 1976. Contracting for optimal delivery time in long-term projects. *Bell Journal of Economics*, **7**, 1, 132–149.

CUMMINS, J. M. 1977. Incentive contracting for national defense: a problem of optimal risk sharing. *Bell Journal of Economics*, **8**, 1, 169-185.

DAYANAND, N. 1996. *Scheduling Payments in Projects: An Optimization Framework*. PhD thesis, The Heinz School, Carnegie Mellon University, Pittsburgh, PA 15213.

DAYANAND, N. AND PADMAN, R. 1993. Payments in projects: A contractor's model. Working Paper 93-71, The Heinz School, Carnegie Mellon University, Pittsburgh, PA 15213.

DAYANAND, N. AND PADMAN, R. 1994. A simulated annealing approach for scheduling payments in projects. Working Paper, The Heinz School, Carnegie Mellon University, Pittsburgh, PA 15213.

DAYANAND, N. AND PADMAN, R. 1995. Project contracts and payment schedules - the client's problem. Working Paper 95-23, Carnegie Mellon University, Pittsburgh, PA 15213.

DAYANAND, N. AND PADMAN, R. 1997. On modelling payments in project networks. *Journal of the Operational Research Society*, **48**, 9, 906–918.

DAYANAND, N. AND PADMAN, R. 1998. Heuristics for the client's payment scheduling problem. Draft Working Paper, The Heinz School, Carnegie Mellon University, Pittsburgh, PA 15213.

DOERSCH, R. H. AND PATTERSON, J. H. 1977. Scheduling a project to maximize its present value: A zero-one programming approach. *Management Science*, **23**, 8, 828–839.

ECKERT, T. 1994. Partnering: A creative, cost-effective concept for medium-sized construction projects. *PMNetwork*, VIII(9).

ELMAGHRABY, S. E. 1990. Project bidding under deterministic and probabilistic activity durations. *European Journal of Operational Research*, **49**, 14–34.

ELMAGHRABY, S. E. AND HERROELEN, W. S. 1990. The scheduling of activities to maximize the net present value of projects. *European Journal of Operational Research*, **49**, 35–49.

FISK, E. R. 1992. *Construction Project Administration*. Prentice-Hall, Englewood Cliffs, NJ.

FLEMING, Q. W. AND KOPPELMAN, J. M. 1994a. The "Earned Value Concept": Back to the Basics. *PMNetwork*, VIII(1).

FLEMING, Q. W. AND KOPPELMAN, J. M. 1994b. The "Earned Value Concept": Taking step two - plan and schedule the project. *PMNetwork*, VIII(9).

FLEMING, Q. W. AND KOPPELMAN, J. M. 1995. Taking step three with earned value: Estimate and budget resources. *PMNetwork*, IX(1).

GIBBONS, R. 1992. *Game Theory for Applied Economists*. Princeton University Press, NJ.

GILBREATH, R. D. 1992. *Managing Construction Contracts*. John Wiley and Soms, Inc.

GRINOLD, R. C. 1972. The payment scheduling problem. *Naval Research Logistics Quarterly*, **19**, 1, 123–136.

HAMBURGER, D. 1989. Bottom line project management. *Proceedings of the Annual Seminar/Symposium of the Project Management Institute*.

HARTMAN, F. T. 1994. Reducing or eliminating construction claims: A new contracting process(ncp). *Project Management Journal*, XXV(3).

HEARN, E. E. 1990. *Handbook on Government Contracts Administration.* Hearn Associates.

HENDRICKSON, C. AND AU, T. 1989. *Project Management for Construction.* Prentice Hall, Englewood Cliffs, NJ.

HERROELEN, W. S., DOMMELAN, P. V., AND DEMEULEMEESTER, E. L. 1997. Project network models with discounted cash flows: A guided tour through recent developments. *European Journal of Operational Research*, **100**, 97–121.

HUGHES, G. 1986. Software development contracts – traps for the unwary. *Australian Accountant*, **56**, 7, 76–79.

ICMELI, O., ERENGUC, S., AND ZAPPE, J. 1993. Project scheduling problems: A survey. *International Journal of Operations and Production Management*, **13**, 80–91.

ISAAC, A. 1990. Analysis of risks in bidding lump sum implementation contracts. *Proceedings of the Annual Seminar/Symposium of the Project Management Institute.*

KOHLI, R. AND PARK, H. 1989. A cooperative game theory model of quantity discounts. *Management Science*, **35**, 6, 693-707.

KOHLI, R. AND PARK, H. 1994. Coordinating buyer-seller transactions across multiple products. *Management Science*, **40**, 9, 1145-1150.

LAMBERT, R. A. 1983. Long term contracts and moral hazard. *Bell Journal of Economics*, **14**, 2, 441–452.

MARTIN, M. D. AND WEBSTER, F. M. 1986. Contract type and the measurement of project success. *Proceedings of the Annual Seminar/Symposium of the Project Management Institute.*

MARTIN, M. D. AND WEBSTER, F. M. 1987. Project integration and subcontract management. *Proceedings of the Annual Seminar/Symposium of the Project Management Institute.*

MARTIN, M. D. AND WEBSTER, F. M. 1988. Managing the future through project contracting. *Proceedings of the Annual Seminar/Symposium of the Project Management Institute.*

PADMAN, R. AND SMITH-DANIELS, D. E. 1993. Early-tardy cost trade-offs in resource constrained projects with cash flows : An optimization based approach. *European Journal of Operational Research*, **64**, 295–311.

PADMAN, R., SMITH-DANIELS, D. E., AND SMITH-DANIELS, V. L. 1997. Heuristic scheduling of resource constrained projects with cash flows. *Naval Research Logistics*, **44**, 364–381.

RAIFFA, H. 1982. *The Art and Science of Negotiation.* The Belknap Press of Harvard University Press, Cambridge, MA.

ROSS, S. 1973. The economic theory of agency: The principal's problem. *American Economic Review*, **63**, 1, 134–139.

RUSSELL, A. H. 1970. Cash flows in networks. *Management Science*, **16**, 5, 357–373.

RUSSELL, R. A. 1986. A comparison of heuristics for scheduling projects with cash flows and resource restrictions. *Management Science*, **32**, 10, 1291–1300.

SANDERS, S. R. AND COOPER, T. 1991. Analyzing construction company profitability. *Cost Engineering*, **33**, 2, 7–14.

SCHNIEDERS, J. R. AND JAMEYSON, R. A. 1987. Financial analysis for new formulation projects. *Proceedings of the Annual Seminar/Symposium of the Project Management Institute*.

SHATTOW, M. 1990. Project planning and control for software development. *Proceedings of the Annual Seminar/Symposium of the Project Management Institute*.

TALBOT, B. F. 1982. Resource-constrained project scheduling with time resource tradeoffs: The nonpremptive case. *Management Science*, **28**, 10, 1197–1210.

WEIDMAN, M. R. (Editor) 1992. *Project and Program Management: A Guide to Managing Project Risks and Opportunities*, volume 6 of *The PMBOK handbook series*. Project Management Institute, Drexel Hill, PA, preliminary edition. For trial use and comment.

WHANG, S. 1992. Contracting for software development. *Management Science*, **38**, 3, 307–324.

22 PROJECT MANAGEMENT IN AUDIT STAFF SCHEDULING

Bajis Dodin

Gary Anderson Graduate School of Management
University of California, Riverside, California

22.1. Introduction

Audit staff scheduling is an integral part of audit planning and control. It is also a very important part of the audit planning and control process due to the desire of the management to control auditing costs, plan the impacted time of the auditors efficiently, and meet the clients' deadlines. In fact, efficient audit scheduling can be used to set these deadlines based on the audit or accounting firm's capacity. Through efficient scheduling audit firms can determine the load and schedule of every auditor, over time requirements, travel schedules and costs, and can respond to changes in the information set (input data) easily.

A typical auditing or accounting firm employs many auditors with different levels of experience. An auditor has his/her preferences as to which experience (s)he wishes to accumulate. The firm also has its priorities with respect to serving its clients efficiently on one hand and honoring the wishes of its own and those of its auditors on the other hand. Hence, there is a match to be maximized, or equivalently a mismatch to be minimized between the workload of the firm and auditors. This gives rise to the performance measure to be used in the audit scheduling process.

The workload of the firm over a given planning horizon consists of several audit engagements. Each engagement is characterized by its type (industry), ready and due dates, processing requirements, and preferences of the client. Due to theses characteristics each audit engagement can be perceived as a project, which is known to be a unique undertaking with specific starting and completion dates, and specific processing requirements. Consequently, each audit engagement can be broken down into interdependent, small and controllable units of work, each to be processed by an individual auditor, known as "audit tasks". The dependencies between the tasks can be captured through a precedence relations structure. Hence, each audit engagement can be modeled as a PERT/CPM network. The precedence can be of the classical finish-to-start type with zero lags, or can allow for overlapping between the tasks, for example with a minimum gap between the start of an activity and the start of its direct predecessors. The networks of the several engagements can be combined together to model the load of the firm as an activity network. Hence, emerges the match between audit scheduling and project management. Therefore, managing the engagements is similar to managing multi-projects with many renewable resources.

The issues facing the audit firm is to determine the working schedule of each auditor, while accommodating their preferences to the maximum possible extent; and also to determine the schedule of each audit engagement while accommodating the objectives of the clients also to the maximum possible extent. It is the desire of the firm to have a scheduling tool which can integrate all of the above segments on one hand, and can be used to respond to changes in the information set, and answer the "what if" type questions on the other hand. As a side benefit and according to Blocher (1979), efficient audit staff scheduling reduces the rate of turnover among auditors; hence leading to reduction in the setup cost of hiring and training. It also leads to reducing the overall cost of the audit; and hence improves the competitive advantage of the firm, Bernstein (1978). Therefore, it is clear that effective audit scheduling is very important. However, it is very complex and difficult to accomplish. Some of these difficulties will be explained in the subsequent sections. They lead to the formation of a complex combinatorial problem.

This article deals with the above problem and develops the desired scheduling tool. It is organized as follows. In the next section the evolution of the research on the audit staff scheduling problem is presented. It is shown that the problem first moved from the stage of audit planning by inspection (trial-and-error) to audit loading using the methods of PERT/CPM or linear programming. Then, at a later stage, it moved from audit loading to audit scheduling, as it is known in project management. Within the stage of audit scheduling it moved also in several states, where in each state more realism was added to the scheduling tool or new solution methodology was developed. In Section 3 the latest mathematical model for the problem is presented; it is the 0 – 1 linear program developed by Dodin and Elimam (1997). This section ends with the classification of the problem using the proposed classification of Herroelen, Demeulemeester and De Reyck (1998). Section 4 deals with the solution methods and their efficiencies where computational experience is provided. Finally, in Section 5 concluding remarks are stated.

22.2. Evolution of the research on audit staff scheduling

Affiliation between audit scheduling and project management started shortly after the introduction of PERT and CPM. Burgher (1964) was the first to extend the PERT method to audit planning. His attempt was expanded in 1977 by Rackie et al to include extending PERT/Cost for audit planning and control. Similarly, network analysis was also extended to internal audit planning, Cirtin (1977). In all instances PERT was used to provide planning guidelines, and it was not used to schedule activities to resources or vice versa. It is assumed, in all of these studies, as it is typically the case in PERT/CPM analysis, that the resources (auditors) are not limiting.

However , in 1978 it was realized by Chan, Lam and Cheng that audit plans and deadlines generated from PERT/CPM are not feasible. Most of the infeasibility emanated from the limiting capacity of the auditors. They developed a heuristic that uses the total slack time of the activity to assign audit tasks to available auditors. Tasks were assigned on the basis of the task with the least slack is assigned first. Naturally, this heuristic results in different completion time than that obtained from PERT/CPM, and it develops an audit assignment based on the available capacity.

In Chan et al's heuristic audit tasks are assigned to auditors based on thc availability of the auditors. The issue of compatibility (match) between the auditor and audit task is not a factor in the assignment process. Consequently even though the audit assignment is feasible, it may not be the most desirable. Similarly, such an assignment may not lead to the one with the least completion time for the audit engagement, or the one with highest utilization (e.g. with the least idle time) for the auditors. It further does not necessarily accommodate the preferences of the auditors and/or the clients. In the sense of optimization, none of the above was an optimization planning tool.

Optimization tools were extended to audit scheduling. The first attempt was made by Summers (1972). He was the first author to formulate the audit planning problem as a linear program. His main concern was to not overburden nor to underutilize the auditors. His cost function includes elements that are hard to measure or quantify. Typical examples are the impact of the auditors past experience or gains from the current assignment on the cost. This difficulty gave rise to the goal programming models developed by Killough and Sounders (1973); Kornbluth (1974); Bailey, Boe, and Schnack (1974); and Welling (1977). In these studies the components of the cost function in Summers' study are unbundled and treated as separate goals, but the decision variables and constraints continue to center on determining the number of hours each auditor should spend on each audit engagement.

The difficulty in obtaining the weights used to scale the various objectives in the goal programming models gave rise to the interactive multiple objective linear programming model developed by Balachandran and Steuer (1982). A series of runs of the model helps in determining the weights preferred by the department manager. The final set of weights are used to determine the number of hours an auditor will spend on each engagement without being over or under utilized.

The difficulty present in the objective function of Summers' model gave rise also to the interactive integer linear programming model developed by Balachandran and Zoltners (B&Z) (1981). Integer programming allows for the accommodation of several of the real world considerations; it allows modeling the preferences of the auditors and the clients as constraints. The objective function is based on minimizing the total mismatching cost between auditors and audit tasks without over or under utilization of the auditor. Knechl and Benson (1991) used linear programming to determine the frequencies and plans of internal audits.

Examination of the above models reveals that they are *loading* but not *scheduling* models. They specify which tasks the auditor will be processing; hence leading to the total number of hours an auditor will be working on each of the audit engagements. None of these publications has dealt with the issue of sequencing the load of the auditor; how to resolve the issues of precedence between audit tasks; or the possibility of having the auditor process more than one task at a time, especially if they belong to two different audit engagements. Similarly, none modeled the setup time and/or setup cost as an auditor changes audit engagements, especially if the audit engagements are in different locations; or how to deal with the deadlines and delays. It is clear that these factors have cost implications that are of primary concern in audit planning and control. Finally, none of these models specify when each task will start and when it will end. Therefore, the issue of sequencing and scheduling the load was resolved through inspection.

The Chan and Dodin (1986) study was the first to deal with the issue of scheduling in an optimization manner. They showed that audit staff scheduling is a complicated form of multi-project resource constrained project scheduling. They used PERT networks to develop an efficient 0-1 programming model for the audit scheduling problem. Their model adds the element of time to that of B&Z; it also adds the precedence relations between the tasks, potential delays in processing the audit engagements; and the fact that an auditor cannot process more than one task at a time. Solution of the integer program yields the schedule of the auditor, audit tasks and audit engagements. It shows when each audit task starts, when it ends, and who the processing auditor is.

In 1991 Dodin and Chan applied their model to real world audit engagements and examined the impact of various objective functions on the operating schedule and the utilization of the auditors. Also in 1991 Drexl formulated the audit scheduling problem as a resource constrained project scheduling problem. He used a hybrid approach involving branch and bound and dynamic programming to find a least expensive schedule under which each engagement is completed by a given due date.

Salewski and Bartsch (1994), in response to findings of a survey they conducted for over 200 of the biggest certified public accounting firms in Germany, attempted to accommodate many of the real world scheduling issues. Due to the complexity of these issues they developed a hierarchical planning model consisting of three levels: medium term, medium-short term, and short term. They used genetic and greedy randomized algorithms to solve the hierarchical model.

Dodin and Elimam (1997) expanded the 0-1 model of Dodin and Chan to accommodate the possibility of overlapping between dependent audit tasks. They also captured the setup time and cost in audit scheduling. Setup time and cost emanate from the possibility of having an auditor change locations as he/she moves between different engagements. However, the addition of these considerations resulted in a complex formulation which is hard to solve by exact algorithms for large size audit scheduling problems.

The difficulty in solving the 0-1 model developed in Dodin and Elimam, which is known to be strongly NP- complete, Garey and Johnson (1975) and (1979), or the three level hierarchical planning model developed by Salewski and Bartsch gave rise to new research centering on solving these models through heuristic search techniques. Salewski, Schirmer and Drexl (1996) demonstrated that a procedure known as RAMSES, developed for solving the multi-mode resource constrained project scheduling problem, is capable of producing "good" solutions in short computational time for the three hierarchical planning model of the audit scheduling problem. Dodin, Elimam and Rolland (1998) developed a tabu search (TS) procedure and applied it to the 0-1 program developed in Dodin and Elimam. They showed that TS could generate close to optimal solutions to large audit scheduling problems in a relatively short processing time.

Rolland, Patterson and Dodin (1998) improved on the tabu search procedure by developing a new heuristic procedure known as adaptive reasoning technique (ART). Both procedures, TS and ART, were applied to the 0-1 formulation developed by Dodin and Elimam. In case of ART the integer programming formulation is augmented with additional non-redundant constraints through memory adjusting techniques. It was shown in this paper that under certain conditions TS is a special case of ART. Computational experience showed the dominance of ART over TS. It also showed that ART consumes more computer CPU time than TS.

In the next section the 0-1 program developed by Dodin and Elimam is presented. This formulation captures many of the real world considerations. They include the auditor utilization and preference factors; clients preferences; precedence and lead-lag relations between audit tasks; setup costs and times; the fact that an auditor cannot process more than one task at a time; delays and delay costs for audit engagements; and each audit task must be completed within its early and late completion times, and the possibility of processing the audit task at different rates.

22.3. Mathematical formulation

First in this section, we introduce the notation used in the 0-1 integer linear programming (ILP) model of the audit staff scheduling problem.

22.3.1. Definitions and notations

22.3.1.1. Constant parameters:

A_{ij} = The set of audit tasks, which can be processed by auditor *i* after s/he completes task *j*.

a_{jk} = Portion of activity *j* to be completed before the start of activity *k*, where *j* precedes *k*, and $0 < a_{jk} \leq 1$.

A_j = The set of tasks that directly succeed task *j*.

B_j = The set of tasks that are connected to task *j* by a direct precedence relationship

c_{ij} = The mismatching cost between auditor *i* and task *j*.

d_g = Due date of engagement *g*.

e_j = The earliest completion time of task *j*.

FS_{jk} = The minimum number of time units that must transpire from the completion of *j* prior to the start of *k*; if $FS_{jk} = 0$, then we have the usual CPM/PERT

g = 1,2,..., G, audit engagement index, where G is the set and number of engagements.

h = 1,2,..., H, time period index where H is the planning horizon.

H_i^- = Minimum number of periods auditor i is available for auditing within H.

H_i^+ = Maximum number of periods auditor i is available for auditing within H.

i = 1,2,..., I, auditor index, where I is the set and number of auditors.

I_j = The set of auditors nominated for auditing task *j*.

j = 1,2,..., J audit task index, where J is the set and number of all audit tasks in all engagements.

J_i = The set of audit tasks which can be processed by auditor *i*.

l_j = The latest completion time of task *j*.

p_g = Cost of having engagement *g* late (tardy) for one period.

s_{ijk} = The cost of having auditor *i* process task *k* after completing task j.

SS_{jk} = The time lag required for the start of task *k* after the start of task *j*. The overlapping relationship between two related activities, *j* and *k*, is given by the ratio a_{jk}. Activity k can start any time after the completion of $a_{jk}\, t_{ij}$ periods of work on activity *j*. Hence, $SS_{jk} = \lceil a_{jk} t_{ij} \rceil$ where $\lceil q \rceil$ is the smallest integer greater than or equal to *q*. The precedence relationship becomes strict when $a_{jk} = 1$.

t_{ij} = The time it takes auditor *i* to process audit task *j*.

22.3.2. Decision variables:

$$x_{ijh} = \begin{cases} 1 \text{ if auditor } i \text{ completes task } j \text{ at the end of period } h; \\ 0 \text{ otherwise.} \end{cases}$$

$$z_{ijk} = \begin{cases} 1 \text{ if auditor } i \text{ processes task } k \text{ after task } j; \\ 0 \text{ otherwise.} \end{cases}$$

Using the above notation and decision variables definitions, we next present the calculations of some parameters used in the ILP formulation. Most of these calculations are derived from the analysis of the corresponding PERT network.

Audit scheduling in an audit firm normally involves more than one audit engagement and perhaps in different locations. Each engagement is characterized by its arrival time, due date, preferences of the client with respect to auditors and audit teams, and other audit requirements. Consequently, each engagement can be looked at as a project. Audit planning and control of the engagement requires first developing the work breakdown structure (WBS) of the engagement. The WBS specifies the audit tasks, precedence and lead/lag relations, processing times as a function of the auditor's experience (efficiency), and mismatching cost. Once the WBS of each engagement is developed the audit process is modeled as an activity network similar to that of PERT/CPM.

Time analysis of the network of all the engagements enhances the task of scheduling. Starting with the arrival time of the engagement, then the earliest and latest completion times of each activity are calculated. The earliest completion time, denoted by e_j, is determined by using the following modification of Bellman's equation:

$$e_j = \max_{k \in B_j} \{e_k - FS_{kj} + \min_{i \in I_j} [t_{ij}]\}$$

where

$$FS_{kj} = t_{ik} - SS_{kj}$$

As for the latest completion time, l_j, it depends on the due date. In auditing, we distinguish between two types of due dates; the first, denoted by d_g, allows for lateness but with a penalty. The second due date is an absolute completion date after which delay is not allowed; i.e., penalty cost is infinite. For all practical purposes, the second due date is set equal to the planning horizon, denoted by H. Given H, then the following expression is used to calculate the latest completion time of task j:

$$l_j = \min_{k \in A_j} \{l_k - \min_{i \in I_k}[t_{ik}] + FS_{jk}\}$$

Normally, $d_g < H$ for all $g \in G$, and d_g is larger than any early completion time[1]. Therefore, it is always the case that $e_j < l_j$ for all $j=1,2,..., J$. The wider the intervals $[e_j, l_j]$ the more flexible the scheduling process; i.e. the more feasible schedules the problem has.

Determining the intervals $[e_j, l_j]$ is helpful in the planning process, but it does not resolve the issues addressed in the previous section. It does not tell us exactly when each task starts and when it ends, and who is the processing auditor. However, these intervals improve the efficiency of the 0-1 integer program, which is used to derive the detailed schedule.

---

[1]Note that e_j is the earliest of all early completion times due to the non-uniqueness of the processing time of task j.

The last issue addressed in this section is the issue of feasibility of the firm's capacity. The capacity of the firm within H is bounded from below by $\sum_{i \in I} H_i^-$, and from above by $\sum_{i \in I} H_i^+$.

Consequently, the capacity of the firm is feasible if:

1. One or more realization of the total work content, independent of the processing auditor, of all G engagements must fall within the above two capacity bounds[2].
2. $\sum_{j \in J_i} t_{ij} \geq H_i^-$ for each auditor $i = 1, 2, ..., I$

This condition implies that

$$\sum_{i \in I} \sum_{j \in J_i} t_{ij} \geq \sum_{i \in I} H_i^- .$$

Furthermore, a tighter feasibility condition is given by:

$$\sum_{j \in J} \min_{i \in I_j} [t_{ij}] \geq \sum_{i \in I} H_i^- .$$

Verifying the feasibility of the capacity; networking based on the WBS; and the calculations of the intervals $[e_j, l_j]$ and the set A_{ij} are normally conducted before the formulation of the problem. Now, we are in a position to state the 0-1 audit scheduling formulation .

22.3.2. Integer program formulation

22.3.2.1. Objective function. It minimizes the total cost of:

- Mismatch between the auditors and audit task. This cost is calculated according to the discussion presented in Balachandran and Zoltners (1981).
- Setup cost: It consists of two segments. The first is limited to the direct travel cost of the auditor between the locations of the engagements; this segment is independent of the auditor rank. The second captures the cost of the travel time between the engagements and the cost of other setup times. This segment depends on the value of the auditor's time.
- Delay cost; if an audit engagement is completed after its due date d_g, then a delay cost is incurred. The delay cost is either agreed upon in the contractual stage between the firm and the client, or it is determined by the firm to reflect its preferences to the clients, value of the engagement and some other factors.

---

[2] These bounds may include regular and overtime working hours.

$$min\, f(\underline{x}, \underline{z}) = \sum_{i=1}^{I} \sum_{j\in J_i} \sum_{h=e_j}^{l_j} c_{ij}\, x_{ijh} + \sum_{i=1}^{I} \sum_{j\in J_i} \sum_{k\in A_{ij}} s_{ijk}\, z_{ijk} + \sum_{i=1}^{I} \sum_{g=l}^{G} \sum_{h=d_{g+1}}^{H} p_g(h-d_g)\, x_{iMgh} \quad (1)$$

where M_g is the last task in engagement g.

22.3.2.2. Constraints. The model contains five sets of constraints:

1. Availability of auditors and the requirement that each auditor must work no less than H_i^- periods and no more than H_i^+ periods within the planning horizon H. Therefore, for each auditor i we have:

$$\sum_{j\in J_i} \sum_{h=e_j}^{l_j} t_{ij}\, x_{ijh} \geq H_i^- \quad (2)$$

and

$$\sum_{j\in J_i} \sum_{h=e_j}^{l_j} t_{ij}\, x_{ijh} \leq H_i^+ \quad (3)$$

2. Each audit task must be processed by one auditor; therefore for each audit task j

$$\sum_{i\in I_j} \sum_{h=e_j}^{l_j} x_{ijh} = 1 \quad (4)$$

3. An audit task can not start before its preceding requirements are satisfied. Consequently for every task j

$$\sum_{h=e_k}^{l_k} \sum_{i\in I_k} (h - FS_{kj})\, x_{ikh} - \sum_{h=e_j}^{l_j} \sum_{i\in I_j} (h - t_{ij})\, x_{ijh} \leq 0 \text{ for each } k \in B_j \quad (5)$$

4. An auditor can not process more than one task at a time. A task j is processed by auditor i in period h if it is completed in period r where

$$h \leq r \leq h + t_{ij} - 1, \text{ and } e_j \leq r \leq l_j$$

Therefore, let $u_j = max\ \{e_j, h\}$ and $v_j = min\ \{l_j,\ h + t_{ij} - 1\}$, then for each auditor i:

$$\sum_{j\in J_i(h)} \sum_{r=u_j}^{v_j} x_{ijr} \leq 1 \text{ for each } h = 1,2,\ldots, H \tag{6}$$

where $J_i(h)$ is the set of tasks auditor i can process in period h.

5. Setup costs: Whenever an auditor changes an audit task a setup cost isincurred. The setup cost depends on the auditor and on "the from" and "the to" tasks. Therefore, as auditor i completes task j s/he might be assigned to a task $k\in A_{ij}$:

$$z_{ijk} - \sum_{r=\theta k}^{lk} x_{ikr} \leq 0 \qquad \text{for each } k\in A_{ij} \tag{7.a}$$

where $\theta_k = \max\{e_k, e_j + t_{ik}\}$

$$\text{and} \qquad \sum_{k\in A_{ij}} z_{ijk} - \sum_{r=ej}^{lj} x_{ijr} = 0 \qquad \text{for each } i \text{ and } j \tag{7.b}$$

The analysis of constraint set (7.b) indicates that the auditor, after completing his/her assignments, must exit the project. Consequently, for each auditor i we add a dummy activity, ∂_i, at the end of the project. This activity is also added to the appropriate sets A_{ij}, but not to the set J_i. To guarantee that ∂_i is the last activity auditor i is assigned to, and that this task is scheduled after the completion of all the engagements, we introduce the following constraints for each $i\in I$:

$$\sum_{h=e_{M_g}}^{l_{M_g}} \sum_{b\in I_{M_g}} h\cdot x_{bM_g h} - H\cdot x_{i\partial_i H} \leq 0 \text{ for each } g\in G \tag{7.c}$$

Analyzing constraint sets (7.a) and (7.b) further indicates the possibility of cycling in the changeover decision variable z. To eliminate cycling, we introduce the following constraint sets:

$$\sum_{j\in J_i} z_{ijk} \leq 1 \text{ for all } i\in I, k\in J'_i \tag{7.d}$$

$$\sum_{k\in J_i} z_{ijk} \leq 1 \text{ for all } i\in I, k\in J_i \tag{7.e}$$

$$\sum_{r=e_k}^{l_k} r \cdot x_{ikr} - \sum_{r=e_j}^{l_j} r \cdot x_{ijr} \geq t \cdot z_{ijk} - M(1 - z_{ijk}) \text{ for each } k \in A_{ij} \tag{7.f}$$

where $J_i' = J_i \cup \partial_i$ and M is a large number

Constraint sets (7.a) and (7.b) are dependent on the set A_{ij}. Theoretically, the set A_{ij} consists of all the activities in the set J_i. However, early and late completion times, along with precedence, can be used to reduce the size of such set. First, a task $k \in A_{ij}$ only if $e_j + t_{ik} \leq l_k$. Then, the interval for the completion time of each element k ε A_{ij} is determined. The earliest completion time of k, denoted by θ_k, is given by $\theta_k = \max\{e_k, e_j + t_{ik}\}$.

The latest completion time of task k $\in A_{ij}$ is l_k, which is determined as stated above.

Study of the above ILP leads to the classification stated in the Appendix. This classification is based on that proposed by Herroelen, Demeulemeester and De Reyck (1998).

22.3.3. Complexity of the ILP

The number of constraints, decision variables, and feasible solutions depends on many factors. It depends on the planning horizon H; number of engagements; size of each engagement represented by the number of audit tasks; precedence and lead-lag- relations; number of auditors; and the gaps between the earliest completion time derived from the PERT analysis and the two due dates d_g and H. The number of constraints and decision variables can be reduced by using the subsets I_j and J_i instead of I and J, and by using the intervals [e_j, l_j] instead of H. The number of feasible solutions is impacted by the subsets I_j and J_i and the width of the intervals [e_j, l_j].

Analyzing the worst case scenario reveals that the number of constraints can be bounded from above by

$$[M(M-1)/2] + \cdot|J| + \cdot|I|(3+H+|G|) + 2\sum_i |J_i|^2 + \sum_i |J_i|$$

where M is the number of rows/columns in the adjacency matrix of the corresponding PERT network. The above quantity can be reduced to

$$2{*}I+J+\sum|B_j|+I{*}H + 2\cdot\sum_{ij}|A_{ij}| + 3\cdot\sum_i |J_i|. + |I|(1+|G|)$$

For instance, in the case study presented in Dodin and Elimam which consists of two engagements with a total of 19 tasks, four auditors and a planning horizon of 12 weeks, the upper bound on the number of constraints is 1239. This number was reduced to 876.

Similarly, the number of variables is bounded from above by

$$I^*J^*H + \sum_i \left(J_i^{\,2} - J_i \right)$$

which can be reduced to

$$\sum_{j=1}^{J} l_j \left(l_j - e_j + 1 \right) + \sum_{ij} A_{ij}$$

Again for the case presented in Dodin and Elimam, the upper bound on the decision variables is 1422 which can be reduced to 520.

22.4. Solution procedures and computational experience

Theoretically, the above ILP can be solved using the exact methods represented by the latest software packages. For instance, Dodin and Elimam applied LINDO (1991) to solve the ILP for a case example consisting of two engagements with a total of 19 tasks, 4 auditors and a planning horizon of 12 weeks. Dodin, Elimam and Rolland used AMPL(1993) to solve the same example. However, in practice we may not be able to apply these methods to large audit scheduling problems. It was demonstrated earlier that the audit scheduling problem is a combinatorial optimization problem classified as strongly NP-complete; in fact, according to Salewski, Schirmer and Drexl it is strongly NP-hard. Consequently, in these cases, using efficient heuristics to derive quality feasible schedules in short search time becomes necessary; see for instance Alvarez and Tamarit (1989), Davis and Patterson (1975), and Kurtulus and Davis (1982). Quality schedules mean scheduling with objective function value close to the optimal value.

Salewski and Bartsch used genetic and greedy randomized algorithms to generate good operational schedules. Also, Salewski, Schirmer and Drexl extended a heuristic procedure known as RAMSES, used to solve the multi-mode resource constraint project scheduling problem, to generate suboptimal operational schedules for the audit scheduling problem in a short computer processing time.

Dodin, Elimam and Rolland introduced tabu search methodology whose promises, if compared with the more traditional search procedures are numerous. First, the solution technique is assumption free. Therefore it is not necessary to make limiting restrictions on the shape of the feasible region, nor on the form of the objective function. Furthermore, TS is capable of escaping local optima through the diversification process. Glover (1989) and (1990) explains the basics of TS and its potential applications.

The TS procedure developed by Dodin, Elimam and Rolland is grounded in intelligent memory structures. It starts with a feasible schedule generated by a greedy forward loading (FL) dispatching procedure. Then a sequence of intensification moves based on add and drop changes are used to improve on the latest schedule. As soon as the improvements on the existing schedule stabilize, a diversification move is initiated. Diversification moves, based on a long-term holding time, are used to sharpen the improvements of the feasible schedules by taking the search process into other segments of the feasible region.

TS was applied to audit scheduling problems with different sizes and characteristics. The tests show that its application results in relatively large improvements in the values of the objective function over the initial feasible schedule. The tests also show that its application requires very little CPU time. Table 22:1 gives the sizes of the problems tested. Seven realizations of each problem were tested; each realization provides a different characterization of the problem. The columns headed by TS in Tables 22:2.a, b, and c show the performance of the heuristic for the 21 problems tested. Table 22:2.a shows the actual value of the objective function; TS results should be first compared with the initial values listed in the same table under FL heading. Table 22:2.b shows the corresponding percentage of the improvement; and Table 22:2.c shows the required CPU time in second.

At a later stage Rolland, Patterson and Dodin tried to improve on the performance of TS with respect to the audit scheduling problem. Their efforts led to the development of a new heuristic search procedure, known as ART (Adaptive Reasoning Technique). This procedure is also grounded in the intelligent memory structures. It is a constructive memory adaptive technique that was proved to provide significant improvement in solution quality over other heuristic techniques. It was first used by Rolland, Patterson and Pirkul (1997) to solve the problem of capacitated minimum spanning tree.

ART, like TS, starts with an initial feasible schedule generated by the greedy forward dispatching procedure. Then it repeats using this procedure to develop a new feasible schedule but after it augments the current ILP with additional non-redundant constraints. After each iteration the additional non-redundant constraints are dynamically added and dropped. The addition and deletion of these extra constraints drive the FL procedure to search new feasible solutions. The memory function of ART is used as a discovery mechanism to find combinations of additional non-redundant constraints to redirect the FL procedure.

ART was applied to the same set of problems used in assessing the performance of TS. Performance of both, TS and ART, was also contrasted with that of GRASP (greedy randomized adaptive search procedure) after it was modified to solve the above ILP. GRASP was initially developed by Feo, Bard and Venkatsaman (1991), and it is used to solve many combinatorial scheduling problems. Results of this comparison are given in Table 22:2. Table 22:2.a shows the terminal value of the objective function for each of the three heuristics, where Table 22:2.b shows the corresponding percentages of the improvement over the value of the initial feasible schedule obtained by FL. Table 22:2.c shows the corresponding CPU time requirements. It is clear that both TS and ART out-perform GRASP. It is also clear that ART out-performs TS; however TS requires far less CPU time.

ART is expected to yield better solutions (on the average) than both TS and GRASP. Similarly, TS is expected to outperform GRASP. ART yields better results than TS since ART is based on modifiying the behavior of the primary rule used to generate feasible schedules, whereas the neighborhood based TS only attempts to correct the errors in the final solution produced by the primary rule. ART and TS generate better solutions than GRASP since in GRASP no memory exists, and no feedback occurs from one iteration to the next; i.e. iterations are independent of each other.

Table 22:1. Sizes of the Tested Problems.

Problem	Number of Engagements	Number of Auditors (I)	Number of Tasks (J)	Planning Horizon (H)	I*J*H
1	2	4	19	12	912
2	2	4	35	35	4900
3	5	6	75	45	20250

Table 22:2. Performance of Three Heuristic Procedures.

a: Value of the Objective Function

Problem	Instance	FL	TS	GRASP	ART
1	a	8900	7900	8900	7200
	b	8900	7300	7300	7000
	c	4900	3600	4900	3000
	d	22900	21600	21900	21000
	e	17000	14800	15500	15100
	f	26900	20900	24500	13200
	g	5560	4560	4560	4480
2	a	20400	15800	19400	15600
	b	18400	15800	16400	14800
	c	20400	16400	19400	15600
	d	20400	16400	19400	15400
	e	26200	22200	25000	18700
	f	36400	31800	26800	29800
	g	17400	13400	16400	11500
3	a	73000	64500	59500	62100
	b	61500	49700	53700	47200
	c	70000	62100	66600	58800
	d	100000	56600	67200	57300
	e	88400	84000	86700	82300
	f	181000	150000	143000	157500
	g	91000	48500	63000	41500

b: Average Percentage of the Improvements

Problem	Instance	TS	GRASP	ART
1	a	11.24	0.00	19.10
	b	17.98	17.98	21.35
	c	26.53	0.00	38.78
	d	5.68	4.37	8.30
	e	12.94	8.82	11.18
	f	22.30	8.92	50.93
	g	17.99	17.99	19.42
2	a	22.56	4.90	23.53
	b	14.13	10.87	19.57
	c	19.61	4.90	23.53
	d	19.61	4.90	24.51
	e	15.27	4.58	28.63
	f	12.64	26.37	18.13
	g	22.99	5.75	33.91
3	a	11.64	18.49	14.93
	b	19.19	12.68	23.25
	c	11.29	6.29	16.00
	d	43.40	32.80	42.70
	e	4.98	1.92	6.90
	f	17.13	20.99	12.98
	g	46.70	30.77	54.40
Average		18.85	11.63	24.38

c: CPU Time Required for Convergence

Problem	Instance	FL	TS	GRASP	ART
1	a	0.08	3	77.05	15.97
	b	0.02	3	68.46	15.88
	c	0.07	3	68.38	17.69
	d	0.01	3	68.97	16.58
	e	0.06	3	62.24	16.09
	f	0.08	3	68.56	17.46
	g	0.07	3	68.43	15.68
2	a	0.23	15	2489.63	212.48
	b	0.24	13	2111.27	223.47
	c	0.23	13	2117.52	214.12
	d	0.22	10	2104.76	227.92
	e	0.26	13	2263.89	252.23
	f	0.27	20	2102.10	250.17
	g	0.25	11	2102.67	203.09

3	a	2.90	52	15165.57	1295.30
	b	3.00	54	10744.11	1244.79
	c	2.95	43	15473.71	1454.99
	d	2.90	55	13624.76	1377.31
	e	3.00	52	17217.85	1249.91
	f	3.00	34	14924.44	2390.06
	g	3.00	50	15412.73	1458.23
Average		1.09	21.71	5635.10	579.50

22.5. Conclusions

This paper investigates the problem of audit staff scheduling. It was found that research on audit staff scheduling went into three stages. The first stage centered on the issue of loading audit tasks to auditors. It used PERT networks as well as optimization models such as linear, goal and multi-objective programming. The second stage dealt with issue of scheduling as it is known in production and project management. In this case detailed information, such as when an audit task starts, how long it takes, when it ends, and who the processing auditor is; is determined. This stage relied heavily on the use of 0-1 integer programming to capture the many real world considerations. The third stage dealt with the development of heuristic search techniques to solve the complex models developed in the second stage.

It was shown that the audit scheduling problem is a multi-project, multi-mode resource constraint project scheduling problem (MMRCPSP). Hence, it is a complex combinatorial optimization problem and can be classified as NP-hard. Consequently the problem benefited from research in the area of project management. Therefore, the problem is first analyzed as a project represented by a project network. The network analysis helps in constructing efficient integer program. Then the integer program is solved by applying any of the heuristics developed for the MMRCPSP. Three heuristic procedures have been tried and have been shown to provide good operational schedules. These procedures are the TS developed by Dodin, Elimam and Rolland; ART developed by Rolland, Patterson and Dodin; and RAMSES developed by Salewski, Schirmer and Drexl.

It is interesting to note that none of the research on audit staff scheduling dealt with the issues of uncertainty or probability in audit scheduling. It is possible, as it is the case in project management, that the duration of some audit tasks can be known only in a probabilistic sense. This is the case in auditing a new technology or business, or in dealing with a new client. Consequently, audit staff scheduling can also benefit from the tools developed for the probabilistic side of project management. This is the subject of future research.

Appendix

Classification of the Problem:

The problem addressed in this article can have the following classification:

m, 1T, °|°, min, e_j, °, l_j, °, mu, °|reg (min. cost)

where these symbols are defined as follows:

α_1= m	numbcr of resource types (auditors)
α_2 = 1T	both renewable and non-renewable resources
α_3= o	renewable resources are available in constant amounts

β_1= o	no preemption allowed
β_2= min	min overlap in precedence relations (start to start)
β_3= e_j	ready times for the activities = ej
β_4 = o	activities have arbitrary integer durations; depends on the processing auditor
β_5 = l_j	deadlines imposed on activities, it is ℓj
β_6 = o	resource is consumed in constant discrete amount
β_7= mu	multi-mode of activity execution; which are the various auditors
β_8= o	no cash flow

δ_1=reg (min. cost) is the performance measure, which is to minimize the total cost

References

ALVAREZ-VALDARÉS, R. and J.M. TAMARIT, 1989, Heuristic Algorithms for Resource-Constrained Project Scheduling, in: R. Slowinski and J. Weglarz (Eds.), *Advances in Project Scheduling*, Amsterdam, pp. 134-143.

BAILEY, A.D. Jr., W.J. BOE, and T. SCHNACK, 1974, The Audit Staff Assignment Problem: A Comment, *The Accounting Review*, July, pp.572-574.

BALACHANDRAN, B.V., and A.A. ZOLTNERS, 1981, An Interactive Audit-Staff Scheduling Decision Support System, *The Accounting Review*, October, pp. 801-812.

BALACHANDRAN, K.R. and R. STEUER, 1982, An Interactive Model for the CPA Firm Audit Staff Planning Problem with Multiple Objectives, *The Accounting Review*, January, pp. 125-140.

BERNSTEIN, P.W., 1978, Competition Comes to Accounting, *Fortune*, July 17, 1978, pp. 88-96.

BLOCHER, Edward, 1979, Performance Effects of Different Audit Staff Assignment Strategies, *The Accounting Review*, July, pp. 563-573.

BURGHER, P.H., 1964, PERT and the Auditor, *The Accounting Review*, January, pp 103-120.

CHAN, K.H., S.F. LAM, and T.T. CHENG, 1978, Audit Scheduling with Limited Resources, *CA Magazine*, Nov. issue, pp. 38-44.

CHAN, K.H. and B. DODIN 1986, A Decision Support System for Audit Staff Scheduling with Precedence Constraints and Due Dates, *The Accounting Review*, January, pp. 125-140.

CIRTIN, A., 1977, Network Analysis and Internal Auditing, *Internal Auditor*, Vol. 34, No. 1, pp. 27-31.

DAVIS, E.W. and J.H. PATTERSON, 1975, A Comparison of Heuristic and Optimum Solutions in Resource Constrained Project Scheduling, *Management Science*, Vol. 21, pp. 944-955.

DODIN, B. and H. CHAN, 1991, Application of Production Scheduling Methods to External and Internal Audit Scheduling, *European Journal of Operational Research*, Vol. 52, pp. 267-279.

DODIN, B. and A.A. ELIMAM, 1997, Audit Scheduling with Overlapping Activities and Sequence Dependent Setup Costs, *European Journal of Operational Research*, Vol. 97, pp. 22-33

DODIN, B., A.A. ELIMAM, E. ROLLAND, 1998, Tabu Search in Audit Scheduling, Forthcoming in *European Journal of Operational Research*, Vol. 104, No. 1 (January 1998)

DREXL, A., 1991, Scheduling of Project Networks by Job Assignment, *Management Science*, Vol. 37, No. 12, pp. 1590-1602.

FEO, T. A., J. F. BARD, and K. VENKATRAMAN, 1991, A GRASP for a Difficult Single Machine Scheduling Problem, *Computers & Operations Research*, Vol. 18, pp. 635-643.

FOURER, R., D.M. GAY, and B.W. KERNIGHAN, 1993, *AMPL: A Modeling Language for Mathematical Programming*, The Scientific Press, San Francisco.

GAREY, M.R. and D.S. JOHNSON, 1975, Complexity results for multiprocessor scheduling under resource constraints, *SIAM J. Computing*, No. 4, pp. 397-411.

GAREY, M.R. and D.S. JOHNSON, 1979, *Computers and Intractability- a Guide to the Theory of NP- Completeness*, W.H. Freeman, San Francisco.

GLOVER, F., 1989, Tabu Search-Part I, *ORSA Journal on Computing*, Vol. 1, No. 3, pp. 196-206.

GLOVER, F., 1990, Tabu Search-Part II, *ORSA Journal on Computing*, Vol. 2, No. 1, pp. 4-32.

HERROELEN, W., E. DEMEULEMEESTER, and B. DEREYCK, 1998, A Classification Scheme for Project Scheduling, Forthcoming in *Handbook on Recent Advances in Project Scheduling*; Edited by J. Weglarz.

KILLOUGH, L.N. and T.L. SOUNDERS, 1973, A Goal Programming Model for Public Accounting Firms, *The Accounting Review*, April, pp.268-279.

KNECHl, W.R. and H.P. BENSON, 1991, An Optimization Approach for Scheduling Internal Audits of Divisicns, *Decision Science*, Vol. 38, No. 9, pp. 1245-1263.

KORNBLUTH, J.S.H., 1974, Accounting in Multiple Objective Linear Programming, *The Accounting Review*, April, pp. 284-295

KURTULUS, I.S. and E.W. DAVIS, 1982, Multi-Project Scheduling: Categorization of Heuristic Rules Performance, *Management Science*, Vol. 28, No. 2, pp. 161-172.

RACKIE, C.C., J.L. KROGSTAD, G. GRUDNITSKI, and D.W. BRYANT, 1977, PERT and PERT/Cost for Audit Planning and Control, *Internal Auditor*, Vol. 36, No. 4, pp. 57-68.

ROLLAND, E., R. PATTERSON and B. DODIN, 1998, A Memory Adaptive Reasoning Technique for Solving the Audit Scheduling Problem, Working Paper 1998-1, Anderson Graduate School of Management, University of California at Riverside.

ROLLAND, E., R. PATTERSON and H. PIRKUL, 1997, Adaptive Reasoning Technique and Greedy Search for the Capacitated Minimum Spanning Tree Problem, *Proceedings of the 2nd Meta Heuristic International Conference*, Antibes, France, pp 73-81.

SALEWSKI, F. and T. BARTSCH 1994, A Comparison of Genetic and Greedy Randomized Algorithms for Medium-to-Short-Term Audit-Staff Scheduling, Working Paper No. 356, Department of Production and Logistics, Institute for Management, Christian-Albrechts University at Kiel, Germany.

SALEWSKI, F., A. SCHIRMER and A. DREXL, 1996, Project Scheduling Under Resource and Mode Identity Constraints. Part II: An Application to Audit Staff Scheduling, Working Paper 388, Department of Production and Logistics, Institute of Management, Christian-Albrechts University at Kiel, Germany.

SHRAGE, L., 1991, *Linear, Integer, And Quadratic Programming with LINDO*, Fourth Edition, Boyd & Fraser Publishing Company, New York .

SUMMERS, E.L., 1972, "The Audit Staff Assignment Problem: A Linear Programming Analysis, *The Accounting Review*, July, pp. 443-453.

WELLING, P., 1977, A Goal Programming Model for Human Resource Accounting in a CPA Firm, *Accounting, Organization and Society*, No. 4, pp. 307-316.

Subject index

Name index

CPSIA information can be obtained
at www.ICGtesting.com
Printed in the USA
LVHW081210200420
654119LV00011B/1003

9 780792 382683